The Alphabet of Galen.
Pharmacy from Antiquity to the Middle Ages

The Alphabet of Galen is a critical edition and English translation of a text describing, in alphabetical order, nearly three hundred natural products – including metals, aromatics, animal materials, and herbs – and their medicinal uses. A Latin translation of earlier Greek writings on pharmacy that have not survived, it circulated among collections of 'authorities' on medicine, including Hippocrates, Galen of Pergamun, Soranus, and Ps. Apuleius.

This work presents interesting linguistic features, including otherwise unattested Greek and Latin technical terms and unique pharmacological descriptions. Nicholas Everett provides a window onto the medieval translation of ancient science and medieval conceptions of pharmacy. With a comprehensive scholarly apparatus and a contextual introduction, *The Alphabet of Galen* is a major resource for understanding the richness and diversity of medical history.

NICHOLAS EVERETT is an associate professor in the Department of History at the University of Toronto.

The Alphabet of Galen.
Pharmacy from Antiquity to the Middle Ages

A Critical Edition of the Latin Text with English Translation and Commentary by

NICHOLAS EVERETT

UNIVERSITY OF TORONTO PRESS
Toronto Buffalo London

ISBN 978-0-8020-9812-2 (cloth)
ISBN 978-0-8020-9550-3 (paper)

Library and Archives Canada Cataloguing in Publication

Everett, Nicholas
The Alphabet of Galen : pharmacy from antiquity to the Middle Ages : a critical
edition of the Latin text with English translation and commentary / by Nicholas Everett.

Includes bibliographical references and index.
ISBN 978-0-8020-9812-2 (bound). – ISBN 978-0-8020-9550-3 (pbk.)

1. Materia medica – Early works to 1800.   2. Medicine, Greek and Roman – Early works
to 1800.   3. Pharmacy – Early works to 1800.   4. Alfabetum Galieni – Criticism and
interpretation.   5. Alfabetum Galieni – Translations into Latin.   6. Alfabetum
Galieni – Translations into English.   I. Title.

RS79.E94 2012   615.1   C2011-908386-8

This book has been published with the help of a grant from the Canadian
Federation for the Humanities and Social Sciences, through the Aid to Scholarly
Publications Program, using funds provided by the Social Sciences and
Humanities Research Council of Canada.

University of Toronto Press gratefully acknowledges the financial assistance of the Centre
for Medieval Studies, University of Toronto in the publication of this book.

University of Toronto Press acknowledges the financial assistance to its publishing program
of the Canada Council for the Arts and the Ontario Arts Council.

Canada Council   Conseil des Arts
for the Arts   du Canada

ONTARIO ARTS COUNCIL
CONSEIL DES ARTS DE L'ONTARIO

University of Toronto Press acknowledges the financial support of the Government of
Canada through the Canada Book Fund for its publishing activities.

For Anna-Sophia and Elias

Contents

❧

List of Plates

～

Acknowledgments

The history of medicine presents many challenges for the historian studying scripts and lay-out of early medieval books who discovered a rough little manuscript in the Vatican library with an intriguing alphabetized format, and who became fascinated with its contents. Challenges to historical periodization quickly surface: the separation of 'ancient' and 'early medieval' medicine easily falters (as the *Alphabet* demonstrates), distinctions between ancient and medieval are often problematic, and from a modern medical viewpoint, ancient/medieval medicine and pharmacy continued into the nineteenth century and beyond. The challenges to understand the natural world described in the *Alphabet* require knowledge of disciplines not usually part of a medievalist's training, and their study continued the discovery.

Reliance on the work of scholars more experienced in the history of ancient and medieval medicine will be apparent in the notes, but I have been particularly inspired by the combination of insight and scholarly rigour in the work of Carmelia Opsomer, John Riddle, John Scarborough, and Jerry Stannard on ancient and early medieval pharmacy. John Scarborough has served as an e-mentor on pre-modern pharmacy, generously sharing his expertise in exemplary scholarly spirit. Cloudy Fischer encouraged this project at an early stage and like many others I have benefited greatly from his numerous publications on medical texts and their manuscript traditions. Contact and subsequent discussions with Eliza Glaze and Monica Green revived levels of energy needed to complete the manuscript; Eliza Glaze generously read chapter 5 and suggested many improvements. My thanks to the two anonymous readers selected by the University of Toronto Press, whose advice and comments improved the manuscript.

Colleagues at the University of Toronto, in the Department of History, the Centre for Medieval Studies, the Department of Pharmacology, and Trinity College, inspire by word as much as example, and I have benefited enormously from their help and advice so generously given when sought. This book was written amid the teaching and administrative responsibilities of a large public university, and I thank the many undergraduate and postgraduate students who have patiently endured my distracted musings on the history of medicine. Teaching with the *Alphabet* reaffirmed the text's uniqueness, and its worthiness of a proper edition with translation. But teaching also served to control inversely the amount of commentary given here: the myriad comparisons with other texts, possible influences, issues of Greek-Latin translation, its manuscript tradition, and so on – these are best left for the classroom and for specialist articles. The University of Toronto Press is to be justly commended for publishing a critical edition of a Latin text, and I thank Suzanne Rancourt for her encouragement and patience, even when process worked against us.

The bibliography mostly stops at 2008: a first sabbatical in 2007 provided teaching relief necessary to make completion possible, and the final book manuscript was submitted in 2009. I tried to include recent material that came to my attention, even if it was not fully incorporated into the discussion (e.g., Ferraces Rodríguez 2009, Petit 2009, Totelin 2009): it is reassuring to realize that the research behind this book is part of larger process of discovery by scholars re-examining ancient and medieval medical texts.

For the permission to reproduce images, I thank the Biblioteca Nazionale of Naples (front cover, plates 4–8), the Biblioteca Apostolica Vaticana (plate 2), the Österreichische Nationalbibliothek of Vienna (plate 1), Georges Fontès of Instants de Saisons (http://isaisons.free.fr/index.htm), and those at the (now defunct) site www.szak-kert.hu (plate 3). The interlibrary loan service at Robarts Library was superbly efficient, and special thanks to Renata Holder, of the Gerstein Science Library, for many kindnesses in accommodating a medievalist. Colleagues near and far who have kindly answered inquiries or helped with bibliography for their respective fields are thanked in the notes where appropriate, but deserve listing here: as well as to the previously mentioned scholars, thanks to Julien Barthe (Chartres), Virginia Brown† (Toronto), Paul Cohen (Toronto), Gerard Duursma (*ThLL*), Ernst Gamillscheg (Vienna), Maria Teresa Gigliozzi (Rome), Dorothea Kullmann (Toronto), Rebecca Laposa (Toronto), Michael McVaugh (Chapel Hill, NC), Francis Newton (Duke), Roel Sterckx (Cambridge), Cindy Woodland (Toronto), and

Roger Wright (Liverpool). Sincere thanks to the colourful characters and staff at the Remarkable Bean, in The Beach, for providing a much-needed second office.

Familial support from Eumundi, Maroochydore, and New Orleans never wavers and defeats the distance. This book is dedicated to Anna-Sophia and Elias, because we learned our alphabets together. The greatest debt of all is to Dita, my A to Z.

Signs, Symbols, and Abbreviations

❧

~	'possibly,' used for tentative plant identifications (see below)
#	entry number in the *AG*
ch.	chapters 1–5 in the present book
Alex. Tral.	Alexander of Tralles *Therapeutica*, ed. Puschmann. (Latin), *Practica Alexandrri Yatros* (1504)
Cael. *Acut.*	Caelius Aurelianus, *De morbis acutis*, ed. Bendz
Cael. *Chron.*	Caelius Aurelianus, *De morbis chronicis*, ed. Bendz
Cael. *Gyn.*	Caelius Aurelianus, *Gynaecia*, ed. M.F. Drabkin and I.E. Drabkin
Cass. Fel.	Cassius Felix, *De medicina*, ed. Fraisse
Celsus	Aulus Cornelius Celsus, *De Medicina*, ed. Marx, trans. Spencer
CLA	*Codices Latini Antiquiores*, ed. Lowe
CMG	Corpus Medicorum Graecorum
CML	Corpus Medicorum Latinorum
De observ.	*De observantia ciborum*, ed. Mazzini
Diosc.	Dioscorides, *De materia medica*, ed. Wellmann
Diosc. Lat.	Dioscorides, *Materia Medica*, ed. Stadler (books II–V), ed. Mihaescu (book I)
DTh.	*Diaeta Theodori*, ed. Sudhoff
Dyn.Vat./SGall.	(Ps. Hippocratic) *Dynamidia*, ed. Mai (ex MSS Vat. Pal. lat. 1088, Vat. Reg. lat 1004) and ed. Rose (ex MS St Gall 762)
Ex herb. fem.	(Ps. Dioscorides) *Ex herbis femininis*, ed. Kästner
Galen *Simp.*	Galen, *De simplicium medicamentorum temperamentis et facultatibus*, ed. K[ühn] 11, 379–892; 12, 1–377

Garg. *Med.*	Gargilius Martialis, *Medicina ex oleribus et pomis*, ed. Maire
Isidore	Isidore of Seville, *Etymologiarum siue Originum libri XX*, ed. Lindsay
K.	Galen, *Claudii Galeni Opera Omnia*, ed. Kühn
LSJ	Liddell and Scott, *A Greek-English Lexicon*, rev. Jones
Marc. *Med.*	Marcellus (Empiricus) of Bordeaux, *De medicamentis*, ed. Niedermann
Med. Plin.	*Plinii secundi iunioris qui feruntur de medicina libri tres*, ed. Önnerfors
MGH	*Monumenta Germaniae Historica*
MGH AA	*Auctores Antiquissimi*
MS, MSS	manuscript(s)
Orib. *Syn.*	Oribasius, *Synopsis.* I–II ed. Mørland, III–IX ed. Bussemaker et al.
Orib. *Eup.*	Oribasius, *Euporista*, ed. Molinier
n.	note
Paul Aeg.	Paul of Aegina, *Practica (Lib. III)*, ed. Heiberg
Phys. Plin.	*Physica Plinii Bambergensis* (*Plinius Valerianus*), ed. Önnerfors
Pelagonius	Pelagonius, *Ars veterinaria*, ed. Fischer
Pliny	Pliny the Elder, *Natural History*, ed. Jones, Rackham, and Eichholz
Ps.	Pseudo-
Ps. Apul.	Pseudo-Apuleius, *Herbarius*, ed. Howald and Sigerist
Scrib.	Scribonius Largus, *Compositiones*, ed. Sconocchia
saec.	*saeculum*, century (used for dating, especially manuscripts)
sp.	species
Theod. *Eup.*	Theodorus Priscianus, *Euporiston libri III*, ed. Rose
Theophrastus *HP*	Theophrastus *Enquiry into Plants*, ed. Wimmer, trans. Hort
ThLL	*Thesaurus Linguae Latinae* 1900–
TLG	Thesaurus Linguae Graecae (database), University of California, Irvine
Varro	Marco Terentius Varro, *De re rustica*, ed. Davis Hooper
Vitruvius	Vitrivius Pollio, *De architectura*, ed. Granger

The Identification of Plants

The identification of plants in ancient and medieval sources is fraught with problems, and the *Alphabet of Galen* is no exception. Descriptions are brief, generic, and in cases so rudimentary that we can never be sure we have the right species or even genus, and it is always possible that the *AG* is describing a species that is now extinct or has evolved. The binomial botanical identifications given in each entry are primarily taken from consultation of André 1985, Beck 2005, and Halleux-Opsomer 1982 – disagreement among these or uncertainty is recorded in the notes to that entry – and also from scholarly literature where relevant, particularly the work of Alfred C. Andrews, John Riddle, John Scarborough, and Jerry Stannard. The symbol ~ is used in cases of considerable uncertainty (e.g., #90–2). A concise guide to the problems of pre-Linnaean identification is Reveal 1996, which contains many helpful references to encyclopedias, dictionaries, and guides to plant identification. Deciding on one common English name among many also common is often an arbitrary act. I have registered two English names where the literature seems to be evenly split in using one or the other, and have consulted Grigson 1974, the polyglot dictionaries of Perdok, ed. 1968, and Váczy 1980, and the information given in reliable databases on the web such as www.naturalstandard.com, and the International Plant Names Index, www.ipni.org.

Warning

The medical remedies recorded and discussed in this book are intended solely for historical analysis. Any use of them for therapeutic purposes could be dangerous.

Plate 1. Vienna, Österreichische Nationalbibliothek, Med. gr.1 fol. 3v. Sixth century. Seven physician pharmacists (clockwise from top middle): Galen, Dioscorides, Nicander, Rufus, Andreas, Apollonius Mus, Crateuas. Galen's centrality and chair reflects his dominant authority in the Greek tradition represented by this alphabetical and lavishly illustrated recension of Dioscorides' text, commissioned by the noblewoman Anicia Juliana in 512 AD. Yet Galen's pharmacological works were largely unknown in the Latin West until the eleventh century, in part explaining the attribution of the *Alphabet* to Galen.

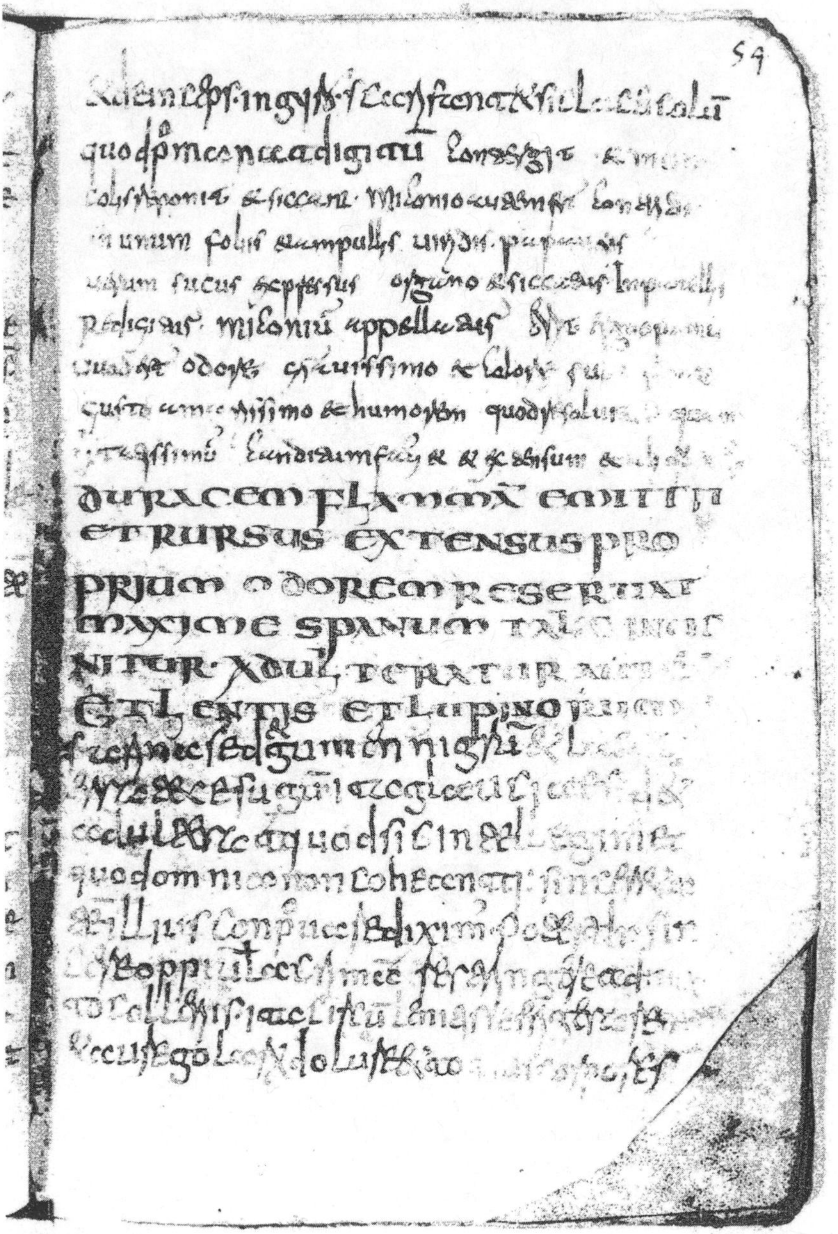

Plate 2. Vatican City, Biblioteca Apostolica Vaticana, Pal. lat. 187, fol. 54. Late seventh/early eighth century. The earliest surviving manuscript of the *Alphabet*. It took three scribes to get through the entry for opium (*AG* #199): see ch. 1.C, and ch. 5.C, MS *V.*

Plate 3. Accuracy in the *AG*. a) Dittany of Crete *AG* #88 (*Origanum dictamnus* L.), 'and on the small branches near the top there appears a small, pink flower' (photo courtesy of Georges Fontès, Instants de Saisons, http://isaisons.free.fr). b) Water parsnip #158 (*Sium angustifolium* L., *Sium erectum* Huds., Coville), 'produces a small white flower' (photo courtesy of http://www.szak-kert.hu). Neither flower is noticed by Dioscorides or Pliny: see ch. 1.I.

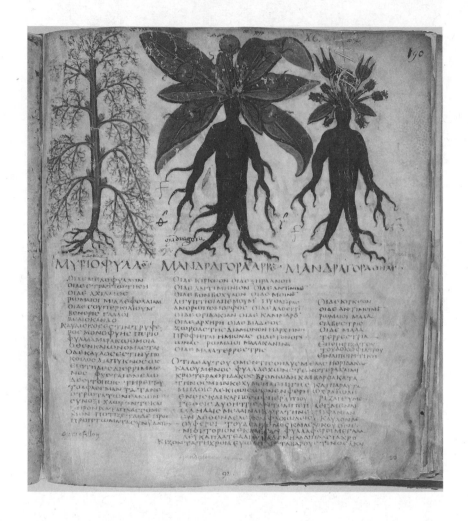

Plate 4. Naples, Biblioteca Nazionale, Gr. 1, fol. 90. Sixth–seventh century. Milfoil (*Achillea millefolium* L.) *AG* #192, and mandrake (*Mandragoras* sp. L.), *AG* #181. Mandrake 'reduces all pain and alertness,' because of its tropane alkaloids: see ch. 2.B. This illustrated (and alphabetized) manuscript of Dioscorides was written in Italy in the sixth or seventh century, around the same time as our earliest surviving copy of the *AG* (Vat. Pal. lat. 187 [*V*]). See ch. 1.D.

Plate 5. Naples, Biblioteca Nazionale, Gr. 1, fol. 48. Sixth–seventh century. Dog's-tooth grass (*Cynodon dactylon* Pers.) *AG* #36; and catmint (*Calamintha* sp. Lmk), two types, *AG* #196.

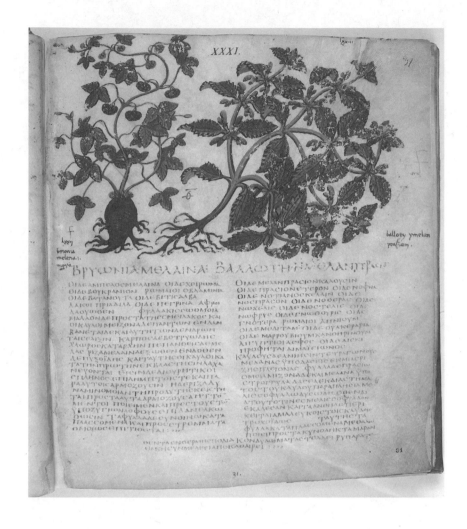

Plate 6. Naples, Biblioteca Nazionale, Gr. 1, fol. 31. Sixth–seventh century. Bryony (*Bryonia dioica* Jacq.) *AG* #47; and black horehound (*Ballota nigra* L.) *AG* #191. The latter is given the name *marrubiastrum* in the *AG*, a term which endured in Italy: see ch. 4.E.

Plate 7. Naples, Biblioteca Nazionale, Gr. 1, fol. 44. Sixth–seventh century. Centaury (*Centaurea centaurium* L.) *AG* #55 (larger and smaller types).

Plate 8. Naples, Biblioteca Nazionale, Gr. 1, fol. 87. Sixth–seventh century. Cardamom (*Ellettaria cardamomum* White and Maton) *AG* #71; and colocynth (*Citrullus colocynthis* Schrader) *AG* #54, 266.

THE ALPHABET OF GALEN

1

Introduction to the *Alphabet of Galen*

ॐ

A. Introduction

The subtitle of this book, *Pharmacy from Antiquity to the Middle Ages*, encapsulates the *Alphabet of Galen*'s place in the history of medicine: a handbook of ancient Greek pharmacy transmitted in Latin to the Middle Ages under the name of a famous doctor. Despite its name, the *Alphabet of Galen* has no relation to the renowned and influential physician of the ancient world, Galen of Pergamum (c. 129–217 AD). The name was a later attribution, applied to a Latin text describing 300 natural products (or 'simples') for medicinal use, arranged in alphabetical order, which in content and style dates to before Galen's time. The earliest surviving manuscript of the text presented here dates from the seventh century AD, and the *Alphabet* circulated in Europe until the twelfth and thirteenth centuries, when it was pillaged for larger and more ambitious pharmacopoeia. While the origin and date of the text ultimately remain mysterious, the *Alphabet of Galen* is a precious survivor of a rich tradition of ancient scientific literature from the third century BC to the first century AD that has mostly been lost. It salvages unique descriptions of plants and of mineral and animal products, and their methods of preparation, including (for example) the first recorded butter churn and the earliest mention of a quill pen. The *Alphabet of Galen* also espouses a distinct pharmacology of drug 'properties,' based on a combination of sensory perception and Greek scientific thought, to explain how each natural product heals certain illnesses. Its simple, technical Latin presents some distinctive features that contribute significantly to our knowledge of Latin and Greek philology, including 20 neologisms (many of which are technical terms) and 20 hitherto unknown uses of the *-aster* suffix (to a known 50 or so).

This book has two main goals. Firstly, it presents the first critical edition of the *Alphabet of Galen*, based on all surviving manuscripts dating from the seventh to the eleventh centuries, accompanied by an annotated English translation, and prefatory discussions exploring its pharmacology (chapter 2), possible sources (chapter 3), language and lexicography (chapter 4), and manuscript tradition (chapter 5). Specialist readers may turn to those chapters for details, and use the alphabetized subheadings in the text, also listed in the Table of Contents, to facilitate quick reference. This first chapter draws upon those findings and is intended to introduce the text and its general character, and to discuss its possible date and origin. The last-mentioned task necessarily invokes specialist discussion of philological matters and manuscript tradition, and again the subheadings can be used to avoid unwanted detail.

The book's second goal, reflected in its subtitle, is to contribute to our understanding of the intellectual heritage of the ancient Mediterranean world and its transmission to medieval Europe. The fall of the Roman empire, the establishment of regional kingdoms under barbarian rule, and conversion to Christianity all fundamentally changed the social, political, and cultural landscape of Western Europe. But always weathering these changes were quiet continuities of ancient medical traditions, repeatedly copied onto parchment and preserved in medieval manuscripts, so that study of the art of healing bridged the two different worlds. Pharmacy was by far the most useful, and consequently the most valued, branch of ancient medicine transmitted to the Middle Ages. By examining the content and transmission of one text, this study hopes to illuminate how ancient ideas of pharmacy continued to define medical discourse and practice in Europe long after Greek and Roman civilization had gone. To this secondary purpose may be added a plea for future students of medical, ancient, or medieval history to investigate other medical texts of uncertain date and origin to redeem their worth and place in history.

Because the issue of dating the text is a constant theme, the chronological shorthand used, often assumed by historians to be self-evident, should be made clear:

- the terms 'ancient Greece' and 'ancient Greek' (in relation to culture or history) generally refer to the period from the sixth to the fourth century BC, though 'ancient Greek' may be used to describe (written) Greek language up to the sixth century AD.
- 'ancient' and 'Antiquity' refer to the period from the third century BC to the second century AD, with 'imperial' specifically meaning the first two centuries AD, the height of Roman imperial expansion.

– 'late antique' and 'Late Antiquity' refer to the period from the mid-third to the mid-sixth century AD.
– 'medieval' and the 'Middle Ages' can be used to cover the period from the mid-sixth to the fourteenth century, but 'early medieval' and 'early Middle Ages' mean from the mid-sixth to the ninth century.

B. Character of the *AG* and Its Date of Composition

Though there is considerable variation throughout, the *Alphabet of Galen* (hereafter *AG*) keeps to a fairly standard format for presenting its information:

a) a brief physical description of plant or product,
b) its origin or location and/or ranking by type,
c) a bald statement that the plant or product has X or Y pharmaceutical 'properties' (*uires*, lit. 'powers,' 'forces'),
d) applications and uses, citing specific complaints or ailments.

The format, and often the content, of the *AG* is similar to that found in the most influential and voluminous ancient pharmacopoeia, *The Materials of Medicine* (Περὶ ὕλης ἰατρικῆς, or *De materia medica* in Latin) of Pedanios Dioscorides of Anazarbus (c. 40–90 AD), and to a much lesser extent, to the information on natural products for medical use given by his contemporary, Pliny the Elder (23–79 AD), scattered throughout his monumental encyclopedia, the *Natural History* (*Naturalis Historia*). The similarities misled some medieval and modern commentators to assume that the *AG* was derivative of these works, but the similarities are owed to the *AG*'s use of the same sources, among which may have been the lost work of Sextius Niger (c. 20 BC–40 AD: see ch. 3.D). The 300 products treated in the *AG* include 220 from plants, 61 from minerals, and 19 from animals. It is, therefore, very much a work on the 'materials of medicine' rather than a 'herbal,' for it even omits some of the more common plants considered to have medicinal or dietary benefit, particularly fruits, which had long been treated in ancient literary traditions.[1] But 300 is still an impressive number: comparable late antique or early medieval works in general treat far fewer items, such as the Ps. Apuleius *Herbarius* (c. 350–450 AD, with 131 plants), Gargilius Martialis's (fl. 220–60 AD) *Medi-*

---

1 Commonly used substances include garlic, lettuce, apple, quince (cydonian apple), onion, mallow, plum, citron, peach, cherry, hazel, pistachio, walnut. Puzzling or unidentified: *alcimonium AG* #11, *amomis* #28, *eridrium* #93, *nascapthon* and *elamulae* #198; *basilisca* #272 n.2.

cine from fruits and vegetables (60 plants), the Ps. Hippocratic *Dynamidia* texts (book 1 dating c. 500–700 AD, with c. 80–90 plants), and Ps. Dioscorides *Ex herbis femininis* (c. 500 AD, with 71 plants), and the similar Ps. Dioscorides *Curae herbarum* (c. 500 AD, 61 plants). The *AG*'s numbers are closer to ancient Roman writers such as Celsus (c. 25 BC–50 AD, c. 250 plants, 47 minerals, 10 animals) and Scribonius Largus (c. 43 AD, 242 plants, 36 minerals, 27 animals).[2] In the next generation, the encyclopedic works of Dioscorides (c. 700 plants, plus another c. 300 animal and mineral products) and Pliny (similar numbers) seem to reflect both the confidence and the wider world of plants and drugs ushered in by Roman imperial expansion and consolidation.[3] Earlier Greek medical writers also seem closer to the *AG*'s range: the Hippocratic corpus (fourth–first century BC) mentions between 200 and 450 species of plants, and Nicander of Colophon (fl. 130 BC) named around 300 plants. But we cannot draw any conclusions here: Galen covered 475 items in his *On the properties of simples*, and early Byzantine writers like Aetius of Amida (c. 500 AD), Alexander of Tralles (c. 525–605), and Paul of Aegina (c. 640s) recorded between 400 and 600 drugs in their respective works.[4] The most we can say is that in terms of numbers, the *AG* looks more ancient, and more Greek, than early medieval.

The plants and products of the *AG* have an ancient, eastern Mediterranean slant, as do the geographic references. The two most frequently named places for the origin of medicines are India (15 references) and Egypt (14), two *loci classici* for exotic drugs and plants known to Greek and Roman medicine.[5] The next most popular locations include Pontus (the area around

---

2   For editions of these works, cited frequently in this book, see the list of abbreviations and bibliography.

3   Scarborough 1986, 59; Nutton 2004, 171–9; idem 2008. On Pliny see also Morton 1986 and Stannard 1965.

4   Scholars arrive at different tallies for the Hippocratic corpus: Riddle 1985, xvii, counts 130 'substances'; Stannard 1961, 499, counts c. 225 plants; similarly, von Staden 1989, 18, c. 250 plants and 40 each of animal and mineral products. Scarborough 1984, 214 and n.27 augments this to 450 species of plants. See further G. Aliotta et al., eds 2003. The plants in the Hippocratic corpus are conveniently listed with botanical identifications in Totelin 2009, 353–61, who also notes (ibid., 2) that the corpus contains over 1,500 recipes. Nicander: Scarborough 1984, also noting there are 550 species in Theophrastus (c. 300 BC). Scribonius: Scarborough 1986, 63. For Byzantine writers, Alexander (495 drugs), Aetius (418 from plants, another 200 from minerals), Paul (600 from plants, 80 non-botanicals), see Scarborough 1984, 227, 225, 228.

5   On the ancient drug trade, Scarborough 1982; Nutton 1985; idem 2004, 171–86; Innes Miller 1969. An excellent discussion of exotic ingredients in the Hippocratic corpus is Totelin 2009, 141–96.

the Black Sea, 11 references), Crete (9), Arabia (9), Macedonia (6), and Ethiopia (6). After this, we have references to Cyprus (4), Syria (3), Italy (3), Cyrene (3), Judea (3), Cappadocia (3), and Armenia (3). References to locations in the western Mediterranean are fewer, including Spain (3), Africa (4), and Germany (1, for its better soap). The exception of Gaul (8 references), however, and the curious references to Italy (see below, ch. 1.F) should give us pause. Nonetheless, the geography reflects a Greek viewpoint: reference is made to the islands of Chios (#151, 186, 228), Melos (#7), Zakynthos (#11), and Lemnos (#276), and to cities such as Alexandria (#94, 163, 261), Colophon (#252), Corinth (#182), and Nicomedia (#119), as well as to the areas or provinces of Arcadia (#201), Attica (type, #25, 179, 207), the Hellespont (#8), Cilicia (#156), Media (#28), Pamphylia (#103), Phrygia (#29, 169), and possibly Lydia (#210). Also the quality of some references suggest eastern provenance, or at least traditions of Greek learning: knowledge of an obscure Thracian river called 'Pontus in Sintia (#167 n.1); or of the river Gage in Lycia (#166); that 'Syrian' spikenard derives its name from the direction of the mountains in India where it grows (#236); an erroneous nod to the 'lands of Ammon, once a king of ancient Cyrene' (#4), and so on. Casual references in the text also evoke antiquity: the form of a flower is likened to 'the masks with gaping mouths used in comedies' (#157); mention is made of 'provinces' ('every province' #153, those of *Gallia* and *Italia* #239), while entries for the plant *silphium* (#147, 250), and an ancient recipe for the Egyptian incense known as *cyphi* (#76), both smack of the ancient Mediterranean world BC. Similarly, the complete absence of any religious references, or indeed of any superstitious practice (particularly striking for a medical text), points to the rationalist traditions of ancient Greek science (see ch. 2.I).

The pharmacology of *AG* also suggests an ancient date. The *AG*'s particular use of drug 'properties' and its concentration on four types (heating, cooling, loosening, and binding [i.e., astringent]) point to traditions of Greek scientific cosmology, and have affinities with concepts of pharmacy found in Hippocratic writings and the use of properties by Dioscorides (see ch. 2.D–E). The pharmacology is certainly pre-Galenic, in that it contains no trace of Galen's elaborate theorizing on drug properties, such as various degrees of intensity or distinguishing active and passive faculties, and so forth.[6] This is not a particularly decisive clue to the text's date,

---

6 Galen wrote several long and theoretical works on drugs, principally *On the composition of drugs according to ailment*, ed. K.[ühn]12, 378–13.361, *On the composition of drugs according to type*, K.13, 362–1058, and *On the properties of simples*, K.11, 379–12.377. It is now recognized that for the first two Galen borrowed heavily from previous works (often verbatim), and their influence in the West was limited to a few

since Galen's pharmacology had little impact in the West until the eleventh century – one of the reasons why this *Alphabet* could be attributed to Galen (see below, ch. 1.D, and plate 1). More importantly, the *AG* does not subscribe to the theory of the 'four humours,' a theory particular to one or two Hippocratic texts and which Galen refined and promoted, so that this 'humoralism' became the dominant medical theory from Late Antiquity onwards and enjoyed a disturbing longevity until early modern times and beyond (ch. 2.G). Instead, the *AG* reflects an earlier, pragmatic tradition of pharmacy of simples based primarily on what was known to work. Its notion of 'properties' is largely theoretical window-dressing to explain how something works, often by linking notions of elemental 'forces' of nature (inside and outside the body) to sensory perception: the taste, smell, touch, or effect of a plant or substance, on which the *AG* can be more explicit than most ancient or medieval texts (see ch. 2.C).

The *AG*'s pragmatic, empirical approach is similar to that of Dioscorides, but Dioscorides broke with the tradition represented by the *AG*'s alphabetical order and arranged his drugs according to their 'affinities' in treating the same ailments. As John Riddle has shown, Dioscorides' arrangement of products according to their common properties pointed the way for investigation into why various different substances should produce the same pharmaceutical effect. Had science continued down this route, it might have led to questions about the active ingredients within each simple that caused similar reactions regardless of their form in nature (plant, animal, mineral): Dioscorides realized, for example, that there was a common substance shared by shelled aquatic animals and limestone (which we now know is calcium oxide). Dioscorides may have pointed the way, but his contemporaries 'saw him point and described the end of his finger.'[7] The tendency after Dioscorides was to make medicines from many different ingredients in the form of compound drugs, that is, polypharmacy, a process not really reversed until the sixteenth century, when

compound recipes. The work on simples was largely an abridgment of Dioscorides, excising the latter's descriptions, rankings of types, advice on preparations, etc., in favour of relating properties to Galen's larger theoretical framework. Even Galen seems to have recognized its limited usefulness: he only provided grades of intensity for a third (161) of the 475 botanical simples he surveyed. Nonetheless I have included *On the properties of simples* among the *Comparanda* to the Latin text (see ch. 3.H) to highlight the different traditions. Excellent, complementary expositions of Galen's pharmacy are given by Riddle 1985, 165–78, Scarborough 1984, 215–20, and Nutton 2004, 241–7. See further Harig 1974, and the papers in Debru, ed. 1997, esp. Touwaide 1997. On the longevity and dominance of Galenism, Temkin 1973.

7 Riddle 1985, 166. On the Latin translation of Dioscorides' *De materia medica*, see ch. 3, n.1. A useful survey of ancient and medieval pharmacy is Schmitz 1998, 12–186.

Paracelsus declared that a simple, or single natural-product drug, was itself a mixture of compounds. The *AG*, therefore, belongs to an ancient, pragmatic tradition of treating simples as the basic 'materials of medicine,' while its alphabetical order and elemental categories of properties may point to a date before Dioscorides (below, ch. 2.D–E, and ch. 3.D–E.).

The language of the *AG*, however, gives a more complicated picture in terms of securing its date and origin. Its simple technical Latin is classical in terms of grammar and syntax, yet the text is also replete with Grecisms, odd linguistic features (the *-aster* suffix, the adverb *uiscide*), and some vocabulary that point toward a late antique date (see ch. 4.B–F). These elements of vocabulary may well be later additions to the text, which still leaves the greater weight of the evidence on an antique, probably imperial, date for the text's original composition. But the question of language presents another conundrum: how much the Latin text, as it stands from A to Z, is a translation of a pre-existing Greek text of similar, alphabetized format. The evidence is inconclusive, but does suggest a bilingual environment of composition (see ch. 4.F). If the bulk of Latin translations of Greek medical lore is anything to go by, then the period from the fourth to the sixth century AD is a strong possibility for the date of the text in its extant format, whatever its original date of composition or language.

To these inferences we must add another two, related problems for dating the *AG* which are dealt with below: firstly, whether the prologue and epilogue found attached to the text in later manuscripts (from the eleventh century) were part of the original composition; and secondly, whether their dedication to a certain 'Paternus'/'Paternianus' signifies an authentic relationship with another Latin text on drugs of equally obscure origin, the *Liber de dynamidiis*, which is also addressed to this same Paternus/ Paternianus, and which circulated alongside the *AG* in some manuscripts. In lieu of a future discovery (a manuscript, a dedication, a literary reference) that adds to our evidence, all we can say with certainty is that the *AG* consists of a body of ancient knowledge on natural product simples centred on the Mediterranean world that, in the state presented in this book, circulated in medieval Europe from the seventh to the thirteenth century under the name of a doctor whose fame was second only to Hippocrates.

C. The History of the *AG*

Somewhere in seventh- or early eighth-century Italy, several scribes took turns in writing out a text entitled *Alfabetum Galieni* onto small pages (c. 14 × 9.5 cm) for a pocket-sized manuscript. The scribes, some with extremely shaky penmanship and grasp of Latin, employed different styles of script, though curiously all used abbreviations common to Irish and Anglo-Saxon manuscripts. They clearly struggled to render the strange Greek terms and exotic names of plants and faraway places in the vulgar Latin of their day. Curiously enough, it took three scribes to get through the entry for *opium* (#199, and plate 2): the first needed to stop after mentioning that harvesting involves wiping off the poppy's milky fluid with a finger; the second suddenly broke off when it came to describing the best opium and its ignition; the third, whose characteristic uncial hand began the manuscript (fols. 8–24v), continued the description of burning opium's 'lasting flame,' that Spain is a major producer (a puzzling observation), and that opium is adulterated with flour made of lentils and lupine: then he too suddenly stops writing and the first scribe returns. A case of opium in the scriptorium? And why the insular abbreviations?

Now preserved at the Vatican library, the manuscript (Vat. Pal. lat. 187, our *V*: see ch. 5.C) is of an unusually early date for medical Latin texts, and is also the earliest known copy of the *Alphabet of Galen*, which survives in another seven manuscripts dating from the late ninth to the eleventh or twelfth century. Hardly a huge number, but still quite significant for the transmission of medical works in early medieval Europe,[8] and many of these manuscripts are among our most important for the transmission of ancient Greek and Roman medical knowledge to the medieval Latin West. Within these manuscripts the *AG* kept literary company with Hippocrates, Galen, Oribasius, Theodorus Priscianus, Ps. Apuleius, and a host of treatises, tracts, letters, and recipes often attributed to ancient doctors or grouped together to form a coherent thematic whole.[9] The *AG* served as a chief source for the creation of new works by eleventh- and twelfth-centu-

---

8   The Latin translation of Dioscorides, for example, survives in three manuscripts, and another two or three fragments: Riddle 1980, 20–3: see also ch. 3, n.1. The most widely copied work of Galen in Latin in the early Middle Ages, *De medendi methodo ad Glauconem* (see ch. 1.E–F below), survives in about 19 manuscripts prior to the twelfth century, as does Ps. Apuleius: Beccaria 1956, 455, 441, et passim.

9   For example Monte Cassino V 79 (*M*), Vat. Barb. Lat.160 (*B*), Chartres 62 (*F*), Lucca Governativa 296 (*L*): for details see ch. 5. The essential catalogue of Latin medical manuscripts (saec. IX–XI) is Beccaria 1956. On their character, see Wallis 1995, 112 n.30, whose judicious count of types of text within described by Beccaria reveals that pharmacological texts (over 324) far outnumber other genres. The next most popular were for general medicine (151, including introductions 25, compendia 93, aphorisms 18, medical philosophy 15), diagnosis and medical semeiotics (112), hygiene and preventative medicine (96, of which seasonal regimen 64, dietetics 28), diseases and cures (65), surgery (58, mostly phlebotomy 46), humoral physiology and pathology (54), gynecology and embryology (44).

ry authors when renewed interest and activity in pharmacy produced larger, more ambitious compendia. In Salerno we can witness the *AG*'s eclipse between Gariopontus (d. 1050), who used the text in his *Passionarius*, and the slightly later *magna opera* of Salertanian pharmacology, Platearius's *De simplici medicina* and the *Circa Instans*, which seem to have ignored it.[10] But elsewhere the *AG* continued to be used until the late thirteenth century. The eleventh- and twelfth-century compilers of the Alphabetical Dioscorides drew their information on minerals from the *AG*, and the text was cited verbatim by Rufinus of Pisa for his encyclopedic herbal (c. 1287). That Rufinus, and his modern editor, erroneously assumed that the *AG* was Dioscorides' *De materia medica* demonstrates the esteem in which the *AG* was held, since Rufinus cited this 'Dyascorides' as his foremost authority among the five or six sources (his 'summi phylosophi' or 'sapientes antiquores') which he sets out for each entry in his herbal.[11] The text was also cited with approval by Simon of Genoa in his herbal of c. 1292.[12]

The *AG* next surfaces in its *editio princeps* published in 1490 by Filipo Pinzi in Venice, among a collection of Galen's works in Latin edited by Diomedes Bonardus. Throughout the sixteenth century this version was reprinted in the *Opera omnia* of the Latin Galen published by the house of Junta (1522, 1528, 1565, 1586), in which it appears in the 'spuria' section of Pseudo-Galenic Latin works. In the following century the text, unaltered, was incorporated into R. Charterius's edition of Galenic and Hippocratic works (Paris 1679), but C.G. Kühn dismissed it as one text among many 'spuria' to be excluded from his monumental, 20-volume edition of Galen's oeuvre (1821–33). Modern scholarship overlooked or mistook the *AG* to be derivative work culled from Dioscorides, Pliny, or Ps. Apuleius's *Herbarius*,[13] but there were some notable exceptions. One of the first

---

10 Gariopontus's *Passionarius* was published as *De morborum causis accidentibus et curationibus ad totum corporis aegritudines praxeon, libri V* (Basel, 1531). See Glaze 2005 (and forthcoming edition).
11 *The Herbal of Rufinus* ed. Thorndike 1946. Thorndike admits that he 'failed to find the version of Dioscorides that [Rufinus] used,' noted that the entries are briefer, and that it contained a 'large number of simples ... not treated in the Latin text [of Dioscorides]' (xxviii). Many of the examples he gives (ibid. n.27) are indeed from the *AG*: e.g., *agresta* (= *AG* #291), *anabula* (= *AG* #252), *bdellium, cassia fistula* (= *AG* #39), *cinamum* (=*AG* #77), *difriges* (= *AG* #89), etc. Yet others derive from another source: e.g., *alchanna, anacardi, banbacium, berberi, ciutium, cima, colocasia, conobium, eruscus*, etc.
12 Simon of Genoa, *Clavis Sanationis*, in *Opus pandectarum Mathaei Sylvatici* (Turin, 1526): 'De simplicibus medicamentis ad Paternianum.'
13 Rose 1874, 34, 38. Singer 1927, 47. Puhlmann 1930, 397. Sigerist 1934, 33. MacKinney 1935–6, 408–12, noted its peculiar character, but assumed (following Rose) that it was

scholars to recognize the *AG*'s original and unique descriptions of plants was Ernst Meyer in his monumental *Geschichte der Botanik* (1854–7), who offered a list of botanical identifications for some of the *AG*'s entries. The editors of the *Thesaurus Linguae Latinae* (1900–) recorded aspects of the *AG*'s unique vocabulary, even comparing manuscripts for readings (our *V*, *L*, *M*, and *C*: see ch. 5), and Jacques André noted its independent character in preparing his dictionary of ancient Latin plant names (1985). But it was not until Carmelia Halleux-Opsomer's article of 1982 that the *AG* received attention in its own right, and was declared to be an independent and unique witness to a rich tradition of ancient pharmacological literature now lost, despite the supposed 'early medieval' version in which it survived.[14] Halleux-Opsomer's study was inspired and insightful, but was forced to rely upon the Junta text (*J*), which, based upon late manuscripts (unidentified and probably now lost), contains many corrupt and misleading readings. The collation of all known surviving manuscripts, particularly our earliest dating from the seventh to the tenth centuries (*V*, *F*, *L*), allows us to reconstruct the text on more secure grounds. As a general rule, the later the manuscript, the closer its text to the printed Junta version (particularly *C*): later manuscripts also show a greater distance from ancient Greek and Latin vocabulary and scientific traditions. Yet the later witnesses should not be left out: they display local colour and facilitate understanding of how pharmacological lore of this sort was accessed, updated, and transmitted over time (see comments in ch. 5.B–C).

D. The Prologue and Epilogue

Our title *Alphabet of Galen* is justified by its predominance among the manuscripts. Though some manuscripts lack one of these words, 'Galen' or 'Alphabet' is always in the title, with one exception where it is simply called the 'Book of drugs' (*Liber pigmentorum L*). Our earliest manuscript witness (*V*) named an addressee called 'Paternus' in its title *Galieni*

---

derivative of Dioscorides. The text was claimed to be a product of the school of Salerno by Salvatore de Renzi 1852, 40–1, who noted its use by Gariopontus in the latter's *Passionarius*, and further suggested that Gariopontus was himself the author. Meyer 1857, IV:484–91, corrected this mistake, but the attribution was still accepted, e.g., Pazzini 1967, 9–11. On Gariopontus, see now Glaze 2005, 53–76. A re-evaluation of De Renzi's work is Jacquart and Paravicini Bagliani, eds 2008.

14 Meyer, 1857 IV: 492–3. Halleux-Opsomer 1982a. Reliance on the Junta text vitiated their commendable attempts to identify the *AG*'s contents in places: e.g., see *AG* #2 n.1, 10 n.1, 47 n.1, 93 n.1, 98 n.1, 120 n.1, 153 n.3, 230 n.1, 234 n.1, 237 n.1, 275 n.1, though cf. #273 n.1, 289 n.1.

alphabetum ad Paternum, as does another early witness from the ninth or tenth century (*Alfabetum Galieni ad Paternum F,* the text of which is closely related to *V*).[15] Two eleventh-century manuscripts, which call the work the 'Book of Galen on herbs, aromatics, [stones,] and animals,' change the name of this addressee to 'Paternianus' and 'Paterninus' (*P* and *B* respectively). More importantly, they preface the text with a prologue and epilogue addressed to this mysterious figure: the epilogue also appears as a preface in *C*, again of the eleventh century.[16] This prologue and epilogue were included in the *editio princeps* of the *AG* edited by Diomedes Bonardus, which sported the title of *Liber Galieni de simplicibus medicinis ad Paternianum*, and was reprinted with that title (exchanging *medicaminibus* for *medicinis*) in the Juntine editions of the sixteenth century (*J*).[17] As it happens, another Pseudo-Galenic work on pharmacy, called the *Liber de dynamidiis*, also addresses this same Paternianus (Paterninus, Paternus), raising the possibility that our text was originally one part of a larger commission or series of works. There are, however, some grounds for doubting whether the prologue and epilogue to the *AG* were part of its original composition, and for suspecting that they, and the prologue to the *Liber de dynamidiis*, were part of a later process of packaging complementary works together as part of a compendium of 'Galenic' works, a medieval anthology of 'the essential Galen' in Latin.

It is possible that the name *Paternus* or *Paterninus/Paternianus* is merely a literary pseudonym, as the epilogue may hint (*pater carissime Paterniane*, though *frater* in *JC*). There is no mention of a person of similar name in Galen's corpus of works, nor any testimony to a Paternus/Paternianus from Galen's time or after who even remotely fits the bill.[18] But this is hardly conclusive: more troubling are aspects of vocabulary in the prologue. Firstly, the stated purpose to describe 'omnia smigmata tam metallica quam et aromatica et omnem herbarum nationem' is a little at odds with the language of the *AG* itself. The word *smigmata* (sing. *smigma/smegma*), from σμῆμα, meaning 'soap' or 'unguent,' is a rare term and

---

15 Full incipits and explicits for each recension are given in ch. 5.C.

16 The prologue is termed *prefatio* in B (and likewise in *J*), while both *B* and *P* term the epilogue a *prologus*, though it was intended to end the text.

17 Though no surviving manuscript contains any form of *simplex* or *medicamen/medicinus* in its title, it was cited as such by Simon of Genoa in his herbal of c. 1292: above, n.12.

18 Only two Paterniani appear in the sources for the period 260–395: a *Paternianus notarius* sent to Illyricum by Valentinian (c. 374), and a *praefectus legionis* of 283–4 in Aquincum (Pannonia Inferior). Martindale et al., eds 1971–92, *PLRE* I: 670–1. The most appealing candidate is Paterius the philosopher, mid- to late fourth century, who taught after Iamblichus in Athens, ibid.

was mainly used in medical contexts to mean specifically a cleansing agent or detersive lotion.[19] It only appears once in the text of the *AG* (#213), with this very meaning of cleansing agent, though two early manuscripts (*V*, *L*) read here *migmata* ('mixtures'), a term we find used in two other Latin medical texts (both late fourth-century), but in its proper sense of 'compound drug' or mixture,[20] whereas the *AG* deals with simples. For the same reason, the reading of *J*, *sinigmata* (presumably from σύμμιγμα, 'commixture'), cannot stand: apart from being fairly rare even in Greek, the meaning is wrong, and no manuscript supports it.[21] On the other hand, the genre of the preface encouraged pretensions of Greek learning, and the use of *migma*, *smigma*, or even *synigma* to mean 'drug/medicament,' in order to state the grandiose intention 'to describe every drug/medicament derived from minerals as well as spices [etc.],' may be serving that end, even if the term was inappropriate *stricto sensu*.

But other terms are also jarring. The use of *aromatica* for 'aromatic herbs' (perhaps simply meaning 'spices' here) seems to be a late antique mutation of *aromata*, the preferred term among ancient Greek and Roman authors for this category of plants, and the *AG* uses *aromata* on one occasion with this meaning precisely (#190). *Aromaticus* is also found on one occasion in the text of the *AG*, but the direct parallel with Dioscorides shows that it is a translation of a Greek source using *aromata* (hence both authors probably used the same source).[22] The use of *natio* for every 'species' or 'type' of plant (*omnis herbarum natio*) is unique, and though the meaning of *natio* can mean 'type' and makes perfect sense, it is not used elsewhere in the text, *species* being the common term throughout.[23] There

---

19 Diosc. II.4, V.108 (both for teeth), 118 (skin lotion); similarly Pliny 20.12, 22.156, 24.43 (skin/facial lotion), but note 34.34 (floating impurities containing copper). cf. Dan. 13.17. LSJ records that the term σμηκτικός was used by Diphilus Siphnius (saec. III BC) to mean 'purgative.'

20 *Med. Plinii* 3.33.1 'ad commendanda migmata ingentes pecunias poscunt [medici]'; Marc. *Med.* pref. 7 'uersiculis quoque lusimus migmatum et specierum digestione compositis.'

21 σύμμιγμα (from συν + μείγνυμι 'to mix [together]') is found only in Plutarch, Posidonius, Chrysippus, Zeno Citeius (in one instance via citation from Galen), and in a Ps. Galenic work (*Definitiones Medicae*), but never in a strictly pharmaceutical sense.

22 *AG* #53 *expressis aromaticis*: cf. Diosc. I.27 'ἐκ τοῦ κροκίνου μύρου τῶν ἀρωμάτων ἐκπιεσθέντων καὶ ἀναπλασθέντων.' On Dioscorides and sources, see ch. 3.B. *Aromata*: e.g., Celsus 3.21.18, Pliny 14.109. A category of plants in Isidore 17.8.1, 'quaequae fragrantis odoris India uel Arabia mittit,' etc. *Aromaticus/-ca*: *ThLL* lists Vegetius, Caelius Aurelius, Pelagonius, Marcellus Empiricus, Theodorus Priscus, Ps. Apuleius, plus nonmedical use in Sedulius and Venantius Fortunatus.

23 Pliny applies *natio* for 'types' of honey and resin, 22.109, 24.32. Halleux-Opsomer

is also no mention of animal products in the preface. Finally, the use of *exemplatio* really rings alarm bells: the word is seemingly unattested until the ninth century, but is obviously derived from *exemplum*, *exemplar*, and (postclassical) *exemplarium*, meaning 'copy,' 'draft,' 'transcript,' and is used as a verb, *exemplare*, meaning 'to copy out,' from the fifth century onwards.[24]

The claim that there was a previously existing text that has (now) been arranged in alphabetical order to avoid 'unfurling the entire roll' (*ne ... totum uolumen euoluat*) begs questions that are difficult to answer. The mention of a roll might suggest a date before the end of the fourth century AD, when the codex completely takes over as the dominant format of texts, a process that begins as early as the first century AD, and which Christians took up with far greater alacrity than pagans and traditions of secular literature.[25] However, throughout this transition, and even well after it, the word *uolumen* could be applied to codex-form books also, and even charters. We find authors such as Theodorus Priscianus (late fourth century) and Caelius Aurelianus (fifth century) using the term *uolumen* to describe their medical works, which may represent tradition rather than rolls.[26] Nonetheless, reference to avoiding the need to 'unfurl the entire roll' by alphabetizing its contents does suggest reading practices of the ancient world.

Alphabetization, unfortunately, provides no chronological clues. The practice dates back to third-century BC Alexandria, was railed against by Dioscorides for works on pharmacy, but remained common practice, as known for Pamphilos's lost work on medicinal plants (c. 150 AD), and witnessed in Galen's own treatise on *Simples* (which relied heavily on Dioscorides), Oribasius's abridgment of Dioscorides' text (c. 360s), two lavish sixth-century manuscripts of Dioscorides, and the works of Aetius

1982a, 68, suggested a neologism for *herbatio* (*J*) but both manuscripts (*B*, *P*) read *herbarum nationem*. She points out (ibid. 69 n.31) that the same classifications appear in Dioscorides' prefaces to books I and V, but the order of the first is different (plants, minerals, aromatics), and the latter is altogether different in scope and tone.

24 As found in Augustine, Sidonius Apollinaris, Gregory of Tours (*ThLL*): its meaning of 'to adduce as example' dates a little earlier. Halleux-Opsomer 1982a translates *J* (*in simplario tractantur*) as 'qui sont traités dans l'art des simples,' suggesting that *simplarium* is a neologism, but the manuscripts rule this out.

25 The classic study is Roberts and Skeat 1983; important updates are Gamble 1995 and Hurtado 2006.

26 Theodorus, *Eup.* I (pref.), 'in his igitur voluminibus non studium tenebo gloriae,' (1.4) 'hoc igitur volumine,' ed. Rose, 1, 4. Caelius *Medicinales Responsiones*, pref. 'ut isto volumine omnis diaeteticarum cura compleatur,' cited in Flammini 1998b, 170.

of Amida (mid-sixth century) and Paul of Aegineta (seventh century).[27] In any case, the prologue confesses the existence of a previous work, which is consonant with all other indications, including the reference to a *volumen*, that the content of the *AG* dates to Antiquity. The issues of vocabulary mentioned above may be owed to scribal traditions anytime from Late Antiquity to the eleventh century, the century in which the manuscripts that include the prologue and epilogue were written. The epilogue presents no troubling issues, except for the question of who were the *antiqui et receptissimi auctores* consulted (see ch. 3). Halleux-Opsomer perceptively noted that the phrase *examinatissima diligentia* is also found in Augustine's *Confessions*, and apparently nowhere else:[28] again, an indicator, however slight, of late antique ambience.

A stronger indicator of a late antique date for the *AG*'s prologue and epilogue is their similarity in style to other prologues or prefaces that are thought to date from this period in their extant form, many of them likewise attributed to Hippocrates or Galen. Among these are the Ps. Hippocratic *Letter to Maecenas* and *Letter to King Antiochus*, which have some resonances with our prologue, as does the preface of a work on the healing properties of betony attributed to Antonius Musa (doctor to the emperor Augustus), a work that is sometimes attributed to Hippocrates in manuscripts.[29] Their simple advice on how to maintain health, their layman's language, and their brevity made these letters 'the most frequently transmitted medical texts from Late Antiquity.'[30] But many of these also may

---

27 Galen, *Simp.* K.11, 379–892; 12, 1–377. On Pamphilos, ibid. VI. *proem.*, K.11, 792–6 (and see ch. 3.E). Oribasius, ed. Raeder, 1928–33. Juliana Anicia manuscript (Vienna, Nationalbibliothek, Med. gr.1) ed. (facsimile) Unterkircher, 1965. Naples, Biblioteca Nazionale, Codex ex Vindobonensis Graecus 1. ed. (facsimile) Bertelli et al. 1992. Aetius of Amida, ed. Olivieri, 1938. Paul of Aegineta, ed. Heiberg, 1921–4. Alphabetization, Daly 1967. See also ch. 3.D.

28 *Conf.* 7.6, 'examinatissima diligentia nosse.' Halleux-Opsomer 1982a, 70. But note also Theodorus, Eup. III (*Gynaecia* pref.), 'et ego quidem te scientia iuvabo ... exerce ... tua diligentia,' ed. Rose 1894, 224; and also Caelius Aurelianus, *Gynaecia*, 'qui sepe exigunt diligentiam medicine (*scil.* morbi feminarum),' in Flammini 1998b, 160.

29 *Ep. Hipp. ad regem Antiochum* 'ut mathesin quoque optime scias, optimum duxi etiam rationem salutis tuae tibi notissimam fieri,' 'ad te scribendum putavi,' in Sconocchia 1998, 123 (from Paris lat. 6837, i.e., our *P*). On the Maecanas and Antiochus letters see Opsomer and Halleux 1985, and Sabbah, Corsetti, and Fischer 1987, no. 320–34. Antonius Musa, 'Cum proposuerim mihi medicamenta dicere,' 'proponendum putavi' 'non esse necessarium putavi,' in Zurli 1992c, 435–6. Also Ps. Apuleius, *De taxone*, 'plurimis exemplis experti sumus,' ed. Howald and Sigerist 1927, 229. Cf. the 'Letter to Marcellinus,' ed. Zurli 1998 (II), 381 (see MS *L* in ch. 5.C).

30 Langslow 2000, 75.

date from an earlier, imperial or late imperial period. On firmer ground is a comparison with the prologue written by Vindicianus, a famous doctor of late antiquity who, as it happens, was a friend of Augustine.[31] While most of Vindicanus's work has been lost, one popular tract which survived (and is found in two manuscripts of the *AG*, *M* and *B*) is the *Letter to Pentadius*, his nephew, a short work which provides an elementary account of physiology and the theory of the four humours. The *Letter* opens with a prologue similar in tone to that of the *AG*:

> I well know that you, dearest nephew, are learned enough in Greek letters to be able to approach this discipline. Nevertheless, lest I omit anything you asked me to set down, I have translated into Latin (*latinaui*) the fundamentals (*intima*)[32] from the medical books of Hippocrates. And because you are worthy of such, I faithfully transmit and will give to you the books of your grandfather, my father, from which you may learn how the entire world works, and also may know how much knowledge our family possessed. With this present book, however, I will undertake to explain to you the nature and constitution of every body.[33]

The rhetorical posturing and air of intimacy is similar, though that was also a requirement for a dedicated prologue across several genres of literature.[34] The distinct verb (*latinare*, literally 'to Latinize'), found in only one other late antique author, Caelius Aurelianus, points to a cultural context that well matches the *AG*, that of the transference of Greek medical learning in the 'Golden Age' of medical Latin from 370 to 450 AD.[35] Vindicianus's advocacy of humoral theory, however, underlines the *AG*'s more antique vintage. Nonetheless, the prologue and epilogue of the *AG* exhibit aspects of vocabulary that point to a late antique environment in which we find similar treatises and letters that transmit Greek medical learning to a

---

31 Augustine on Vindicianus, *Conf.* 4.3.5, 7.6.8: *Ep.* 138.3. See also Cilliers 2005. On Latinity, Önnerfors 1993, 280–7.

32 Another reading almost equally attested in the manuscript tradition is *syntomata*, as recorded by Rose (next note), 485, 6. *latinare/* Caelius Aurelianus *latinauimus*, see Zurli 1992b.

33 Vindicianus, *Epistula ad Pentadium* ed. Rose 1894, 485–6.

34 Surveyed in Santini and Scivoletto 1990, 1992, and Santini, Scivoletto, and Zurli 1998. On Greek medical epistolography, see Wiedemann 1976, 21–35. See also Langslow 2007.

35 Langslow 2000, 63. Caelius, *Chron.* V.77 (Bendz ed., conjectures *latin<iz>auimus*): and *Acut.*, *Pref.* (see Urso 1998, 182) for further emphasis on translation from Greek: likewise the preface of Ps. Oribasius, 'Libri quinque de simplicium uirtutibus,' Flammini 1998c, 302 (note 'singulari adiutorio *possint subvenire* [simplicia medicamenta]').

Latinate audience, claiming either Hippocrates or Galen as their authority, or indeed as their author. With this in mind, we can now address the other text dedicated to Paternianus.

E. The *Liber de dynamidiis*

A text entitled the *Liber de dynamidiis*,[36] which appears right next to the *AG* in at least three manuscripts (*F*, *C*, *P*), and was included in another (*B*), also addresses Paternianus.[37]

> This is a description of the properties of all drugs which concern knowledge of the art of medicine. If those who practise the art of medicine fully know these, they shall always assist men with these aids and cures. Because if one is truly assiduous, and becomes fully versed in this discipline and its doctrine, he will easily be considered with favour.[38] But those who ignore the doctrine will be all the more frustrated; for the more they neglect this, the

---

36 *Liber de dynamidiis*. Also published among the 'spuria' in *J*, fols. 18–19. Besides our *F*, *M*, and *B*, Beccaria 1956 lists another five manuscripts of this *Liber*: Paris, lat. 7028, bis, fols. 136v–137v, 144–6, saec. X: Vendôme, Bibl. Municipale 109, fols. 87a–87vb, saec. XI, see Sigerist 1943: 78–9; Glasgow, Hunterian Museum V.3.2, fols. 22v–24v, saec. X; London, British Museum, Sloane 475, fols. 34v–35, saec XII; London, British Museum, cod. Additional 8928, fols. 5v–6, saec. X. The *Liber de dynamidiis* is often confused, in manuscripts and in print, with another text dubbed by the Juntine editors as the *Liber Alter de Dynamidiis* (*J*, fols. 19–35v: incipit: 'Libellum quem roganti tibi promisi'), which is a hotchpotch of introductory medical material, including brief and simplified summaries of humoral theory, the different categories of drugs or *dynamidia*, Ps. Hippocrates *Epistula ad Maecenas*, and *Epistula ad regem Antiochum* (above, n.29), before switching its main focus (fols. 23–35) to recipes addressing specific ailments listed head to toe. Some of the 40 or so drugs mentioned (fols. 19v–23) appear in the *AG*, but the focus is upon their relation to humoral theory. The Juntine editors stated that the work was derived from the works of Aetius of Amida ('… erroribus tamen plurimis scatens'), but this remains to be demonstrated.

37 *J*, fol. 18. Incipit: 'Verum haec uirtutis omnium medicamentorum.' Last line: 'Nunc vero dynamidiam eorum nominum exponere institui, quae multis generibus ex dissimilibus rationibus constat: sicut infra ostenditur.'

38 The phrase 'facile recipitur in crasiam' is unclear and probably corrupt: the word *cras*[*s*] *ia*[*m*] is unattested elsewhere. Halleux-Opsomer 1982a, 76, suggests it is a corruption of *gratia*, and translates the phrase as 'on sera facilement reçu en faveur.' I follow the same sense but suggest 'crasia' may be a deviation of *cras*, 'thereafter.' Another possibility is *crasia* as a form of Gr. *krasis*, 'balance' (of humours, hence 'health'), hence 'balance can easily be achieved.' But this requires rewriting the entire sentence, the subject of which is the person studying (*non negligens … plenus fuerit*). There is, however, great variation among the manuscripts for the whole prologue, as noted by MacKinney 1935–6, 408–9.

more they are a danger to the living. For they put together medicines that are mixed up or use substitutes that do not match the ailment and still less the wounds. It is therefore necessary to commit this doctrine to memory first, lest the healer should make an error in application. In the beginning, therefore, I composed a treatise written for Glauco[39] concerning all elements and humours, urine, the pulse, and inflammations. In the second work I set forth the knowledge, quality, and defects of all drugs [*omnium pigmentorum intellectum et qualitatem et defectum posui*] for you, most noble Paternianus. Now I have attempted to explain the property [*dynamidia*] of all of those named, which comprise many different classes and different natures, just as is explained below.

What follows from this is a list of 21 different classes of drug (using Greek terms: see below), a simple statement of the 10 things from which drugs are made (animals of sea, animals of land, plants, seeds, wood, stones, flowers, juice, sap, and 'mineral substances' [*rebus metallicis*]), a short statement on how the *qualitas* and *potentia* of substances can be determined by four senses (see ch. 2.C), some fairly banal notes on use of contraries and combinations of substances, and a concluding statement that some ailments are incurable by drugs, iron, or fire (i.e., surgery and cautery).

Despite the address to Paternianus, the language and substance of the *Liber de dynamidiis* differ greatly from the *AG*. The *AG*'s prologue and epilogue do not mention properties, and its text does not subscribe to humoral theory, which governs the *Liber*'s conception of pharmacy. Moreover, the 'second work' is described as an account of the 'intellectum et qualitatem et defectum' of 'omnium pigmentorum': if this was intended to refer to the *AG*, the choice of words is odd, for they have no resonance with the *AG* itself, and one might expect mention of the *AG*'s alphabetical order. There is, however, one slight, and slightly odd, connection between the two texts. Of the 21 drug types listed by the *Liber de dynamidiis*, the *AG* mentions seven.[40] Two of these, *styptica* and *diuretica*, are so com-

---

39 That is, *De medendi methodo* (translated variously as *Method of Healing* or *Therapeutics*), which was dedicated to Glauco, who is otherwise unknown to us: it was under the title 'ad Glauconem' that the work was largely received in Latin translations: see Kudlien and Durling, eds 1991, and Sabbah, Corsetti, and Fischer 1987, nos. 275–9. Given that the manuscripts of the translation date from the ninth century, the *terminus ante quem* is Cassiodorus, *Institutes* I.31.2 (ed. Mynors 1937: see below ch. 1.F). Fischer 2003, 111, suggests that it was available from at least the first half of the fifth century.

40 *J*, fol 18v. All 21 categories are given a short description: I include here the description (in brackets) of only of those categories that appear in the *AG*: *Hypnotica, Peptica, Eustomacha, Styptica* [quae stringunt partes relaxatas], *Lea, Horectica, Colleta, Diuretica*

mon in pharmacological literature that their appearance throughout the *AG* is hardly surprising; but the other five (*metasyncratica, thermantica, paregorica, anaplerotica, cathartica*) appear in only one entry, that for oil (*oleum* #203), while *paregorica* appears again in the entry immediately after, that for eggs (*oues* #204). Moreover, the entry for oil contains yet another category not used elsewhere but also not mentioned by the *Liber*, that of *tonoticum*, which seems to be of late antique currency (#203 n.6). There are no obvious answers here: the inclusion of these categories in two entries may be owed to later tradition, though arguably not directly from the *Liber*, given the *AG*'s inclusion of *tonoticum*, and given the otherwise total lack of the *Liber*'s categories throughout the *AG*.

This is not to say that the categories listed in the *Liber de dynamidiis* could not be *applied* to the properties and actions described in the *AG*.[41] It is this complementary aspect, I would argue, that caused these otherwise different texts to be collected together and associated with Galen. The manuscript tradition suggests as much: the *AG* circulated with Galen's *De medendi methodo (ad Glauconem)* in three manuscripts (*F, M, B*) two of which (*F, B*) also include the *Liber de dynamidiis*, and another manuscript (*C*) places the *AG* and the *Liber de dynamidiis* together, under the assumption that they are both works of Galen ('Explicit liber Galieni [=*AG*]. Item eius degmatici [sic] uiri libellus' [= *Liber de dynamidiis*]). In one manuscript (*F*), all three texts are placed together to suggest a composite whole not unlike the schema set out in the *Liber de dynamidiis* prologue. Moreover, the *Liber de dynamidiis* and the *AG* were not the only texts attributed to Galen and packaged along with the *De medendi methodo*. Another text which appears frequently alongside these texts (and is found in our *M* and *B*) is a Ps. Galenic work called the *Liber Tertius*, intended as a 'third book' of Galen's *De medendi methodo*. According to Klaus-Dietrich Fischer, this is a somewhat fragmentary Latin translation of an earlier, more exten-

---

[sunt quae urinam provocant], *Cathartica* [sunt, quae ventrem et vulnera soluunt, et purgant], *Picra, Glycea, Thermantica* [sunt excalfactoria, quae calore suo omnem perfrictionem, et stricturam calefaciunt], *Pysctica, Paregorica* [sunt, quae mitigant ne malum crescat: non persanant], *Trachea, Staltica, Catulotica, Anaplerotica* [sunt, quae cicatrices crassant; quae altera vulnera complent et lympidant], *Metasyncretica* [confirmatoria, quae persanant], *Malactica, Toxica*. On *metasyncretica*, a Methodist term, see ch. 2.E.

41 For example, though not specifically called *malactica* ('quae omnem duritiem emolliunt et soluunt' *Liber*), plenty of substances in the *AG* soften or relax indurations (*duritiem emollit, laxat*, etc.); both honey (*AG* #179) and salt (#239) 'fastidium tollit/auertit,' which would match the *Liber*'s *horectica* ('quae auferunt fastidium'), and so on.

sive Greek text dating to the imperial period.[42] When it was translated into Latin, or first attributed to Galen, remains unknown.

In this sense, the *Liber Tertius*'s obscure history from Antiquity to surviving manuscripts (from the late eighth century onwards) parallels that of the *AG*, as with many other Ps. Galenic works that appear alongside them: *De catharticis, Definitiones medicae, Epistulae de febribus, Introductio siue medicus, De urinis*, and so on.[43] Like the many Ps. Hippocratic works on similar subjects, we may never know their ultimate origin or source, but their attribution to the 'big two' physicians of the ancient world and their collection into one volume were a product of Late Antiquity, when learned medicine, like Christianity, became 'defined in relation to a fixed series of books, a canon of orthodoxy.'[44] This did not mean a static reception of earlier, authoritative older texts, or simple abridgment: the Latin versions of *De mendendi methodo* books I–II, for example, were not simply translations, but reorganized Galen's text into thematic chapters, excised much of the theoretical discussion, and elaborated upon obscure passages by providing glosses. When we look about for evidence of reading and studying Galen in a Latinate environment, one place stands out above all others.

F. A Road through Ravenna?

There was undoubtedly an active medical school in early medieval Ravenna in which the study of Galen's work formed part of the curriculum. A ninth-century manuscript (Milan, Biblioteca Ambrosiana, G.108 inf.) pre-

---

42 Fischer 2003, 101–32, and edition of text at 283–346: on possible date, 131–2, 288. Fischer invites comparison with material transmitted in the *Liber Byzantii* (on which see Fischer 1998), and notes its transmission along with Ps. Soranus, the *Liber Aurelianus* and the *Liber Esculapii*: see Sabbah, Corsetti, and Fischer 1987, no. 59–60. Although the *Liber Tertius* mentions some of the substances found in the *AG*, these were common to pharmaceutical lore, and there seems no relation between the two texts.

43 Respectively, Sabbah, Corsetti, and Fischer 1987 nos. 269–70, 547–8, 273–4, 281, 288–9, providing references to editions and studies. Of these, *De urinis* appears in our (manuscript) *B*, *Definitiones* in *F*, *De febribus* in *M*: on this last, see Flammini 1998a. The *Definitiones medicae* are a partial translation of a Greek work that dates before Galen (probably saec. I: ed. K.19, 346–462): see Kollesch 1973, 144–5. It is very similar to another Ps. Galenic work, *Introductio seu medicus*, ed. K.14, 674–797, for which see now the edition of Petit 2009. Another anonymous (Greek) medical text dating from the first or second century AD is the *De morbis acutis et chroniis*, ed. Garofalo 1997.

44 Nutton 2004, 309. Likewise, Wallis 1995. To this I would also add developments in law: see Everett 2009.

serves, along with some Hippocratic works, a copy of lectures on works of Galen (*De sectis, Ars, De pulsibus ad Teuthram*) given by a certain Agnellus, 'professor of medicine' (*yatrosophista*), also termed 'chief physician' (*archiatrus*), plus another commentary on the *De medendi metho-do*, which seems to be by someone else.[45] The lectures of Agnellus were recorded 'ex voce' and edited by a certain Simplicius (*medicus*), 'in Ravenna.' Unfortunately, Simplicius gives no indication of the date, but Agnellus's method and commentary are remarkably similar in parts to another commentary which survives from sixth-century Alexandria, where there was an established curriculum based upon a set canon of Galen's works, reconstructed largely thanks to ninth-century Arab authors.[46] Unfortunately, we have no other testimony to Agnellus or Simplicius that might help us date their activity: any time from the mid-sixth century to 640, when Alexandria fell to the Muslims, seems possible, or at least before 751, the fall of the exarchate of Ravenna to the Lombards. Other evidence points toward the sixth century. A charter from Ravenna recording a land sale in 572 was witnessed by a Eugenius, an official of the state treasury, who declared himself to be the 'son of Leontius, physician of the Greek college [*schola*],'[47] possibly testimony to instruction in 'Greek' medicine in the Byzantine capital of Italy. We know also that Latin translation of Oribasius was prepared in or around Ravenna – the translator alludes to

---

45 On *archiatrus*, Nutton 1988.

46 This simplifies a complex philological problem: Agnellus's commentary on *De sectis* is attributed to a certain Gessios (*iatrosophista*) in a late manuscript (Vat. Pal. lat. 1090, saec. XV), and is remarkably similar to another Greek commentary by John of Alexandria (*iatrosophista*, saec. VI–VII), also transmitted in much later manuscripts (saec. XIII–XIV). See Palmieri (ed. *De sectis*) 1989, (ed. *De pulsibus*) 2005, and also Palmieri 1993, 1994, 1997, 2000, 2001, 2002. An edition of *De sectis* with English translation is Westerink et al., eds 1981. The commentary on *De medendi methodo* (on which see Palmieri 1981) closely resembles another commentary (in Greek) on this work by Stephanus (of Athens or Alexandria, c. 550–650?), on which see Dickson, ed. and trans. 1998. Stephanus also wrote a commentary on Hippocrates' *Prognosticon* (ed. and trans. Duffy 1983), the text (in Latin translation) which precedes the Galenic lectures in Milan G.108 inf., and wrote another commentary on *Aphorisms* (ed. Westerink 1984). Some questions remain about whether 'Agnellus' is a corruption of an eastern/Alexandrian name (Angeleo, Anquilaos: see Irvine and Temkin 2003) and whether the text itself is merely a Latin translation of an Alexandrian commentary: parts of the same syllabus were translated into Syriac as early as the sixth century. See further Temkin 1935, Beccaria 1971, and the papers in Garofalo and Roselli, eds 2003. On Ravenna's 'school,' see Mazzini and Palmieri 1991. On Galen's *De medendi methodo* and commentaries, Peterson 1974.

47 *PItal*. 35, ed. Tjäder 1982, II, 226, 'Eugenius Pal[atinus] sac[rarum largitionum] filiu[s] Leonti medici ab schola greca.' The term *schola* ('corporation,' 'guild,' 'community') may have a military application here: Brown 1984, 77, 185.

the city several times – and probably in the Ostrogothic period, though we cannot be sure.[48] Evidence such as this has prompted arguments for Ravenna as the location responsible for several other Latin medical works that reflect translations from Greek, such as the Latin Alexander (of Tralles), Rufus of Ephesus (*De podagra*), and Hippocratic works in various guises, such as the *De observantia ciborum*, as well as Latinate compilations such as the *Physica Plinii*. Whatever the merits of these arguments, the *AG* shares none of the linguistic affinities found in these works which point to north Italy, at least, as a place of origin.[49] But there is a good chance that our earliest manuscript of the *AG*, Vat. Pal. Lat. 187 (*V*), which attributes the work to Galen, was written in or near Ravenna, given its script and orthography (see ch. 5.C). Curiously, a beautifully illustrated (and alphabetized) version of Dioscorides (in Greek) dates to this same period, and may also be a product of Ravenna.[50]

In any case, the four works commented upon in the Milan manuscript (Ambrosiana G.108 inf.) served as the introductory material for what seems to have been a 24-book curriculum: from the introductory works one moved on to anatomy, physiology, pathology, therapeutics, dietetics, and hygiene.[51] A striking omission in this curriculum, however, is

---

48 See Mørland 1932. 'quem accepi a Martyrio arciatro Ravenna' (895, 12–3), 'epithima diaspermaton quam accepi a Maximo pimentario Ravenna' (903, 17), 'pimentari autem Ravenna aqua mittunt' (906, 115); 'quod quidam ex Mestre loco, nostris temporibus noscitur adinuenisse' (916, 27–8). Note also the reference to 'eastern' (Alexandrian?) physicians, 'sed cum aliis medicis orientalium in meridie constitutis quod ab his didicimus tradimus scriptum' (198, 22–5). On the linguistic evidence, which points to northern Italy rather than Ravenna specifically, see Adams 2007, 472–89 (and next note).

49 Many of these attributions (e.g., Mazzini 1981, and idem, ed. 1984 for *De observantia*: see also ch. 3.F) rely on historical assumptions about Italy's role in preserving classical traditions that are no longer tenable: rightly sceptical is Vázquez Buján 1984a, 1984b, 1998. The more recent linguistic analysis of the texts mentioned above (plus the Hippocratic *De conceptu*) by Adams 2007, 497–511, suggests only a 'common geographical origin' (ibid. 507) of northern Italy. None of the morphological or grammatical features discussed by Adams appears in the *AG*. On the Latin Alexander, Langslow 2006, 25.

50 Naples, Biblioteca Nazionale, Gr. 1 (6th–7th cent.): see plates 4–8. Bertelli, in Bertelli et al., eds, 1992, 125–31, suggests Ravenna c. 590–600, but there is no decisive proof: other centres of Greek culture (e.g., Rome, Naples) are also possible.

51 Latin titles given in Westerink et al. 1981, vii. I give here the English titles and schema from Nutton 2004, 413 n.28. Introductory tracts: *On Sects, Art of Medicine, Synopsis on the Pulse, Method of Healing for Glaucon*. Collection 1 (anatomy) *On Bones, On Muscles, On Nerves, On Veins and Arteries*, (physiology) *On Elements, On Temperaments, On the Natural Faculties*. Collection 2 (causes and symptoms) *On Affected Places*. Collection 3 (on the pulse) *On the Differences between Fevers, On Crises, On Critical Days, Method of Healing, On the Preservation of Health*.

any of Galen's works dealing with pharmacy per se. In the East, study of the other 20 books of the Galenic curriculum could easily and naturally progress to Galen's works on pharmacy, which Muslim authors took up early and with great enthusiasm. But in the Latinate West, where access to Galen was restricted to these introductory works in translation, there was a large gap to be filled with respect to the most commonly sought aspects of medicine: an authoritative guide to pharmacy. Given the manuscript tradition outlined above, the *AG* was obviously used to fill that gap. Its attribution to Galen and its attachment to other Galenic material may have happened anytime in the centuries after Galen, or even during his time: Galen himself complained of forged works with his name on them, and for this reason compiled a bibliography of his works.[52] But in looking for where and when the *AG* may have been consulted as a Galenic work on pharmacy in late antique or early medieval Europe, Ravenna stands out as a possible link in the chain of transmission.

Arguably, however, this took place after Cassiodorus's career as a functionary in Ravenna (c. 500–40). For after his monastic retirement in Calabria, Cassiodorus advised his monks at Vivarium to learn 'the properties of herbs and study the mixture of drugs carefully,' and to this end recommended that, in the absence of being able to read Greek:

> … you have first the *Herbal* of Dioscorides, who discusses and sketches accurately the plants of the fields. After this, read Hippocrates and Galen translated into Latin, that is, the *Therapeutics* of Galen addressed to the philosopher Glaucon, *and a certain anonymous work that has been collected from various authors*, then Caelius Aurelius' *On Medicine*, and Hippocrates' *On Plants and Cures*, and various other works written on the art of medicine that with the Lord's aid I have left to you in the recesses of our library. (My emphasis)[53]

The description of the 'anonymous work' has some resonance with the epilogue of the *AG*, which declares that *auctores antiqui receptissimi* have been consulted. But the overall context suggests that this was another introductory work to complement the *De medendi methodo*, and that Cassiodorus

---

52 Ed. K[ühn].19, 8–48.

53 Cassiodorus, *Inst.* I.31.2, ed. Mynors 1937, (emphasis) 'et anonymum quendam, qui ex diversis auctoribus probatur esse collectus.' The similarity was noted by Halleux-Opsomer 1982a, 80. For identifications of these works (Dioscorides or Ps. Dioscorides *Ex herbis feminis*; *On Medicine* as Caelius Aureli[an]us's *On acute diseases* and *On chronic diseases*, often transmitted together; Hippocrates *Plants and cures* as extracts from *Diet*), see Riddle 1981; Nutton 2004, 300.

did not know of the *AG*, or at least of its attribution to Galen, though it may well have been one of the 'various other works' in his library.

Nonetheless, the *AG* contains some intriguing references to Italy. Firstly, we are told that the spurge tree grows in Mauretania and 'as some would have it, in the regions of Italy that border Gaul' (*AG* #94). As discussed elsewhere (ch. 3.D), Dioscorides complained that Sextius Niger erroneously located the spurge in Italy, and Pliny noted that an inferior variety grew in Gaul, which seems to point to the same type of confusion between spurge and ground-olive. Secondly, the *AG* (#135) lists two types of indigo, one found in Italy which consists of 'scum collected from using purple dye which floats upon the surface and sticks to the sides of the cauldrons.' Both Pliny and Dioscorides refer to this type without mentioning Italy, and though Pliny further notes that it is chiefly found in purple dye shops (*in purpurariis officinis*), he does not specify their location, the centres of production in his day being Rome, Tyre, and Sidon.[54] The *AG*'s account – mixing indigo with sand of silver oar, shaping it into fingers or lozenges, and that it creates wrinkles – is independent of the traditions used by Dioscorides and Pliny, and as with the spurge reference, could be read either way as someone writing outside of Italy or within. Thirdly, in its account of salt, the *AG* (#239) mentions that apart from collection on seashores, it is mined or processed, 'as is the case in Gaul but especially in Italy and various other provinces.' Again the *AG* departs from Dioscorides and Pliny, who mention nothing similar, and though Pliny's long description of salt types (33.73–99) mentions three or four locations in Italy, they are not favoured or related to 'mined or processed' types. There are other signs: for white hellebore the *AG* used the word *ueratrum*, which according to Pliny was the Italian name (#285 n.1); and the entry for hyssop (#297 n.1) seems to be referring to the Italian type rather than the eastern variety found in Dioscorides.

There are, however, many places where we could expect more information if the text was of Italian origin. For example, Pliny notes that Italians call purslane (#32) *inlecbra*, but the *AG* does not mention such, even keeping to the Greek term *andrachne* rather than the common Latin name of *portulaca*. Likewise, the type of cyclamen described is germane to the eastern Mediterranean, whereas the variety described by Pliny is found in Italy (#63 n.1). We find similar absences concerning Italy when we compare the *AG* with Diocorides on, say, types of wolfsbane in Italy (#98 n.1),

---

54 See Jensen 1963; Elliott 2004.

or lees from old Italian wine (#112 n.1), and so on.[55] This type of deduction can be misleading, however: both Dioscorides and Pliny state that lovage (#148 n.1) was so-named (*libysticum*, λιγυστικόν) because it came from Liguria, where the natives call it *panaces*. But lovage was found in many places around the eastern Mediterranean, whence it was imported into Italy at an early date.

G. Self-Medicating in Late Antiquity

So far we have mostly looked at technical issues to determine the date and origin of the *AG*, and many of these point to Late Antiquity as the period in which an earlier text of the *AG* was dressed up with a prologue and epilogue and associated with other texts attributed to Galen to form the pharmacological component in a packaged, introductory reading kit of the 'essential Galen' in Latin. There is, however, a cultural aspect to consider which has not received much scholarly attention: several other late antique works on pharmacy all claim to have been written to enable laymen to practise medicine on their own, without having to rely on doctors. In the preface to the Ps. Apuleius *Herbarius*, for example, the author claims to have cobbled together 'a few properties of herbs and cures for the body because of the verbose stupidity of the practice [*professio*] which we prefer to call the profits rather than cures of doctors,' who are no more than 'fortune hunters' (*lucripetae*). Now his fellow citizens, acquaintances, and travellers may treat themselves, even if doctors will resent his work.[56] Similarly, the author of the *Medicina Plinii* claims that the ignorance and profit motive of doctors led him to collect 'remedies for health' (*ualetudinis auxilia*) and fashion them into 'a sort of abridgment' (*uelut breuiario*), so that he and others could avoid doctors, nor will anyone owe him a fee or 'have to pay for the consultation' of the book.[57] Admittedly, complaints

---

55 Other examples: #272 *basilisca*, a type of man orchis: the only other text to mention this name is Ps. Apuleius, who adds that it is called *regia* in Italy: the description of cyclamen, #63 n.1, is the same as that of Dioscorides, whereas the description of Pliny seems to be of *Cyclamen hederifolium* L. or *Cyclamen europaeum* L. found in Italy.

56 Ps. Apul. *pref.* ed. Howald and Sigerist, 15. 'nostra litterata scientia inuitis etiam medicis profuisse uideatur.' Ps. Apul. prol. ed. Howald and Sigerist, 15. Voigts 1978. On the preface, see Zurli 1992a; Maggiulli 2000. It is uncertain whether the preface was original or was added in the fourth or fifth century: Maggiulli and Buffa Giloito 1996, 19–32, argue for a fifth-century addition, but not convincingly. The perceived borrowings from Apuleius's *De Platone* (e.g., *lucripetas* from *lucricupidinem*, etc.; Maggiulli, ibid., 24) seem strained. Their identification of plants, ibid., 170–3, is most welcome.

57 *Med.Plin. Prol.* ed. Önnerfors, 4. On the prologue, see Segoloni 1990a, and Buffa Giolito 2000.

about doctors and the need for self-sufficiency, a core Roman value, are found in Celsus and Pliny, who contrasted Roman pragmatism with Greek quackery, and the two works just mentioned relied heavily on Pliny. But the theme of handbooks or abridgments for independent lay use seems to be a keynote for fourth- and fifth-century works on pharmacy. Marcellus of Bordeaux's 'manual' of domestic medicine, which adds local Gallic folk remedies, chants, and charms to the recipes of Scribonius Largus, attempted to create a 'little book of things that work' (*libellus de empiricis*) so that his sons could treat themselves 'without the intervention of a doctor.'[58]

But even doctors caught the trend. Oribasius wrote two handbooks for laymen, a *Synopsis* for his son Eustathius, and another for Eunapius the philosopher (c. 347–410), and abridged both Galen's *Simples* and Dioscorides' text into a condensed, alphabetically arranged listing of 554 drugs (equalling 100 pages of a modern printed edition).[59] We mentioned Vindicianus's letter to his nephew, but he also wrote another to the Emperor Valentinian (364–75) with advice on drugs for constipation.[60] Vindicianus's student Theodorus Priscianus (fl. 364–75) wrote a tract on gynecology that pays much attention to pharmacy, including fertility drugs from a *medica* named Victoria. But Theodorus also composed a handbook, the *Euporiston Phaenomenon*, focusing on *euporista* (literally 'common,' 'easy to procure'), which he defined as 'easy and natural remedies that are beyond dispute' and which mother nature had placed in common plants (*uiles herbae*), minerals, and other things, so that the sick 'need not head off straightaway to Pontus or scour the deep interior of Arabia to obtain storax or castoreum or other riches from distant parts of the globe.'[61] The entries for storax and castoreum in the *AG* (#254, 50) betray its more antique origins, but its format and brevity well match the late antique ethos of pharmaceutical self-help, and an attested need for short, authoritative guides to what was known to work, without theoretical discussion, and without the need for doctors, who, as Roman imperial presence receded, became fewer across the landscape of early medieval Europe.[62]

---

58 Marcellus, *Med. pref.*, ed. Niedermann 1968 I, 2. 'gratulemini [filii] super hac re labori studioque nostro, quod uobis absque medici intercessione opem necessariam curationemque praestabit.'
59 Oribasius, *Med. Coll.* XI–XIII, ed. Raeder II, 80–180. See esp. Scarborough 1984, 221–4.
60 *Ep. ad Valentinianum*, in Marc. *Med.* ed. Niedermann 1968, I, 46–53.
61 Theodorus, *Eup.* I *pref.*, *Gynaecia*, I.1, V.13, ed. Rose, 4, 222, 233. See Fraisse 2003a. On Theodorus's Latinity, Önnerfors 1993, 288–300. Apollonius Mys (c. 50 BC–25 AD) wrote a popular work called *Euporista*: von Staden 1989, 543.
62 On doctors, see Nutton 1981, 1988, 2004. The same trend can be seen in dietetics with Vindicianus's letter to Valentian (in Marcellus, *Med.* 1., ed. Niedermann, 46–52), and Anthimus's *On the observance of foods* to the Frankish King Theuderic (511–34), ed.

H. Evaluating the *AG*'s Pharmacy

The *AG* belongs to a long tradition that is not only still with us today, but is thriving. Over 85 per cent of the world's inhabitants still use plants as their primary form of health care. Despite the triumphs of modern combinatorial chemistry in isolating, extracting, and synthesizing the chemical compounds of natural products, over 50 per cent of modern pharmaceuticals are still derived from the earth's biodiversity, and only 60 per cent of the 11,500 different natural products known are replicated by synthetic compounds. New techniques to tap nature's resources (random screening, phylogenetic analysis, environmental genome sequencing, and so on) and the subsequent identification of unique bioactive molecules have resulted in the discovery of important new drugs.[63] Journals devoted to scientific research on the biochemistry and potential medicinal properties of natural products have doubled in volume in the last five years alone. Large, printed pharmacopeias for pharmacognosy have been replaced by websites and journal search engines.[64] Additionally, natural product pharmacy as 'alternative' medicine in the developed world is big business, and grows more popular each year, forcing governments and the health-care industry to examine and regulate standards, practices, and sales with greater stringency. Texts recording past uses of natural products have certainly played a part in modern drug discovery: atropine, aspirin, ephedrine, reserpine, quinine, digoxin, digitoxin, tubocurarine, morphine, and codeine, all compounds originally derived from plants, were all discovered through scientific investigations of folklore claims, and there have even been calls for the re-examination of ancient and medieval texts for new leads.[65] While that

---

with Eng. trans. Grant 1996. On the doctors named in sources after 600 in Italy, see Cosentino 1997; Pilsworth 2009: southern Italy, Skinner 1997, 127–36 et passim.

63 Cox 2007, citing the examples of palytoxin, taxol, unique COX-2 inhibitors, artemisin, anti-spasmodic isperenoids, prostratin, and BMAA. On discoveries from natural products over the last 25 years, see Newman and Cragg 2007.

64 The *Journal of Ethnopharmacology*, for example, doubled the number of papers it published from 2003 to 2005, and similar trends can be seen in *Planta Medica*, *Journal of Natural Products*, and *Phytotherapy Research*, not to mention articles on natural products in more traditional journals such as *Life Sciences* and *Biochemical Pharmacology*. For serious criticism and caution, see Gertsch 2009. Websites for the American Society for Pharmacognosy, the Society for Medicinal Plant and Natural Product Research, the European Scientific Cooperative on Phytotherapy, and the International Society for Ethnopharmacology contain many useful links to search engines and resources for research. A particularly good website is the naturalstandard.com.

65 Holland 1996, 1. Adding the first three mentioned is Gilani and Rahman 2005.

task may be worthwhile, what can be said here is that many of *AG*'s prescriptions for therapeutic use hold up remarkably well against the findings of modern science, while many seem to have no biological or physiological basis and remain puzzling.

The *AG*'s treatment of 'simple' (that is single-product) drugs, and its preference for empirically verified results over theory, make it easier to evaluate than many other ancient or medieval pharmacy texts. Pharmacy based upon humoral theory as found in Galen and later exponents of Galenism, or compound pharmacy[66] (whether based on humoral theory or not) which involved the mixture of different ingredients (from half a dozen to over 60, as in some ancient recipes, diligently copied by medieval scribes), may well have little more than placebo effects, as has been claimed for nearly all pre-modern pharmacy.[67] This is unfair, despite the undeniable therapeutic power of placebos, which has challenged their place in modern clinical trials (from the 1970s onward) as a standard 'control' procedure for establishing the casual relationship between a therapy and its outcome.[68] The *AG*, however, does not present us with the heuristic muddle of different ingredients mixed together, nor is it governed by any dominant or overriding theory. As such, its 'simples' can often be measured against modern phytochemical and biochemical research.[69] To

---

66 Compound pharmaceuticals are found among our earliest medical writings (e.g., Ebers papyri, see #76 n.1, and in Hippocratic writings, see Totelin 2009), and in most authors of medical works. They often appear as singular texts in nearly every early medieval manuscript of medicine, often derived second-hand and adapted from ancient authors such as Scribonius, Pliny, Galen, Celsus, etc., who themselves borrowed from earlier works. Examples of collections are Marcellus of Bordeaux (who borrowed heavily from Scribonius), and those edited by Sigerist 1923, and Jörimann 1925, and the *Lorscher Arzneibuch*, ed. Stoll 1992. An insightful overview is Halleux-Opsomer 1982b; see also Hirth 1980, and the relevant essays of Stannard 1999a and 1999b. Wallis 1995, 112, notes that of the 324 pharmacological works among early medieval manuscripts listed by Beccaria 1956, 137 of these were recipes, with another 24 on weights and measures.

67 As claimed by A.K. Shapiro and E. Shapiro 1997, building on an earlier study of A.K. Shapiro 1959, but their accounts of pre-modern pharmacy are too general and unduly dismissive. A more historically sensitive and sympathetic account of pre-modern pharmacy is Mann 2000. Stimulating approaches to understanding two compound drugs are Norton 2006 (mithridatum), Nockels Fabbri 2007 (theriac). On later medieval (11th–14th cent.) theories of compounds, McVaugh 1975.

68 Modern use of placebos is now being questioned: the criticisms of Hrobjartsson and Gotzsche 2001, among others, have unleashed a flurry of questions and responses: see Louhilala and Puustinen 2008; Spiegel and Harrington 2008. Interestingly, the neurological perspective is at the forefront of upholding the therapeutic value of placebos: see Oken 2008; Diederich and Goetz 2008.

69 This is not to deny that the *AG* refers to mixing products for different therapeutic use

take a well-known example commonly used today, aloe vera does indeed 'bind' wounds (*AG* #5), for aloe latex contains anthraquinone glycosides (aloin, aloe-emodin, and barbaloin) and acemannan (a mannose polymer) that accelerate wound healing, as well constituents such as bradykininase and magnesium lactate, which have anti-inflammatory properties and antipruritic effects.[70] The anthraquinone glycosides do indeed 'loosen the bowels,' and aloe has been used as a laxative into the modern era, but modern research warns that prolonged use contains some risk of liver damage, despite the *AG*'s claim that aloe is 'good for stomach and liver ailments.' The *AG* also claimed that aloe 'reduces bile and phlegm,' or can be mixed with colocynth and scammony to treat fevers – both claims that rest upon erroneous notions of bodily fluids and pathology (nothing more than symptomology here), and which require reconstructing imaginative scenarios of chance efficacy to understand how they came to be accepted. Another example is the *AG*'s entry for comfrey (#255), which says that its bitter taste and stickiness make it useful for treating 'coughing up blood' and that it 'vigorously binds and agglutinates.' According to recent research, comfrey has analgesic, anti-inflammatory, and anti-exudative actions, and its extract is used in topical preparations for the treatment of muscle and joint complaints. For comfrey contains mucilage, tannin, and allantoin, the latter of which stimulates tissue regeneration, hence its use for both external injuries and gastric ulcers.[71] Most of this fits the bill given in the *AG*: the 'coughing up blood' being a common symptom (that is, hematemesis) of a gastric ulcer. However, the discovery of pyrrolizidine alkaloids in comfrey also suggests it should not be used for internal medication.[72]

That some of the *AG*'s prescriptions can be supported by modern research (see ch. 2.B) should not surprise us. Traditions concerning natu-

---

(e.g., #5 as below, 8, 99, 103, 109, 136, 152, 153, 188, 194, 203, 240, 288: note also concerning lye #150). With the exception of multi-ingredient products like cyphi (#76), and the many references to mixing a product into eye-salves, caustics, antidotes, purgatives, and a few others (e.g., #94, with rue to reduce side-effects of thirst; #156, with myrtle oil to tame frizzy hair, etc.), most mixes are limited to one or two ingredients, most commonly with water, wine (see #292), vinegar, honey, or oxymel (see #5 n.2) to create a topical or potable solution. On the whole one detects a sceptical stance toward mixing ingredients, in keeping with the warnings for detecting adulteration (e.g., #28, 44, 97, 199, 279). On the manufacture of different forms of drugs (lozenges, pills, plasters) in antiquity, Gourevitch 2003.

70 Boon and Smith 2009, 25–32.
71 Staiger 2005. On history of its use, Englert, Mayer, and Staiger 2005.
72 Trease and Evans 2002, 389.

ral product drugs, stretching back over millennia, endured because they worked, as John Riddle has championed for Dioscorides' work and for the history of contraceptives.[73] Traditions, however, were not immune to corruption, and they could also completely fail. For example, Crateuas the rootcutter (fl. c. 130 BC) noted that meadow saffron (or autumn crocus, *Colchicum autumnale* L.) is good for treating gout, and he was right: modern science has discovered that meadow saffron contains an active substance, colchicine (an alkaloid compound named after meadow saffron), that attacks the chemical chain reaction in the formation and deposit of uric acid on the joints.[74] But oddly enough, later commentators, such as Dioscorides, Pliny, and the *AG* (#80) made no mention of it being used to treat gout, despite enumerating other therapeutic uses – especially odd for Dioscorides, who admired and used Crateuas as a source.[75] Irrespective of the value or frailty of traditions, we should not overstate the *AG's* acceptability to modern science or medicine. Based on a fairly primitive understanding of the human body and of pathology, many of its prescriptions are fanciful, vague, useless, and possibly downright dangerous – the warning at the start of this book should be heeded, the *AG* should not be consulted for actual use!

I. Evaluating the *AG's* Botany

The botanical descriptions given in the *AG* can be extremely crude and unhelpful, yet the same has been said of the far more prolix descriptions of Dioscorides and Pliny, and it is generally accepted that users of these texts had to have prior knowledge of the plants discussed in order to identify them.[76] Most ancient authors claimed to be relying on observation, though clearly they often copied out descriptions from earlier works, despite emending or adding details as they saw fit. Another difficulty is that we

---

73 Riddle 1985, 1992a, 1997; the subject can be found throughout his work, some of which is collected in idem 1992b. This is the place to acknowledge my debt of inspiration to Riddle's works.

74 Riddle 1996, 10. See also Lee 1999. Riddle 1996 also notes that Dioscorides' observation (IV.93) of stinging nettle (*Urtica dioica* L. + sp.) as a diuretic has been confirmed by science: it contains a phytoagent that can be an effective treatment for prostatic hypertrophic prostrate gland, a symptom of which is urinary retention. Both the *AG* (#288) and Pliny 21.92–4 fail to mention any diuretic effects.

75 Diosc. I.29, II.127, III.125, etc. On Diosc. and tradition, see Scarborough and Nutton 1982; Riddle 1985, 18–20. On Crateuas, see Wellmann 1897.

76 Riddle 1981, 45, idem 1985, 27–32, 176–217. On Pliny, Stannard 1965, Morton 1986, Scarborough 1986.

can never be sure exactly which species of a particular plant the author refers to, and it is also possible that a particular species has become extinct or evolved over the last two thousand years. There is no evidence that the *AG* was ever illustrated, as found in the (Greek and Latin) tradition of Dioscorides, Ps. Apuleius, the Ps. Dioscorides *Ex herbis femininis* and *Curae herbarum*.[77] However, the epilogue of the *AG* claims that 'ancient and most popular authorities' were consulted, many entries begin with the statement that a plant is '(well-)known to everyone' (*nota omnibus*), which probably refers to previous written sources (see ch. 4.H), and the text frequently mentions that 'some say,' or 'others say,' explicitly stating in two entries that 'others describe' (#92, 99; see ch. 3.A). Comments such as these suggest that fuller descriptions and details on a particular plant could be found by the reader elsewhere, but were out of place in what was intended to be a handy reference work.

Despite its brevity, however, the *AG* can be more accurate than Dioscorides or Pliny in some places. For example, the entry for dittany of Crete (#88) accurately describes its many branches, how they are heavily noduled, 'around which sprout little leaves that are similar to mint, but are paler and more narrow, and on the small branches near the top there appears a small, pink flower.' Dioscorides' description, which was no lengthier, said it resembles pennyroyal, 'though has larger and woolly leaves and some sort of woolly on-growth. It bears neither fruit nor flower' (III.32). Pliny's account (25.92) says much the same, and both authors must have meant another plant, or had never laid eyes on dittany.[78] The *AG* describes the water parsnip (#158) as producing 'a small white flower,' which indeed it does, but again this is not mentioned by Dioscorides or Pliny (see plate 3). The same can be said for descriptions of animals and minerals, such as the careful description of blister beetles (#65), or that of gypsum (#126), compared to the meagre notice of this last from Dioscorides.

J. Conclusion

Presenting a text which adds significantly to our knowledge of ancient medicine and its transmission to the early Middle Ages, yet remains of uncertain date and origin, is hardly satisfactory. I can only hope that this critical edition of the *AG*, and the care taken to present the evidence and

---

77 Riddle, 1980, 1981. Bertelli et al., eds 1992. Collins 2000.
78 Noted by Meyer 1856, IV: 493, 'mit weitläuftiger eigenthümlicher Beschreibung.' Cf. Diosc. II.127, Pliny 22.84. On dittany, see Andrews 1961a, 78–9.

arguments for determining a date and origin, will spur further investigation and expertise to provide more certainties than managed here. New evidence – the discovery of a new manuscript, or a literary reference to the *AG* – may also change the picture. But a search for certain origins may also be misguided for a reference work that, whenever and wherever its initial composition, was augmented, updated, and corrected over many centuries, possibly as many as eight (first century BC to seventh century AD), and as the surviving manuscripts show, continued to be altered for another five centuries with interpolations, omissions, and occasionally flagrant rewriting of the text to change its meaning. This was not simply carelessness. Ancient (Western) medicine had long treated texts as simply vehicles for ideas to be transmitted, challenged, debated, and selectively pillaged, in comparison with Indian and Chinese traditions, in which text is considered scripture to be memorized (as with Āyurveda), or as a 'classic' to be commented, glossed, and edited (as with the *Ben Cao*: see ch. 2.C). Early medieval readers were even more cavalier than ancients about ancient 'authorities' of secular literature, for ancient authors belonged to an alien, and increasingly remote, religious and cultural milieu. As a result, ancient medical texts 'could with impunity be subjected to radical and unabashed re-working, dismemberment, and de-authorization,'[79] as we find among our early medieval manuscripts.

Nonetheless, we can be certain of the following. The *Alphabet* presented here circulated in Western Europe from the seventh to the twelfth and thirteenth centuries, and was considered a genuine work of Galen. With 300 entries on plant, animal, and mineral products, it had a more comprehensive range than most late antique or early medieval texts of a similar nature, and its alphabetical format facilitated quick reference. Its ancient origins are betrayed by its similarity in style and content to authors such as Dioscorides and Pliny, its pharmacology that relies on concepts (drug properties and elemental forces) common to ancient Greek science, its lack of subscription to humoral theory, and the overall ambience of the text's references to the ancient Mediterranean world, with an eastern or Greek viewpoint. The *Alphabet*'s language suggests it was redacted in a Greek-Latin bilingual environment, and it used Greek and Latin sources that are unidentified. For this reason it transmits technical vocabulary not recorded elsewhere, and likewise unique descriptions of products and their preparation. Other aspects of its vocabulary point to late antique

---

79 Wallis 1995, 125, who builds on the ideas of Keil, Temkin, and Baader concerning Christian opposition to pagan conceptions of 'knowledge' or 'wisdom.'

additions. Sometime between the death of Galen (c. 217 AD) and the eleventh century a prologue and epilogue were attached to the *Alphabet* and it was associated with other works attributed to Galen (*Liber de dynamidiis*, and a Latin adaptation of authentic work of Galen, the *Therapeutics for Glauco*) as part of an ensemble of introductory texts for medical instruction. Testimony for such a setting is found in sixth- or seventh-century Ravenna, but clearly the *Alphabet*, whether attributed to Galen or not, was circulating in Late Antiquity, and well matches late antique interests and activity in translating Greek medical traditions into Latin, and in providing handbooks of practical pharmacy. Finally, the *Alphabet* is a philological gem: it adds 20 new words to the lexicon of ancient and medieval Latin, and 20 hitherto unrecorded uses of the Latin *-aster* suffix, by far the most ambitious use known.

Despite its pseudonymous deceit, it is not difficult to understand why the *Alphabet* gained the authority of Galen's name, and why it endured from Antiquity when so many other texts from the same tradition have been lost. Anyone reading an ancient treatise or compound recipe which refers to an unknown simple drug could quickly find in the *Alphabet* a short description (appearance, habitat, different and best types, preparation methods), what 'properties' it has, and what it can be used for. A copy of the *Alphabet of Galen* would certainly have helped the English bishop who wrote (around 750) to a fellow bishop at Mainz to complain that 'we have some medical books, but the foreign drugs [*pigmenta ultramarina*] we find prescribed in them are unknown to us.' Such 'foreign' drugs continued to be prescribed in medical manuscripts circulating in Carolingian Francia a century later, and that region's increasing trade with the Mediterranean world made them available.[80] The *Alphabet of Galen* also retained currency because it contained many practical, simple remedies that had stood the test of time. Without much theoretical ado, a reader learned which plants, minerals, or animal products could agglutinate wounds, help with indigestion, gas, stomach cramps, parasites, diarrhoea, coughing, sore throats, and inflammation, or which could be used in eye-salves, antidotes, plasters, and compresses, and what substances promoted regular urination and menstruation. The *Alphabet*'s lack of magic or superstition may have disappointed some readers, but it was not above popular beliefs or con-

---

80 Cyneheard of Winchester to Lull of Mainz, MGH Ep. I, 403, Ep. no. 114. The manuscript Vienna 751 (saec. IX med.) reads *sigmenta*, which seems corrupt. On Latin medical texts in Anglo-Saxon England, see Cameron 1982 and 1983, and Adams and Deegan 1992. See also ch. 4.D (concerning *uiscide*). On medical manuscripts in Francia, McCormick 2001, 708–15.

cerns, such as mixing verdigris with the urine of an infant boy (#136), how much skink can one drink before its aphrodisiac qualities turn fatal (#267), or what to use for make-up, hair-dyes, or depilatories (ch. 2.J). These last further expose the *Alphabet*'s ancient roots, but as is clear from the very survival of the *Alphabet*, a dividing line between antique and medieval medicine is largely a modern imposition. Medieval readers continued to seek inspiration from ancient medical texts, all the more if they had weighty names attached. The *Alphabet*'s attribution to Galen may well be considered a symbol of the political and cultural caesura – between East and West, Greek and Latin – that came after the fall of Rome. But the persistent use of the *Alphabet of Galen* in the centuries to come is also proof that there was much continuity in pharmacy from Antiquity to the Middle Ages.

2

Pharmacology

ॐ

A. Introduction

This chapter explores the particular type of pharmacy espoused by the *AG* and attempts to place it in historical context. The *AG* certainly relied on past, empirical tradition of 'what works,' but goes further than that in attempting to provide explanations for what works by appealing to the notion of 'properties' within natural products. These properties are correlated to sensory perception: the taste, smell, perceived effect of the product on the body (e.g., soporific) are considered, explicitly and implicitly, as a guide to determining the properties of that substance. Interestingly, the *AG* has more to say on sensory perception than most ancient or medieval pharmacy texts. The *AG*'s use of properties is similar to that of Dioscorides, but its more limited number of categories, and its predominate fourfold paradigm (heating, cooling, binding, and loosening properties) may reflect an earlier stage in the development of the concept between the Hippocratic writings (fourth to first century BC) and Dioscorides (first century AD). Discussion then turns to explore other aspects of the *AG*'s pharmacy, including its references to humours, its few explicit theoretical statements, and the striking absence of superstition or magic. Firstly, however, a brief discussion about the chemical properties of natural products and their medical use will help to provide a general context for explaining the type of prescriptions we find in the *AG*.

B. Natural Products and Pharmacy

Plants

Plants manufacture different chemicals for different reasons: to regulate

growth cycles, to maintain their metabolism, and to protect against harmful bacteria, fungi, insects, and herbivores. The different chemical compounds found within medicinal plants may act individually, additively, or
in synergy to improve human health. Phytochemical research increasingly
stresses the synergistic element, and the interplay between constituent
groups, as the key to a plant's medicinal efficacy. The active constituents
of most medicinal plants are their secondary metabolites, of which there
are three main groups: nitrogen-containing substances, terpenes, and phenols. Of the nitrogen-containing substances, alkaloids are among the more
powerful because they directly affect the central and/or peripheral nerv-
ous system, as with the coniine in hemlock (#49), solasodine in hound's
berry (#84), aconitine in wolfsbane (#98), or scopolamine in henbane
(#295) and mandrake (#181: see plate 4), colchicine in meadow saffron
(#80), and of course morphine (along with over 60 different alkaloids) in
the opium poppy (#199), to mention the best-known.[1] Terpenes are complex arrangements of hydrocarbons (known as arenes or aromatic hydrocarbons) found in the essential oils within the plant, often used today in
the manufacture of fragrance and flavour concentrates. These oils are often
highly antiseptic and bactericide, but also have antipyretic, carminative,
counterirritant, expectorant, anti-tussive, antispasmodic, contra-allergen-
ic, anti-pruritic, anti-inflammatory, and immunostimulant properties.[2]
Phenols add hydroxyl groups to aromatic hydrocarbons to form complex
chemical compounds, two categories of which, flavonoids and tannins, are
particularly prevalent in the plants of the *AG*. Flavonoids have significant antioxidant and anti-inflammatory effects, such as quercetin found
in nettle (#288) and caraway (#60), or rutin found in rue (#227), bramble
(#230), elder (#107, with isoquercitin), safflower (#82, with luteolin), and
shepherd's purse (#283, with diosmin and hesperidin), and the unusual fla-
vonoid combinations in squill (#264) and violet (#287). Tannins are large
molecules that are highly solvent in water, and while they have been shown
to contain potential anti-inflammatory, bactericide, and anti-parasitic

---

1   On different alkaloids in plants and their medicinal use, Roberts and Wink, eds 1998,
    and van Wyk and Wink 2004, 372–8. Other important nitrogen-containing substances
    include non-protein amino acids (NPAAs), cyanogenic glucosides and glucosolinates,
    which are all found in the plants of the *AG*.

2   E.g., (monoterpenes) menthol in mint (#185), thujone in wormwood and sage (#22,
    102), 1,8-cineole in cardamom (#71), azulene and chamazulene in camomile (#72), ter-
    pinen-4-ol in juniper (#139); (triterpenes) ursolic acid in elder and oregano (#107, 206),
    squalene in olives (#203), glycyrrhizin in licorice (#122). The many monoterpenoids
    and sequiterpenoids (camphene, β-phellandrene, α-zingiberine, *ar*-curcumene) in ginger
    (#253) explain the *AG*'s comments: cf. Boon and Smith 2009, 134: 'stimulating carmina-
    tive that aids digestive function and tonifies the gastrointestinal system.'

qualities, their diverse use is often owed to one property: they react with proteins to form an inert, protective layer. This effect is often described in the *AG* (and elsewhere) as 'binding' (*stringens*, hence 'astringent') in reference to their tightening effect on body tissue and membranes – just as the tannins in wine cross-link with the surface proteins within the mouth to create a puckering sensation until reversed by a body enzyme.[3] Tannins can be used internally, for mouth or gum complaints (as with tamarisk #193, mulberry #194, blackberry #195, bramble #230), and also against diarrhoea, since they form a surface lining on the intestine to protect it from further irritation (e.g., the tannin-rich pomegranate #183, also grape vine #291 and betony #290). Used externally on burns, abrasions, and wounds, tannins bond with the proteins of damaged surface cells to form a tight protective layer which limits contamination and loss of serum (e.g., rhubarb #231, rose #232, comfrey #255). By also reacting with the dead proteins, tannins neutralize potential sources of food for harmful bacteria.

Of course there are many other important chemical constituents in the plants of the *AG* that have preventative and therapeutic effects (polysaccharides, sterols, saponins, aldehydes, polyacetylenes, various types of acids, and so on). The renewed scientific focus on botanical sources for new drugs, resuscitating and refining the tradition of 'weeds and seeds' pharmacognosy that comprised most pharmacy before the Second World War, has produced a range of authoritative guides which can be consulted.[4] Again, it needs to be stressed that the chemicals in plants are usually multifunctional compounds carrying more than one pharmacologically active chemical group, and that these occur in complex mixtures. As such, plant medicines have subtle effects on several different biochemical pathways and receptors in the body and mind that contribute, directly and indirectly, to restoring health. Moreover, the combination of compounds can differ greatly between plant organs, within developmental periods, and even diurnally. The *AG*'s awareness of these differences, and its often startlingly accurate recommendations,[5] build on millennia of trial and error.

---

3   See Bates-Smith 1973; McManus et al. 1981.

4   Excellent guides are Wichtl 2004; Van Wyk and Wink 2004; Ross 1998– (still in progress); Harborne et al., eds. 1993; Boon and Smith 2009 (though limited to 55 plants). Also helpful are the updated editions of large pharmacognosy manuals such as Trease and Evans 2002, Parfitt, ed. 1999 (*Martindale: The Complete Drug Reference*), Osol, Pratt, and Gennaro 1973– (Dispensatory of the United States of America), and more recent projects such as the *European Scientific Cooperative on Phytotherapy* (ESCOP) monographs ed. 1996–9, or the *German Commission E Monographs* ed. 1998. Less useful is the *Physicians Desk Reference for Herbal Medicines*, revised annually.

5   E.g., fennel, 'especially used in eye-complaints' *AG* #110, contains anethole, used to treat blepharitis and conjunctivitis in humans and animals.

Minerals

The considerably large number of mineral substances in the *AG* comprise products from copper (#1, 3, 20, 37, 48, 52, 66, 89, 145, 178, 257), zinc (#45, 210), lead (#46, 182, 237), iron (#20, 276), sodium (#9, 10, 38, 110, 239), calcium (#11, 12, 27, 83, 67, 150), antimony (#241), sulphur compounds (#7, 8, 242), and a host of different earths (#125, 127–32) and stones (#161–78) with medicinal properties. These are mostly prescribed for their astringent, drying, caustic, cicatrizing, and detersive properties, particularly in external applications (salves and ointments) to treat dermatologic conditions, and they are still employed today in many skin-care and cosmetic products.[6] They are also used to used to remove abnormal growths (#8, 52, 66, 67) or excrescences (#10), and as ingredients for eye medicines (*collyria*), or for curing eye problems, such as thyite stone (#175), having 'properties that bite severely, and it therefore helps with dim eyesight and diminishes corneal abrasions,' or azurite (#37), which is 'particularly used for drying up watery eyes and nourishing the eyelids.' Some mineral products are also recommended as an internal astringent for treating dysentery or diarrhoea, which is unusual among ancient literature, and probably very dangerous.[7] Other minerals are said to have 'loosening' and relaxant properties (#163, 176, 249), while others have specific (and sometimes puzzling) uses beside those already mentioned, such as square stone preventing conception (#168), rock crystal stimulating lactation (#174), pumice stone for toothpaste (#213), lignite stabilizing loose teeth (#166, also 167), and flamestone inducing sleep when smeared on the forehead (#173). Occasionally, as with some plant entries, the text omits any uses, or simply states that the mineral has the same properties as another (#1, 178).

Animals

In Greek and Roman authors, Dioscorides and Pliny included, animal products were often prescribed for illnesses thought to be caused by the animals themselves, particularly 'the hair of the bear that bit you' variety (scorpion as an ingredient to treat scorpion stings, eating lizards or

---

6   Cosmetic use: e.g., Chian earth (#129), tightens skin and 'makes it shine.' Astringent and cooling properties: #20, 89, 241. On minerals in Dioscorides, Riddle 1985, 145–64; in Pliny, Healy 1986. An excellent guide to minerals in modern cosmetics, skin products, and food is Leung and Foster 1996.

7   E.g., rock alum #48, Samian earth #125, gypsum #126, lodestone #165. As Riddle notes (1985, 153), Dioscorides seems only to have recommended Sinopic red ochre for internal use (V.96, with egg). cf. *AG* #276.

mice when bitten by such, or mad-dog liver to treat rabies) but also across different species (hippopotamus testicle for reptile bites, pickled tuna for dog bites), and frequently as antidotes for poisons and snake bites.[8] Curiously, there is none of this in the *AG*, another aspect of its comparatively rationalist approach (see below, ch. 2.I). The majority of animal products are described as emollients that possess softening, loosening, or warming properties, or some combination of these, and are prescribed for treating indurations, swellings, or pain in the joints, such as castoreum (#50) being good for chronic pain in the tendons.[9] Many are mentioned as ingredients for eye-salves, and some have more specific uses, such as butter (#41) for vaginal pains, genital swellings, and dryness of the lungs, egg (#204) for inflamed eyes and hoarse throats, skink (#266) for an aphrodisiac, cuttlefish (#248) for toothpaste, and donkey dung (#115) for toothaches. Some animal products have astringent properties, such as rennet (#87), fish glue (#140), and stag-horn (#86), and the astringency of blood from woodpigeons (#245) is particularly good for eye problems.

C. Sensory Perception

Far from simply repeating traditions known to work, the *AG* justifies its recommendations on a more theoretical level by making a connection between two concepts: the effect of the plant or substance on the senses (particularly taste and smell, but also touch), and the particular medicinal 'properties' within that plant or substance. Interestingly, the *AG* has more to say than most ancient or medieval texts on how taste and smell indicate the medicinal properties of a plant. But this sensory paradigm also reveals another, fundamental theoretical underpinning of the *AG*'s pharmacology: the notion that an ailment should be treated by a drug considered to have characteristics that are opposite to the symptoms being treated. The 'theory of opposites' in medicine stretches back to prehistory, was pervasive throughout most pre-modern pharmacy, and is still around in household and alternative medicine today: eating hot chillies to cure a cold, for example. The *AG* contains many obvious instances: the onion-like bulb

---

8   On ancient and medieval beliefs about animal products, see MacKinney 1946; Scarborough 1979. Examples in Diosc. are found throughout book II: scorpion 11, hippopotamus 23, pickled tuna 31, mad dog 47, lizard 66, mice 68. See esp. Riddle 1985, 132–41. On animal products in Pliny (mainly books 28–32), see Bodson 1986.

9   Emollients: butter #41, wax #59, marrow #184, propolis #215, fats #214, greasy wool fat #296. On castoreum cf. Diosc. II.24: tendons but also reptile bites, emmenagogue, abortifacient, against narcolepsy, 'deadly' drugs, etc.

of squill (#264) has a bitter, acrid taste that slightly warms the mouth, and therefore is said to have sharp and heating properties, considered to be useful for conditions caused by excess cold or moisture, such as coughs or dropsy; the cool texture of harvested barley grains (#240) suggested a cooling and moist effect that was useful for treating fevers;[10] and likewise the cold feel and metallic taste of many minerals, stones, and earth suggested they were good for cooling and soothing inflammation or slowing down movement within the body, as when gypsum (#126), considered to have 'vigorously cooling properties' (*uires refrigerantes fortiter*), restricts the flow of blood when smeared on the forehead.[11]

In much pre-modern medicine, the faculties of taste and smell, the 'chemical senses,'[12] were considered to be key methods for determining pharmaceutical effects of a plant or product, or in modern technical terms, organoleptic testing. While authors such as Theophrastus (fourth century BC) attempted to use taste and odour to classify plants, and philosophers of the stature of Plato and Aristotle questioned the relationship between these senses and specific sensations, ancient and medieval authors on pharmacy rarely comment on how or why a particular taste or smell suggested a heating, cooling, astringent, or loosening effect. We can appreciate the difficulty of attempting to do so: not only do these two senses continue to defy precise measurement,[13] any correlation with a predictable effect on the body was tenuous at best. Some of the most toxic plants have little or no discernible taste or odour. Only on one occasion, for example, did Dioscorides directly correlate an astringent taste with an astringent action.[14] Galen explicitly stated that astringency and other 'secondary' properties (sharpness, sweetness, sourness, and so on, as opposed to the four primary properties of wet, cold, warm, and dry) were determined by taste, but he never elaborated on how this was done. It was up to later authors such

---

10 As far back as Hippocratic writings: Stannard 1961, 517.

11 Earths #125–31; stones #161–2, 168–70, 172, 174, 177, 241. Note #89 (cools 'just like snow'). Lettuce was commonly considered a cooling agent (Pliny 20.67, 25.121, 21.58; Diosc. II.136) but surprisingly does not appear in the *AG*.

12 Moncrieff 1967, but since then the fields of psychophysics and neuroscience have developed increasingly sophisticated understandings of the biochemistry of chemoreception, best accessed through journals such as *Chemical Senses* (1974–) and *Attention, Perception and Psychophysics* (1965–).

13 On olfaction in history, see Cain 1978; new directions with psychometrics, Wise, Olsson, and Cain 2000. On taste in history, Bartoshuk 1978; new directions Schiffman 2000.

14 I.82 (mace); Riddle 1985, 37. But note the perceptive comments of Scarborough in Beck 2005, xvi–xvii, on the emphases in Diosc. books I, II, and III.

as Aetius of Amida (sixth century) to expand upon the idea.[15] Curiously enough, the *Liber de dynamidiis*, which as we have seen (ch. 1.E) was also dedicated to Paternianus and shared a similar manuscript tradition to the *AG*, is quite rare in stating explicitly that the type and strength (*qualitas et potentia*) of drugs can be determined by sight, touch, taste, and smell: taste reveals whether they are bitter (*amara*), sweet (*dulcis*), salty (*salsa*), sharp (*acria*), astringent (*stypica*), or mild (*lenia*), while smell indicates whether they are heating (*thermantica*), astringent (*styptica*), harsh (*austera*), or putrid (*putrida*). The banality of these statements aside, the language used (*austera, putrida, salsa, thermantica*) further confirms the impression that the *Liber* has a separate origin from the *AG*, and was only later associated with it.[16]

More common in medical texts is the sort of general comment, made several times by Pliny, that the bitter taste of a plant or substance, or its penetrating aroma, is evidence of its medicinal properties.[17] The *AG* also employs this notion, but goes even further in explicitly stating that the bitterness or pungency of a substance is the medicinal agent at work: the *acrimonia* of hogweed (#269), a term which can be translated as 'pungency,' 'sharpness,' or 'harshness,'[18] loosens the bowels; the 'pungency' (*acrimonia*) of fleabane (#79) wards off snakes, but also helps with their bites, and chronic genital disease. Likewise the pungency and bitterness (*acrimonia et amaritudo*) of gentian (#123) gets rid of snakes, expels the foetus (see ch. 4.H), and is a good antidote to poisons; the bitterness (*amaritudo*) of wild lupines (#153) not only 'kills intestinal parasites but also expels the foetus'; while the 'bitterness and pungency' of wild gourd (#54) 'forcefully agitate

---

15 Galen, *Simp.* I.2 (water), VI. *proem.* (plants) (K.11, 392–3, 791–4). Harig 1974, 68. See also Barnes 1997, van der Eijk 1997, von Staden 1997, and Debru 1997. On Aetius and taste, see Scarborough 1984, 224–5.

16 *J*[unta], fol. 18v–19. Sight shows the similarity of the leaves and flowers, its texture (*materiam*) and colour; touch reveals softness (*mollities*), hardness (*durities*), and its degree of heaviness and lightness. On the vocabulary of touch and texture in medical texts, see Boehm 2003; on the five senses in the Ps. Hippocratic *Regimen*, Jouanna 2003.

17 Examples discussed by Scarborough 1986. Pliny also links bitterness with astringency, e.g., 24.127 'adstringit vehementer cum amaritudine.' Note also Ps. Antonius Musa, *De herba vettonica* (*pref.*), ed. Howald and Sigerist 1927, 5: 'itaque potest effectus sanitatis eleganter, non ut aloe uel absintium, quorum non est tanti sentire auxilium, quam ut amaritudinem sentias ... odore est tam amabilis, ut ignorantem cogat suspicari herbam utilem corpori.'

18 Its Greek equivalent, δριμύτης, and the adjective δριμύς (likewise Lat. *acer, acris*), had the same semantic range, and are just as difficult to translate. So too in other authors, e.g., Pliny 24.128 'cum quadam acrimonia dulcem' (sarcocolla), Caelius Aurelius, *Acut.* 2.41 'omnis acrimonia tumoribus incongrua,' ibid. 2.139.

the entire body with shaking' as well as purge the bowels. We learn that the bitter (*amara*) almond is used more frequently (than the sweet type) 'for medical purposes because it is more astringent.'[19] The connection between taste and property can be quite explicit, as with daucos (#91), which has sharp and heating properties 'so that when tasted they [the properties] immediately bite and heat the tongue.' By the same logic, the more bitter or sharp the taste, or more pungent the aroma, the more powerful the medicine, as the *AG* often states when declaring the 'best' type.[20] Other authors often note that a wild variety is stronger than the domesticated kind, but oddly enough there are almost no examples of such in the *AG*, which regularly states that the wild and cultivated varieties have the same strength.[21] Nonetheless, the *AG* does occasionally mention that different species, determined by colour, are better than others.[22]

Sensory perception, however, needs to be understood in the broadest sense of including also the perceived effects on the body. For experience taught that taste and smell could be completely unreliable: many toxins, for example, have no taste or odour, and whole classes of chemicals supplying the active substances of drugs cannot be detected through organoleptic or other sensory perception. Mandrake (#181), for example, is said to have cooling properties, but these were deduced from its anodyne, analgesic, and soporific effects (all duly noted), which resulted in a deadening effect or loss of sensitivity, interpreted as a loss of innate heat in the body. The same is said for other potentially toxic plants with similar effects (hemlock #49, henbane #295, opium #199). That most toxic plants were considered to be 'cooling' (because deadening the senses or consciousness) in the ancient world is underlined by the origins of the word *pharmakon*, which can mean 'drug' or 'poison.'[23] Likewise, a plant may be considered to be 'astringent' even though it has a sweet, rather than bitter,

---

19 #24; Diosc. I.123 simply says the bitter type is more effective.
20 More bitter or pungent as more powerful: #4, 39 50, 54, 71, 97 (where reduced bitterness, hence effectiveness, results from adulteration), 118, 123, 137, 146 ,180, 199, 202, 231, 236, 252. But cf. #19, 235, where these characteristics suggest the opposite.
21 Wild and cultivated with same strength: #40, 138, 183, 297. Cf. #153, the larger lupine is better than smaller wild variety. Elsewhere, however, 'wild' does suggest a powerful effect: #54, 96, 151 (fig), 195, 199, 203 (cucumber). Cf. Diosc. *pref.*: stronger properties are found in plants which grow in mountainous and rugged terrain, weaker properties in those on plains, wetlands, and poorly aerated places.
22 Colour examples: #4, 97, 137, 207, 235, 236, 247, 252. Of course many other differences are cited (e.g., #19) besides the many concerning place of origin (e.g., #15, 16, 58, 259, 285, etc.).
23 See Artelt 1968.

taste or odour, because of its 'binding' or 'contracting' effect on the body, whether it be for external uses, such as drying up teary eyes (#170), agglutinating wounds (#73) or applied in plasters (#25), or used internally as an anti-diarrhoeal (#183), and so on.[24]

Sweetness in taste or smell is associated in the *AG* with softening, loosening, and warming actions. The organoleptic associations are made clear in some cases. Honey (#179) was thought to have warming and soothing properties, so it is not surprising that the best wax (#59), a common emollient, was that which had a 'sweet and honey-like fragrance,' and likewise the best storax (#254), which 'can gently loosen, for it warms and gently softens indurations,' has a sweet, honey-like fragrance also.[25] That sweetness, heating, loosening, and even 'sharp' properties were associated with diuretics, emmenagogues, and purgatives should not surprise us either when we consider how those effects were perceived to provide relief from fluids and elements stored in the body which must be evacuated to maintain health.[26]

The use of sensory perception as a means of explaining, evaluating, and categorizing the medicinal properties in natural products can be found in our earliest accounts of pharmacy: Ayurvedic herbalism (dating back to 2000 BC) linked health to the balance of six fundamental *rasas*, a word which encompasses the meanings of taste and smell, but also mood, emotion, and even the soul's relationship to the divinity.[27] Equally ancient Chinese traditions of plant medicine classified herbs according to seven categories of taste that relate to function (acrid/pungent herbs have dispersing and moving properties, sweet herbs are for tonifying and harmonizing, sour for stabilizing and binding, bitter for sedating, purging, drying, and so on).[28] It is intriguing that our earliest surviving copy of a Chinese treatise on pharmacy, the *Shen Nong Ben Cao Jing* (*Divine Husbandman's Classic of Materia Medica*), dates from the second or third century AD and describes 365 herbs for medicinal use: a date and number similar to

---

24 See also #2, 61, 7, 170, 278.
25 Note #189 '[roots] sweetly fragrant have a taste that heats the tongue.'
26 Loosening/warming/heating, #120, 122, 223, 225, 253, 258, 263, 293; sharp/heating/diuretic, #14, 15, 26, 29, 35, 73, 139, 270, 271, 272, 279; loosening/heating/sharp/purging, #14, 16, 55, 62, 63, 94, 144, 250. On purging and movement of fluids, see Maurel 2000.
27 See Zimmermann 1987, 1989; Zysk 1996; Singhal and Patterson 1993, 33–7.
28 On Chinese herbal pharmacology, see Chen and Chen 2004, and Bensky, Clavey, and Stöger 2004. On Chinese medical thought and philosophy, see Maciocia 1989.

the *AG*.[29] While there is no direct relationship between the two texts,[30] it is not surprising that the links between sensory perception and medicinal effect are similar, such as henbane classified as 'cooling' because of its analgesic and anodyne properties (cf. *AG* #295), or the sweetness of grapes associated with 'heating' properties (cf. *AG* #291), and so on.[31] These near-contemporary Chinese parallels follow the same methodological arc of beginning with empirical observation, applying organoleptic considerations, and only then relating these to a larger, philosophical or cosmological framework and the place of the human body within it. The key to understanding the larger theoretical framework of the *AG* is found in its concept of properties, to which we now turn.

D. Drug 'Properties'

The *AG* employs the term *uis* (pl. *uires*) literally 'strength[s]' or 'force[s],' a direct translation of the Greek term δύναμις, to describe the active element in a natural product that has therapeutic effect; both words are best translated into English as 'property.' The *AG*'s use of properties is similar to that of Dioscorides, but is more restrictive in its number of categories, and largely clings to a dominant paradigm of four properties that are often paired (heating and loosening, cooling and binding/astringent). It is possible that this more restricted, fourfold paradigm of properties reflects an earlier stage of pharmacy that developed sometime in the period between the Hippocratic writings (third century BC) and Dioscorides (first century AD). Before discussing that prospect, however, it is worth comparing the *AG*'s use of *uires* to its closest counterpart, the δυνάμεις of Dioscorides.

The categories of properties employed by the *AG* are similar to those found in Dioscorides, but very rarely do the two texts agree when it comes to specifying the properties of a given subject – another reason for ruling

---

29 The dating of the *ben cao* texts is difficult because of later interpolation and corruption, and the earliest version we have survives through a revision by the Taoist scholar T'ao Hung-ching (452–536). See Makato 2005; also the overviews of Unschuld 1985, 101–16, idem 1986, 10–83, and Needham 1954–, VI, pt.1 220–38. Updates on research are available online through the International Dunhuang Project: http://idp.bl.uk. My thanks to Roel Sterckx, Cambridge University, for references and for discussion of these texts.

30 On imperial and late Roman contact with China, see Leslie and Gardiner 1996. Archaeological evidence from the Persian Gulf suggests that trade between the Sassanian Persia and China increased in Late Antiquity: Daryaee 2003.

31 Cited and trans. Unschuld 1986, 22. On henbane, compare Diosc. IV.68, Celsus 3.18.12, 3.27.2C, 6.6.9A, 6.7.2D, 6.9.2, and Pliny 25.35–7.

out any direct influence between these authors (see ch. 3.B). Nonetheless, some of the most common terms are directly comparable:

	Diosc. δύναμις	AG uis
warming/heating:	θερμαντική	excal[e]faciens/ calefaciens/calefactoria
cooling:	ψηκτική	refrigerans
astringent/binding:	στυπτική, στυφελή, σῦψις	stringens, styptica
loosening:	λυτική	relaxans, laxans
softening:	μαλακτική	emolliens, remolliens
diuretic:	οὐρητική, διουρητική	diuretica
sharp/sharpening:	δριμύς	acris (and adv. acriter)
dispersing:	διαφορητική	euaporans

Other properties mentioned by Dioscorides are conveyed in the AG by verbs or other constructions: the property of 'drying' (ξηραντική) is rendered with siccat[ur]; that of hardening (σκληρυντική) with firmat/confirmat; relieving flatulence (ἐκκριτικὴ φυσῶν) with inflationem sedat/dissipat or inflationibus prodest; thinning (λεπτυνική) with attenuat, leuis/leuiter facit; staunching (σταλτική) with ad profluuium facit/statuit/stringit; gluing [wounds] (κολλητική) with glutinat; and so on.[32] But Dioscorides also employed various constructions besides adjectives to convey either properties or actions, and was far from uniformly consistent. Although those listed above were the main categories of property recorded in the AG, it also used more specific properties on occasion, such as 'biting properties' (uires mordentes, #175, 283), 'causing perspiration' ([uires] sudorem suscitantes, #271, equivalent to Diosc. 'διαφορητική'), or 'good against poison' (aduersus uenena facientes [uires], #272), 'glutinous properties' (#73), and even 'staining properties' (inficientes, for dyeing) in the case of shoemaker's black (#20), and so on. As we shall see (ch. 4.C), the AG's use of present participles to characterize drug actions has its closest parallels in Celsus and Pliny, and may reflect a stylistic response to Greek adjectives ending in -ικός. This is certainly consonant with the comparisons to Dioscorides' adjectives given above, and the AG's own claims to use Greek sources (ch. 4.F).

---

32 For comparisons of some terminology (madefacere, humectare, siccare, exsiccare, desiccare, refrigerare, infridigare, calefacere) among other Latin pharmacy texts, see Santamaría Hernández 1994.

Like Dioscorides, the *AG* often links the properties and the therapeutic actions with a causal conjunction such as 'therefore': 'X has heating properties, *therefore* it is good for healing Y.' For this the *AG* most frequently uses the word *unde*, translated in this edition as 'therefore,' but it literally means 'from which place,' 'whence,' or 'from which'; hence the selection of this particular word tends to underline the causal link between property and action. Only occasionally does the *AG* use the similar *sic* ('thus,' 'therefore,' 'so') in this sense. Oddly, the usual Latin words for 'therefore' such as *igitur, ergo,* and *ideo* are mainly employed as purely grammatical conjunctives, often connecting one section to the next: most commonly for *igitur* to link the description of the plant/substance to its properties ('Y has red leaves. It *therefore* [*igitur*] has X properties'); and similarly, *ergo* to commence the 'ranking' comments ('the best type is therefore [*ergo*] Y'),[33] though there are exceptions for both of these. Dioscorides likewise used conjunctives to link drug property and use, particularly 'therefore' (ὅθεν), but also other constructions such as '[X properties] operating on' (ποιοῦσαν πρὸς), '[X properties] are suitable for' (ἁρμόζουσαν πρὸς), and so on.[34] Also, like Dioscorides (particularly his books III and IV), in many cases the *AG* does not provide properties, or even actions (such as a substance being 'warming' or 'loosening'), but only mentions medicinal uses.[35] But it also does the opposite, and mentions only properties or actions without any indication of application or medicinal use.[36] Dioscorides rarely did this. Nevertheless, we find the same 'open-ended' statements concerning drug actions ('it has cooling properties,' 'it is an astringent,' 'it softens') in ancient authors such as Celsus and Pliny,[37] and certainly in the more terse herbals that circulated in the early Middle Ages, such as the Ps. Hippocratic *De observantia ciborum,* the *Dynamidia* texts, and the *Diaeta Theodori.* Such bald statements obviously left much room for individual judgment on the part of the physician or interested party for how and when the substance should be used.

---

33 *Sic: AG* #16, 34, 50, 52, 54, 97, 125, 234, 287; *igitur,* passim, though there are exceptions, e.g., #196, 235; *ergo,* e.g., #180, 199, 209, etc.; *ideo*: prol. and epil., #16, 150, 179, 188, 194.

34 E.g., 'ὅθεν,' Diosc. III.75: 'ποιοῦσαν πρὸς' III.62 (cf. *AG* #26); ἁρμόζουσαν πρὸς, III.59; also γὰρ ('for') III.146 (cf. *AG* #138); 'ψυκτικοῦ ὄντος' ('since it cools') III.86. Riddle 1985, 34, cites I.20, I.39.

35 E.g., #56, 65, 78, 79, 99, 101, 103, 104, 109, 112, etc.

36 E.g., #46, 49, 64, 84, 85, 98, 111, 146, 176, 180, etc. Of course some properties (e.g., 'diuretic') required no comment.

37 E.g., Pliny 27.133 (Cretan alexanders can) 'calfacere [et] extenuare,' but 'warm and diminish' what? Cf. *AG* #271.

E. Four Main Properties, Implicit Theory, and Greek Cosmology

Terminology aside, the *AG* also departs from Dioscorides in two noticeable ways: it uses fewer categories of properties, and it cleaves to a dominant paradigm of four major categories that are mostly paired: heating/loosening (*excalefaciens/relaxans*), and cooling/astringent (*refrigerans/stringens*). But like Dioscorides, and as noted above, these are mostly applied in an 'allopathic' manner of opposites to the condition or symptoms: cooling (properties) against heat (inflammation, fever, pain, etc.), or warming (properties) against cold (chills, phlegmatic conditions, loss of sensation), and so on. It is not difficult to understand this in the context of sensory perception discussed above: astringent and cooling actions reduce, tighten, slow, sedate, or stop abnormal developments, while heating and loosening actions help disperse, dry, relax, soften, and reinvigorate body tissue, organs, damage, or disease.[38] Curiously enough, the same pairing of heating-loosening and cooling-astringent is found in another Latin author, Celsus (c. 25 BC–50 AD), who summed up his own survey of foodstuffs with a quick summary of 'foreign drugs' (*peregrina medicamenta*) by stating that 'Generally, those [substances] which are powerful to repress inflammation and cool harden the tissues: those which are heating disperse inflammation and soften.'[39] The paradigm is much the same, but Celsus did not really employ the concept of 'properties' for drugs. In the preface to his survey of seventeen different drug types, he stated that all drugs (*medicamenta*) have 'special powers' (*proprias facultates*), and that he intended to give an account of the 'names, strengths, and their mixture in compounds' (*nomina et uires et mixturas eorum*). But Celsus used neither the term *facultates* nor *uires* in his survey (nor *dynamis*, for that matter), and used entirely different language from anything we see in Dioscorides or the *AG*, despite his heavy reliance on Greek sources, particularly Hippocratic writings, and the frequent deployment of Greek vocabulary or synonyms throughout his work.[40] Writing a generation after

---

38 There are, of course, exceptions: e.g., spikenard (#236) has 'gently styptic and slowly warming properties,' when these two categories seem opposed elsewhere. On other occasions the logic is difficult to determine, such as rock crystal (#174) having astringent properties, but somehow stimulating lactation.

39 Celsus *Med.* 2.33.6: 'Fereque quae vehementer reprimunt, et refrigerant, durant: quae calfaciunt, digerunt et emolliunt.'

40 The 17 drug types he discusses in *Med.* 5.1–16 mainly rely on verbs to describe function. I list them here: suppress bleeding (*sanguinem supprimunt*); agglutinate wounds (*glutinant vulnus*); subdue inflammation (*reprimunt*); mature abscesses and promote

Celsus, Pliny used similar language (*facultates, uires, effectus*) to describe the 'powers' of plants and other products (below, ch. 2.F), but not in any consistent way that suggests a developed sense of 'properties' as the determinative force, as we find in his contemporary Dioscorides.

It remains unclear how Dioscorides arrived at his concept of 'properties,' and what his models were.[41] In the preface to his work, Dioscorides specifically states that properties were determined by the empirical observation of a drug's effect on the body or medicinal application (δοχιμασία), and not on any theoretical considerations. Indeed he heavily criticizes his predecessors for 'prattling about causation,' which led them to considering different particles (ἄναρμοι ὄγχοι) of a drug: a veiled reference to the school of Asclepiades of Bithynia (c. 129–40 BC), which he explicitly criticizes elsewhere in his preface. From what we know about this school's atomist doctrines, Asclepideans were more concerned with diet rather than drugs to maintain the flow of bodily corpuscles that prevented illness, caused by too many or bad particles blocking the pores or channels in the body. But Asclepiades' followers, founders of the Methodist school of medicine, developed a complex pharmacology to fit their therapeutic paradigm of 'constricting' and 'relaxing' the channels, a therapy described as 'metasyncrisis,' for which there were 'metasyncretic' drugs, that is, those which change the conditions of the pores, particularly prescribed for chronic conditions that resist milder treatments.[42] As noted in the introduction (ch. 1.E), the term 'metasyncretic' (for properties) appears once in the *AG* (#203, olive oil), alongside four other categories for properties using Greek terms not found

suppuration (*concoquunt et movent pus*); 'open the mouths, as it were, in our bodies [*aperiunt tamquam ora in corporibus*] called in Greek stomum'; have a cleansing effect (*purgant*); corrode, eat away, or 'are erodents' (*rodunt*); consume, devour, corrode, eat away, or 'are exedents' (*exedunt corpus*); scorch, parch, burn, singe, or 'are caustics' (*adurunt*); 'generally induce scabs on ulcerations in the same manner as when burnt by a cautery' (*fere crustas ulceribus tamquam igne adustis indicunt*); loosen such scabs (*crustas has resolvit*); 'are most powerful to disperse whatever has collected in any part of the body' (*ad discutienda uero ea quae in corporis parte aliqua coierunt maxime possunt*); epispastics (*evocat et educit* cf. 5.17.2); relieve that which is irritated (*levat id quod exasperatum est*); 'make the flesh grow and fills in ulcerations' (*carnem alit et ulcus implet*); soften, or 'are emollients' (*molliunt*); cleanse the skin (*cutem purgat*). On Celsus's use of Greek sources, many of which remain unidentified, the fundamental study was Wellmann 1913, but see Von Staden 1996, 394–408; Serbat 1995, liii–lxviii; and for pharmacy in particular, Pardon 2003.

41 Nutton and Scarborough 1982, 191–2; Riddle 1985, 33.

42 On Asclepiades' pharmacology, Scarborough 1975; on Methodist pharmacology, see Tecusan 2004, 12–14, and for *metasyncrisis*, Fr.[agment]·1 (Aeitus, *Med.* 3.185), Fr. 180 (Galen, *Therapeutics* 4.4–5), Fr. 207 (Galen, *Simp.* 5.24–5), and the next two notes.

elsewhere in the text, but are found in the *Liber de dynamidiis* (among its 21 categories of drugs), which circulated alongside the *AG* in some manuscripts (*F, C, P*, and separated in *B*). As suggested, this transmission may partly explain the incorporation of these terms into one entry of the *AG* (and the repetition of *paregoricus* in the next, #204), but not entirely, given the use of *tonoticum* in this same entry also (see #203 n.6). While the *AG*'s emphasis on 'loosening' and 'binding,' and its allopathic theoretical stance, might vaguely echo Methodist notions of 'constricting' and 'relaxing,' these concepts were fairly common in pharmacological lore, as were heating and cooling, which seem subordinate to constricting/relaxing in Methodist drug therapy. In any case, Dioscorides used the term *metasyncretic* (though sparingly) as category of drug property or action,[43] and later Methodists such as Soranus of Ephesus (fl. 89–117 AD) carefully consulted Dioscorides for their drug lore.[44] If the *AG* reflects Methodist principles in any way, it is most likely owed to these common traditions, rather than a direct tradition of Methodist pharmacology.

There was, however, another school of thought in which preoccupation with diet promoted the concept of properties, namely, the Hippocratic tradition, and it is within this tradition that we may find the origin of the *AG*'s reliance upon four main properties. The Greek term *dynamis* was applied as widely in Greek culture and science as the modern English equivalents of 'power' or 'force,' but we find it used in a specific sense among Hippocratic writings, in which foods are described as hot, dry, cold, or moist by nature because of the *dynamis* within each. A *dynamis* was not a fixed property but a potential property, for each body also has its own *dynamis*, against which the foreign, introduced *dynamis* will react: hence wine may possess this or that property, but its effect is also due to the properties within the body of the patient who drinks it at a particular time, thus explaining why different people have different reactions to wine. Moreover, preparation of a food or drug, or the use of additives, will alter its properties. It is this, the combination of factors, which seems to have been in dispute in ancient medical thought, rather than determining the precise property, or more precisely the potential property, of each substance.[45] No extant Hippocratic treatise addresses the topic of *dynamis*

---

43 Diosc. I.38, III.35, IV.153, V.6, V.119.
44 On Soranus's pharmacology, Scarborough 1991b, esp. 215. On Soranus and Methodism, see Hanson and Green 1994; Lloyd 1983, 168–200; van der Eijk 1999.
45 On *dynamis* in Greek thought, see Miller 1952, idem 1959; in the Hippocratic corpus, Plamböck 1964. On its importance in the Ps. Hippocratic *On Ancient Medicine*, see Schiefsky 2005, esp. 229–35.

directly, but across those Hippocratic writings which mention foodstuffs, and by extension drugs, we find reference to four primary properties of warming, cooling, drying, and moistening.[46] The notion of four properties is an extension of the basic principles behind Greek philosophical or scientific thought that stretched back to the pre-Socratics and beyond: that the world was made up of four primary elements of fire, water, air, and earth, which were intrinsically linked with four primary qualities of hot, cold, dry, and wet.[47] These categories were applied to medicine as early as Empedocles (fl. 460 BC), and there are glimpses of matching drug properties with the elements in the Aristotelian corpus which may have provided a 'blueprint' for assumptions found in Hippocratic writers.[48] But the four elements were not universally accepted as applicable to medicine, and some Hippocratic texts, such as *On Ancient Medicine*, specifically argued against them. Nonetheless the number four held great attraction for Hippocratic authors.[49] The most obvious example of this is the four-fold schema in the influential treatise known as the *Nature of Man*, which develops the theory of the 'four humours,' the theory that later, largely thanks to Galen, came to be seen as *the* Hippocratic theory of medicine. Galen refined and bequeathed this (four-)humoral theory to become the dominant medical theory up to the seventeenth century and beyond, and it was integral to his pharmacology and own notions of drug 'properties.'[50] Neither the *AG* nor Dioscorides subscribed to humoral theory – in both authors 'humours' are merely bodily fluids, or fluids in general (see below, ch. 2.G).

Hippocratic writings promoted no uniform theory of pharmacology, or of even of properties, yet somehow 'all extant writers on pharmacology after the Hippocratics took for granted the role of *dynameis*.'[51] To lay all credit with Hippocratic writers is mistaken: Theophrastus (c. 370–288 BC)

---

46 See Stannard 1961; Scarborough 1983. The concept of properties per se was rarely used for plants, as is clear from the listing of all plants in the corpus and their attributes in Aliotta et al., eds 2003. It also seems to have been irrelevant to compound medicines within the corpus, Totelin 2009.

47 Lloyd 1964.

48 Scarborough 1983, 308–11.

49 Schöner 1964.

50 Lucid expositions of Galen's pharmacy are given by Riddle 1985, 165–78, Scarborough 1984, 215–20, and Nutton 2004, 241–7. On Galen's debt to Hippocratic pharmacy, Jouanna and Boudon 1997. A succinct account of humoral theory that was widely copied in the early Middle Ages was given by Isidore, *Etym.* 4.5.3. On its easy application and adaptability, Stannard 1985; Nutton 2004, 82–6. On its longevity, Temkin 1973.

51 Nutton and Scarborough 1982, 191. No uniformity: Scarborough 1983.

wrote of plants with 'medical powers/properties' (φαρμακώδεις δυνάμεις) in book 9 of his *Historia Plantarum*, though he did not classify or categorize them.[52] Nonetheless, the recurring theme of four (elements, humours, properties) found in Hippocratic and even earlier Greek writings may reflect the origins of the *AG*'s concentration on four main properties. It also may suggest that the *AG*'s categories belong to an earlier stage of pharmacy than that represented by Dioscorides. Unfortunately, we cannot know, owing to the loss of evidence. As discussed in the next chapter (3.E), Dioscorides alone names 19 Greek authors from this period whose writings he used which are no longer extant, Pliny used many of the same if not more authors, and Galen cites yet another ten or so writers from the first centuries BC and AD whose work survives only in fragments or has been completely lost. Nonetheless, there are two other indicators of an early date for the *AG* to which we now turn: use of the term *uis,* and its references to 'humours.'

F. *Uis* vs *uirtus*

We have noted elsewhere (ch. 4.C) the singularity of the *AG*'s Latin terminology for its four main categories of astringent, heating, cooling, and loosening properties (*stringens, excalefaciens, refrigerans, relaxans*), terms which are not found anywhere else as adjectives for 'properties' per se, with the closest analogies being found in ancient writers such as Pliny and Celsus, rather than late antique writings. The same can be said for the *AG*'s preferred use of *uis/uires,* rather than *uirtus/uirtutes,* which it only uses occasionally, mostly in entries that show signs of interpolation by the use of different terminology from the rest of the text.[53] But *uirtus* was the favoured term for drug 'property' from Late Antiquity onwards, witnessed as early as Gargilius Martialis (c. 260) and found in most of our Latin works on pharmacy, including *De observantia ciborum,*[54] the *dynamidia* texts, and significantly, the Latin Dioscorides – none of which, however, can be dated securely. The *AG*'s use of *uis* arguably points to Latin

---

52 Theophrastus, *HP* 9.8.1. There is some doubt that Theophrastus authored book 9: Scarborough 1978 defends its authenticity.

53 *AG* #1, 68, 151, 153, 239. Probable interpolation: #150, 240 (cf. *humectam/humectantia*), 179 (*calida*) 203–4 (*thermantica, metasyncratica,* etc. see ch. 1.E, 2.E), 292 (*humida, frigida*). As 'strength,' #100, 103.

54 *De observ.* employs the term *uirtus/uirtutes* for properties of foods, uses *uis* on one occasion for 'property' ('humidam et frigidam uim habet' [*olus agreste*], 72), but elsewhere uses it in the sense of 'force': *uis aceti* 86, *uis uentorum* 2, *uis* of rubbing with oil 101.

antiquity, when its original meaning of 'force' or 'power' became cognate with other terms to describe medicinal properties in plants. We see this usage as early as Cato the Elder (243–139 BC), who extolled the medicinal *natura* or *uis* of the cabbage, and it continued into the first century AD. Scribonius (fl. c. 47 AD), a contemporary of Celsus, lauded the *uis medicamentorum* in the preface to his book of compound drug recipes, but like Celsus does not employ the notion of 'properties.'[55] Nor does Pliny, in whom the terms *uis, natura, facultas, potestas, effectum* are interchangeable, but noticeably, at the beginning of book 25, where Pliny provides a potted history on those who have written on plants, he criticizes Greek authors for simply listing the 'powers and property' (*potestates uimque*) of plants without providing a botanical description. Pliny also voices agreement that there is 'nothing which cannot be achieved by the power [*uis*] of plants but the properties [*uires*] of most are still unknown.'[56] Understandably then, *uis* was the term which fourth-century authors such as Marcellus and the *Medicina Plinii* took from Pliny to describe drug properties, and likewise was the preferred term in Ps. Apuleius (e.g., *uis calfactoria*), though that author also used on occasion *uirtus*.[57] Clearly the terms *uis* and *uirtus* were interchangeable as a translation of *dynamis* as early as Gargilius and probably earlier, but *uis* was the preferred term for the earliest reception of the concept, as is also found in other works besides Ps. Apuleius which circulated in the same manuscripts as the *AG* such as

---

55 Cato, *On agriculture* (ed. Hooper 1999), 157.1, 'ualidam habet naturam et uim magnum habet,' 157.12 'brassica erratica maximam uim habet,' etc.. On Cato's terminology, see Boscherini 1970, and idem 1993, who defends doubts about Cato's authorship of this section of *De agricultura*. Maggiulli 2000, 150, reads Cato as evidence of the 'ascendancy of Ps. Hippocratic humoral theory' found in *Regimen*. Scribonius, *pref.* 6. See Sconocchia 1985.

56 Pliny 25.15 'nihil non herbarum ui effici posse, sed plurimarum uires esse incognitas.'. Interestingly, Pliny gives three reasons for this: because this knowledge is primarily among illiterate country folk; because those who learn it refuse to teach others; and because many think the knowledge is the restricted preserve of women.

57 Marcellus, e.g., 8.2 (*ui[s] medicamenti*), 26.60, etc. *Med. Plin.* 10 ('[farina opopanacis] uim habet mirifice excalefactoriam'). Ps. Apul. *pref.*, 'paucas uires herbarum et curationes corporis'; also 3.4 'et ceteris pro cuiusque uiribus sic dabis,' 29.4 'sucus pro uiribus cuiusque potui datus,' 93.18 'Puleium quam uim medicamenti secum habet.' *uis calfactoria* 19.4. As force: *uis ueneni* 19.1, the *uis* of dittany of Crete (= *AG* #88) kills snakes, even by fragrance 62.3. *uirtus*: 11.4, 12.2, 12.59, 126.1, 131, and in the two *precationes* attached to 119 'ut uenias ad me cum tuis uirtutibus,' and 126 'cum gaudio uirtus tua praesto sit.' On the whole, the text avoids using either *uis* or *uirtus* by concentrating on specific applications and ailments – an absence underlined by the interpolations from Dioscorides.

Sextus Placitus's *Liber Medicinae ex animalibus* (in manuscripts *L*, *M*, and *B*) and Ps. Antonius Musa's *De herba vettonica* (in *L* and *B*).[58] The date of all of these texts, however, is uncertain, as is that of the Ps. Hippocratic *Letter to Maecenas*,[59] though it may be the earliest text to mention the terms *dynamis* and *uis* together. After describing various types of disease and some treatments for them, the *Letter* concludes with an exhortation to consult the work of Terentius Euelpistus, 'in the last book of which you shall read about the *dynames* of plants,' which are (apparently) altered by the phases of the moon, so by knowing these phases 'we ought to monitor the property and power [*uim et potestatem*] of plants when considering and compounding medicines.'[60] Terentius's work does not survive, but Celsus called him 'the most famous oculist of our times,' and cited some of his compound drugs and salves, as did Scribonius, and later Galen. Nonetheless, this further suggests that Greek concepts of plant 'properties' were circulating in first-century Rome, and *uis* was the preferred Latin translation. Isidore of Seville (d. 636) recognized this six centuries later in his unique description of *dynamidia* as a genre of medical book, a word and concept not known to Antiquity:

> *Dynamidia* are concerned with the power [*potentia*] of plants, meaning both their force and their potential [*uis et possibilitas*]. For the force within plants used to cure ailments is called *dynamis*, and so the term *dynamidia* is used to name writings that describe the medical uses of plants.[61]

---

58 Sextus Placitus, *Lib. med.* 30.4 'Columbae stercus habet uires multas'; Ps. Antonius Musa, *Lib. de herba vettonica, pref.* '[medici] non totis perspexerunt uiribus.' Their date is very uncertain. Sextus used Pliny, but it is unclear whether he used Marcellus (c. 400) (Sabbah, Corsetti, and Fischer 1987 no. 480–5: hence Langslow 2000, 68) or was used by him (Howald and Sigerist 1927, xxii: hence Opsomer 1989, lxii): Ps. Antonius may have been used by Marcellus, but is thought to date before Ps. Apuleius (late fourth century) (Sabbah, Corsetti, and Fischer 1987 no. 26–34; Langslow 2000, 68; Opsomer 1989, lxiii).

59 The version in Marcellus, *pref.* (ed. Niedermann 1968, 26–32) is the earliest: another (B) version dates from the early ninth century. Its transmission is linked with the similar *Epistula ad Antiochum regem* (ibid., 18–24). It has been argued that both texts derive from a Latin translation of a single Greek letter attributed to Diocles of Carystus (saec. IV BC): see Opsomer and Halleux 1985; Sabbah, Corsetti, and Fischer 1987, nos. 321–40; and Sconocchia 1998.

60 Marcellus, *pref.* ed. Niedermann 1968, 32. Marcellus himself also mentions several times that phases of the moon affect the properties of plants (1.43, 8.24, 8.41, 16.101, 25.13, 31.18), but the idea is foreign to the Hippocratic corpus, Dioscorides, the *Dynamidia* texts, and the *AG*.

61 *Etym.* 4.10.1 The passage has aroused comment: MacKinney 1935–6, 412–3; Mazzini 1984, 31–2, idem 1992; Maggiulli 2000.

Isidore mentions three other genres of medical books (*aforismus, prognostica, butanicum herbarum*), the first two of which recall titles of two very popular Hippocratic works. Hence he probably had in mind the texts entitled *Dynamidia* that have come down to us in two slightly varying versions and which seem to be extensions of the Hippocratic *De observantia ciborum* (that is, a Latin translation of book 2 of *Regimen*). Oddly enough, the *Dynamidia* texts exclusively use the word *uirtutes* for properties, but that may be the result of their late manuscript tradition (from the ninth century onwards).[62] In any case, the choice of *uis* for the concept of healing 'property' within plants, combined with the adjectives used for their categories (*excalfaciens, refrigerans, stringens, relaxans*), further suggests an ancient, rather than late antique, date for the *AG*, as do its references to humours, some theoretical statements in the text, its lack of magic or superstitious elements, and its references to cosmetics.

G. Humour, Bile, and Phlegm

The *AG* refers on several occasions to *humor*, but mostly in its original, loose sense of meaning body 'juices' or fluid, as found in Dioscorides, Celsus, and Pliny.[63] That *humor* simply meant moisture or fluid seems clear from its use to describe the composition or constituents of different earths and stones.[64] But elsewhere it is applied to bodily fluids, as when spurge (#94) is said to 'contract humours as it purges, and can heat watery humours,' or that lodestone (#165) 'draws out all humours through the bowels,' or that the astringency of rock alum (#48) is excellent for staunching 'discharges of blood and humour and stopping diarrhoea.' There are, therefore, many humours rather than strictly four, as indicated when honey (#179) 'reduces humours by thinning them,' when the turnip (#233) 'nourishes harmful [*inutilem*] humour,' and when the heat of stinging nettle (#288) 'breaks down all humours.' In one case, however, humours are paired with phlegm as fluids that can be reduced by the warming action of mercury (#160), yet on several occasions phlegm is paired with bile as

---

62 Earlier versions certainly existed to which Isidore had access: see the comprehensive studies of Ferraces Rodrígez 1999, 19–126.
63 Riddle 1985, 42–3. Celsus: e.g., 4.22.4 diuretics 'are beneficial by directing humour to another part.' On Pliny and Hippocratic thought, see Beagon 1992, 225–32; Nutton 1986; Byl 1994.
64 #127, 130, 131; stones #161, 170 (*humorem quasi lacteum*), 171, 175; plants #97, 181, 199, 201, 252, 254; as liquid 276. As both moisture and saliva, #147.

two things to be reduced by a drug.[65] This would appear to reflect earlier Hippocratic treatises, which ascribe particular importance to these two fluids above others, some (like *Affections*) suggesting that they are the origins of all disease, others (like *Airs, Waters and Places, The Sacred Disease, and Diseases 1*) seeing them as two polarities within the human constitution, an imbalance of which causes disease, and is to be addressed by diet and attention to climate.[66] The same can be said of the references to black bile, such as when dodder (#103) not only purges phlegm 'with equal force through the mouth and through the bowels' but also 'dissolves black bile' (see also #55). It has been noted that before the Hippocratic treatise *The Nature of Man* raised the importance of black bile to the level of one of the four humours, black bile was considered in the earlier treatises as merely another type of bile distinguished by its colour, and was not regarded as a humour.[67] This seems to be the case in the *AG*, which refers also to 'red' bile on two occasions (#159, 292). Moreover, that the *AG* used both the terms *bilis* and *cholera* for bile[68] suggests these references were simply picked up from several works, and that there was no overarching theoretical commitment to bile as a specific substance with specific attributes, as confirmed by an entry for bile itself (#116), understood here as gall (from animals), which used the eminently Latin word *fel*, rather than the Greek term χολή (as used by Dioscorides II.78). In any case, Hippocratic physiology promoted the idea that juices, humours, and biles could be moved about the body to either cause or cure illness, and that all bodily fluids, including menstrual blood and urine, needed to be constantly evacuated to prevent their stasis or coagulation within the body. This concern seems to inform the *AG*'s consistent focus on purgatives, diuretics, and emmenagogues, as well as explain the few references to 'drawing out,' 'dissolving,' or 'reducing' humours, bile, phlegm, and other fluids. Humoral theory, as expounded by *Nature of Man*, played no part in the *AG*'s understanding of physiology or therapeutics.

---

65 #5, 16, 82, 159. That traveller's joy (#81) 'reduces phlegm downward' echoes Diosc. (IV.180), 'ἄγει κάτω φλέγμα καὶ χολήν'.

66 Nutton 2004, 79.

67 Nutton 2004, 83–4.

68 *Cholera*: #5, 16, 82, 103, 159, 292. Those suffering from *cholera* are called *cholerici*, #94. It is difficult to determine the meaning of 'nigram bilem prouocat' in #55: as it is listed immediately after mention of purging the bowels, combined with the sense of *prouocat* for urine and menstruation throughout the text, I have translated it as 'forces out,' but it could also mean 'stimulate the production of.'

H. Explicit Theorizing in the *AG*

Only on few occasions does the *AG* reveal some sort of reasoning or theory behind its prescriptions, such as when mandrake (#181) is said to have cooling properties, 'and for the same reason [*et hac ipsa ratione*] also astringent properties,' or that opium (#199) 'reduces all types of fatigue in the body, and for the same reason [*illa uidelicet ratione qua*] we find it also induces sleep.'[69] There are, however, three instances in the *AG* where peculiar, theoretical statements are made that further reveal the text's orientation, and originality. Two of these instances appear in the entry on wine (#292). After the initial claim that all wines are warming, we are told that wine:

> supplies the body with both strength(s) and movement [*uires atque motum*], because it heats, repairs, and restores any part of the body which is constrained by cold or pain.

While similar sentiments about wine can be found in other authors, the precise language used here that links heat with strength and restoration is not found elsewhere.[70] The second theoretical statement appears as an authorial interruption after the recommendation that wine should be mixed with water:

> ... And here something needs to be said about water itself. All water has cold, soft, and moist properties, and hence is administered many times when treating acute fevers. But if it alone is applied to malignant wounds the result is disastrous, for all wounds to the body themselves are soft in nature. Then the malignancy thrives and grows, finding an agreeable environment in which it can nourish itself and develop further.

There seems to be no direct source for this statement, the closest parallel being the Hippocratic treatise *Wounds*, which opens by extolling the

---

69 Also #149, 281. More commonly, *ratio* is used to mean 'method' (of collection, preparation, confection, adulteration), e.g., #3, 97, 126, 136, 156, 201, 208, 280, 296. Also as 'test' in #149.

70 Cf. *De observ.* 44, '[uinum nigrum] siccat consumens humorem ex corpore calore suo.' Diosc.V.6.10 names the shared properties of wine as 'warm[ing], easily digested, wholesome, stimulate the appetite, nutritious, soporific, strengthen[ing], and give a good complexion.'

virtues of wine for nearly all wounds, then notes that the physician's task is to achieve a dry state for wounds, the state closest to being 'natural,' by expelling any wet or moist 'humours': Galen repeats these precepts of *Wounds* in his *Therapeutics*.[71] While the three texts are following same logic, their focus and wording are quite different, ruling out direct influence. That wounds are 'soft in nature' may reflect Hippocratic thought or simply common sense, and the idea that disease thrives on excess moisture was common to ancient medical thought, acutely evident in the notion that drinking too much water results in coughs, dropsy, and in humoral terms, phlegmatic conditions.[72] Nonetheless, the interruption is a welcome, personal voice in an otherwise impersonal text written in sober, technical prose.

One other comment is worthy of note for its deductive reasoning. We are told that honey (#179)

> alone is something of a compound mixture in itself, because it is collected from many different flowers and the juice of different plants. For this reason it can be useful as a fast-acting agent for all ailments, hence it is mixed with all types of antidotes.

No other ancient source makes a comparable statement, which reveals at least some degree of thought on how honey is made, and how the diversity of natural elements within honey qualifies it as a compound drug. Modern research is only now beginning to understand the potential of antibiotic compounds in honey, such as methylglyoxal, for medicinal uses, particularly for topical application in treating wounds and burns. A 'honey compound' has even been touted as an improved means for administering honey's healing properties.[73] As noted (in the translation) for the reference to 'heavenly dew' in this entry, we may well be hearing faint echoes of ancient Greek voices now lost to us.

---

71 *Wounds*, 1, ed. Littré VI, 400. Galen, *Therapeutics* 4.5 (K.10, 278, trans. in Tecusan 2004, 483). Here I follow Scarborough 1983, 318 n.58, in using the title *Wounds* rather than the traditional English rendering of *Ulcers* for Περὶ ἕλκων.
72 See Skoda 1994. There is nothing similar in Diosc. V.10, Celsus 2.18.12–13, or Pliny 14.59–70. The *De observ.* contains two entries on water (44, 106), the first ridiculously short ('aqua frigida et humida est') and preceding the entry on wine; the second warning against water being too cold or too hot because it deprives the body of 'humor' and fills it with cold spirit, which causes deterioration. For the *De observ.* on wine, see ch. 3.E.
73 Comparisons: see notes to #179. Medicinal rediscovery, Zumla and Lulat 1989. An update is Cimolai 2007. Honey compound: see Osman, Mansour, and El-Hakim 2003.

I. The 'Doctrine of Signatures' and the Absence of Magic in the *AG*

Two final points should be made concerning the pharmacology of the *AG*: the possibility that the appearance of a plant or substance influenced the author's understanding of its medicinal effects; and the complete absence of religion, superstition, or any aspect of the supernatural in the text. The notion of sympathy, or 'doctrine of signatures' as it came to be called by sixteenth-century spiritualists and chemists, is evident in our earliest sources for pharmacy.[74] It was based upon a likeness, whether real of fancied, between a pathological condition and the appearance of a natural substance. An obvious example in the *AG* is that the double-bulbed root of the man orchis resembles testicles, and is an aphrodisiac (#272). The connection is not made explicit – most authors avoided doing so, these associations being a little too folksy for men of learned, literate traditions – and there may be more implicit in the text, such as the use of red-coloured plants or minerals for staunching bleeding (#83) or stimulating menstruation (#219). Likewise, sympathy could easily extend to the theory of opposites, so that hard or soft plants, fruits, or other substances might be thought beneficial for treating ailments with 'soft' or 'hard' features respectively (unhealed wounds, indurations, etc.). However, the overall 'rationalist' stance throughout the *AG* suggests that any such connections were unintended, underlined by the absence of superstition in the text, especially when compared to Dioscorides and Pliny.[75]

While Pliny included much superstition and magic in his accounts of pharmacy,[76] Dioscorides was quite reluctant to do so, and a quick comparison with Dioscorides' attitude further highlights the *AG*'s rationalist or secular nature. For where Dioscorides was prepared to admit superstitions into his account of certain products, usually (but not always) distancing himself with the comment 'some say' or a similar phrase, the very same products in the *AG* contain no such information. For example, Dioscorides (II.126) reports that the root of plantain is carried 'by some' as an amulet 'to disperse scrofulous swellings in the glands,' but the comparatively short entry for plantain in the *AG* (#217) sticks to common-sense applications of its leaves and juice, and there is no mention of amulets throughout the entire text – striking for an ancient or medieval medical

---

74 For its history and relation to scientific inquiry, see Court 1999, Bennett 2007, Pearce 2008. On its use in ancient and medieval sources, see Stannard 1982a (in Pliny), and 1985 (medieval).

75 E.g., cf. *AG* #290 (betony) with Pliny 25.84, 101, and Diosc. IV.1.

76 Stannard 1965, 1977, 1982a, and Scarborough 1986, 1991a.

work.[77] Similarly, both texts describe leopard's bane as resembling a scor-pion in appearance (IV.76, #21), but Dioscorides adds that 'they say its root, when brought before a scorpion, paralyses it, and that it stirs again when white hellebore is set before it,' a legend repeated by Pliny (27.4–10), who includes a host of others. One of the few occasions in which Diosco-rides did not distance himself from a superstitious practice was to state that squill 'wards off evil when hung whole on front doors' (II.171), but the *AG* (#264) restricts its comments to squill's sharp and heating effects being useful against coughs and dropsy.[78]

The *AG* comes closest to peddling superstitions when it recommends the placing of lupine flour and horehound in the navel to expel intestinal parasites (#153) or smearing honey externally on sore throats (#179); when it states that gypsum smeared on the forehead restricts the flow of blood (#126), that flame stone 'is said' to induce sleep when smeared on the fore-head (#173), and that some people mix worm-like verdigris with the urine of an infant boy, which goldsmiths find good for soldering (#136). None of these are very wacky, and topical applications may invite the placebo effect. The only other possibly superstitious element mentioned in the *AG* is the charge that the (unidentifiable) plant known as *basilisca*, which the *AG* classifies as a type of man orchis (#272), has properties that are 'use-ful against poisons, so that anyone who has this plant with them is safe from all types of snakes.' Hardly superstitious (it may simply mean that the antidote is then ready to use if needed), and all the more reasonable when we compare it with the only other known source for this plant, Ps. Apuleius. As discussed in the next chapter (ch. 3.F), that text shares some verbatim description and the statement about carrying it to keep safe from snakes. But Ps. Apuleius also mentions three types of snakes with noxious, magical powers, and a superstitious ritual for harvesting the plant. Since the similarities between the two texts seem owed a common source, rather than direct influence, we have further indication of the *AG*'s avoidance of

---

77 Lloyd 1983, 129, argues that Hippocratic writings do not refer to amulets and in general avoid the magical: Totelin 2009, 123 n.54 counts three amulets in the gynecological trea-tises, and his approach to the recipes underlines the extent of both symbolism (particu-larly in remedies concerning reproduction or contraception) and the theory of opposites in the choice of ingredients.

78 Other examples: asparagus (Diosc. II.125) grows where beaten rams' horns are buried (cf. *AG* #23). The root of sorrel (II.114) is used by 'some' as amulets against scrofulous swellings by hanging it around the neck (cf. *AG* #154). Cf. also his comments on black hellebore (IV.162) with #285. Contra Riddle 1985, 88, this reference to gatherers of black hellebore praying to Apollo and Asclepius is not the only mention of a deity in Dioscorides: see *AG* #76 n.6.

superstitious or magical elements, and its concentration on drug properties and therapeutics.

Other appeals to magic or superstition common to medical texts, such as magic words or 'word sympathy,' numerology, and astrology, find no place in the *AG*.[79] The only reference to a deity in the *AG* is that the 'ancients' used to burn the incense called *cyphi* (#76) 'for their gods.' The origin of the incense, and Dioscorides' similar entry for such, reveal that the 'ancients' in question are the Egyptians. But curiously, our earliest manuscript (*V*) adds the phrase 'and whose doctors unwholesomely prepared rites' (*pistelentius iscaribus* [= *sacribus* for *sacris*? *escaribus* from *escae*? cf. *ThLL* V.2.856] *laborabant*). The word *pestilentius* (properly *pestilenter*) obviously indicates disapproval, but the appearance of this phrase in only two manuscripts (*C*, a later recension, differs slightly) suggests it is a later interpolation by a (Christian) scribe. It has been said that both Pliny and Dioscorides wrote at a time when superstitious and magical thought were on the rise, while Greek rationalism was on the retreat. If Dioscorides really was 'the model of a rational Greek'[80] in his resistance to magic in his day, then the *AG* was even more rational; or it was written in an earlier era, as the fourfold paradigm of drug properties may also suggest.

J. Non-Medical Uses

The *AG* includes information on natural products that are strictly outside the purview of medicine, yet were obviously considered worth recording. Compared with Dioscorides or Pliny, references to non-medical uses are kept at a minimum, and are mostly restricted to cosmetics – a subject that again betrays the text's ancient date and context.[81] Several references are made to hair dye: ampelitis earth (#127) is used to dye red and white hair black, while sage leaves (#102) can also be to used as a black hair dye. Myrtle (#188) is used as hair dye, and also promotes 'flowing hair,' and for this reason it is also mixed with rockrose (#156) to tame 'frizzy' hair. Loss of hair can be countered by oil of mastic tree (#151), and the radish

---

79 For examples of these see Stannard 1977, 1985. Magic in ancient medicine has been studied extensively: a brief guide is Nutton 2004, 268–70, 406. See also Scarborough 1991a.
80 Riddle 1985, 138. Rationalism to magic, Lloyd 1979, 1983. On magic in Cassius Felix and Theodorus Priscianus, Fraisse 2003b.
81 For Dioscorides, Riddle 1985, 88–92. Noticeably, the *AG* does not discuss perfumes as do these other ancient authors, though it does remark on the pleasing fragrance of some substances: e.g., #15, 30, 59, 61, 73, 77 ('more powerful than other aromatics'), 102, 137, 180, 202, 253, 254, 260, 262, 270, 279, 287, 293.

(#234) helps with 'baldness of the head' (*alopeciae capitis*), and also *alopecia* in general (#33), while depilatories are mentioned twice (#8, 280). We are informed that ampelitis earth (#127) is used for eye make-up, yet the entry for *purpurissum* (#211), which we know from Pliny was a type of cosmetic, only mentions that it was used by dyers to obtain a rosy colour. Presumably these are dyers of textiles, which also gain a mention in the entry for indigo (#135), and Sinopic red earth (#276) is said to be a powerful dye to achieve redness. Aphrodisiacs are mentioned six times, mainly by referring to engendered activity (*uenerea*).[82] Such uses or traits never stand alone as attributes but are complementary to the medicinal applications recorded. The care taken to record the ingredients and measures taken to prepare the incense *cyphi* (#76) is owed to its use in compound medicines for its sharp and loosening properties, while the fact that pumice stone (#213) is good for sharpening quill and reed pens (see ch. 4.C) is complementary to its use as a toothpaste and as an ingredient in cleansing lotions.

Despite the *AG*'s inclusion of animal products, and in contrast to Dioscorides and Pliny, there are only a few references to veterinary medicine, which had close links with medicine, particularly in the traditions of animal husbandry among the Romans, who surpassed the Greeks in at least this branch of medicine.[83] That lye made from ashes of brushwood (#150) is applied to castrated sheep seems included to demonstrate its efficacy for healing cuts, and likewise lye from oak wood (#149) is also good for treating humans and sheep, but it should be tested on sick sheep first by having them drink it. The reference to dogs eating dog's-tooth grass (#36, and see plate 5) to purge themselves seems gratuitous (neither Dioscorides nor Pliny mentions it). References to warding off animals, such as the many mentions of putting snakes (and/or vipers) to flight,[84] or that the tamarisk (#193) kills lice and drives away gnats and fleas, are obviously another aspect of how pharmacy could be used counter threats to health.

K. Conclusion

The pharmacology of the *AG* offers a significant contribution to our knowledge of ancient and medieval thought on pharmacy. Its distinct

---

82 #92 (*amatorium*), 100 (*coitum ... ad uenerias inspirat*), 124 (*ad uenerea percitat*), 196 (*ad uenerea mire*), 266 (*ad uenerea facit*), 272 (*uenerea prouocat*). On types and terminology, see Denniger 1930; for Greek works, Faraone 1999.

83 On the links between veterinary and human medicine, see Scarborough 1969, 171–3; Fischer 1988a, Adams 1995. In Dioscorides, Riddle 1985, 77–82.

84 #79, 88, 108, 123, 201, 206, 221, 272 (on this last see ch. 3.F): separate from treating snake bites: #155, 191, 271, 284, 301.

use of properties and Latin vocabulary to describe them, its explicit and expressive association of these with sensory perception, and the practical, non-sensational nature of its prescriptions – often quite accurate according to modern scientific research – add considerably to our understanding of ancient medical traditions and their transmission to the Middle Ages. But pharmacology also strongly indicates an 'ancient' date for the text. The *AG*'s use of *uires* suggests an early reception in Latin of the concept of 'properties' (δυνάμεις) as found in Dioscorides. Yet its more restricted number of properties, and its predominate fourfold paradigm of paired properties (heating/loosening, cooling/binding), may well reflect an earlier stage in the development of this concept, sometime between Hippocratic texts and Dioscorides. Unfortunately, the loss of pharmacological writings from this period, by all indications a well-established and even flourishing genre (see ch. 3.D–E), prevents any firm conclusions. Other aspects also imply an ancient date. The *AG*'s references to humours, bile, and phlegm are similar to those found in Dioscorides, Pliny, and Celsus, and suggest a date well before the ascendancy of humoral theory championed by Galen. Inclusion of non-medical uses such as cosmetics and aphrodisiacs also smacks of the ancient world, while the almost total lack of magical or superstitious elements, in contrast to the *AG*'s two closest analogues of Dioscorides and Pliny, may well suggest earlier, more rationalist traditions of Greek science. Its few explicit theoretical statements also seem to reflect this earlier tradition, and reveal a critical intelligence able to theorize about therapeutics and evaluate claims. Whenever and wherever the *AG* was originally compiled, it is one of the few Latin texts on pharmacy which circulated in Late Antiquity and the early Middle Ages that allow us to see clearly the connection between theory and practice in the use of natural products to treat medical conditions.

3

Sources Compared and Lost

❧

A. Introduction

In the *AG*'s epilogue, the author states that, besides relying on memory and experience, he has consulted 'ancient and most popular authorities' to compile his list of drugs. Whether or not the epilogue is authentic (ch. 1.D), the text constantly refers to 'some say' or 'yet others say,' and on two occasions 'some describe' (*aliqui describunt* #92, 99). None of these sources is named, and there are no obvious candidates among our surviving sources. But from what we can tell, many of them dated to the first century AD or earlier. To see this clearly, we must first untangle the issue of the similarities of the *AG* to Dioscorides, which then allows us to address four related topics: the *AG*'s occasional similarity to Pliny; the possibility that the *AG*, like Dioscorides and Pliny, drew from the now lost work of Sextius Niger; some intriguing links with (what we know of) some other lost sources dating from the second century BC to the second century AD; and finally the shared linguistic echoes with two other works from Late Antiquity, namely, the *Diaeta Theodori*, pseudonymously named after a less-famous doctor, and Ps. Apuleius *Herbarius*, pseudonymously named after a Roman novelist. At the end of this chapter I include some brief guidelines for using the listed sources for comparison (*Comparanda*) underneath the critical apparatus at the foot of the Latin text of the *AG* for each entry.

B. Dioscorides

Only one surviving text from Antiquity, Dioscorides of Anazarbus's *De materia medica* (Περὶ ὕλης ἰατρικῆς), can be shown to have direct affinities with the *AG*, in terms of both content – of botanical description, medicinal 'properties,' and medicinal uses – and in the order in which these are presented. Parallels with Dioscorides abound. Some are noted in the English translation, but these neither are comprehensive nor indicate Dioscorides as the source, even for near-verbatim passages (allowing for issues of Greek-to-Latin translation),[1] and in a few cases, near-verbatim whole entries. For it is clear that the similarities between the two texts derive from their use of similar sources belonging to a rich tradition of botanical and medical literature now lost to us.[2] There are three principal reasons for presuming this which we explore here. Firstly, many entries for the same product differ entirely from Dioscorides. Secondly, despite direct echoes between the two texts on a given substance, other aspects always differ considerably, be it the colour of the flower, the shape of the leaf, the preparation methods used, the location of the 'best' types, the properties recorded, or the prescribed medical uses, so that rarely does a direct echo extend beyond a phrase or just a few words. Thirdly, the *AG* describes a dozen drugs not mentioned by Dioscorides (who recorded over a thousand), nearly half of which are found in Pliny, who also did not know Dioscorides' work.[3] Other points are discussed elsewhere (see ch. 4.E).

1 It should be stated here that our concern is with the original Greek text of Dioscorides, not the Latin translation edited by Stadler et al. which comes down to us in ninth- and tenth-century manuscripts and is often dated to the sixth century (without compelling reasons). This Latin text is largely a slavish, word-for-word translation in highly vulgarized Latin that shows no affinity with the Latin of the *AG* (see ch. 4.E for some examples). The earliest use of Dioscorides by a Latin author is Gargilius Martialis (fl. 222–35), but we do not know if he used the Greek or a Latin translation. As Ferraces Rodrígez 1994 has shown, the *Ex herbis femininis* and the *Dynamidia* texts both drew from a translation of Dioscorides that has not come down to us, reminding us that the Latin text edited by Stadler et al. was certainly not the first in the chain of transmission from Greek to Latin. Moreover, the examples cited by Ferraces Rodrígez (spurge, mandrake, wild gourd, squill) prove that the *AG* did not use this earlier, hypothesized translation either. On Dioscorides' transmission in Italy see Touwaide 1992, and more generally, Riddle 1980, 1985. It is also worth noting that another work on simples attributed to Dioscorides (Περὶ ἁπλῶν φαρμάκων, ed. Wellmann 1914, III:151–317), but of doubtful authenticity (see Riddle 1980, 134–9, idem 1985, xxvi–ii), has no relation to the *AG*.

2 On lost botanical and pharmacological works of antiquity, see Stannard 1968, and below ch 3.E. There is much (still) in Wellmann 1889 and 1897. For the following discussion, the comparative references to Dioscorides and Pliny can be found among the *Comparanda* at the foot of the Latin text of the *AG*.

3 Not in Dioscorides: soap *AG* #259, lye #150, skirret #273 (if it is not parsnip), blackberry #195, *folium* #111, squared stone #168, rock crystal #174; and others known to Pliny (on which see also below), *purpurissum* #211, sulphur #249, *syricum* #237, flame stone #173, batrachid (frog) stone #176.

At no point does the *AG* name a source or author, despite the frequent references to 'some say … but others say' in several entries, a trait it shares with Dioscorides, though Dioscorides sometimes named authors, Pliny even more so. For example, both the *AG* and Dioscorides report that 'some describe' the seed of *dorycnion* (#92) as something used for 'aphrodisiac[s]' (*amatorium*), revealing that both authors have lifted that phrase from an earlier work, rather than having themselves independently reviewed the work of those 'some.' Their common debt to earlier tradition can be detected in other instances, such as the comment both authors make at the close of their entries for Melian earth:

> We recommend that which is fresh, quite soft and delicate, and which crumbles apart quickly when wet and contains no rubble. We require that all clays [*creta*] be tested this way, and likewise for all those which we have mentioned above. (*AG* #131)

> This one [Melian earth] and in general all soils must be chosen free of stones, fresh, soft and crumbly, and they should easily dissolve when water touches them. (Diosc. V.159)

The entries have other similarities (ashen colour, alum-like taste, resultant dry tongue), but also differ: the *AG* omits entirely any mention of properties or medicinal uses, which Dioscorides provides, while also noting that painters use it to enhance and prolong the life of colours. The *AG*'s singular use of the word *creta* possibly points to another, Latin tradition (it is the term used by Pliny for medicinal earths), for the *AG* consistently uses the Greek word *ge* for 'earth' in all its entries on such (#125, 127–32). When tracing the myriad sources used by Dioscorides, Max Wellmann was so shocked by the debt of that author to previous works that he exclaimed in print, 'Dioscorides is nothing other than a compiler of the first century after Christ!'[4] Wellmann's comment is important, if unfair, in that it serves to highlight the debt of surviving authors to a literary chain of previous works now lost.

The most compelling evidence for the *AG*'s independence from Dioscorides is that many entries differ entirely, such as that for honey (#179), where the comments about different types of bees or that honey is a compound medicine in itself are not found in Dioscorides or in any other

---

4   Wellmann 1889, 548; cited by Riddle 1985, 18, who shows that Wellmann tempered this
    view as his critical edition of Dioscorides progressed.

source (ch. 2.H); or the entry for moon foam (#12), which states that it is made from 'heavenly dew' placed upon the stone called *specularis*, a stone recorded only in Pliny, Petronius, and Lactantius. Compare Dioscorides' entry, given here in full:

> Selenite, which some people have called *aphroselinos*, because it is found at night when the moon waxes, occurs in Arabia. It is white, transparent and light. Filed down, it is given as a drink to epileptics and women use it as a protective amulet. It also seems to make trees fruitful when attached to them.[5]

The only element in common is that of treating epilepsy, and the mention of amulets merely highlights the gulf between the two texts, for as we have seen (ch. 2.I) the *AG* contains no such references to magic or superstitious practices of any kind, very unusual for an ancient or medieval medical text. Similarly, Dioscorides' entry on gypsum consists of a simple sentence: 'Gypsum has astringent, adhesive, anti-hemorrhagic, and antiperspirant properties. But if drunk it kills by suffocation.' The description in the *AG* (#126), with its method of preparation in a *clibanum*, comparison to Samian earth, and so on, clearly stems from a different tradition. The same can be said for many of the mineral entries: the *AG* is often more discursive, which partly explains their interpolation into texts such as the lapidary of Damigeron, and even into later medieval versions of the Latin Dioscorides, for the earliest recensions of such, dating from the ninth and tenth centuries, lack the entries for stones; hence the broken tradition was healed by pilfering from the *AG*.[6] Other entries for minerals show a close affinity to Dioscorides' text, and there are others not found in Dioscorides but known to Pliny (discussed below).

While it is easy to point out other big differences between the *AG* and Dioscorides – such as the entry for butter (#41), with its unique mention of a butter churn, or that for pumice stone (#213), mentioning quill pens (see ch. 4.C), and so on – it is difficult to qualify the smaller differences and similarities without the price of tedium for the reader, and descent into madness for the researcher, so a few examples will suffice. The descriptions of costus root (#64) are similar, but the *AG* mentions Cretan and Egyptian varieties, and criticizes the Syrian, while Dioscorides says nothing of the first two, and only mentions the Syrian type in passing. Shoemaker's black

---

5   V.141, trans. Beck 2005, 400.
6   Text of Damigeron ed. Halleux and Schamp 1985. Cf. the entries for Thracian stone *AG* #167, Diosc. Lat. V.129. Also blister beetles *AG* #65, Diosc. II.61, where the *AG* gives more information.

(#20) is in part almost verbatim in similarity, but the *AG* includes a different preparation (rusted iron and vinegar) and different properties (cooling, rather than burning). Mountain germander, herb ivy, and ground pine are likewise treated in the same entry (as types of *chamaepitys*, #142), but the order is inverted, and so on. Common tradition must also explain the similarity of some striking phrases, such as the flower of lonchitis (#157) resembling 'one of those masks with the gaping mouth used in comedies'; that 'Syrian' spikenard (#236) is named after the direction it faces on a mountain in India; that Thracian stone is found in the river Pontus in Sintia (#167), an obscure location. Likewise the less exotic similarities shared by the two texts: the doubling of cassia to equal cinnamon as an ingredient (#75), Ethiopian ebony is better than Indian (#95), the best castoreum comes from two beaver testicles joined together (#50), cankerwort leaves are hairy like those of bindweed (#106), and meadow saffron is also known as wild onion (#80). The same can be said for medicinal uses, such as stinging nettle taken daily with honey eases coughs (#288), juice of the root of yellow flag is 'amazing' for clearing eyesight (#30), or the digestive uses of aloe (#5).

At times the *AG*'s Latin comes close to being a direct translation of Dioscorides (as with *-astrum*, ch. 4.E), but in other instances the texts vary a great deal. The same can be said for many opening phrases of an entry, not just the 'X is a plant known to all/well known' formula (on which see ch. 4.H), but also the description, such as that for storax:

Styrax est quasi lachryma arboris quae et ipsa stryax appellatur, similis tota cydonia. (*AG* #254)

στύραξ δάκρυόν ἐστι δένδρου τινὸς ὁμοίου κυδωνίᾳ. (Diosc. I.66)

The simple difference that the 'tree itself is [also] called storax,' however, already points to a different tradition, for the *AG* concentrates on the wood-dust obtained from larvae exiting the branches as an essential element of preparation, while Dioscorides merely mentions it in passing as one of several possible ingredients for adulterating storax, and on the whole the descriptions differ considerably in content and format. Hence when we carefully comb through the seemingly almost verbatim descriptions of azurite (#37), salt froth (#38), parsnip (#222), and bitter vetch (#133), we find differences that, combined with the evidence presented above, rule out direct influence, even in cases where the *AG* seems to follow Dioscorides blindly, such as when both liken the seed of shepherd's-purse (#283) to that of garden cress (*cardamum*), for which the *AG* fails to provide an entry. Both texts, therefore, are heavily dependent on a common tradition, despite both authors' claims to first-hand experience.

C. Pliny

That Pliny, Dioscorides, and the *AG* all used similar sources is clear from some entries, such as the coupling of *pompholyx* and *spodos* as two different forms of zinc oxide (#210), or that adulteration of scammony (#252) with spurge is detected by its heating effect in the mouth, the changing taste of agaric (#16), the two different types of hypocist (#298), the three types of calamine (#45), and so on. As mentioned above, the *AG* describes several mineral products not mentioned by Dioscorides but known to Pliny, namely, *purpurissum* (#211, for which the *AG* offers an alternative name, *fuscum*, not recorded elsewhere), sulphur (#249, including the 'live' type), *syricum* (#237), flame stone (#173), and batrachid (frog) stone (#176). Here and there the two texts contain other information not found in Dioscorides, such as likening the appearance of ammoniac salt to *schiston* (#239), or the three types of copper pyrites (#89), and there are other elements that point to an exclusively Latinate tradition, such as the name of *malicorium* for pomegranate rind (#183), *caricae* as a type of fig (#109, though in other respects the entry greatly resembles that of Dioscorides), that the Greeks call Assian stone (#163) *sarcophagon*, and so on. Some locations also reflect a Latinate tradition, as when Celtic spikenard (#238) is said to grow in the Alps 'and in the region of Noricum,' when Pliny says Pannonia and the 'Norican Alps,' and Dioscorides has 'Alps and Liguria'; or that scammony (#252) grows in Colophon and its resin is called *colophonia*, of which Dioscorides is unaware, but Pliny, Scribonius, and Caelius Aurelianus knew; and that there is an 'Alexandrian' mustard seed (#261). There is some information in the *AG* that seems partial to Italy (ch. 1.F, ch. 4.E), but on no occasion does the *AG* replicate the type of specialist knowledge on Italy's flora and fauna that Pliny so affectionately included, such as *inlecbra* being the Italian name for purslane (#32); the *AG* does not even use the common Latin name of *portulaca*, but rather the Greek *andrachne*. Also only very rarely does the *AG* resemble Pliny in language and expression. The closest parallel is the entry for axe weed (italics highlight the similarities):

> *Pelecinus herba est quae in segetibus nascitur*, plures et tenues ramulos habet, foliola pusilla et *granula in folliculis terna uel quaterna*, subrufa et amara ualde. Quae possunt cum aliqua acrimonia uiscide stringere, propter *quod et in antidota mittitur* et menstrua prouocat. (*AG* #219)

Pelecinon in segetibus diximus nasci, fruticosam cauliculis, foliis ciceris. semen in siliquis fert corniculorum modo aduncis ternis quaternisve, quale git novimus, amarum, stomacho utile. additur in antidota. (Pliny 27.121)

While the differences are apparent, Pliny is closer to Dioscorides in likening the leaves to those of the chickpea, and describing the pods as 'like little horns,' yet the latter does not comment on the seed growing in bunches of three or four, and all three texts offer different medicinal uses. Clearly in this case the *AG* was using similar, but not the same, sources as these two first-century authors. Let us now look at one possible source.

D. Sextius Niger: The Possibility

In the introduction to his monumental work, Dioscorides criticizes the efforts of his more recent predecessors for describing in great detail 'materials that are common and well known to everyone,' and setting down 'slapdash accounts of the properties of drugs and the method of testing them.' Among these he names the early first-century (AD) authors Julius Bassus, Niceratus, Petronius, Sextius Niger, and Diodotus, none of whose work survives, save only through citations from later authors. But then Dioscorides singles out the work of Sextius Niger (c. first century BC–first century AD) in particular:

> For instance, Niger, who is considered prominent among them, supposes that the resinous juice of the spurge [τὸ Εὐφόρβιόν φησιν ὀπὸν] is the juice of the spurge olive [χαμελαία] which grows in Italy, and that perfoliate St John's wort [τὸ ἀνδρόσαιμον] is the same as crispate St John's wort [ὑπερικόν], and that aloe is dug up in Judea [ὀρυκτὴν ἐν Ἰουδαίᾳ γεννᾶσθαι], and propounds wrongly many other notions similar to these that are contrary to manifest facts, proving unequivocally that his account is based not on personal observations but on writings that he has misunderstood. Niger and the rest of them have also blundered regarding organization: some have brought into collision disconnected properties, while others used an alphabetical arrangement [κατὰ στοιχεῖον], separating materials and their properties from those closely connected to them.[7]

Curiously enough, the contents of the *AG* largely matches Dioscorides'

---

7   Diosc. I. pref.. Translation my own navigating Beck 2005, 2, and Scarborough and Nutton 1982.

descriptions, or rather criticisms, of Niger's lost work *The materials of medicine* (Περὶ ὕλης ἰατρικῆς, the same title used by Dioscorides), which we only know from Dioscorides' and Pliny's comments on it.[8] To begin with the last comment about organization: although it is unclear whether Niger was among the 'some' who employed alphabetical arrangement, Dioscorides' statement is itself one of the earliest attestations of the use of alphabetical order for classifying anything other than books, which began in the libraries of the Ptolemaic kingdom. Wellmann believed that Niger, and probably Crateuas (c. 120–63 BC), were the target of Dioscorides' comment, but the proof is lacking.[9] Of course, alphabetical arrangement for pharmacological texts was also used after Dioscorides (e.g., Galen, Oribasius, and even for Dioscorides' own text). In the *AG*'s preface the author claims to have put everything into alphabetical order himself, but he may have been inspired by one of his sources. In any case, the reference there to having to unfurl an 'entire scroll' (*totum uolumen*) suggests a pre-fourth-century date at least.

But it is Dioscorides' comments on spurge, St John's wort, and aloe that link Niger's lost text with the *AG*. Dioscorides complained that Niger confused the juice of spurge (εὐφόρβιον) with the juice of 'spurge olive [χαμελαία] growing in Italy.' The *AG* (#94) states that spurge is the sap of a tree that grows in 'Mauritania and, as some would have it [*ut aliqui uolunt*], in the regions of Italy that border Gaul,' and is the only source (besides Niger) to mention Italy. Dioscorides' own entry for spurge claims that it grows in 'Autololia next to Mauretania.' Pliny locates it in the same region (Mount Atlas), and has seen a treatise on the plant by King Juba II of Numidia (and Mauretania), who discovered it. But toward the end of his account Pliny also notes an inferior variety from Gaul which 'comes from the spurge-olive' (*ex herba chamelaea*) – hence Pliny reflects

---

8   Wellmann 1897, 21. Wellmann's article, and the evidence for Sextius's work he collected in an appendix to his edition of Dioscorides, vol. III, 146–8, are complemented by the attempt of Capitani 1991 to fill out the context of Sextius's career. Additionally, see Riddle 1985, 14–19; Marganne 1982; Scarborough 1986, 69–76.

9   Wellmann 1897, 18, cited the alphabetized section in Pliny (22.73–91), which mentions several Greek authorities, and argued that these references were more likely derived from a Greek text such as that of Sextius Niger: likewise Scarborough 1986, 70 n.84, citing Pliny 27.4–142, as probably from Niger. But Pliny did not specifically name Sextius as his source for these entries as he does in other places. Only one plant (*sium*, water parsnip) of the dozen or so listed by Pliny in 22.73–91 appears in the *AG* (#158), and only a few appear in Dioscorides; while the longer section of 27.4–172 is clearly from an alphabetized Greek source, only a dozen or so of its c.115 plants appear in the *AG*, with very different descriptions. On alphabetization, see Daly 1967, 35, and ch. 1.D.

the same 'confused' tradition found in the *AG* and Niger.[10] The *AG* and Pliny share other information not given by Dioscorides: that it resembles frankincense and that it severely dries out your throat (the *AG* uniquely offers a method for mitigating this). On the other hand, Pliny and Dioscorides share similar descriptions of the harvesting process (incision at a distance by javelins, collection of sap in washed animal stomachs), and its use against snake bites (by pouring it into an incision in the skull) – aspects not recorded by the *AG*. The issue is further complicated by the fact that all authors contain another entry for spurge under the name of *tithymallos*, for which both Pliny and Dioscorides describe seven types with specific names.[11] The *AG* merely states (perhaps resignedly) that 'Spurge has almost as many names as there are types': but uniquely it also distinguishes separate methods for collecting its sap (from crushed branch tips then collected in shells) and its juice (from pounding and pressing the leaves), demonstrating a different tradition. Nonetheless, the *AG*'s mention that 'some' think *euphorbium* is found in Italy may well be referring to Niger.

The second complaint of Dioscorides, that Niger thought perfoliate St John's wort (τὸ ἀνδρόσαιμον) to be the same as crispate St John's wort (ὑπερικόν), is reflected in the *AG*, which gives an entry only for *hypericum* (#299), and its description matches that of Dioscorides for the perfoliate variety (ἀνδρόσαιμον, III.156 = *Hypericum perfoliatum* L.), in that both describe its flower as small and yellow ('golden' *aurosum AG*, 'quince-yellow' μήλινα Diosc.). The family of St John's wort (Guttiferae) is one of the more difficult in modern botanical taxonomy. Dioscorides broke new ground in carefully distinguishing four different types (III.154–7), though he mentioned similar medical uses for each (hip ailments, burns, diuretic, emmenagogue), the same uses found in the *AG*'s singular entry. But Dioscorides' description of the crispate variety (ὑπερικόν, III.154 = *Hypericum crispum* L.) also shares aspects with that of *AG*: it has leaves like rue, and a 'capsule the size of barley,' but its flower is 'like a gilliflower' (λευκοΐῳ ὅμοιον). Hence the *AG* was guilty of the same conflation of the perfoliate and crispate varieties for which Dioscorides criticized Niger, as were no doubt many other authors prior to Dioscorides' exactitude; again suggesting an early date for the *AG*.[12]

---

10 Diosc. III.82. Pliny 25.77–9, 'sed iubae volumen quoque extat de ea herba et clarum praeconium.' None of these authors mentions anything similar when describing spurge olive itself: Diosc. IV.171, Pliny 24.133, *AG* #144.

11 Diosc. IV.164, Pliny 26.72, *AG* #280.

12 Pliny 26.85 also likens the leaves of the *hypericon* to rue, and says its black seed 'ripens at the same time as barley,' but he does not mention the flower, nor that of the 'other' type

Finally, Dioscorides criticized Niger for claiming that aloe is 'mined in Judea' (ὀρυκτὴν ἐν Ἰουδαίᾳ γεννᾶσθαι). The *AG* (#5) noted that 'Some say it is gathered from rocks [*de petris collecta*] in Judea, Asia, and even India.' Since Sextius wrote in Greek, it is plausible that he wrote something akin to Dioscorides' ὀρυκτὴν ... γεννᾶσθαι' (lit. 'obtained by digging'), and that in translation to Latin this became *de petris collecta*, to mean 'pulled from the ground.' But Pliny reflects a similar confusion when he claims that 'Some have reported that in Judea beyond Jerusalem there is a mineral aloes [*metallicam eius naturam*],' which was the most inferior type.[13] As with spurge (above), Pliny, the *AG*, and Niger all partake of the same tradition criticized by Dioscorides in his preface, and again the 'some say' of the *AG* may be referring to Niger. The *AG*'s account certainly has nothing in common with that of Dioscorides (III.22, except the likeness to squill, a common epithet), who says that aloe grows in India, whence its juice is exported, and also 'in Arabia, in Asia, along certain coastal areas, and on islands, as for instance on Andros.' Scarborough and Nutton have suggested that Dioscorides may have described Socotrine aloe (*Aloe perryi* Baker), from the island of Socotra due east off the Somalian coast, introduced to the Roman West sometime in the first century AD, documenting the extension of Roman trade with the Far East.[14] If so, the aloe of the *AG* may predate this commerce. Pliny notes that 'the best' is imported from India and that it also grows in Asia, and other aspects of his description resemble that of Dioscorides, but some medicinal uses are shared with the *AG*, which, as with mining/digging up/harvesting 'in Judea,' confirm a common tradition, one shared with Sextius Niger.

There are other faint echoes of this shared tradition. According to Pliny (28.120), Niger said that more than a *drachma* in weight of skink flesh, mixed with a *hemina* of wine and drunk, was dangerous, as does the *AG* (#266, a *cyathus* of wine), but not Dioscorides (II.66). Pliny also says (32.26) that Niger recommended the Pontic castoreum as the best type, which we find in the *AG* (#50), where Dioscorides (II.24) is silent – though both the latter and Pliny (who cites Niger here as his source)

---

of *hypericon* 'called by some *caro*,' which is obscure. Note also AG's *scopa regia* #275, traditionally another Latin name for St John's wort: the description is brief, but the likening of its leaves to lettuce is unique, perhaps ruling out any of Dioscorides' four types.

13 Pliny 27.14–20, at 15.
14 Scarborough and Nutton 1982, 210–12, and Scarborough 1982. On trade, Innes Miller 1969, and Raschke 1978. Note that Celsus, 1.3.25–6, merely mentions aloe. The earliest known illustrations of aloe are found in frescoes at Pompei: see Feemster Jashemski and Meyer 2002, 88.

deny the legend that the beaver removes his own testicles when being pursued. However, there are omissions to suggest also that Niger was not consulted: Niger (says Pliny 20.129) wrote on garden cress (*nasturtium*), which gets only a fleeting mention in the *AG* (#283 n.1, 289 n.3); but also on mallow (*malua*, Pliny 20.226, also Gargilius *Med.* 5); on the yew (*taxus*, Pliny 16.51); on cheese (particularly from horse milk, Pliny 28.131), and on the salamander (Pliny 29.76). None of these are mentioned in the *AG*.

E. Other Lost Sources BC to AD

What all this confirms, rather than merely suggests, is that the *AG* used sources which, like Sextius Niger, predate Dioscorides and Pliny and conveyed the same traditions found in all three surviving works. It also confirms that the many echoes of Dioscorides found in the *AG* derive from their use of the same or similar sources, and not from direct knowledge of one another – the differences far outweigh the similarities. A pre-Dioscoridean date for the *AG* is also suggested by other entries, such as its claim that the root of wind rose (#13) is called agrimony (*eupatorium*), which is completely erroneous, for agrimony is another plant entirely. Dioscorides goes out of his way to correct the confusion of 'some people' between agrimony (εὐπατόριος), the wind rose (ἀργμώνη), and the poppy anemone (ἀνεμώνη) by providing careful descriptions of each, and pointing out the differences (Diosc. IV.41, II.176, 177). The *AG*'s author had obviously not read Dioscorides on this plant, yet those who did (Galen, Oribasius, Alexander of Tralles) took note of Dioscorides' distinctions: our author knows nothing of this tradition, perhaps because he wrote before it.

Which authors may the *AG* have used for its descriptions and prescriptions? For a start, we should consider that besides Sextius Niger, Dioscorides mentioned another nine authors in his preface alone: Julius Bassus (c. 1–50 AD), Niceratus (first century AD), Petronius Musa (first century AD), Iollas of Bithynia (third century BC), Heraclides of Tarentum (first century BC), Crateuas the root-cutter (first century BC), Andreas the Physician (third century BC), Diodotus (first century AD), and Asclepiades (first century BC). To these we can add the other eight or so authors he mentions in the text itself, while scholarly detective-work reveals that Dioscorides also used Nicander of Colophon (fl. c. 135 BC), and perhaps also Diocles of Carystos (fourth century BC) and Apollodorus (third century BC) – tallying 19 lost authors all up, though there were undoubtedly many more.[15] Pliny, whose *Natural History* was published thirteen years after Dioscorides' work appeared (64 AD), also mentions many of the same authors as his sources (Iollas, Heraclides, Crateuas, Andreas, Julius Bassus, Niceratus, Asclepiades), plus others such as Apollodorus (fl. 280 BC), King Juba II (see #94 n.1), and contemporaries like Xenocrates of Aphrodisias (fl. 70 AD), who wrote works entitled *Useful things from plants* and *Useful things for man from animals*. But Pliny also mentions a whole host of Greek authors dating from the sixth to the first century BC, many of whom are just names to us. It has been argued by some scholars that Pliny knew these authors through later doxographical collections of writings, possibly by Solon of Smyrna (third–second century BC), which he may have known directly or through Sextius Niger, whom he relied upon heavily for his drug lore.[16] Moreover, Pliny also had access to (now lost) Latin works, such as that on medicinal plants by Valgius Rufus (consul 12 BC), to whom Horace dedicated poetry, and those of Antonius Castor, a centenarian doctor whose botanical gardens in Rome contained many medicinal plants which Pliny inspected personally. But Pliny also used Celsus, Varro, and Cato, whose works partly survive: there is no correspondence between these and the *AG*. Though the text that has come down to us was written in a Latinate environment, the *AG* seems primarily Greek in origin and orientation (see ch. 4.F).

In his own pharmacological works, Galen employs a host of Greek authors on pharmacy who span the period of the first centuries BC to AD and whose work survives only in fragments or only through Galen's extensive quotation of them. Of these we can rule out as possible sources

---

15 Diagoras (late 3rd cent.), Diosc. IV.64, Erasistratus of Ceos (3rd cent. BC) IV.64, King Juba II (50 BC–23 AD) III.82, Mnesidemus (4th cent. BC) IV.64, Philonides (1st cent. AD) IV.148, Thessalus of Tralles (1st cent. AD) I.26, Peteesius (? according to Scarborough 1986) V.98: also possible is Theophrastus (371–c. 287 BC) on plants, since Dioscorides names him twice, though Wellmann 1889 and others suggest that Dioscorides probably knew Theophrastus through intermediaries such as Sextius and Crateuas. On Dioscorides' use of Nicander, Scarborough 1977, 1979. On Andreas, von Staden 1989, 472–7.

16 Philistion of Locri (4th cent. BC), Chrysippus of Soli (c. 280–c. 207 BC), Praxagoras (fl. c. 310 BC), his pupil Pleistonicus, Dieuches (4th cent. BC), Medius, Simus, Cleophantus, Pythagoras (6th cent. BC), Glaucias (3rd cent. BC), Dionysius, Chrysermus (1st cent. BC?), Tlepolemus, Dalion, Sosimenes, Evenor (4th cent. BC), and Hicesius (1st cent. BC). On Pliny's sources: Wellmann 1924, and notably Scarborough 1986, arguing for Solon of Smyrna, and who notes that Pliny mainly kept to authors of the first century AD. There may well be two Xenocrates: fundamental is Wellmann 1907.

those concerned with compound drugs, and perhaps those who wrote in verse, like Servilius Damocrates and the two Andromachi (all first century AD), though they were known to Pliny. Although Pamphilos (c. 150 AD) wrote a work on plants and their *dynameis* that was arranged in alphabetical order, it was also full of magical lore from Egypt and elsewhere such as incantations and 'ridiculous' stories.[17] But at least two of the seventeen or so authors cited by Galen are intriguing candidates. Heras of Cappadocia (20 BC–20 AD) composed a work called either *Narthex* ('Medicine chest'), or the 'book of properties.'[18] While it is clear from the bulk of Galen's citations that it contained many compound medicines, Galen also mentions Heras in relation to 80 or so natural products, and most of these appear in the *AG*, including exotic or mysterious items like *cyphi* (#76), *alcyonion* (#11), (Lemnian) *spragis* (#276), and the hypocist (#298). Of course these were the simples most commonly written about, which explains their inclusion in Dioscorides also, and there are some simples that the *AG* did not mention.[19] Nonetheless, the range of simples, contained in a work known to be concerned with their 'properties,' represents the tradition from which both Dioscorides and the *AG* drew, and to which they both belong.

The second author of interest cited by Galen also reflects a common tradition rather than a direct source for the *AG*. Statilius Crito was a doctor to the Emperor Trajan (c. 100 AD) who had made his fame as a historian,

---

17 The key study is Fabricius 1972, esp. 180–236. See also Touwaide 1997. On Pamphilos, Galen *Simp.* VI. proem. K.11, 792–6, and von Staden 1989, 439. We can probably rule out Asclepiades Pharmacion (1st. cent. BC–AD: Fabricius 1972, 192–8), highly esteemed by Galen, because traces for his cure for earaches (Galen, *Comp. med. sec. loca,* III. 3, K.12, 610) find an echo in Dioscorides' entry on opium (IV.64.4); the *AG*'s entry for opium (#199), although noting it cures earaches, does not combine it with other ingredients. Archigenes (fl.100 AD; Fabricius, ibid. 198–9) wrote on compounds and his remedies included amulets (see ch. 2.I).

18 Galen seems to acknowledge two titles in *Comp. med. per genera* I.13 (K.13, 416), 'καὶ ὁ Ἥρας ἐν τῷ βιβλίῳ τῶν φαρμάκων, ὅ τινες μὲν ἐπιγράφουσι νάρθηκα, τινὲς δὲ τόνον δυνάμεων,' though a little later mentions 'καὶ Ἥρας μὲν ἐν ἔγραψε βιβλίον τῶν δυνάμεων,' ibid. I.15, (K.13, 441). Fabricius 1972, 183 n.8, suggests emending τόνον ('intensity') to τόμον ('book') in the first citation, as synonymous with the βιβλίον of the second, on the basis of *Meth. med.* II.11 (K.11,137), 'κατὰ τὸν Ἥρα γέγραπται τόμον.'

19 *cyphi* (K.14, 207.6), *alkuonion* (12, 398.10), *spragis* (14, 207.8), hypocist (13, 39.12). Not in the *AG*: βησασᾶ (K.12, 941.18), ἡδύσαρον (13, 1042.15), κίκινον (13, 782.11), ναρδόσταχυς (12, 942.1), πολύτριχον (12, 430.13), σκόρδιον (14, 206.11), σπλάγχνον (13, 1042.17), φρυκτή (13, 766.2), χάρτης (13, 339.14–15), plus particular varieties or species not recorded by the *AG*, ἀριστολόχια δακτυλῖτις (13, 544.7), ἶρις Ἰλλυρική (12, 941.15), Ἀμμωνιακὸν θυμίαμα (13, 544.7), etc. See Fabricius 1972, 183–5, 209–12.

recording the Dacian wars of Trajan's campaigns north of the Danube. But Crito also wrote a work on *Cosmetics* (Κοσμητικά), or the 'art of dress and ornament,' and two other works on pharmacy, one entitled *Pharmakitis*, the other on either compound drugs or simples: again Galen seems to give two different titles for one work.[20] The *Cosmetics* contained a recipe for a depilatory containing yellow orpiment, quicklime, Selinus earth, and fine meal (or starch, ἄμυλος), which Galen dutifully recorded. John Scarborough has noted that the only author before Crito to mention yellow orpiment as a depilatory was Dioscorides, and that no other author recommends mixing it with quicklime. As it happens, the *AG*'s entry for orpiment (#8) mentions that 'when mixed with quicklime and smeared on the body it removes hair. In Greek this mixture is called *psilotron*, but is known in Latin as *acilea*.' This Latin term is unattested elsewhere, and presumably derives from *acies/acer* ('battle line'/'sharp edge,' etc.: perhaps *acilea* is a slang term for 'razor'?), reflecting the mixture's extreme causticity, as was noted by Galen. The word *psilotron* for depilatory (from ψιλόω, 'strip,' 'bare') can be found as early as Theophrastus, but Dioscorides did not use that word to describe orpiment's use 'to remove hair' – further proof of the *AG*'s independence from this author.[21] While Galen did include Crito's recipe under the rubric of 'Hair Removers' (Ψίλωθρα τριχῶν'), it is very unlikely that Crito was the source for the *AG*'s notice, given the absence of Selinus earth and fine meal in its own *psilotrum*, and nothing is said about depilatories in the *AG*'s entries for these two products (#130, 6). The noble Roman audience of Crito's *Cosmetics* may well have used the term *acilea* for caustic depilatories using orpiment and quicklime, or Crito may have used that term himself in the *Cosmetics*, or in his other lost work. It may have been beneath Galen to record such local slang.

The examples of Heras and Crito, and the traditions they shared with the *AG*, highlight the richness of pharmacological literature of the third century BC to Galen's time and beyond which has been lost to us. Herophilus of Alexandria (c. 335–250 BC) taught that 'drugs are the hands of the gods,' and his followers embraced pharmacology as an essential branch of medicine, so that they did not (according to Celsus) 'treat any kind of dis-

---

20 Fabricius 1972, 190–2, 261. On Crito's career and works, see Scarborough 1985, and 2008. Simples or compounds: Galen, *Comp. med. per genera* 5.6 (K.13, 786), 'τῆς τῶν φαρμάκων συνθέσεως'; ibid, 6.1 (K.13, 862), 'περὶ τῶν ἁπλῶν φαρμάκων αὐτῇ λέξει.' On *pharmakitis* as a genre of recipe books, Totelin 2009, 98–102.

21 Diosc. V.104 'ψιλοῖ δὲ καὶ τὰς τρίχας.' Theophrastus, *HP*, 9.20. Crito's *psilotron*: Galen, *Comp. med. sec. locos*, I.4 (K.12, 453): Scarborough 1985, 395–6.

ease without drugs.' Likewise, their rivals, the early Empiricists, extolled the clinical importance of pharmacy, but very little of the drug lore of either of these schools and their practitioners survives.[22] Such loss makes it difficult to measure the *AG*'s debt to these previous traditions, but the parallels with Dioscorides, Pliny, and Sextius Niger, and shared traditions with Heras and Crito, suggest a degree of antiquity that helps explain why the *AG* has no direct correlation with other works on pharmacy which circulated in early medieval Europe. We now look at the two examples in which a correlation can be made, but can be explained by common (lost) sources or later interpolation.

F. Two Linguistic Echoes: The *Diaeta Theodori* and Ps. Apuleius

Of the many Latin medical texts that have been dated to the early medieval period, only two show a direct, verbatim relationship to the *AG*, namely the *Diaeta Theodori* (*DTh.*) and Ps. Apuleius *Herbarius*. The *DTh.* appears in a few manuscripts that date from the eleventh or twelfth century, including Vienna 2425 (our *W*), in which we find a highly edited version of the *AG*.[23] The *DTh.* is quite different in character from Dioscorides, Pliny, and the *AG*: it opens with a short and somewhat confusing statement on the worth of determining what constitutes the maintenance (*officia*) of health, and what harms us, then launches immediately into discussion of the *uirtutes* of barley and wheat, taken mostly from the second book of the Hippocratic *Regimen* (that is, the Latin *De observantia ciborum*), and likewise on different kinds of milk, vegetables (*legumina*), animals, birds, fish, herbs, greens (*olera*), eggs, cheese, a long section on wine, water, and finally on bathing, vomiting, and exercise, finishing off with some potions for maintaining health to be drunk in March and April.

---

22 On Herophilus and Herophileans, von Staden 1989, esp. 14–19, 400–1, 418–19 (citing Celsus *Med.* 5 proem.), and 450–2. On Empiricists and pharmacy, Deichgräber 1965, fr. 136–9, 157–63, 192–240, 252–5, 259, 262, 267–74.

23 Sabbah, Corsetti, and Fischer 1987, no. 200–4. The false attribution to Theodorus (scil. Priscianus) is found in most manuscripts. Different recensions and editions have appeared, but that of Sudhoff 1915 seems the more reliable for being based on three manuscripts, including Brussels, Bibl. Royale 1342–50 (upon which were based the *editio princeps*, publ. Venice 1502, and that which appeared among the Physica S. Hildegardis, publ. Strasbourg 1533), London, British Libr., Harley 4986, saec. XI, and Vienna 2425 (our *W*: see ch. 5.C). The text has received little attention, but see Baader 1970, 15–17, whose discussion of its late Latin terminology demonstrates its dissimilarity from the *AG*. On Theodorus, see Conde Salazar and López de Ayala 2000, and in the same volume, Fraisse 2000.

Although we find many of the same expressions used in the *AG* in terms of drug actions,[24] to a degree of similarity not shared with any other text, they are almost never used in respect to the same substance, and the two texts mostly record opposite effects or properties (cold/hot: good/bad for stomach, etc.) – another example of the colossal diversity between pharmacological texts, the sort that calls into question whether there was any real understanding (sensory, theoretical, or otherwise) behind some of these works. Nonetheless, the *AG*'s entry for wine (#292) has several connections with the entry for such in *DTh.*: they share the opening statement that 'all wines are warm in nature,' share two almost verbatim phrases concerning wine that is 'seasoned with raisins' (*ex passo*) and violet-wine,[25] and also refer to wines seasoned with wormwood (*uinum absinthiatum DTh.*), boiled must (*defretum AG, defricatum DTh.*), marjoram, and mastic, though the effects ascribed to these by the two texts differ considerably. Also the *DTh.* discusses many more adulterated wines, and continues with a catalogue of regional wines (Sorrentinum, Chium, Asianum, etc.). Hence although there is no direct relation between the two texts, the shared phrases and terminology demonstrate that both texts have borrowed from similar traditions concerning the medicinal potentials of foodstuffs, as we see in the *De observantia ciborum* and the Ps. Hippocratic *Dynamidia* texts (which regrettably do not list wine).[26] Two examples will suffice:

On new wine:

> *De observ.* 347: uinum nouellum magis egeritur quam uetus ex eo quia prope dulce est
> *DTh.* 268: uina recentia uetustis calidiora sunt, plus nutriunt

---

24 For example: *uentrem stringit, inflationem facit, urinam et menstrua prouocat, lumbricos occidit, partum expellit* (note *nauseam stringit*, 190). It also refers to *uirtutem uiscidam* (of sea water 391), see ch. 4.D; and both texts refer to the barley products of *alfita* and *ptisana* (13, *AG* #240, not found in the *De observ.*). *DTh.* uses *uirtus* to denote properties, *uires* for 'strength' (*acquirit uires, continet uires* 14, 23).

25 'Conditum ex passo similiter plus tamen humactat et stomacham iuuat' (277. Cf. *AG* 'similiter et urinam deducit, corroborat stomachum': 'uiolatium urinam procurat ... etc.' 282 nearly verbatim. Likewise the nearly verbatim comment that 'aqua omnis frigida natura [*uirtus AG*] est et humida' (387), but the rest of the entry differs considerably, underlined by the following two chapters on 'rain water' and 'on the different natures of different waters,' this last taken, as the text informs us, 'ex libro decimo Oribasii' (423).

26 Note Wallis 1995, 116, '[the *DTh.*] updates for medieval Western readers the list of wines in the original Hippocratic text.'

AG #292: uinum recens calidum minus est uetusto et uentrem procurat

On white wine:

> *De observ.* 345: uinum album austerum calefacit, non tamen siccat, magisquae
> per urinam quam uentrem egeritur.
> *DTh.* 269 uina alba dulcia nutriunt et urinam procurant.
> *AG* #292: uinum album diureticum est et toto corpori aptum est.[27]

Even if we could date the *DTh.* more precisely, it would merely underline the difficulty of dating the original compilation of the *AG*, for this
particular entry on wine (#292) differs from the rest of the text in terminology (*uirtus calidus, frigidus, humidus, thermanticus, uentrem/urinam
procurat,* etc., language found only in the entry immediately before this
on the grapevine, *AG* #291), and in its theoretical statements about wine
and water (ch. 2.H). As we shall discuss in the next chapter (ch. 4.C–D),
the *AG*'s Latinity shares some characteristics with Ps. Hippocratic Latin
works (*sed et,* and *uiscidus* as 'powerful'), but not enough to draw a direct
link with that tradition. In content as in language, the *AG* stands apart
from any known Latin medical work.

There is, however, one other instance in the *AG* of a direct echo with a
another source. The *AG*'s entry for man orchis (#272) mentioned a third
type called *basilisca,* attested only in Ps. Apuleius (130), that remains unidentified. The Ps. Apuleius description differs considerably: it opens by
stating that it grows where a basilisk (*serpens basiliscus*) has been present,
then describes three different types of basilisk serpent (golden-hued, spot-
ted with golden head, blood-red with a golden head), attributing different levels of deadliness to them – whatever they see or touch bursts into
flames, disappears, or is destroyed, so that only bones remain, attributes
that are similar to those given by Pliny.[28] But it contains two descriptions
that are almost verbatim to the *AG*:

> a) … herba talis similis salicis foliis oblongioribus et angustis, guttis aspersis
> nigrioribus, radix eius pedis ursi similis. (Ps. Apul. 130. 14)

---

27 The texts share other echoes, such as the comment concerning eggs: 'eo quod naturae
fit animantis' *AG* #204: 'eo quod natura forte sit et animantibus' *DTh.* 349; 'fortia ideo
[oua] quia natiuitas est animantes' *De observ.* 328.
28 Pliny 8.78, possibly a source, but his description differs considerably. On Pliny and
snakes, see French 1986, 105, and idem 1994, 270–82.

… folia sunt priori satyrio similia, sed grauiora et leuiora, punctis asperis nigrioribus, radix eius pedis ursi habens similitudinem. (*AG* #272)

b) Si quis homo eam secum habuerit, ab omne generatione serpentium erit tutus. (P. Apul. 130.12)

… siquis secum habuerit ab omni genere serpentum erit securus. (*AG* #272)

Besides both texts mentioning its golden-yellow flower (*florem crisococcum* Ps. Apul.: *flore aureo uel croceo AG*), in every other respect they differ entirely, particularly the comment in Ps. Apuleius that 'anyone who wants to pick [basilisca] should be clean and mark around it with gold, silver, deer horn, ivory, a bear's tooth, a bull's horn and place honeyed fruit in their tracks.'[29] Clearly, therefore, the two texts were drawing from a similar source rather than directly influencing one another.

G. Conclusion

The closest analogous text to the *AG* in terms of style, content, and format is that of Dioscorides, but it is evident that the *AG* did not know Dioscorides' text or use it directly, for three reasons: the *AG* contains completely different descriptions for some products; where the entries for a product are similar there are too many differences in description (colour of flower, length/shape of leaf, etc.), preparation methods, origins, and ranking of best types to suggest any direct influence; the *AG* contains drugs not mentioned by Dioscorides, some of which are mentioned in Pliny, a contemporary of Dioscorides. Nor was Pliny a source for the *AG*, but it is clear that all three texts drew from the same traditions dating from the first century BC to AD. One of these may have been the lost work of Sextius Niger, which Pliny and Dioscorides acknowledge as a direct source: the faults of Niger which Dioscorides criticized in his preface seem to be echoed in the *AG*. On the other hand, we know from Pliny and Dioscorides that Niger said many other things that are not recorded in the *AG*, which suggests the latter may have used the same sources as Niger (Solon of Smyrna? Xenocrates?) rather than Niger himself. The range of writers

---

29 'Qui eam leget, munus circumscribateam auro, argento, corno ceruino, ebure, dente apruno, corno taurino et fruges mellitos in uestigio ponat,' Ps. Apul. 130.16. See also ch. 2.I.

on pharmacy from the second centuries BC to AD demonstrates the rich literary tradition to which the *AG* belongs, but we cannot see from which texts it drew specifically. While the two late antique (or early medieval) texts of Ps. Apuleius and the *Diaeta Theodori* contain at most one or two verbatim lines of text found in the *AG*, these passages can be explained by a shared tradition (Ps. Apuleius) or later interpolation from the *AG* itself (*Diaeta Theodori*), and do not help us to establish the origin or date of the *AG*. Hence the findings on the *AG*'s sources support those on its pharmacology (ch. 2) and suggest that the *AG* is pre-Galenic, and may date as early as the first century AD, possibly even BC. As we shall see in the next chapter, there are also linguistic reasons for dating the text to this period, though other aspects point to Late Antiquity. Finally, it should be reiterated here that, despite its title, there is no sign that the *AG* was influenced by or had any connection with Galen's work, with the possible exception of the reference to a type of toothache (*haimodia*) cured by purslane (see *AG* #32 n.2, and ch 4.F), which, again, seems owed to common traditions.

H. The *Comparanda*: An *apparatus comparationum*

The comparisons (**cf.**) listed at the foot of the apparatus criticus to the Latin text for each entry are intended as an aid to modern scholarship on ancient and medieval pharmacy, and should by no means be misunderstood as an *apparatus fontium*, that is, a list of the sources which the author of the *AG* may have used for that entry. With these one can compare the different treatments of a simple by different texts, all of which reveal the *AG*'s singularity and its often more informed comments. I have only included references which comment on the nature, properties, or description of the drug that might be usefully compared to the *AG*'s orientation and purpose, hence Dioscorides (Gr.), Pliny, Gargilius Martialis's *Medicines from vegetables and fruits*, Galen's *On the properties of simple drugs* (Gr.), Ps. Apuleius, the Ps. Hippocratic *Dynamidia* (St Gall and Vatican recensions), *Ex herbis femininis*, and Isidore's *Etymologiae* are always listed, while references to Celsus, Scribonius, Caelius Aurelianus, Theodorus Priscianus, and others are restricted to citations which say something about the drug, rather than simply as an ingredient in a compound medicine. Similarly, references to the Latin Oribasius, Latin Alexander of Tralles, and others heavily dependent on Dioscorides or Galen are kept to a minimum and are given only when they contain additional information that relates to the *AG*'s contents. Full references to many of these can be found in

Opsomer's *Index de la pharmacopée de Ier au Xe siècle* 1989, which lists all occurrences of thousands of simples (*medicamenta*) in 69 different Latin texts, excluding Pliny, Gargilius, and the Ps. Hippocratic *Dynamidia*. Correspondence with simples mentioned in the Hippocratic corpus can be achieved by using the alphabetical arrangement in Aliotta et al. 2003, and the indices in Totelin 2009.

4

Language, Latinity, and Translation

৵

A. Introduction: Language and Dating the *AG*

The *AG*'s anonymity, uncertain date, and obscure origins are obstacles to determining its place in the history of Latin literature, but it certainly deserves one. This relatively short text adds 20 new Latin words to our lexicon, a quarter of which are Greek terms unattested elsewhere. It contains over 20 -*aster* suffix adjectives hitherto unattested, by far the greatest usage of this suffix for any single text. It uses sources lost to us, employs and translates Greek terms and expressions, and its alphabetical format is a milestone in the history of Latin scientific literature. We shall first discuss the language of the *AG* with a view to its implications for dating the text, before addressing its two of its unique linguistic features: the words *uiscidus* and *uiscide*, and the uses of the -*aster* suffix, some of which offer interesting parallels with Greek constructions in Dioscorides. We then turn to the *AG*'s use of Greek, and end by listing some of the text's terminology that presents problems of meaning and translation, followed by a list of new and interesting words.

The *AG* comes down to us in manuscripts that date several centuries after its composition, and given that each manuscript can often present different orthography, grammar, and syntax, assessment of the text's Latin and its place in the history of the language is fraught with uncertainty. Nonetheless, there is sufficient consistency among the manuscripts, despite their individual departures from the norms of classical Latin, to suggest that at least parts of the original text were redacted in clear, grammatically correct technical Latin that may date as early as the first century AD, as some features seem to suggest. On the other hand, certain other philological features, particular words and distinctive terminology, suggest a time frame from the fourth to the sixth century, but these may also be owed to later interpolation. As primarily a reference work, the *AG*'s technical nature does not easily lend itself to dating. The literary culture of Late Antiquity is renowned for its verbosity, mannerism, and abstract expression, a 'jewelled style' in which the medium becomes an important part of the message. But technical literature, including late antique medical Latin writings (and Greek, for that matter), maintained a more pragmatic style than most literary genres, and often presents a mixture of 'high' and 'low' registers, a combination of normative and substandard features, that challenges conventional periodization for the development of Latin.[1] As a reference work, the *AG* has no literary pretensions, and all of these problems are compounded by its use of unknown sources, so we cannot measure its departures from an earlier work. As explained in the next chapter (5.A), there is enough uniformity of language among our manuscript witnesses to suggest that the *AG* was written and circulated in a Latin that adhered to classical conventions, hence the analysis that follows does not consider the vulgar Latin spellings or grammar found in some manuscripts (particularly *V*) and recorded in the variants (ch. 5.B), but instead focuses on the text as presented in this edition.

B. Grammar

The *AG*'s Latinity shows none of the usual signs of vulgar Latin, in terms of either substandard grammatical forms or proto-Romance features, that we find in late antique literary and legal texts.[2] Neither erroneous use of verbs (the muddling of conjugations, periphrastic verb paradigms) nor corrosion of the declension system is evident, as we find in some late antique medical works such as the Latin translations of Hippocratic works and *Dynamidia* texts, Ps. Apuleius, and the Latin Dioscorides.[3] For exam-

---

1   In general Clackson and Horrocks 2007, 229–304, and on technical Latin in relation to medicine, Langslow 1989, 1991, 1991–2, 1994, 2003, and esp. 2000. The study of Latin regional variation by Adams 2007 offers new perspectives for late Latin. Jewelled style: Roberts 1989. Note also Haverling 2008. Late antique prose: Everett 2003, 19–35.

2   Here I draw on previous research: Everett 2003, 130–62; 2009. Excellent guides to vulgar Latin are Herman 2000, and especially Wright 1982, (ed.) 1991. An extensive guide to medieval Latin is Stotz 1996–2004. On genres, Mantello and Rigg, eds 1996.

3   Good overviews of medical late Latin are Baader 1970 and 1971, Vázquez Buján 1984a and b, Mazzini 1991, Önnerfors 1993. Essential is Langslow 2000, which focuses on issues of technical language and the influence of Greek. On Hippocratic Latin, Mazzini 1984 and 1985, Vázquez Buján 1998, and Fischer 2002. Invaluable are studies of individual texts or authors such as those of Svennung 1935 (on Palladius), Mørland 1932 (on Lat. Oribasius), and Adams 1995 (on Pelagonius). On Lat. Dioscorides, see Mihaescu

ple, there is no shortage of synthetic passive verbs, their loss usually a tell-tale sign of late Latin, as is the loss of deponent verbs, but these too are found in the *AG*. Passives are appropriate for technical literature because the treatment of objects is often the subject: things are beaten, smashed, pressed, purified, cooked, burned, cooled, stored, and so on.[4] Comparison with later, truly 'vulgarized' technical literature, such as the artisan recipes preserved in an eighth-century manuscript at Lucca, which rely on second person active verbs or the imperative governing objects ('take the hide and you put it in lime,' etc.), merely underlines the antiquity of the *AG*.[5] The *AG* also sports passive infinitives (*curari, purgari*), and one perfect passive infinitive (*sumpsisse*).[6]

Likewise in favour of an early date is the competent use of accusative-and-infinitive constructions for subordinate clauses (*aiunt lachrymam esse arboris* #11), rather than the use of conjunctives with finite verbs (*aiunt quod/quia lachryma est arboris*) for verbs of thinking, saying, or perceiving,[7] and the retention of correct case for verbs that take the dative, such as *prodesse* and *resistere*, or that take the ablative, such as *constare* and *utor*.[8] The one example of the infinitive with *habet* to

1938, and the studies of Ferraces Rodrígez 2000, 1999, 2009a and b, which also examine the *Dynamidia* texts and *Ex herbis femininis*: on the latter see also Maltby 2008. A proper study of Ps. Apuleius is needed: a start is Maggiulli and Buffa Giolito 1996, 36–53.

4   The frequent use of *inuenitur, dictur, additur, mittitur* aside, other examples include *colligitur, reponitur, coquitur, dignoscitur* (44), *auellitur, funditur, suspenditur, capitur, foditur, aspergitur, purificatur, frangitur, bibitur, conciditur, tunditur, exprimitur, uritur, conficitur, comburitur, abraditur, limatur, illinitur, resoluitur, tingitur, fingitur, defunditur, accenditur, siccatur, comminuitur, subigitur, redigitur, contrahuntur, efficiuntur, intelliguntur, falluntur, incenduntur, clysterizantur, dissipantur, agglutinantur, luxuriantur, operantur*, and so on. Passives are also used in simple expression (*Sed hic spernitur quod* [*humorem emittit*] #171, *somnifera creditur* #181, *sed tactu facile deprehenditur* #235). On passives, Stotz 1996–2004, IV:171–92. Deponent verbs (apart from the frequent *nascitur* #2, 4, 11, etc.): *sequitur* #28, *insequuntur* #51, *patiuntur* #103, *utuntur* #76 and 104, *uescuntur* #156. On the historical development of deponents, Flobert 1975; Stotz 1996–2004, IV:334–51.

5   Hedfors, ed. 1932. See Everett 2003, 155–61. Linguistic analysis: Adams 2007, 465–72.

6   Passive infinitives: *solidari* #151, *puluerizari* #76, *curari* #150, *lauari* #292, *purgari* #36, *inueniri* #135, 257; *sumpsisse uidetur* #12. Stotz 1996–2004, IV:172–92.

7   Examples: *peritissimum esse te probaui* pref., *esse dicitur/dicuntur* #21, 46, 58, 65, 97, 155, 203, 47 (*facere*) etc., *debet esse* #228, 238, *prodesse potest/uidetur* #12, 179, 269, 292, etc.; *prodesse comprobatur* #6, 245, *esse putant* #190, etc.

8   *prodesse* #12, 179, 245, though also with the preposition *ad* before the object, #6, 125, 269. *resistere* #16. Note also *praecipitur* #239, 292, *recedit* #75, *constat* #209, *utor* #76, though incorrectly applied to accusative in #104, which may be scribal error.

express obligation, common in vulgar Latin (even of early date),[9] suggests it is later scribal corruption, for throughout *oportet* and *debere* usually serve this purpose.

One aspect that might be considered to be post-classical, if not quite vulgar Latin, is the occasional use of the demonstrative pronouns *ipse* and *ille* as adjectives (*ipsa folia, illa comula*), which points forward to the definite article in Romance. It is also used occasionally (and superlatively) in combination with *hic* (#5, 63, 201, *hoc ipso* 88, *hac ipsa* 181). But overwhelmingly *ipse* and *ille* are used correctly, including the genitive form (*illius, ipsius*) in place of a noun in subordinate phrases, and notably in the retention of the neuter form of *ille* (*illud*). If anything, the adjectival sense is one of emphasis ('the root itself') or clarity ('the seed of this very same plant,' 'the juice from the root itself').[10] In any case, even technical Latin of classical date relied on demonstratives more than literary Latin as a simple means of precision.[11] Moreover, we find other classicisms such as the use of *prae* for 'because of' (#79).

Before moving on to discuss aspects of the *AG*'s vocabulary, Mazzini's survey of the grammatical features of the Latin translations of Hippocratic texts noted two aspects which we find in the *AG*: the frequent use of comparative adjectives; and the use of the conjunctive *sed et* for *et*. But this last is not, as Mazzini suggested, evidence of late Latin, for we find it Cicero and a few times in Celsus, and it was used extensively by Pliny.[12] As for the comparatives, which Mazzini believed were owed to translation from Greek, the *AG* uses them constantly in relation to other plants or substances as part of its descriptions (e.g., *foliola … similia menthae,*

---

9 'et ea quae sinus collectionem habet euaporare' #39. *Habet* with infinitive is found as early as the first century AD: Thielmann 1885. Herman 2000, 56–8; Stotz 1996–2004, IV: 174–7.

10 Adjectival: *ipsa folia* #188, *ipsa radix* #189, 201, *ipse matrix* #202, *ipse autem orobus* #207. Clearly emphasis: *illa caerulea* #25, 30, *nigra tamen illa stringit* #188, *illa uidelicet ratione* #199, 281 (but cf. #201, *hac ratione*), *circa illam comulam densam* #221. With *hic*: *has ipsas laminas* #3, *haec ipsa mixtura* #8, *ipsa aqua recocta* #17, *ipsum lac* #41, *coctura ipsa* #45, *harum ipsarum* #199, *his ipsis pilulis* #209, *haec ipsa* #213, *ipsum chalcanthum* #51, *hic ipse succus* #201, *haec ipsa* #202, *hoc ipsum semen, haec ipsa herba* #217, *haec ipsa radix* #234, *hanc ipsam lachrymam* #278. Emphasis: *succus illius* #49, *sapore ipso* #239, etc. passim.

11 Trager 1959. Stotz 1996–2004, IV: 122–30.

12 Mazzini 1984, 21, 22–3. *Sed et*: Pliny 2.5,173, 191, 209, 6.54, 128, 7.114, 169, et passim. Celsus *proem.* 74, 5.17.1, 6.15.3. Celsus clearly preferred *sed etiam*, which serves the same purpose. Cicero, *Att.* 11.9.2. *Sed et* is used 27 times in the *AG*: #7, 39, 49, 58, 64, 74, 79, 86, 88, 97, 107, 127, 129, 149, 150, 153, 155, 181, 188, 194, 199, 203, 209, 214, 218, 233, 292.

sed exalbidiora et angustiora #88, ... *similem aristolochiae, sed nigrorem et lentiorem corticem* #123), or when comparing species (e.g., #64, 70, 78, 142, etc.), and in prescribing 'the best' type or part (e.g., *quarum [radicum] eligimus quae sunt grauiores et pleniores et duriores* #61, 131, etc.). Very rarely is a comparative form used without a comparator (explicit or implicit) to intensify degree, though occasionally we find such (e.g., *fuscius* #1, *latiores* #8, *fuscidiora* #210, *infirmiores* #235), and when describing the effect or properties of a substance (*uiribus acrius* #14, *uires aliquatenus acriores* #202, *quidem cachry acrius et uiscidius* #229, though the comparison with the rest of the plant might be understood), and some adverbial forms (*quae instantius stringere uolumus* #125, *si enim frequentius in ore teneatur* #151, *gargarizetur frequentius* #195).[13] Hence, despite the frequency of comparatives in the *AG*, worth noting in itself, it does not employ them to intensify the adjective in the same manner as the Latin Hippocratic texts.

C. Vocabulary

The language used to convey drug action stands apart from late Latin texts of a similar nature. We have already seen (ch. 2.F) how the use of *uires* rather than *uirtutes* for drug properties points to Antiquity. Adding to this impression is the *AG*'s remarkable and quite distinct use of present participles to describe *uires*, particularly the commonly mentioned actions of heating, binding, cooling, and loosening. The term *excalfaci-ens/-entes* for 'heating' properties seems to be unique: the verb itself (*excal[e]facere*) is mostly limited to (above all) Pliny, Scribonius, and a few appearances in Chiron, as are the closest analogues (*uis excalfaciendi, natura excalfactoria, ad excalfaciendum,* etc.), but use of the present participle is unique to the *AG*.[14] Likewise, the participle *stringens/stringentes,* used as a Latin equivalent to *stypticus* to describe an astringent drug property, is unique among medical literature, but the closest analogues (*adstringens, con-*

---

13 An odd case is the description of hartwort (#251), 'grana affert uastiora et oblongiora quam utrumque,' which may reflect translation from Greek. Dioscorides III.53 has nothing similar ('καρπὸς ὑπομήκης, γεγωνιωμένος'). On comparatives, see Stotz 1996–2004, IV: 295–301.

14 Celsus V.18.1. Pliny: *excalfacit*, 21.139, 21.151, etc.; also *ad excalfaciend-*, 21.163, 20.68, 28.136; *excalfactoria natura*, 23.152, *excalfactoria uis*, 21.120, hence also in *Med. Plin.* 3.13, 'vim habet excalefactoriam' and in Marcellus. Mazzini 1984, 19, suggests that the use of *calefacio* for *calefio* is 'tipicamente tardo,' but the evidence points the other way. Later texts used *calidus*, e.g., Diosc. Lat. I.29. On Pliny's therapeutic language, Bonet 2003.

stringens) seem to be Celsus and Pliny.[15] Similarly, the *AG* consistently uses the participle *refrigerans/refrigerantes* to describe a cooling action, a term common to Pliny, but only used sparingly by other earlier writers (Celsus, Chiron),[16] while late antique writers generally used *frigidus* or similar adjectives to describe drug action or effect. The *AG*'s use of *relaxans/relaxantes* for 'loosening' properties also seems unparalleled in other texts. This is complicated by the frequent variant reading of *laxans/laxantes* among the manuscripts, but both terms are extremely rare to describe drug actions. The closest parallels are, again, Pliny and Celsus, in their use of verbs and gerunds to convey 'loosening' actions.[17] It has been suggested that the use of present participles to describe the therapeutic effects of foodstuffs and medicines found primarily in Celsus, and less so in Pliny, was a stylistic innovation to translate Greek adjectives ending in -ικός.[18] This is certainly compatible with the *AG*'s claims to have used Greek sources (below, ch. 4.E–F) and the correspondence with Dioscorides' adjectives for 'properties' (δυνάμεις, ch. 2.D). We do find the verbs *calefacit*, *restringit* (e.g., *uentrem restringit*), and *refrigerat* used to describe drug actions in later texts such as the *De observantia ciborum* (*saec.* V–VI),[19] but the use of the present participle form combined with *uis* is consonant with the age of Celsus and Pliny. Other philological aspects

---

15 *adstringens*: Celsus 5.28.11: Pliny 17.251, 24.13, 31.98, etc.. Also used for taste, *gustu adstringens*, 27.128; *uis adstringendi* 23.13, 24.36, 24.93, 34.127, 35.188, etc.: *uis constringens* 23.100. Also *styptica uis* or *natura* 21.166, 32.11, 24.120.

16 Celsus 2.33.3, 5.18.1. Chiron 394. The only secure date for this enigmatic text is its use by Vegetius (saec. IV ex.–V in.), but some of its terminology (e.g., *veterinarius*) suggests a date considerably earlier: see Fischer 1989b, 77–80; Adams 1995, 6. Pliny refers to *uis refrigeratoria* 22.90, *uis refrigerandi* 23.54, 24.93, 24.110, 27.132, *summa uis in refrigerando* 25.70, *vis ad refrigerandum* 25.140, *natura refrigerandi* 27.76, 35.196, 36.156, etc. and often without *uis* or *natura* (*refrigerat*, *refrigerant*, *refrigeratio*, etc. 20.74, 20.121, 20.213, 21.75, 21.84, 21.121, 21.130, 23.8, etc., and note *refrigerante herba* 22.126).

17 Pliny uses *laxo* (never *relaxo*) in similar ways: 19.78 'vis mira colligendi spiritum laxandique ructum': 23.157 [oleum] nervos laxandos [utile est]: 24.88 'ad laxandas siccandasque fistulas'; 8.129 'herbam quandam ... laxandis intestinis.' Celsus 2.1, the body is *relaxatum* by the heat of summer: see also 2.3, 2.17.3, 3.22.5 (skin), 7.7.8 (eyelid), cf. 7.25.2, 7.30.1; also *laxo*; 2.14.4 *laxaretur*, 7.19.2 *laxet*, 7.7.8 *laxanda sutura*, and so on, as well as *laxus*, *laxior*, etc. Cf. also 7.33.2, 8.10.1. 8.10.7.

18 Langslow 2000, 350-2, mostly forming neuter plurals (Celsus: *reprimenta*, *refrigerentia*, *calfacientia*, *inflantia*, *rodentia*, *calorem/pus/urinam moventia*, etc.. Pliny: *erodentia*, *urentia*).

19 Date according to Mazzini, ed. 1984, 32. The *De observ.* contains some expressions similar to the *AG* (e.g., *urinam prouocat*, *flegma deducit*, etc.), but on the whole its language differs considerably. Likewise the *Diaeta Theodori* ed. Sudhoff 1915 (see ch. 3.F), and Ps. Theodorus *Simp.* ed. Rose 1894 (Theodorus).

point to Antiquity. The *AG* knows that the term for pomegranate rind is *malicorium* (#183), which Pliny said was the preferred term used by doctors, and it was known to other golden age writers such as Petronius and Celsus, but was subsequently forgotten after them (Marcellus took it from Pliny). The use of the term *morbus regius* to mean jaundice suggests a date earlier than the fourth century, when it begins to acquire other meanings such as leprosy or other visible skin diseases.[20]

On the other hand, there are many other philological clues that suggest a later date. With the exception of the term *morbus regius*, the *AG* never uses *morbus* to describe a disease or ailment, instead preferring the more neutral term of *uitium*, meaning 'natural defect' in the body.[21] It has been suggested that the word *uitium* replaced *morbus* from the third century onwards in Latin medical texts for two reasons: *morbus* was etymologically and semantically related to the word *mors*, death, whereas *uitia* were 'disorders' or 'faults' of the body that medical science had some hope of addressing and curing; also, *morbi* were associated with demonic powers in the ancient, pagan world, and visitations of such were usually considered fatal, hence the term evoked terror or impending death.[22] The *AG* also uses terms which appear only in other late Latin sources, though they may also be later interpolations. For example, use of the compound noun *carpobalsam* for the fruit of balsam seems to be a late antique phenomenon,[23] and *propoma*, an aperitif mentioned in the entry for wine (#292), is attested only among four other late Latin writings: Palladius, *Medicina Plinii* (both of the fourth century), the Latin Oribasius, and the Latin Alexander of Tralles (both of uncertain date, but post-fourth and -sixth centuries respectively). It seems unknown to Pliny or Dioscorides, though Galen mentioned it in his tract on barley water, if indeed he wrote it.[24]

Other words, however, beg the question of whether the *AG* is simply the first attestation of such in Latin, such as the verb *masso* (#279), 'to chew' (from the Greek μασ(σ)άομαι), found only in Theodorus Priscianus

---

20 See #196 n.5.

21 *uitia capitis* #12, *longinqua uitia neruorum* #50, 261, *indurata uitia neruorum* #203, *(omnia) uitia thoracis* #75, 91, 258, *pulmonum uitia* #109, *uitia oculorum* #110, *ueretri uitia* #151, 152, *stomaci uitia* #153, *uitia oris* #194, *plurima uitia* #249. The term *aegritudo* is used only in #109. Totally absent is any reference to disease as a consequence of immorality, as we find in pagan as well as later Christian authors: see Mazzini 1998.

22 Migliorini 1993, 99–105, though her focus is classical sources. There is some dispute over the etymology of *mors/morbus* (Migliorini, ibid., 101 n.31), but Isidore 4.5.2 (*mortis uim*) had no doubts.

23 #202 n.4. Another questionable case is *asclusus*, see #39 n.2.

24 Galen *De Ptisana* 57 (K.6, 828). See Hartlich, ed. 1923.

(c. 400), but fairly common in Greek, and the word *bolarium* (#125, from Greek βωλάριον, little lump, a diminutive of βῶλος) is only recorded in the second-century AD poet Septimus Serenus, but this case is complicated by the lack of this passage in the earliest manuscripts of the *AG*, raising the spectre of interpolation. A good illustration of the difficulty of using such evidence to date the text is word *oletanus*, 'foul-smelling,' 'stinking' (#95, 221). It is not found elsewhere, but clearly derives from the verb *oleto*, which we find only in the first-century author Frontinus (*De aqueductu*, 97), and a letter of Ambrose of Milan (c. 381 AD), in reference to a heretic (*se oletauit perfidia*). Similarly, some alternative names given for plants are found only in later sources, such as *apollinaris* for mandrake (#181), known to *Ex herbis femininis* and the *Dynamidia* (both possibly fifth-century), the use of *columbina* as an alternative for vervain (#224) and known to Ps. Apuleius (or at least the synonyms that were attached to that text in the fourth or fifth century), and the alternative name of *pulicaris* given for fleawort (#216), used by a handful of late Latin texts, the earliest of which was Theodorus Priscianus[25] and Caelius Aurelianus (both c. 390–450): its placement at the end of the entry, however, may suggest it was added later. Similarly, use of terms such as *tonoticum* (#203 n.6) have a late antique flavour, but as discussed elsewhere (see ch. 1.E, ch. 2.D), this particular entry on oil, and the following one on egg (#204), are the only entries among the *AG*'s 300 that use five of the Greek terms listed as 'drug types' in the *Liber de dynamidiis*, suggesting it is a later interpolation. We can only guess about the significance of several other terms not recorded elsewhere but which do not seem later additions: *fuscum*, an alternative name for *purpurissum* (#211), a type of sulphide of arsenic called 'flower' (*flos* #242), *lamna* as another name for orpiment (#8), the verb *subuiridicat*, 'to turn a pale green colour' (#45), and so on (see ch 4.I).

An intriguing example of vocabulary that suggests a later date for the *AG* is the reference to writers sharpening their quills or reed pens (*pennas uel calamos*) with pumice stone (#213). The earliest attestation of the use of the *penna* ('feather,' and usually a goose feather) as a writing instrument is found in a famous passage by chronicler known as Anonymus Valesianus (writing c. 530s), who describes King Theoderic the Ostrogoth

---

25 Other references shared only with Theodorus are the use of *atreton*, #253 n.2, and mention of 'juice of donkey dung,' #115, n.1. On Theodorus's Latin, see Önnerfors 1993, 288–301; Langslow 2000, 53–6 (who suggests Greek was Theodorus's first language), et passim; Fischer 2000a, 2000c. Mørland 1952 details the affinities of Theodorus with the Latin Oribasius that raise unresolved questions about the relation of these two texts. On the language of the Latin Oribasius, Adams 2007, 473–89.

using a stencil with the word *legi* ('I have read [it]') for signing state documents because he was illiterate. Our next attestation is Isidore of Seville, who, in a chapter of his *Etymologiae* on book makers and their instruments, comments that 'the instruments of the scribe are the reed and the pen. With these the words are affixed to the page, but the reed is from a tree, the pen from a bird.' Confirmation that by Isidore's day quill pens were standard is seen in the Anglo-Saxon world, where several riddles of Aldhelm (639–709) and others in Old English refer to feathers as writing implements.[26] The quill may have been used earlier: it has been alleged that some of the Dead Sea Scrolls, dating as far back as the first century BC, were written with a quill or feather of some sort. But it is thought that the slow but inexorable switch from papyrus to the use of parchment as the primary writing material (fourth to fifth century) encouraged the greater use of the quill, which was more flexible than the reed and better suited the parchment surface.[27] Hence the mention of quill pens in the *AG* hardly serves to date the text's composition, but it does highlight a difference from the two ancient authors who also comment on pumice: Dioscorides, who fails to mention any such use and whose entry on the pumice stone (V.108) differs considerably; and Pliny (36.154–6), who jokes that pumice is used as a depilatory for women, 'and nowadays also for men, and moreover, as Catullus reminds us, for books [*libris*]' – a reference to Catullus 1.2, 22.8, who mentions its use in smoothing the edges of a book roll. In noting the use of pumice by 'writers' to sharpen quill pens, the *AG* seems more early medieval than ancient.

D. *Uiscidus* and *uiscide*

The *AG* exhibits two related linguistic features which contribute to Latin lexicography and may have implications for dating the text: the use of the adjective *uiscidus*, meaning 'vigorous,' 'powerful,' a word which appears only in three or four other late Latin medical writings, and in only a few instances with that meaning; and a related adverb, *uiscide*, meaning 'powerfully,' 'vigorously,' which is unique to the *AG*.

---

26 *Anon. Val.* 14.79, 'per eam [laminam] pennam ducebat.' On this text, Adams 1976.
    Isidore 6. 14. 3, 'Instrumenta scribae calamus et pinna. Ex his enim verba paginis infiguntur, sed calamus arboris est, pinna avis.' Also noteworthy in this respect is Isidore's unique reference to syricum (Syrian minium) to write out the chapter headings in books (19.17.5): see #237 n.1. On Aldhelm and riddles, see Fry 1992.

27 This is the theme found in Wattenbach 1896, 227–8; Stiennon 1973, 159–62; Bischoff 1990, 18, and Brown 1998, 57 ('introduced in the sixth century'), but no early medieval author voices such an opinion. For a tenth-century fountain pen, see Bosworth 1981.

The *AG* always uses the adjective *uiscidus* in the sense of meaning 'powerful' or 'effective' – three times to describe a drug property (*uis uiscida*), seven times to mean 'effective' or 'powerful' with respect to medicinal uses or effect, once to describe a 'powerful' smell (*odore uiscidum*), and once in a manner that suggests a meaning of 'concentrated' to describe the sap of cedar.[28] Thanks to the detective work of Svennung, who found the word used by Palladius (early fifth century) to describe smell (*odor uiscidus*), and who subsequently traced the term using the archives of the *Thesaurus Linguae Latinae* in Munich,[29] we know that the only other attestations of this word are found in Latin medical writings of late antique date, namely, Gargilius Martialis (*Ex pomis*), the Latin Dioscorides, Theodorus Priscianus, and Latin translations of three Hippocratic works (*Prognostics, Seven signs, Airs waters places*). But these texts largely restrict their use of *uiscidus* to describing a taste or smell as 'sharp,' 'pungent,' or 'strong,' and only by extension a 'sharp,' 'pungent,' or 'astringent' property or medicinal effect. A few examples will suffice.

In our earliest datable example, Gargilus Martialis twice couples *uiscida* with *constrictiua* (a rare word for 'astringent') to describe the immature fruit (*poma*) of the medlar. Theodorus uses the term for 'sharp' or 'sour' vinegar (*acetum uiscidum*: likewise in Lat. Diosc., translating ὀξους δριμέως) and once to describe 'more pungent myrrh' (*murra uiscidiore*). The connection between pungency and astringency can be seen in the

---

28 Comparison with Dioscorides' Greek given where possible or similar. As property (*uis*): (*alcimonium*) *uires uiscidas et efficaciter relaxantes* #11, (flower of copper) *uires uiscidas et stringentes* #52, (squared stone) *uires uiscidas et stringentes* #168. Effective/powerful: (deadly carrot) *optimum et uiscidissimum* #281, (hellebore) *uiscidissimum medicamentum et fortissimum purgatorium* #285, (quicklime) *uiscidissima* #67 (similarly Diosc. V.115, 'ἐνεργεστέραν'), (gall types) *quae ipsae sunt uiscidiores* #121 (similarly Diosc. I.107, a type that is ἐνεργεστέραν), (roots of Celtic spikenard) *uiscidissimae* #238 (Diosc. I.8, 'useful and aromatic' [ἡ χρῆσις καὶ ἡ εὐωδία]), (frankincense rosemary) *quidem cachry acrius et uiscidius* #229 (Diosc. III.75 'warming and extremely drying properties' [δύναμιν ἔχει θερμαντικήν, ἀναξηραντικὴν σφόδρα]), (juice of the hypocist) *non dissimilis acaciae sed mollior tactu et effectu uiscidior* #298 (Diosc. I.97.2. '[properties similar to shittah tree but] is more astringent and more desiccative' [στυπτικωτέραν δὲ καὶ ξηραντικωτέραν]). Smell: *odore uiscidum* #147. As 'concentrated': (cedar sap) *solidum, uiscidum* #57 (Diosc. I.77, best is 'thick, translucent, robust, heavy in scent' [παχεῖα καὶ διαυγής, εὔτονος, βαρεῖα τῇ ὀσμῇ]). 'Sticky' is conveyed with *glutinosum* (#94).

29 Svennung 1934, who gives references to same editions cited in the present book. My thanks to Gerard Duursma of *ThLL* for helpful correspondence on this and on *uiscide* (below). Note two authors who we know consulted the *AG*: Gariopontus 1.6 'de sanguine uiscido, id est amaro' (likewise *Alphita*); Simon Januensis, 'uiscidum mordicatiuum, pungens linguam, acuti saporis.' On Gargilius's Latin, Önnerfors 1993, 264–75, and Maire, ed. 2002, eadem 2003.

Dynamidia's suggestion that a wild vegetable which is 'pungent and sour' is astringent.[30] The Latin Hippocratic works, in contrast, present a meaning more akin to 'hardy,' 'tough,' 'concentrated,' as when the treatise known as the *Seven signs* uses it to describe how, just like the bodies of men, those of animals and trees become *sicca et amara et uiscida* with age, in contrast to being *umida, mollia*, and *imbecillia* when young. Unfortunately an original Greek text is lacking for *Seven signs*, but in *Airs* (c.10) we find bile described as 'concentrated and thick' (*uiscidum et pingue*) to translate the Greek superlative παχύτατον,[31] and similarly in the *Prognostic* (c.14) 'concentrated spittle' (*sputum ... constitutum uiscidum*) for the Greek ἄκρητον, meaning 'pure,' 'strong,' 'vigorous.'

The closest analogues to the meaning of 'powerful' found in the *AG* appear on two occasions in the Latin Dioscorides. In general, the Latin Dioscorides uses *uiscidus* most often to translate Dioscorides' use of δριμύς, meaning 'sharp,' 'pungent,' 'piercing,' 'acrid,' and so on (equivalent to the Latin *acer/acris*), and on one occasion it employed the noun *uiscitudo* for 'sharpness.'[32] Here and there *uiscidus* is used for taste and smell,[33] while two times it is employed to describe 'humours,' and once to describe a 'woody' habitat of fish thistle.[34] Only in two instances, however, the Latin Dioscorides uses the *uiscidus* to qualify in a more general sense the power or efficacy of a drug in a manner that comes close to the *AG*'s meaning: dittany of Crete is 'more effective' (*uiscidior*) than cultivated pennyroyal in their otherwise shared medicinal uses, a direct transla-

---

30 'uiscidum et austerum ... restringit.' *DynVat* II.1/*SGall* 55. Note also the preface (*SGall*), 'species [herbarum] que cum uideantur uiscide. facile egeruntur.'

31 *Airs* ed. Grensemann 1996, from Glasgow, University Library, Hunter 96 (T.4.13), saec. VIII/IX.

32 *uirtus est [erucae agresti] diuretica et uiscidior ortino* Diosc. Lat. II.126 (δριμύτερον πολλῷ τοῦ ἡμέρου II.140); *uiscidior tantum est nec uirtute acrior* III.39, *magis uiscidior* V.124 (both translating δριμύτερον/ δριμύτητα). Note also *mixtus confectionibus uiscidis uirtutes earum extinguit* I.29 (for ἀμβλῦνόν τε τὰς τῶν ἑλκούντων φαρμάκων δυνάμεις I.30). Noun: *uiscitudinem caret et fit melapticu* (scil *malactica*) I.62. ἀποβάλλει τὸ πολὺ τῆς δριμύτητος καὶ γίνεται μαλακτικόν I.61. Similarly, *odorata cum uiscitudine* III.73, *pro uim uiscitudinis eius* III.82. The only other attestation of a *uirtus* is *Dieta Theodori* 391, 'aqua maritima uirtutem habet uiscidam et termanticam.'

33 Taste: *uiscidum et mordet* Diosc. Lat. I.12 (for δριμύ καὶ δηκτικόν I.14); *stiptico sapore sentitur et uiscido* I.16 (στυπτικός ὑπόδριμυς I.18); *uiscida gustu non suabi* I.28 (ὑπόδριμυν I.28). Smell: *odore uiscidu* I.26 (δριμύς I.26).

34 Humours: 'unhealthy,' *uiscidos humores nutrit* Diosc. Lat II.70 (from γεννητικὸς κακοχυμιῶν II.87); *humores uiscidos et absteros temperat* II.69 (ποιοῦσα πρὸς τὰς δριμύτητας II.86). 'Woody' places, *locis uiscidis* III.11 (ὑλώδεσι τόποις III.12).

tion of ἐνεργεστέρα;[35] and a third type of opium poppy is 'wild, quite small and *medicinal*' (*agrestis, minior et uiscidus*), to translate the comparative form of φαρμακώδης.[36]

The *AG*'s use of *uiscidus*, therefore, somehow bypasses the meanings of acrid, sharp, or pungent, for which it uses other words (*acer, cum acrimonia, stringens, acidus/subacidus*), and keeps to the meaning of 'powerful' as found only twice in the Latin Dioscorides, and as hinted at with the meaning of 'strong' or 'concentrated' in the Latin translations of the Hippocratic *Prognostic* and *Airs*. This singular focus of meaning is repeated in the *AG*'s use of the adverb *uiscide* to mean 'powerfully,' used to intensify drug actions, particularly astringent actions, and also used to intensify colour, smell, and taste (e.g., pepper '*powerfully* bites and warms the tongue').[37] This adverb is unattested in any other Latin source, with the remarkable exception of (possibly) Gildas and (certainly) Anglo-Saxon authors and glossaries of the eighth and ninth century, in which it is given the Old English equivalent of *tochlichae/tohlice/tholice*, that is, 'toughly.' The semantics differ slightly in that the Anglo-Saxon usage may be more owed to *uiscatus/uiscum*, 'sticky,' 'adhesive.'[38] Nonetheless, we know that Anglo-

35 *uirtus est illi tanta, quanta et puleio, et paulo uiscodior est* III.33 (ποιεῖ δ' ἅπαντα, ὅσα καὶ ἡ ἥμερος γλήχων, ἐνεργεστέρα δὲ πολλῷ III.32). The (erroneous) translation of πολλῷ with the phonetically similar but semantically opposite *paulo* is also found in Hippocratic Latin translations: Mazzini 1984, 27.

36 Lat. IV.60, 'ἀγριωτέρα καὶ μικροτέρα καὶ φαρμακωδεστέρα' (i.e., than the two other types), IV.64.

37 Intensifying an astrigent action: *uiscide stringere potest* #2, *uires enim habit acriter et uiscide stringentes* #20, *uiscide stringit* #32, 51, 68, 145, 161, 162, 208, 219, 261, superlative *uiscidissime stringere* #89. Purgative actions: *uentrem uiscide purgant* #158, *expurgat uiscide* #221, also #226, 252. Heating: *potest igitur acriter uiscide et calefacere* #10, also #198, 202, 277. Loosening/relaxing: *uiscide relaxare potest* #163, also #243, 247, *uiscidissime*, #249, 287. Other actions: *amaritudine et acrimonia sua uiscide* [*corpus commouet*] #54, *uiscide dolorem placat* #69. Colour: *exalbidum uiscide* #111. Taste: *gustu linguam uiscide excalefacit* #137, (pepper) *gustum uiscide mordet et calefacit* (209). Smell: *odore graui et quidem uiscide* #201.

38 Gildas, ed. Mommsen, MGH, AA. 13: pref. 14, 'per tot annorum spatia [in]interrupte uiscideque protractum,' though in other recensions it appears as *inscide*. See O'Sullivan 1978, 65, 68. Glossaries, ed. Sweet 1885, 83 (Epinal) no. 163, 104 (*uiscidae*); (Erfurt) no. 2170, 107 (*uiscide*). Aethelwald, MGH AA 15, 530, l.76, 'erant iuncti bitumine / germanitatis uiscide.' Cf. trans. of Miles 2004, 92, 'adhesively joined … in the pitch of brotherhood,' but 'powerfully' may be better. On the date of the glossaries I follow Parkes and Bischoff in Bischoff et al., eds. 1988, 16, 19. Two other attestations of *uiscide* can be dismissed: those in the lapidary of Damigeron (Svennung 1934, 20) derive from the *AG*: the *voce of* Du Cange, *Glossarium*, 'Gloss. MSS ad Alexandrum Iatrosoph.' is obscure. My thanks to David Langslow (cf. idem 2006) and Gerard Duursma of *ThLL* for their advice.

Saxon England possessed many of the same Latin medical writings found on the continent, and the *AG* may well have been there.[39]

The *AG*'s extensive use of the adverb *uiscide* to mean 'powerfully,' and its singular use of the adjective *uiscidus* as 'powerful/effective,' remain unique and unaccounted for. Svenning suggested that the origins of *uiscidus* stem from a semantic combination of *uirus*– meaning 'venom,' 'poison,' or an acrid secretion of plants/animals (e.g., snakes), and that the subsequent adjective *uirosus* refers to an unpleasantly strong or pungent smell or taste, with some semantic connection to *uiscum*, mistletoe, birdlime (*AG* #286), for its pungent and sticky qualities. For Svenning, *uiscidus* is to *uiscum* what *fumidus* is to *fumus*, *gelidus* to *gelu*. But in the *AG* at least, the meaning of *uiscidus* and *uiscide* more appropriately derives from *uis* (force, power), without any of the negative or unpleasant associations of *uirus*, *uirosus*, or *uiscum*. Only these last two are employed in the text, twice each: *uirosus* in the sense of 'sticky,' 'viscous,' 'fetid,' and *uiscum* for mistletoe and 'resin.'[40] It is possible that the author simply developed the term logically from *uis/uires* for pharmaceutical 'powers'/'properties,' whether or not he was aware of *uiscidus* meaning 'astringent.' That he employed *uiscidus* not as it was used by late antique authors, but rather as hinted at in the translations of Hippocratic works and found only twice in the Latin Dioscorides, may also suggest an earlier, imperial date for the text's composition.

E. The *-aster/ -astrum* Suffix

The *AG* presents us with one noun and 20 adjectives that employ the unusual suffix of *-aster/ -astrum*, making welcome contributions to the history of the Latin language.[41] Only around 50 such uses of this suffix, for both nouns and adjectives, are known from Latin sources dating from the third century BC to the late seventh century AD, yet those found in the *AG*, with the two exceptions of *nigraster* and *oleastrum*, are not recorded elsewhere. I list them here with translation:

---

39 Cameron 1982, 1983, 1993. Also Adams and Deegan 1992.

40 *uirosissimus* 'stickiest [sap]' #280, *uirosius* 'more fetid' #60 (neither entry refers to *uires*): mistletoe #286 ('uires acres et quae durities remollire possunt'), *uiscum* 'resin' #70 (no *uires*).

41 A longer version of this section on *-aster* was published in *Galenos* (Everett 2010): my thanks to University of Toronto Press for permission to reprint some of this material there.

acidastrum #202 'particularly sour'
subacidastrum #183 'sour'
  (pomegranate)
cineraster #170 'ash-like' (colour)
cepastrum #243 'onion-like' (smell)
croceastrum #197 'saffron-coloured'
fucastra #212 'dark'
marrubiastrum #191 'black horehound'
melastrum #215 'honey-like' (fragrance)
myrrastrum #271 'myrrh-like' (taste)
nigastrum #190, *nigrasta* #231
  'a little black[-ish]'
oleastrum #265 'wild olive' (smell)
paleastrum #269 'a little chaffy'
pallidaster #169, 172, 286
  'somewhat pale'

puluerastrum #44, 135 'a little
  dusty'
purpurastrum #55, 114, 272
  'purplish'
resinaster #202, 254 'a little
  resinous,' 'resinous'
rubeastrum #200, 243, 'reddish'
similastrum #88 'a little like'
spicastrum #190 'like
  spikenard' (taste)
subcaerulastrum #136
  'blue-green' (eye-salve)
subrufastra #272 'a little reddish'

In Latin literature the *-aster* suffix is a type of diminutive, implying inferiority or incomplete resemblance.[42] There are different theories on its linguistic origins,[43] but we find it in Latin writers as early as the archaic period (Plautus, Terrence, Titinius, Lucilius), throughout the classical and silver periods, and into late Latin: our two latest attestations are Pope Gregory the Great (d. 604), who refers to an iron tool 'which is called a *falcastrum* because of its similarity to a scythe [*falx*],' and the Anglo-Saxon

---

42 The examples given below are collected and discussed by Seck and Schnorr von Carolsfeld 1884; they cite about 70, but many of these are variations (*apiaster, apiastra*, etc.) and many derive from glossaries, hence are fraught with uncertainties of date and provenance. See also Bolling 1897. Thomas 1940 argues against a strictly pejorative sense, and demonstrates its wider use designating similarity or variation. He was the first (and only) scholar to notice the 'grand nombre' of terms in the *AG*, and suggests it 'laisse entrevoir combien fut grand leur développement dans le vocabulaire technique' (522), but only cites four examples (*offucastra, pallidaster, purpurastrum, resinastrum*) without discussion. On *filiaster*, see Watson 1989.

43 Seck and Schnorr von Carolsfeld 1884 suggested adaptation from substantives in *os* + *tro* (e.g., *flustrum*, from *flovostrum*) to other nominal stems; Lindsay and Stolz saw it as an extension of the comparative *-tero-*, derived from *-atus* formations: Bolling 1897 suggested that the botanical terms derived from the qualifying adjectives *silvestris* and *agrestris*.

Aldhelm (d. 709), who refers to a *porcaster obesus*.[44] Aldhelm's example (also used by Ps. Apuleius) reflects the learned, Roman literary tradition in which *-aster* conveyed a pejorative sense, such as Cicero referring to an *Antoniaster* (a would-be Antony), or Augustine referring to Cicero as a *philosophaster* (would-be philosopher), and which is found in adjectival form applied to bodily defects (*calvaster, claudaster, mancaster, surdaster, canaster*). Gregory the Great's example, however, derives from the more basic meaning of approximation or similarity, such as the examples of *filiaster* (stepson, someone like a son, child from illegitimate union), *patraster* (stepfather, like a father), *matrasta* (stepmother, etc.), which we find in inscriptions as early as the first century AD. This same sense of approximation was also used for botanical terms to denote a wild variety of a species or some other variation, hence *oleaster* as the 'wild' olive (*olea*) (Cicero, Pliny), *alicastrum* as specifically 'summer' spelt (*alica*) (Columella), *apiastrum* as a type of wild parsley (*apium*, celery) (Sallust, Varro, Columella), *mentastrum* as a type of mint (*menta*) (Celsus, Columella, Pliny), *pinaster* as a wild or cluster pine (*pinus*) (Pliny, *pinastellus* Ps. Apul.), and so on.

The 41 or so securely datable attestations of its use are fairly evenly spread across time (8 from the archaic period, 10 classical, 7 silver Latin, 11 late Latin), hence offer little help for dating the *AG* and its extraordinary number of examples – by far the most ambitious use of the suffix on record. That three late Latin medical texts use the suffix may or may not be significant: *fuluastrum, -astra* (for *thyrsum, radix*) in Ps. Apuleius, *nouellastrum* (*uinum*) in Marcellus Empiricus, and *crudastrum* (*ficatum porcinum*) in Anthimus.[45] Marcellus and Anthimus, a Greek who worked at King Theoderic's court in Italy, both wrote for a Gallic audience, and Ps. Apuleius's location is indeterminable: Africa and Italy are the best bets. What they all have in common, however, is their use of late Latin with many vulgar traits. This raises the question, which has not been sufficiently addressed let alone answered, of the suffix's survival in spoken Latin for the period after Gregory's *falcastrum*, for the suffix can be found in modern Romance languages, including Italian *-astro* (*biancastro, giallas-*

---

44 Gregory, *Dial.* II.6. (also Isidore 20.14.5). Aldhelm, *vers.* 15. Du Cange records *uirgastrum* for *uirgetum* in a charter of 993: Seck and Schnorr von Carolsfeld 1884, 404.

45 Ps. Apul. 110 'ex interp. Diosc. II.14.7 sqq' according to Howald and Sigerist, but nothing similar is found in the Greek or Latin Diosc. Ps. Apul. also gives three *-aster* nouns as alternative names: *pinastellus* (= *peucedanos*) 95, *porcastrum* (= *portulaca*) 104, and *apiastellum* (= *batrachion*) 8 interp., idem (= *brionia*) 67. Marcellus 8, 'vinum optimum, sed novellastrum.' Anthimius 21, 'ficatum … in subtilibus carbonibus assetur ita ut crudastrum sit.'

tro, nerastro), French *-âtre* (*blanchâtre, jaunâtre, noirâtre*), and to a lesser extent Spanish (*padastro, madrasta, apiastro*). Most of these are owed to early modern antiquarian and humanist coinage, imitating its use in Latin texts (especially words like *poetastro, politicastro*, etc.), but undoubtedly some uses of the suffix have an ancient or medieval pedigree and survived in oral tradition.[46]

Intriguingly, Italy offers the best evidence for continuity of *-aster* via the vernacular: we find *filiastro* in a document of 1211, and Dante used *vincastro*. More compelling, for our purposes at least, is that two of the *-aster* words in the *AG* itself are found in late antique and Renaissance Italy. The only other attestation of *nigraster* (#190, 231) is in the fourth-century Sicilian astrologer-turned-Christian-polemicist, Firmicius Maternus. But of far greater significance is the case of *marrubiastrum* (#191: see plate 6), unattested elsewhere in Latin, but attested in early Italian by the sixteenth-century physician and botanist Pierandrea Mattioli, who noted in his commentary on Dioscorides (published 1563–5) that black hore-hound (*marrobio nero*) 'in Italia è tutto notissimo, e chiamasi da chi marrobiastro e da chi marrobio bastardo.' Significantly, *marrubiastr-um/-o* is not found in any other language, and is obsolete in modern Italian botany. Hence the *AG* uniquely used a plant name that was common parlance in sixteenth-century Italy, suggesting that *marrubiastrum* had deep roots in the peninsula, and that the *AG* was written there. Unfortunately, the only other modern Italian derivatives from the *AG*'s examples (*cinerastro, pallidastro, nerastro*) suggest later humanist coinage.[47] To privilege Italy as the bastion of *-aster* use and survivals would be unwise without a systematic survey. But the evidence over time points to a greater use of the *-aster* suf-

---

46 In Italian it is equivalent to *-accio*, hence *poetastro, politicastro, fratellastro*, etc. See Rohlfs 1966–9, 1127, and the *voce* in the Treccani *Vocabulario della lingua Italiano* I (1986), 320. Individual words can also be checked with the *Opera del Vocabolario Italiano* database at http://www.ovi.cnr.it/. For French, see the useful entry in *Trésor de la langue française* III (1974), 806–8. The survivals in modern French are mainly those for colours. On Spanish, see Pharies 2002, 112–13, who states that of the thirty or so usages in that language, most are of modern coinage, though maintain the same semantic categories as the Latin; an exception is *madrast[r]a*, which can be traced to the thirteenth century. My thanks to colleagues in Toronto, Paul Cohen, Dorothea Kullmann, and Will Robins, to Maria Teresa Gigliozzi in Rome (Treccani and La Sapienza), and to Roger Wright in Liverpool, for advice on these issues.

47 Andrea Mattioli, *Volgarizzamento di Dioscoride* (Venice, 1563–5), 448. *cinerastro*, cf. *cinerazzo*, first attested in *Dioscoride fatto de greco italiano* (Venice, 1542), I.13. *pallidastro* first attested in Luigi Guglaris, *Quaresimale* (Bologna 1676), I.70. *nigastrum*-Ital. *nerastro*, first attested in Antonio Vallisneri, *Opere fisico-medicale* (Venice, 1733), II.4.

fix there than anywhere else, which seems significant for a Latin text that provides by far the greatest number of -*aster* examples, nearly all of them unique.

Comparisons with Dioscorides

Another insight into the *AG*'s use of the -*aster* suffix is afforded by a comparison with similar passages in Dioscorides, which suggests that some uses may have derived from attempting to translate Greek, but not Dioscorides' text. Particularly intriguing are the six cases (*acidastrum, resinastrum, spicastrum, nigrastrum, purpurastrum, oleastrum*) which correspond to the use of present participle constructions in Dioscorides for the same product (references can be gleaned from the *Comparanda* in the *AG*). Hence the description of the best Mecca balsam (#202) as having a 'pungent, sharp fragrance that is *not particularly sour*' (*non acidastrum*) finds an almost verbatim description in Dioscorides, 'vigorous, pure and *not sharp*' (εὔτονος καὶ εἰλικρινὴς καὶ μὴ παροξίζων), which is the only recorded participle form of παροξίζω, 'to have a somewhat sharp smell.' The same entry also records that Mecca balsam becomes red, dry, and a 'little resinous' (*resinastrum*) with age, where Dioscorides simply states that it deteriorates, 'thickening by itself,' using a present middle participle (παχυνόμενος χείρων γίνεται): but the variance between the two descriptions seems to rule out direct influence.[48] On the other hand, the description of the taste of malabar (#190) as 'not salty but rather like spikenard' (*non salsum sed potius spicastrum*) seems a direct translation of Dioscorides (ναρδίζον δὲ τῇ γεύσει <καὶ> μὴ ἁλμυρίζον), using the participle form of ναρδίζω, a verb not attested before Dioscorides, though here notably the other participle μὴ ἁλμυρίζον is rendered with a simple adjective in the *AG*, where one might expect *salsastrum* if the author was translating directly from Dioscorides' text and struggling with participle forms. This same entry uses the term *nigrastrum* to describe the colour of the best malabar, where Dioscorides again uses a present participle and dative construction, 'off-white with a tinge of black' (ὑπόλευκον ἐν τῷ μελανίζοντι). The *AG*'s other use of *nigrastra* for the colour of rhubarb (#231), however, is rendered by Dioscorides with a simple adjective 'black' (ῥίζα μέλαινα). The small flower of centaury is described as 'purplish' (*floscellum purpurastrum*) where Dioscorides uses the participle form of ἄνθος κυανίζον, from κυανίζω, 'to be dark in colour' (and see plate 7). Yet the other use

---

48   Closer to the *AG* is Pliny 12.116, 'rufescit deinde simulque durescit e tralucido.'

of *purpurastrum* (for man orchis, #272) has no resonance with Dioscorides' description of such. Finally, garlic germander (#265) is said to have a wild olive smell (*odore oleastrum*), where Dioscorides said it is 'smelling somewhat like garlic' (ποσῶς τῇ ὀσμῇ σκορδίζοντα, which, incidentally, agrees with manuscript *P*, *aliastro*: cf. Lat. III.121, 'odore et gustu alei habens'), using the only known instance of this verb σκορδίζω, and again as participle.

There are also another six cases where the *AG*'s -*aster* suffix corresponds to an equivalent, simple adjective in Dioscorides, and another four cases where they slightly modify the meaning but are similar. These are best presented in point form, and comparison with the Latin Dioscorides, and Pliny where relevant, further suggests a Greek context for the *AG*:

SIMPLE ADJECTIVES:

– *subacidastrum,* a type of sour pomegranate (#183): ὀξεῖα, 'sharp/sour.' Lat. I.109, 'acra.' Pliny 23.108 *acerbus* (as one of nine types).
– *croceastrum* (*inter*), narcissus flower (#197): ἔσωθεν δὲ κροκῶδες, 'saffron-coloured in the middle.' Lat. IV. 154, 'in medium florem croceum.' Pliny 20.128 *herbeacum* ('grass-coloured' one of type types).
– *pallidaster,* colour of Phrygian stone (#169): ὠχρὸς, 'pale.' Lat. V.149, 'colore uiride.'
– *purpurastrum* (*pallore*), the 'slightly purple shade' of flower of Cretan spikenard (#114), which is similar to but larger than the flower of the narcissus: a direct parallel to 'ἄνθη πρὸς τὰ τοῦ ναρκίσσου, μείζονα δὲ καὶ ἐν τῷ ὑπολεύκῳ διαπόρφυρα': LSJ (citing this instance) trans. 'shot with purple.' Lat. I.7. 'flores narcisso similes factura, colore ueluti purpureo.'
– *rubeastrum* (*extrinsecus*), the colour of sagapenon (#243): ἔξωθεν μὲν κιρρόν, 'outwardly tawny.' Lat. III.91, 'rufus a foris.'
– *subrufastra,* the colour of leaves of man orchis (#272): ἐνερευθῆ, 'a little reddish' (from ἐρευθής, 'red'). Lat. III.138, 'colorata.'

SLIGHT DIFFERENCES:

– *cineraster,* colour of milkstone (#170): μέντοι ἄλλως ἔντεφρος, 'really ash-coloured.' Lat. V.158, 'dictus est pro colore lactis.'
– *paleastrum,* seed of hogweed (#269): ἀχυρωδέστερον, 'more chaffy' (than that of hartwort).
– *puluerastrum,* the feel of adulterated saffron (#44): ἐντρέχειν κονιορτῶδες, 'presence of dust.' Lat. I.24 'puluere plenus est.'

– *resinaster* (remains very), texture of the best storax (#254): λιπαρός, ῥητινώδης, 'fatty [and] resinous.' Lat. I.78, 'pinguis, resine similis.'

These examples convey the impression that the *AG* was using Dioscorides or a very similar Greek source. Against this, however, is that five of the *AG*'s –*aster* usages have no resonance with Dioscorides' text whatsoever (*fucastra, pallidaster* in both #172 and 286, *similastrum, subcaerulastrum*), while another six depart from Dioscorides in significant ways, again best presented in point form:

Departures from Dioscorides

– *cepastrum*, smell of sagapenon (#243): μεταξὺ ὀποῦ σιλφίου καὶ χαλβάνης, '[smelling] between juice of laserwort and galbanum.' Lat. III.91, 'odore inter silpi et galbanu habens.'
– *melastrum*, fragrance of the best bee glue (#215): εὐώδη, 'aromatic.' Lat. II.67, 'odorata.' Pliny 11.16 'odore et ipsa etiamnum graui.'
– *myrrastrum*, taste of seed of Cretan alexanders (#271): <ὄζον> σμύρνης, 'smelling of myrrh.'[49] Lat. III.73, 'uelut murrae habens odorem.' Pliny 27.133, 'odore medicato cum quadam acrimonia iucundo.'
– *puluerastrum* (and smooth, deep blue), indigo from India (#135): (κυανοειδές τε καὶ) ἔγχυλον, (λεῖον), '(deep blue), juicy (and smooth).' Lat. V.117, 'colore quianeu habet et lene.' Pliny 35.46, 'cum cernatur, nigrum, at in diluendo mixturam purpurae caeruleique mirabilem reddit.'
– *purpurastrum* (and whitish), colour of man orchis flower (#272): κρινοειδὲς ἔγχυλον, 'lily-like and white.' Lat. III.138, 'flore lilio simile et candidu.'
– *rubeastrum*, colour of the best omphacion (#200): ξανθὸν καὶ εὔθρυπτον, 'yellow and crumbling.' Lat. V.8, 'rufus et fragilis.' Pliny 12.131, 'rufa.'

The impression gained from all of these comparisons is that the author of the *AG* was using Greek sources that were similar to, if not the same as, those used by Dioscorides, but he did not consult Dioscorides' text, and at times used completely different sources – an impression which is

---

49 Given the *AG*'s description, Wellmann's insertion of 'ὄζον' (δριμὺ γευομένῳ <ὄζον> σμύρνης) may be mistaken, and the correct reading is 'pungent in its taste, which is like myrrh.'

consonant with other findings (ch. 3.B). We shall now look more closely at the use of Greek terms and glosses in the *AG*, which confirm that the text originates in a comfortably bilingual environment, which explains its ease in dealing with Greek sources and a quintessentially Latinate phenomenon like the -*aster* suffix.

F. Greek in the *AG*

As could be expected for a Latin medical text, the *AG* is infused with Greek vocabulary for diseases or conditions (*coeliacis, ictericus, cacoethes, panaritium*), anatomical terms (*parotis, thorax*), treatments (*clysterizare, catismatis*), and, of course, nomenclature of plants, minerals, and animals.[50] In addition, on at least 21 occasions the *AG* specifically refers to 'Greeks' and their language as a source of authority and specialist vocabulary.

To begin with nomenclature, that the author was immersed in the Latin world is confirmed by the large number of Latin names used rather than Greek equivalents: hence *lenticula* rather than *phakos* for lentils (#152), *nepita* not *calaminthe* for catmint (#196, and plate 5), *bryonia* not *ampelos* for bryony (#43), *calx uiua* not *asbestos* for quicklime (#67), *fel* not *chole* for bile (#116), *git* not *melanthion* for black cumin (#120), and so on, including some distinctly Roman-Italian names like *sertula* for king's clover (#262), and *ueratrum* for hellebore (#285).[51] This is in contrast to the Latin Dioscorides, which mostly just Latinizes Greek names.[52] But intriguing in this respect is the *AG*'s retention of Greek names when we know (from Pliny, among others) that Latin names were well-established, such as *asphodelus* and not *albucus* for asphodel (#33), *conium* and not *cicuta* for hemlock (#49), *conyza* and not *pulegium* for fleabane (#79), *helenion* and not *inula* for elecampane (#99), *elelisphacos* and not *salvia* for sage (#102), *erythrodanum* and not *rubia* for madder (#105), and so on.[53] Also telling in this regard is the retention of the word *ge* (γῆ) for the

---

50 The bibliography on the use of Greek in Latin medical texts is quite large, but see André 1963, Mazzini 1978, and especially Langslow 2000, 76–140.

51 *AG* #1, 3, 7, 8, 14, 20, 31, 54, 72, 76, 78, 84, 97, 110, 112, 113, 115, etc. A glance at Dioscorides' names in the *Comparanda* section (see ch. 3.H) makes the choices apparent. Note also #20 n.1, and 96. In many cases the other is given in the text: e.g., *ueratrum/ hellebore* #285 (likewise in #97). Cf. *uella* #289, a 'Gallic' term according to Pliny.

52 See Stadler 1898.

53 Also #42, 36, 85, etc. The author seems to know a Latinized Greek term for foam of soda (*aphronitrum* for ἀφρὸς νίτρου, #9), rather than the more common Latin name of *spuma nitri*, as Martial (14.58, cited by Isidore 16.2.8) recorded: 'Rusticus es? nescis quid Graeco nomine dicar: / spuma vocor nitri. Graecus es? aphronitrum.'

entries on different soils (#125, 127–32: *ge samia, ge ampletis, ge erethria*, etc.), as is clear by their placement among products beginning with 'G' and the uniformity of such in the manuscript tradition, though the scribes were clearly unsure of what to make of it, often mangling it to make odd amalgamations (*gesamia, gemapelitis*, etc.). This certainly suggests that an original Greek text is behind this section, though of course G's placement as the seventh letter is Latinate (in Greek it would be third), and there are other signs of Greek heritage, but also Latin editing.[54] Less significant are the few attestations of Greek measurements (*cyathus, drachma*), for as both Celsus and Pliny acknowledged, Greek weights and measures were commonly used in medical writings, so they gave conversions (though Pliny continued to employ Greek). But the *AG* also uses the Roman measurements of *scripula* (#151) and *librae*, this last for the ingredients to make the incense *cyphi* (#76), where the similar recipe in Dioscorides used a combination of Greek measurements (*xestes, mnai, drachmai*).[55]

On at least 21 occasions the author draws attention to Greek terms as glosses, sometimes in first person voice (*appellamus/dicimus Graece*), sometimes in third person (*Graeci dicunt/uocant/appellant, dicitur/appellatur Graece*), occasionally with an adverb (*graece*), and a few times as a gloss (*id est*). Over half of these are simple translations of Latin terms, giving the Greek equivalent, or specific types of products that are known by a Greek name only.[56] But others suggest a more profound acquaintance with Greek, and with Greek sources now lost to us. For example, the author knew that the adjective in Greek for 'without holes/apertures'

---

54 Not rendering rough breathings (#38, 99, 101, 295–300) may also reflect Greek origins. Other aspects point to Latinate ordering, eg. *phu* #114, and the use of k for chi (χ), #142–4. There are also odd placements such as *qu-/c-olocynthis* #226. Cf. stone entries (#161–78, *lapis*, orig. '*lithos*'?).

55 *drachma* #82, 103, 266; *cyathus* #266. Pliny 21.185 (repeated *Med. Plin.* prol., and Marcellus, *Med.* pref.). Cf. Celsus, 5.17.1C, who gives conversions but sticks with Latin measurements. Curiously, *cyphi* was unknown to Celsus and Pliny, though their contemporary Scribonius mentioned it.

56 Simple translations: *gigarta* #3, *narthex* #4, *polypodium* #90, *cicidas* #121, *lepidas* #145, *rhoa* #182, *libanus* #279. Greek names: *schiston* #7, *eupatorium* #13, *sarcophagon* #163, *stacte* #180, *alphiton* #240, *mastiton* #250, *molybdaena* #256 (= 182). Glosses (*id est*): *agasylis* #4, *calaminis* #210, *pampini* #291, *scotosin* #292. Some examples may belong to later traditions, such as the gloss of *chaumenon* as a synonym for burned copper (#1), unattested elsewhere, and not found in our earliest manuscripts (*VLF*), though its appearance in *BWP* suggests it was comfortably incorporated by the tenth or eleventh centuries at least. The term καῦμα, 'burning heat,' first appears in Latin in the fourth century (Jerome, Ps. Jerome, Fulgentius of Ruspe), but the first medical reference seems to be Latin Oribasius (*Syn.* 6.26 'in pestilentia cauma est in thorace et lingua combusta').

is 'what we call *atreton*' (#253), here used to describe best type of ginger. Dioscorides used the term *ateridonista*, 'not worm-eaten,' and warns that ginger decays easily. The *AG* also notes that 'in Greek we call' grape-seeds *gigarta* (#3), a very rare term in Latin, used only by Palladius and Cassiodorus. Curiously, in his own, similar entry for verdigris, Dioscorides used a different word, στέμφυλα, which has the same ambiguous meanings of *gigarton* as grape-seed or olive-pit.

There are other hints of Greek sources now lost. As seen (ch. 3.E), the *AG* mentions a depilatory made from orpiment called *psilotron* (#8), and also gave a Latin name (*acilea*) unattested elsewhere. The mention of a toothache caused by eating unripe fruit, 'which we call in Greek *haemodia*' (#32 n.2), seems to derive from a lost source. Although Dioscorides also mentions the *haemodia* in his own entry for purslane – and it is the only time he mentions it – he does not qualify it in any way, and says nothing about teeth or fruit. The only author to do so is Galen, which seems to constitute the only link between this author and the *Alphabetum* that circulated under his name. In the entry for purslane in his own *On simple drugs*, Galen notes that it helps with *haemodia*, which is caused by 'contact from acidic juices.' But in his *On the properties of foods*, Galen begins his discussion of the medicinal qualities of sour milk (ὀξύγαλα) by stating that it does not harm teeth in their natural state, but only those affected by a condition of (humoral) imbalance and preponderance of cold, 'and they sometimes have the symptom called *haimodia*, which generally follows [from eating] unripe mulberries, in addition to other astringent and acidic foods.'[57] As is well known, Galen used a host of sources, often verbatim and often without acknowledgment, for his own writings on pharmacy (ch. 3.E), and it would seem here that Galen and the *AG* used a similar source: the lack of any other resonance between the *AG* and Galen rules out direct influence.

The same sense of the *AG*'s reliance on other sources, or simply knowledge of Greek, occurs in other entries where there is no equivalent in Dioscorides or any other text, such as the reference to crushing mandrake in a machine to make 'what we call in Greek *mandragorochylon*' ('mandrake-juice,' #181), or likewise to a resin that 'we call in Greek *strobilina*' (#228), which obviously derives from the word for stone pine (*strobilos*), but is unattested elsewhere.

---

57 Galen *Simp.* VI.43 (K.11, 831), 'αἱμωδίας τέ ἐστιν ἴαμα ἐξηρασμένα ταχέως ὑπὸ τῆς τῶν ὀξέων χυμῶν ὁμιλίας'; idem, *De alimentorum facultatibus libri* III, (K 6.689), 'καὶ τὸ σύμπτωμα αὐτοῖς ἐνίοτε γίγνεται τὸ καλούμενον αἱμωδία, τοιοῦτον οἷον ἐπὶ τοῖς ἀώροις συκαμίνοις ὅσα τ' ἄλλα στρυφνὰ καὶ ὀξέα συμβαίνειν εἴωθε.'

Other references to Greek are puzzling, as when the *AG* states that the salt which is mined is called in Greek *alasorexaton* (#239 n.1), which is etymologically correct ('salt' and 'mined'), but the name is unattested elsewhere, and it may simply represent a corrupt translation of Greek. Similarly, we are told the best type of nutgalls (#121) are the darker, heavier, and more plump kind 'which the Greeks call *cicidas*,' but this is simply the word for nutgall (κηκίς, -ῖδος). Given that κηκίς also means 'gushing/ bubbling forth,' 'ooze,' and 'dye,' the *AG* may be referring to specialist nomenclature, the record of which has been mangled over time.

G. Conclusion

The *AG*'s philological riches add considerably to our historical record for one, relatively short text. Yet finding a precise historical context that matches all of the linguistic phenomena discussed above has proved difficult. While the grammar and some vocabulary (particularly the language for properties, *uires excalfacientes*, *stringentes*, etc.) suggest an imperial date, which is consonant with the *AG*'s pharmacology and sources (chs. 2 and 3), other aspects of vocabulary point to a late antique ambience. These may be explained as interpolations or additions picked up over time, but they had certainly been incorporated by the seventh century, our earliest surviving recension (*V*): their survival in nearly all other recensions suggests that they were integrated much earlier. The use of *uiscidus* to mean 'powerful,' and the unique use of its adverbial form *uiscide*, are employed too extensively to be considered later additions, and they have a late antique flavour. To date this earlier requires us to date the similar use of *uiscidus* in the Latin translations of three Hippocratic works and in the Latin Dioscorides to an early period also. That may be possible, but it also requires us to view the semantic development of *uiscidus* as progressing from meaning 'powerful' to meaning 'bitter,' 'astringent,' 'pungent,' rather than the other way around. This becomes more plausible if *uiscidus* and *uiscide* derived from *uis*, as the *AG*'s usage suggests, rather than a semantic combination of *uirus* and *uiscum*: or it may be that the *AG*'s author developed the term independently, and hence provides no clues for dating the text.

The remarkable number of -*aster* suffixes seems to provide no hints for dating the text either. On six occasions they correspond to present participle constructions in Dioscorides, which suggests that the author of the *AG* deployed them to translate from Greek sources used by both authors. But most examples have no resonance with Dioscorides' text and seem independently coined. Their concentration in the *AG* may reflect a more Italianate environment, but more evidence and more research on the *-aster* suffix are needed to confirm this: in any case the *AG* makes a significant contribution toward both those ends. References to 'as the Greeks say' or 'as we say in Greek,' and the use of Greek terms, including neologisms and the use of *ge* for (medicinal) 'earth,' suggest a deep Greek substratum behind the text that was facilitated by a bilingual author and environment, also reflected in the mix of Greek and Latin names for plants. This too may reflect an imperial as much as a late antique date: there are no significant clues either way. If the *AG*'s Latinity provides no clear answers, then the weight of the other evidence documented in these chapters – pharmacology, sources, and general world view (ancient, eastern Mediterranean) – pushes the date back to Antiquity.

H. Difficulties of Terminology and Translation

In keeping to fairly literal translations of terms, I have tried to replicate the use of simple, technical language of the *AG* and similar works, and not stray into modern specialist language of medicine, botany, or related topics. Hence *thyrsulum* is translated as simply 'stalk' or 'stem,' rather than the less-known 'thyrse,' though a proper modern botanical definition is 'a more or less ovoid ellipsoid panicle, with cymose branches.'[58] As noted throughout the English text and in the section on sources, many ambiguities are resolved, where possible, by referring to Dioscorides' text, or to Pliny where relevant. Medical terms present the greatest difficulties, particularly the identification of particular ailments or diseases. Although a technical vocabulary and specialist terms existed, ancient and medieval authors were not consistent in their use of them, and rarely do we find efforts to systematize the use of terms, or even disputes over their meaning between authors, that would help us identify the particular illness in question.[59] Secondly, the aetiology of disease was symptom-based, in that the physical manifestation of the disease was directly associated with its cause, hence its classification. Fevers, for example, were considered a dis-

---

58 Stearn 1966, 530.
59 On identifying particular diseases see Migliorini 1993, the individual papers in Debru and Sabbah, eds. 1998, and Deroux, ed. 1998. For the Hippocratic corpus, see the papers in Thivel and Zucker, eds. 2002; from Greek to Latin in this corpus, Bozzi 1981. In the absence of microbiology, concepts of germs and contagion relied on notions of 'contact' (physical, approximate, environmental): see Grmek 1984, Bodson 1991, Nutton 1997, Stok 2000.

ease rather than a symptom, and some notions of disease do not match modern understandings: 'dropsy' (*hydropsis*) was thought to be caused by an excess of water (or simply fluid, such as blood) in the body, hence was used to cover a range of illness in which symptoms could include inflammation of the tissues, or the perceived collection of fluid, such as interstitial fluids in various types of edema, or coughs and respiratory problems. Some terminology is dealt with in the translation (erysipelas #68 n.1, *regius morbus* #196 n.5). The following, selective examples are restricted to frequently mentioned terms that have some ambiguity. On the terms for different forms of drugs (*collyrium, pastilla, emplastra*, etc.) and how they are made, see Gourevitch 2003.

Duritias/durities. Translated as simply 'induration.' It appears often in the *AG* as something to be mollified, softened, or loosened by the drug in question. Its range of application in Latin is extensive.[60] Its Greek equivalent, σκληρία, likewise simply meaning 'hardness,' was also used for a range of meanings, and we find it in Dioscorides used a number of different ways, ranging from to breast carcinoma to calluses on the fingers.[61]

Inflationes. Literally 'swellings,' it is mostly used to describe symptoms relating to the stomach or intestines, hence it is translated here as 'gas' or 'bloating.' It is used extensively in classical and late Latin. The equivalent Greek term (in Diosc. and others) is στρόφοι, though Celsus seems to render this as *tormina* 'gripings,' particularly for the symptoms of dysentery.

Eye problems: *asperitudo caligo, cicatrix, lippitudo, vitium*. Many different eye problems are referred to in the *AG*, and as with most ancient and medieval texts, it is difficult to identify the problem and render it with a modern term, though Greek writers were generally more specific in their terminology than Latin authors: see Marganne 1980, Fronimopoulos and Lascaratos 1991, Grmek 1998, and Fischer 2000g. The Greek term for cataract, *hypochusis* or *hypochuma*, literally means 'overflow of humour' (thought to have entered in the space between pupil and lens), translated as *suffusio* by Celsus and early Latin authors. *Suffusio* is used only once in

---

60 Pliny: *duritiae genarum* 28.169; *duritiae anginis* 30.35; *duritiasque et sinus corporis* 26. 127; as *carcinomata* 26.127; as *collectio, contractio* 24.184; cf. *strumae* 22.144, 26. 51; *tumor* 24.184; *furunculi* 28.140. Note Vindicianus *Med* 27, 'saxitate stomachi, id est duritie (*graece* σκίρρωσις).' See also Scrib. 124, 131.
61 Riddle 1985, 56.

the *AG* (#63), yet other terms suggest similar if not the same complaints: 'scar (tissue) on the eyes' (*cicatrices oculorum* #52) may mean cataracts, as could 'swelling on the eyes' (*tumorem oculorum* #152, 262), 'painful discharges' (*oculis ex rheumate dolentibus* #291), 'eye complaints' (*uitia oculorum* #110), or 'teary eyes' (a variety of expressions #6, 37, 44, 45, 58, 170, 199, 213). *Lippitudo*, usually reserved for ophthalmia, is used twice (#118, 217), and *aegilops* (lit. 'goat-eye'), lachrymal fistula, appears once (#188), as does *asperitudo* ('rough eye,' #283), often used by Latin authors as an equivalent for Greek *trachoma*, which may mean conjunctivitis, ophthalmia, or trachoma. *Caligo* (lit. 'cloud,' 'fog,' 'darkness') *oculorum* is often translated as 'dim eyesight' (#56, 96), but it also may refer to the effect or symptom of any of the above-mentioned problems, likewise the ability of some medications to 'clear up vision' (*claritatem oculis praestat* #30, *ad oculorum claritatem faciens* #179). In all cases the translation has tried to remain fairly literal with attention to context and (where possible) comparison with Dioscorides.

Expellit partum and ***prouocat menstrua*.** The phrase *expellit partum* occurs 13 times in the text, and *deiicit partum* twice (#91, 155): these are translated literally as 'expels the foetus' and 'ejects the foetus,' which retain something of the ambiguity in the Latin as to whether this refers to an oxytocic or abortifacient action. There are good grounds for both interpretations. The *AG* also refers to 'partum mortuum expellit' (#88), suggesting the foetus is alive in other instances, and the etymology of *partus*, from *pario*, to give birth, may suggest the full term has been reached – the *AG* called a chicken egg *partus gallinae* (#204) for example – whereas *fetus*, and the Greek equivalent ἔμβρυον, both derive from the notion of swelling/growing inside. If the uses of *partus* in Pliny, Celsus, and other texts is our guide,[62] then *expellit partum* seems to have an oxytocic sense. On the other hand, the considerable force of the verb *expellit* is also used to describe the expulsion of intestinal parasites (*lumbricos*) or bladder and/or kidney stones (*calculos expellit*, e.g., #247). Catmint (#196, and plate 5) is said to 'expel intestinal parasites, maggots, the *partum*, and kills worms in the ears,' while the pungency and bitterness of gentian (#123) can 'expel the *partum*, put all snakes to flight, and is useful against poisons' – both examples strongly suggest an abortifacient action, as do oth-

---

62 Conveniently set out by Mazzini 1993, to which we should add Pliny 27.120, 28.251, 29.47, 30.124, 30.125. Elsakkers 2008 shows that *partus* could mean offspring or neonate in penitentials and other early medieval texts.

ers that pair expelling intestinal parasites with the *partum* (#153, 196, 287). Others, like pennyroyal (#214), are well-attested as abortifacients from ancient to modern times (Scarborough 1989). Similar ambiguities exist in Dioscorides' use of the phrases 'ἄγει ἔμβρυα ('draws down foetuses') and ἐκτινάσσει ἔμβρυα ('expels foetuses'), both of which are often coupled with drawing or expelling the menstrual period (καὶ ἔμμηνα), and/or the afterbirth (καὶ δεύτερα) (Diosc. I.76, II.109, III.3, 31, 72, 102, etc.). In ten cases *expellit partum* in the *AG* is coupled with 'stimulating menstruation' (*prouocat menstrua*: #22, 55, 72, 74, 88, 91, 99, 155, 214), including that rue (#227) 'stimulates urination and menstruation, and if more is consumed it also expels the foetus.' The *AG* contains as many instances of 'stimulates menstruation' (often coupled with *urinam*), without mentioning foetuses. John Riddle (1992a, 1997) has suggested that phrases such as 'stimulates/ promotes menstruation' were often coded shorthand for abortifacients, but other scholars have pointed out that the common concern for regular menstruation in women was about fertility and women's health in general (see Green 2005, 2008a, and 2008b, with many references). The example of rue (#227) suggests that *prouocat menstrua* and *expellit partum* were considered separate actions, but the ambiguity of the latter's meaning only leaves us with even less certainty as to whether *prouocat menstrua* was intended to signify arbortifacient potential.

Notum omnibus. Literally 'known to everyone/all,' which appears throughout. An alternative translation is 'well-known,' which better echoes the Greek γνώριμον (Diosc. II.113), but even the Latin Dioscorides translated this as *omnibus notus est*; likewise, caraway (III.57): 'καρὼ σπερμάτιόν ἐστι γνώριμον'. Diosc. Lat. 'careu semen est omnibus notu,' *AG* #60 'Careum est semen omnibus notum.' The same goes for other forms such as 'bramble is a familiar [plant]' (βάτος ἦν γινώσκομεν, IV.37; Diosc. Lat. 'Batus omnibus nota est'), where the *AG* has 'everyone knows bramble' (*rubum omnes nouerunt* #230), and so on. The phrase 'known to all' better maintains the ambiguity that the 'all' may refer to written sources used by the *AG* (*aliqui describunt* #92, 99: see ch. 3.A). In many cases, the phrase *omnibus notum* is followed by brief descriptions (e.g., for the rose #232, 'which everyone knows,' and so brief as to be useless), suggesting that the author saw no need to repeat what was easily accessible or frequently written about.

Capitellum, -a. Literally 'little head,' it means 'tips' (#82, 199, 260 [var.

W, a name for soap: see ch. 5.C]), 267 [pods], 280. In #271 (*ramulum et capitellum*) it seems to mean 'umbel,' as found in Diosc. III.68, σκιάδιον, though he also uses that for 'flower-head' and other, related parts of the plant (II.139, III.27, 49, IV.36, 173).

Oblongus, -a. Variously translated with its two ancient meanings of the shape 'oblong,' and 'long'/'lengthy.' Sometimes context makes it clear – e.g., the leaves of centaury (#55) are 'magis oblonga quam rotunda,' – though on other occasions it is impossible to tell – e.g., #265 'uirgulas mittens quadratas, rectas et oblongas,' and #275 'radicem tenuem et oblongam et uirgulas oblongas.' Ambiguity is resolved where possible by comparison with Dioscorides and Pliny.

L[a]euis/lenis. Depending on the context, *leuis* is translated as 'smooth' (or even 'polished'), and also as 'light' in weight (not colour). Among the manuscript tradition there was much interchange between *leuis* and *lenis*: this is noted in the critical apparatus where appropriate, but most often choice was based on context, use elsewhere in the text, and Dioscorides' text. In modern botany it means 'smooth, free from unevenness, hairs or roughness.'[63]

Acre/acris. (cf. Gr. δριμύς). Both taste and properties can be 'sharp': for properties it is generally translated as 'severe,' 'keen,' 'penetrating,' 'stinging,' and so on, whereas 'sharp' makes sense as a description for taste (e.g., cheese), and is often associated with astringency, e.g., 'properties that are sharp to the taste and a little styptic' (#95). It is also used to describe physical attributes, such as a seed being 'pointed' (#277). Use of the adverb *acriter* is often applied to drug actions and properties: 'a sharply (that is, severely) astringent property,' though it can also apply to 'loosening' (*relaxare*), hence 'severely [*acriter*] loosening property.' See also ch. 2.C–D.

Effectus duplex. Translated as 'double application,' for it seems to refer to either drinking or eating the substance for different medicinal purposes, as applied to aloe (#5) and egg (#204), though this last also implies a further difference between cooked and raw. We are also told that wheat juice (#240) 'duplex est uel triplex uirtute,' which appears to mean the same thing, for 'it has the same properties as that which we said for the egg,

---

63 Stearn 1966, 454.

hence it is better to drink wheat juice than eat wheat.' I have found no similar examples of this, though the use of *effectus* for the power or efficacy of a medicine is evident in Celsus, Pliny, and Scribonius (see ch. 2.F).

Colours. Achieving precision for ancient and medieval use of words and expressions describing colour is notoriously difficult. Artists, decorators, dyers, tailors, and others certainly required precision with colours, but there are few traces of their specialized nomenclature in the surviving evidence. The studies of André 1949, Skard 1946, and Guillaumont 1950 (and on Greek and Latin, see the papers in Villard, ed. 2002) have shown that we cannot apply modern standards to pre-modern descriptions of colour, which vary considerably from author to author and their location in time and place. The *AG* frequently employs the following terms, which appear to have a range of meanings that defy regularity or precision:

pallidus: white, whitish, pale, yellow, light green (Stearn 1966, 250 '[used] for every pale tint of the artist's palette').
rufus: red, orange, reddish brown.
subrufus: slightly red, reddish, orange, tawny, reddish brown, yellow. Also *croceus* (or *croceatus*, *crocatus*).
subalbidus: white, whitish, pale.
exalbidus: white, whitish, pale.
candidus: white, bright, yellow.
violaceus: blue, purple, violet.
niger: black, dark (cf. wine, #292).

Interestingly, the typical term for yellow, *flauus*, is never used, and instead seems to be intended by *croceus* (*croceatus*), *aureus*, or *candidus*, cf. Stearn 1966, 245. Twice the text refers to the colour of fire (#55, 237), which is notoriously difficult to pin down: Stearn ibid., 250.

I. Neologisms and Rare Words

Besides the 20 -aster suffix adjectives and nouns discussed above, the following words are also only found in *AG*. Both *haemodium* and *alosanthos* (as a compound noun) are attested in Greek but not Latin, hence are included here. 'Neologism' refers to our historical records rather than a new coinage: many of these words may have been common but have not reached us by another source. Rare words are simply those that appear in few Latin sources, most of them from Greek. Numbers refer to *AG* entry #, unless prefaced with p.[age] number.

Neologisms

acilea, the Latin term for *psilotron*, a depilatory, 8, p. 77 ch. 3.E.
alosanthos, from Gr. ἄνθος ἁλός, salt efflorescence, 10.
alasorexaton, a type of salt (mined in Greece), 239.
chaumenon, a synonym for burned copper, 1: possibly a later interpolation, see above, n.56.
elamulae (*elamnolae V*, *lamulae F*, etc.), unidentified plant with bark, 198.
exhumido, -atur, pull out from ground, 5.
exhumido, -ans, moisten (-ing property), 122.
exstaurando, restoring (tissue), 150.
flos, 'flower,' a type of sulphide or arsenic, 242.
fuscum, another name for *purpurissum*, a cosmetic, 211.
haemodium, a type of toothache, 32, p. 105.
lamna, another name for orpiment, 8.
mandragorochylon, mandrake juice (or crusher), 181.
mastiton, alternative (Gr.) name of laserwort, 250.
nuclipinum, stone pine (*Pinus pinea* L.), 228.
oletan-us/a, foul-smelling, stinking 95, 221, from *oleto*, to befoul, defile.
sitatula (*VFLBPC*), for *hirsuta* (*J*), shaggy? 225.
subuiridicat, to turn pale green (*subviridis*), 45.
strobilina, a dry resin (made from stone pine), 228 (see also *nuclipinum*).
uiscide, powerfully, forcefully, passim (above, ch. 4.D).

Rare Words

aegaeum, aegean, 44.
appollinare[m], alternative name for mandrake, 181.
asclusus (*asclosus, assulosus, astulosus*), with/containing splinters, 39 n.2, 53, 119.
atreton, Gr. ἄτρητος, without holes (ginger), 253.
axungia, axle grease, 288.
bolarium -a, Gr. βωλάριον, little lump, 125.
botrytin, type of calamine ('bunched'), 45.
calliblephara, Gr. καλλιβλέφαρον, eye make-up, 127.
cancerosus, cancerous (wounds), 101, 292.

catismata, from Gr. ἐγκαθίσματα, sitz baths, 188.

centuripinus, of Centuripae (mod. Centorbi, Sicily), type of *bulbus errati-cus* (meadow saffron), 44.

cephalalgicus, Gr. κεφαλαλγικός, chronic headache, 12.

cicidas, Gr. κηκίς, a type of nutgall, 121.

clibanum, an iron furnace, 126.

clysterizare, to inject via clyster or syringe, 150.

coeliacis, Gr. χοιλιακός, intestinal problems/pain, 125.

columbina, 'dove-like,' an alternative name for vervain, Gr. περιστέριον, from περιστέρα/*columba*, pigeon or dove, 224 n.2.

cornetum, horned, 44.

coxiosus, Gr. ἰσχιαδικός, suffering from hip ailment, 99, 196, 271, 299.

cribellare, pass through a sieve, 76 var., 133.

cyrenaicum, Cyrenic, 44.

defrustratus, deceptive (fragrance), 118.

exulcerare, ulcerate, cause to fester, 169, also *exulcerantia* 81, *exulcerantes vires*, 42.

flammine[us], flame-coloured, 173.

gargarismum, Gr. γαργάρισμα, a gargle, 109.

ictericus [*cirinus et niger*], Gr. ἰκτερικός, (sufferers of) yellow and black jaundice, 56.

ichthyocolla, fish glue, 140.

ios (s)colex, worm-like verdigris, 136.

languentia (*loca*), enfeebled parts, 261.

malaxare, to soften (stools), 133.

malicorium, pomegranate rind, 183.

masso, from Gr, μασ(σ)άομαι, to chew, 279.

melicratus, water-mead, 103.

nauseosus, nauseous (stomach), 103, odour, 257.

onychiten, type of calamine, 45.

ormini, -us, type of asparagus, 103.

ostracitis, type of calamine, 45.

panaritium, from Gr. παρωνυχία, whitlow or felon (bacterial or fungal infection between nail and skin), 188.

paregoricus, a type of drug, 204.

partus -a, Parthinian (? stone), 44.

planta, young plant, slip, 122.

pulenta, pollen, finely ground flour, var. 152.

pulicaris, an alternative name for fleawort, 216.

quadratulus, squared (stems), var. 265.

racemorum, a cluster or bunch, 28.

ros coelestis, heavenly dew (an ingredient for moon foam), 12, (for honey), 179.

selgium, a type of saffron, 44.

sinapismus, mustard-plaster, 196, 261, 281.

sium, a type of saffron, 44.

smegma, -ta, detersive solution/lotion, prolog., 213.

syriacum, Syrian (saffron), 44.

stagneus (*stanneus*), made of tin (vase), 296.

synanche, sore throat, 194, 195.

uariola, varied in colour, 65.

uascellum, small vase, 146 var.

uastitudo, width/thickness, 44.

uiscidus, powerful, passim (above, ch. 4.D).

Manuscripts

꒰

A. Overview and Editorial Principles

As noted in the Introduction (ch. 1.C), the *AG* survives in eight medieval manuscripts, ranging in date from the seventh to the twelfth century. All have been consulted and collated for the edition presented here, along with the first printed edition of 1490, which was subsequently reprinted several times in the sixteenth century (see below, ch. 5.D). Although eight surviving medieval witnesses sounds like a small catch, this is fairly plentiful for the transmission of medical works before the twelfth century: by comparison Dioscorides (in Latin) survives in only three manuscripts and a few fragments, while the most popular, Ps. Apuleius's *Herbarius*, survives in 19 manuscripts dated before c. 1100. That there once were more circulating is clear from the citations of the *AG* by twelfth- and thirteenth-century authors (ch. 1.C), from the lack of any direct relationship between the printed version of the Renaissance and surviving manuscripts, and from testimony such as the twelfth-century catalogue of books at the monastery of St Amand (c. 1150–68), which lists two copies of the *AG* among 36 different medical works contained in 18 volumes, none of which has been identified with extant manuscripts.[1]

---

1 Given in Becker 1885-7 (repr. 1973), 231–3 (no. 114): but better is Desilve 1890, 153–4, 169–70, who dates the catalogue (*index maior*) to 1150–68: see also Derolez 1979, 230–1. I thank Eliza Glaze for the references and for discussing this catalogue with me: see Glaze 2000, 5:52–6. The wording of its title[s], 'Galienus ad Paternum de qualitate herbarum, aromatum lapidum seu animalium' (x2), is the same which appears in the *incipit* of manuscripts *B* and *P*, and the explicit of *F* (see below). Many catalogue entries, or lists of contents prefacing a manuscript, that mention *Galien[us] ad Patern[-um, -ianum]* etc. may refer to *the Liber de dynamidiis* (see ch. 1.E): e.g., Oxford, Merton College 324, fol. 1v.

The uncertainty surrounding the *AG*'s authorship, date, and origin presents serious problems for determining its text, problems only compounded by its surviving manuscripts beginning centuries after the text was first redacted in some form or other. Henry Sigerist rightly noted that many early medieval medical texts 'were so widely used that they underwent considerable changes and it is probably hopeless to attempt to reconstruct an archetype.'[2] Heeding this advice, the present edition privileges the oldest manuscript, *V* (Vat. Pal. lat. 187), for determining the overall content of the *AG*, but not the form of its language. For *V* was written in a wildly vulgar Latin with extremely corrupt orthography and grammar well on its way to proto-Romance, but is inconsistent in all aspects. Nonetheless, its antiquity (it is an extraordinarily early manuscript for a medical text), and the similarity of its content with our next earliest manuscripts, *F* and *L*, present a common tradition that serves as the principal guide for the text presented here. This common tradition is evident in phrasing, syntax, order of entries, and in particular the many omissions of text. The tradition is also confirmed by features such as the new entry for *ochra* in #207 (n.2), and the editorial comment at the end of the entry for wormwood #22 (n.2) – this last demonstrating that the slightly later and very incomplete *M* is also related to this tradition. *M2*, however, which appears in the same manuscript and contains many interpolations from Isidore, represents another branch of transmission also found in *C*, again heavily interpolated with Isidore. The text of *C* is the closest to the printed Junta edition (*J*), evident in the more inventive passages of both, but *C* was not a direct source for the *editio princeps* of Bonardus in 1490, subsequently reprinted (without any noticeable changes) in *J*: unfortunately we know nothing about the manuscript(s) which Bonardus used. For this reason, *J* has been collated as another witness.

Manuscripts *B* and *P* represent another branch of transmission, and are often identical in terms of variants, though they are not directly related. The Latin of *P* is far better, and in general the scribe strove to create fairly clear and comprehensible text where *B* can be obscure. There are some common elements among *B*, *P*, and *C*, such as the substitution of *maleficorum et venenarum* for *serpentum* (*ab omni genere serpentum ... securus*) in #272, reflecting the later date of these manuscripts. *W* differs from all others in completely re-rewriting the text to create a more succinct version that largely omits description, methods of preparation, adulteration, different varieties, therapeutic uses, and so on, in favour of simply recording

---

2 Cited in Jones 1939, 73.

the name of the product, its location, and its properties. Further details on the characteristics for each manuscript can be found in their individual descriptions below.

While I have tried to reflect a consensus of manuscripts in the choice of readings, most often the uniformity among *FLBP* has been used as the corrective against the wayward grammar and orthography of *V*. True, our other manuscripts all post-date the clean-up of Latin that coincided with the Carolingian reforms of the eighth and ninth centuries, but these too are hardly free of vulgarisms of their own, and most importantly, they show almost no relation to *V* in orthography and grammar (even *F*, which in every other respect closely resembles the text of *V*). Instead, the *AG*'s manuscript tradition presents problems of content, not of form. A recent book on nutrition that rails against modern eating habits reduced its message for marketing purposes to a sensible seven-word motto: 'Eat food. Not too much. Mostly plants.'[3] Remove or alter the punctuation, elide or alter a word, and the meaning could be altered considerably ('Don't eat food too much, especially plants'). This is the sort of dilemma we face with the different recensions of the *AG*, rather than odd spellings and unorthodox grammar, none of which are consistent enough in the manuscript tradition to warrant their consideration as a feature of an 'original text,' an unknowable entity anyhow. Moreover, the norms of standardized Latin, at least in terms of spelling and grammar, were ubiquitous in medieval scribal culture – through the traditions of classical texts, through legal culture and written instruments, and most importantly through the Vulgate Bible – even if individual scribes slipped into vulgar or Romance-type spoken forms when *writing* out text, which can look significantly different from what they read.[4] This, and the signs of original propriety of the *AG*'s Latin discussed in the previous chapter (ch. 4.A–B), justify the decision to standardize the Latin text in this edition. Another control on the text has been to monitor the Greek of Dioscorides because of its similarity of content and format for a given entry (e.g., #52 n.3, 188 n.1, 288 n.2, *tussem ueterem* rather than *tumorem uentris JBPC*, etc.).

None of our manuscripts can be dated precisely: the dates assumed here are based on a consensus among the scholarly authorities cited, with the exception of *V*, for which I argue a slightly earlier date. Likewise, locations can only be approximate: the Beneventan script of *M* (and *M2*), *B*, and *C* all

---

3 Pollan 2008.
4 Everett 2003, 36–42, 45–52, 130–62; Wright 1982, ed. 1991: all with references to further literature.

confirm a southern Italian origin, but only *M* can be ascribed with any confidence to the abbey of Monte Cassino. Manuscripts *V* and *L* are also from Italy, but both present ambiguous features for determining their precise origin: the use of 'insular' abbreviations and very Italianate vulgar Latin in *V*, the dry-point note in Greek characters (on fol. 35) suggesting Mantua for *L*, though other signs point to Lucca. Notes by later medieval or humanist readers on fly-leaves or spare folios, or in the margins, or *ex libris* markers, help to determine the later history of the manuscript and its travels (*V* to Lorsch by the thirteenth century, *C* to Cambridge by the fifteenth century).

A *stemma codicum* would look something like this:

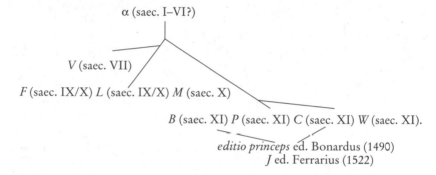

α (saec. I–VI?)

V (saec. VII)

F (saec. IX/X) *L* (saec. IX/X) *M* (saec. X)

B (saec. XI) *P* (saec. XI) *C* (saec. XI) *W* (saec. XI).

editio princeps ed. Bonardus (1490)
J ed. Ferrarius (1522)

B. Variants

The recording of variants has been restricted to those which reflect a different, yet still sensible meaning, and particular care has been taken to record omissions (*om.*), important for determining the manuscript traditions mentioned. Omissions of whole entries are indicated under the apparatus criticus with *deest* (e.g., #47 *deest P*, or #56 *deest ex lacuna V*, because that particular folio is missing in *V*). Damaged, defective, or partly illegible entries are indicated in the variants with *def.*, and sometimes under the apparatus criticus for particularly difficult sections in the manuscript (e.g., #67 *partim def. F*). Also included are some readings that reflect the historical development of Latin, such as the intriguingly deviant orthography of *V*, affording insight into the reception of ancient technical terms in the midst of the Dark Ages, or others such as the use of *vascellus* for *vasculus* that display the development of Latin toward Romance in later manuscripts.[5] In some instances classicizing forms were best used; I followed *J*

---

5 Only *V* consistently uses *-ulus*. On the diminutive suffix *-cellus*, see Stearn 1966, 306.

in rendering the superlative of *acer* as *acerrimus*, though it is most commonly rendered in the manuscripts as *acrissimus* (e.g., #264). I have generally not recorded variants deriving from grammatical variations, such as different case endings and verb forms, which abound in the different recensions, as scribes navigated the meaning of the phrases before them and adjusted the grammar as they saw fit. This is also true for prefixes that can shade meaning (e.g., [*con-/de-/ad-*]*stringit*, [*pro-/ e-*]*manat*), though these are noted where it seemed to matter. The only occasions in which the manuscript *sigla* appear *inside* the lemma parenthesis (]) are those that concern the use of the -*aster/-astrum* suffix, for a clear representation of the chosen reading (e.g., #169, pallidaster *BL*] paledaster *VF* etc.).

Many of the variants reflect confusion or ignorance, especially concerning Greek words or technical terms, and I have noted some of these, particularly for *L* and *P*, which sometimes attempt to render Greek words in Greek characters. Later manuscripts display less familiarity with the ancient world, such as *W* and *P* reading *potiolanum* for *puteolanum* (#211), or their rendering of ('what the Greeks call') *cacoethe* (#70) as *cacimaculas W*, or *kake P*. Confusion over a basic Greek term can be seen in scribal adaptations of *ge* (γῆ) for 'earth'/'soil' (#125, 127–32: see ch. 4.F). By the tenth or eleventh century no one seemed to know what was a *colum* (colander), and preferred the more familiar *sacculum* (#97). But confusion is found in earlier manuscripts also: *F* has three different words for rendering *cachrys* (*acris*, *cacreus*, *corius*) in #229. Occasionally personal interventions appear, such as the Christian disapproval of Egyptian religious practice in *V* (#76 n.6), the comments on local soap in *W* (#259, and below), or the change in *M* from cyclamen's root being 'cut out when the plant flowers' to 'its juice when mixed with wine gets you drunk' (#63).

C. Manuscripts

The manuscripts are discussed in chronological order (as much as their uncertain date allows): likewise the listing of their *sigla* among shared variants. I list them here for quick reference:

V – Vatican City, Biblioteca Apostolica, Pal. lat. 187 (saec. VII–VIII).
F – Chartres, Bibliothèque Municipale 62 (115) (saec. IX–X).
L – Lucca, Biblioteca Governativa cod. 296 (saec. IXex–X).
M and *M2* – Monte Cassino, Archivio della Badia, cod. V. 97 (saec. X).
B – Vatican City, Biblioteca Apostolica, Barb. lat. 160 (saec. X–XI).

C – Copenhagen, Det Kongelige Bibliotek, Gamle Kgl Samling cod. 1653. 4 (saec. XI).
P – Paris, Bibliothèque nationale lat. 6837 (saec. XI–XII).
W – Vienna, Österreichische Nationalbibliothek 2425 (saec. XI–XII).
(*J* – *Editio princeps*. 1490, repr. 1522. See below, ch. 5.D).

The following descriptions are mainly limited to the manuscript's content, date, and its text of the *AG*. Fuller physical descriptions (size, gatherings, prickings, margins, punctuation, etc.) can be found in the literature cited. Many of the texts within these manuscripts, often of composite nature, still require thorough investigation for a more correct identification: where no modern scholarship clarified matters I have relied on titles (here italicized) given in manuscripts.

V – **Vatican City, Biblioteca Apostolica, Pal. lat. 187, fols. 8–66v**. Late uncial and early minuscule, saec. VII–VIII. *Incipit Galieni Alphabetum ad Paternum. I. aer iustus. II agatia*, etc. Text defective, ending abruptly on fol. 66v at 'corroborat simul et sthumachu[m], quod ipsum sinapi.' (that is, #261 mustard seed). Also missing is fol. 25, and therefore also the text for *cucurbita syluestri* (#54), *centaurea* (#55), *chelidonia* (#56), *cedrum* (#57).

Contents:
1) Psalter (Hieronimum), [fol.] 1–6v.
2) Medical recipes: *Ungentum ad uentrem solvendum, Potio ad physicos*, 7.
3) (Ps.) Galen, *Alphabetum Galieni*, 8–66v.

Bibliography: Petrucci 1986, 118. Bischoff 1974, 60 and VIII.2. Lowe CLA I:81.
Our earliest witness, which one historian called a 'filthy little book,' one of several 'dirty and well-worn medical manuscripts of convenient pocket size' that attest to continued interest and trade in eastern spices and drugs from Late Antiquity to the early Middle Ages: McCormick 2001, 713 n.77. Like *F* and *C* (below), *V* prefaces each letter-section with a numbered list of following entries (I aer isutus, II agatia, III aeruca, IV ammoniaco, etc.) which it then repeats in the main text. The letter 'G' is used to symbolize 'VI' (hence XXG, XXGI, XXGII, XXGIII = 26, 27, 28, 29), a trait which Cappelli 1961, 418, dates to the eighth century. Its vulgar Latin is similar to that found in Italian documents of the sixth, seventh, and eighth century (Everett 2003, 132–8; Adams 2007, 457–64; and below). Further sug-

gesting an Italian environment is its use of the term *saliuncola* rather than *saluicula* for Celtic spikenard (#238): according to Pliny (21.43) and Diosc. (I.8), *saliunca* was the local Ligurian name. The script is rapid and somewhat careless, written in two or three different hands (one 'di assoluta rozzezza': Petrucci), and as a result alternates between uncial and minuscule forms with cursive elements. Particularly Italian traits are the 't', with its cross-stroke forming a loop to the left; 'e' and 'c' sporting a little tag to the left; the shoulder of the 'r' extending above the following letters; the ligature '&' used for both soft and hard *ti*; and a superscript 'a' frequently occuring in ligature. Notably, the different scribes all used abbreviations of 'insular' type: hr = *autem*, # = *enim*, ÷ = *est*, b: = *bus*, q with superscript o = *quo*, qnm with stroke above = *quoniam*; p with crossed stem = *per*, p with line above = *prae*; p with vertical dash above = *pri*; p with superscript o and p with squiggly line on stem = *pro*. There are signs of haste: e.g., *driopteris* is listed as third among the four headings for D (fol. 32v), but in the text is *daucum* (#91), not listed among the headings. The censorious comment that Egyptian doctors 'unwholesomely prepared rites' (*iscaribus* = *sacris*? *escaribus*, ex *escarius*? cf. *C. var.*, *casibus*: see #76 n.6, and ch. 2.I) seems to be the addition of an indignant scribe. The *ex libris* on fol. 1 confirms that the manuscript was at Lorsch by the early thirteenth century.

Lowe dubbed the script of the main hand as an 'uncial of the declining stage,' and dated the manuscript to the eighth century, while Bischoff suggested late eighth century. There are, however, good grounds for dating it earlier. Our earliest surviving examples of Ps. Apuleius *Herbarius* all date from the sixth and seventh centuries, were all from Italy, and were all written in an uncial script with similarly barbarous Latin: of these, the script of *V* closely resembles those of two seventh-century fragments now preserved at Halberstadt (CLA VIII.1211), and another (also with Ps. Hippocratic *Ep. ad Maecenas*) now at Munich (CLA XII.1312): others are CLA X.1685 [Leiden, Univ. Bibl. Voss. Q 9, the earliest], III.301, VIII.1050, IX.1312. Moreover, the two (unedited) medical recipes on the recto side of fol. 7 of *V* are written in a north Italian cursive script that resembles the Ravennate papyri of the sixth and seventh centuries. A similar script is found in the palimpsest fragment of a medical recipe (still unedited) in Milan, Ambr. C.105 inf. (fols. 1 and A 9, lower script), which Lowe (CLA III.324) dates to the sixth century. The Psalter which comprises the first seven folios of our manuscript (CLA I: 80a) was written in an 'ungainly' (Lowe) minuscule script with some uncial forms (g and a), uses vulgar Latin spellings that are similar to those of the *AG* text, and also some of the same abbreviations (for *quoniam*, *per*, *pro*, *-tur*, *-bus*), but not all. The use of uncial script, and the presence of the medical recipes, suggest an earlier date for both the Psalter and the *AG*, which seem contemporary.

The following spellings occur regularly in the manuscript, and give some idea of the 'vulgar Latin' character mentioned above: *ramuli* [for *rami*], *hodure* [*odore*], *orina* [*urina*], *hodoniam* [*ideonee*], *ispissus -a* [*spissus*], *agris/agriter* [*acris/acriter*], *sugus* [*succus*], *quoquitur* [*coquitur*], *sablunis* [*sabulis*], *postillus* [*pusillus*], *gliuole* [*glebula*], *sthumachum* [*stomachus*].

F – Chartres, Bib. Municipale 62 (115). fols. 54v–73v. Caroline minuscule saec. IX–X. *Incipit alfabetum Galieni ad Paternum.* – (Yhacantum) adfectum stiptica est. *Explicit qualitas omnium herbarum.*

Contents:
1) Soranus, *Quaestiones medicinales*, [fol.] 1–16.
2) Galen, *Ad Glauconem de medendi methodo I*, 16–36v, 38.
3) Extracts on ages of man, various herbs, and correspondences between humours, elements, and seasons, etc., 37.
4) *De pulsibus et urinis*, 38–40v.
5) Galen, *Ad Glauconem de medendi methodo II*, 40v–53v.
6) Ps. Galen, *Liber de dinamidiis*, 54–54v.
7) Ps. Galen, *Alphabetum ad Paternum*, 54v–73v.
8) *Liber Escualpii*, 74–109.
Various recipes on fols. 41, 44, 109.

Bibliography: Beccaria 1956, 126–9. Wickersheimer 1953, 166–9. MacKinney 1952a, 8 and fig.1. Idem 1935, 403, 408, idem 1937, 125, 206 n.260 and tav. II. Tribalet 1938, 15–23, 32, 90. Diels 1905, I:144, no.175: II, 93–4.

Item 3 partly ed. Wickersheimer 1966, 167–8; item 8 ed. Manzanero Cano 1996. MacKinney suggested that this manuscript originated from the monastery of Fleury, and noted (1935, 403) that the *AG* 'seems to have been an important part of the dynamidic literature of the period.' But Chartres itself is a possible origin: the manuscript is recorded in several notices concerning the library of Notre-Dame of Chartres's collection up to the seventeenth century. It was badly damaged from bombing in the Second World War, but luckily a microfilm was made prior to the war: my sincere thanks to Julien Barthe, Directeur des bibliothèques de Chartres, for generously sending two copies of the relevant folios from the microfilm. As with *V*, numbered chapter headings for the entries are given for some letters (A, B, C, T, V, X, Y) that do not always correspond with the text. The text of the *AG* seems to derive from the same tradition as *VM*, confirmed by the editorial comment at the end of #22 (as above, ch. 5.A). Despite demonstrating considerable orthographic propriety, the scribe continually wrote *uirum* for *uerum* (from around 'E' onwards), though had no trouble with *uero*; and *postilla* for *pusilla*, as found in *V.*

L – **Lucca, Biblioteca Governativa cod. 296 (B.196), fols, 81v–107v.** Minuscule saec. IX ex–X. *Incipt liber pigmentorum.* Aes ustum fit maxime de clavis cupreis ex vetustis – [Yagointhum] et effectum stiptica est valde. *Explicit qualitas omnium harbarum et aromatum vel lapidum et animalicularum.*

Contents:
1) Antonio Musa, *De herba vettonica liber*, [fol.] 1–2.
2) (Ps.) Apuleius, *Herbarius*, 2–18.
3) *Epistua ad Marcellinum*, 18.
4) *De taxone liber*, 18v–19v.
5) Sextus Placitus, *Liber de medicinae ex animalibus*, 19v–27v.
6) *curae herbarum*, 27v–46v.
7) *Incipit de ponderibus medicinalibus Dardantii philosophi*, 35v (following blank folio).
8) *curae ex hominibus*, and *curae ex animalibus fiunt*, 47–81.
9) (Ps.) Galen *Alphabetum ad Paternum*, 81v–107v.
10) Dietic calendar, 108–108v.
11) *Dies aegyptici* (on lunar calendar in Egypt), 108v–109.
12) *Lunas de somnium* (which moons bring joy, danger, etc.), 109–109v.
13) Various recipes (*confectio caustici.* etc.), 109v–110.

Bibliography: Ferraces Rodrígez 2007 (comprehensive). Collins 2000, 156–62, and fig. 36 (fol. 44v) 37 (fol. 48). Beccaria 1956, 285–8. Muzzioli 1954, 42–3. Simonini 1936, 188–90. Sigerist 1934, 48. Singer 1927, 39. Giacosa 1901, 349–53 and tav. of fol. 18v and 43.

Items 1, 2, 4, and 5 are edited in Howald and Sigerist 1927. Item 3 ed. and discussed by Zurli 1998 (II), and see below. Item 6, claimed by Riddle 1980, 128, as a copy of Ps. Diosc. *Ex herbis feminis*, is the *curae herbarum*, ed. Mattei 1995, on which see Ferraces Rodrígez 2004. Item 8 ed. Simonini 1936, 207–12. Item 7 ed. Ferraces Rodrígez 2007, 21–3. Item 8 partly ed. Benassai 1998 (*ex animalibus*), but as shown by Ferraces Rodrígez 2006,

2009b (also 2007, 50–1), these texts (*curae ex hominibus et ex animalibus*) both derive from Pliny (books 28–30), and should be considered a composite. Items 10–13 are saec. XII additions: Johnson 1937, 90–1. Ferraces Rodrígez 2007, 23–4, transcribes and discusses a few additional recipes and texts throughout the manuscript in margins and blank spaces: fols 11–13, (in saec XII hand) 'ad dentium dolore et curatio,' 'alia cura ad dentoum dolore probata'; fol. 39v, 'benediccio ad caseum.' Special thanks Arensio Ferraces Rodrígez for forwarding his work on short notice.

The manuscript is north Italian: the colophon scratched in dry-point on fol. 35v 'Lodericos me scripsit in Mantoa' (facsimile in Giacosa 1901, 52; Ferraces Rodrígez 2007, 23) refers to the restoration of that folio with the Dardantius text written in a slightly later hand. The famous Lucca manuscript, Biblioteca Capitolare 490 (saec. VIII ex.– IX in.), includes the *AG*'s entry for rock alum (*chalcitis*) among its artisanal recipes (see #48 n.1), hence suggesting that an earlier copy of the *AG* was present in Lucca. The script is usually dated to the ninth century: Collins 2000, 158 n.69, notes a tenth-century date proposed by P. Supino-Martini. The manuscript contains some rough illustrations of plants and animals for the Ps. Apuleius, *curae herbarum* and *curae ex animalibus* texts, which Muzzioli 1954, 42–3, suggested were north Italian. The text of Ps. Apuleius (on mandrake) contains an interpolation from Isidore (*Etym.* 17.9.30), suggesting at least a *post-quem* for this recension of that text: Ferraces Rodrígez 2007, 56. The manuscript includes the prefatory letter 'ad Marcellinum' (fol. 18v) attached to *Ex herbis femininis*, first published by Giacosa 1901, 351, and more recently edited by Zurli 1998 (II), collating three other known manuscripts: London, Brit. lib. MS Harley 4986 (saec. XII), London, Wellcome 573 (saec. XIII), Lyons 1283 (saec. XV). Riddle 1981, 57, suggested that this Marcellinus was the same Marcellinus *comes* to whom Cassiodorus wrote (*Variae* 1.17.1–2), and that Cassiodorus wrote this letter also: this is refuted convincingly by Zurli 1998 (II), 383. The *AG* text contains a few interpolations from Isidore, such as *balsami arbor* (*Etym.*17.8.14), and also a badly mangled entry for 'iratios' ('braton multi dicitur' etc.) that does not seem to be from Isidore or Pliny.

M and *M2* – **Monte Cassino, Archivio della Badia, cod. V. 97, pp. 466a–474b (=*M*), 545a–552b (=*M2*).** Beneventan minuscule, saec. X in., in two columns, mostly by one hand. Two copies of text, the second in a later hand (saec. XII), but both cease abruptly among the 'C's, *camelleon* (*chamaemelus* #72) in *M*, and *colceconeis* (*colchicon* #80) in *M2*. *M*, 466a: *In nomine sancte Trinitatis. Incipit Alphabeta* – 474b: *Camelleon radicem vastam et nigram et variam spissam sub rufa morsum sub mutilo.* After the abrupt cessation of text, a thirteenth-century Gothic hand completes the entry 'mordens, si masticetur [etc],' followed by a librarian's note, 'In libro minore requiritur, quem sic incipit aeris usti fuit, et est parvus liber.' *M2*, 545a: *Incipit alphabetum et quarundum herbarum nomina quae ex aliqua sui causa resonant habentes nominum explanationem non tamen omnium herbarum etymologyam inuenies, nam pro locis mutantur etiam nomina.* (= Isidore, *Etym.* 17.9.1, also found as the prologue in C). 552b; *colcecones effemerum hanc ipsam erbam uuluum erraticum dicunt effert ...*

Contents:

1) *Sapientia artis medicinae*, [p.] 1a–b.

2) Hippocrates, *Prognostica*, 1b–3a.

3) *Incipiunt indicia ualitudinum Yppogratis*, 3a–4a.

4) *Quomodo visitare debes infirmum*, 4a.

5) *Cura febrentibus*, 4a–b.

6) Vindicanus, *Epistula ad Pentadium*, 4b–6a.

7) Vindicianus, *Gynaecia*, 6a–8a.

8) *Incipit epistula Yppogratis de fleubothomia*, 8a–10a.

9) *De mensura tollendo sanguinem in magnitudine aegritudinis et fortitudo uirtutis*, 10a–10b.

10) (Ps.) Aristotle, *Problemata*, 10b–12b.

11) *De passionibus unde eveniunt*, 12b–13b.

12) Isidore, *Etym.*, 4.1–2, 13b–20b.

13) (Ps.) Hippocrates, *Epistula ad Antiochum regem*, 20b–23b.

14) *Item alia epistula*, 23a–243a.

15) *Item alia epistula*, 24a.

16) *Item alia epistula*, 24a.

17) *Epistula de ratione uentris vel uiscerum*, 24a–26a.

18) *Item alia aepistula de pulsis et uirinis*, 26a–26b.

19) *De pulsibus et urinis*, 26b–33a.

20) Galen, *Ad Glauconem de medendi methodo I–II*, 38a–89.

21) (Ps.) Galen, *Liber Tertius*, 89a–108b.

22) Theodorus Priscianus, *Euporiston II*, 109a.

23) *Liber Aurelii*, 109a–131a.

24) *Liber Esculapi*, 131a–199b.

25) Commentary on Hippocrates, *Aphorisms I–VII*, 199b–282a.

26) Alexander of Tralles, *Theurapeutica. I–III*, 282a–466.

27) (Ps.) Galen *Alphabetum ad Paternum*, 466a–474b.

28) (Ps.) Apuleius *Herbarius*, 477a–522b.

29) *De herbis*, 475b–476a.

30) (Ps.) Dioscorides, *Ex herbis feminis*, 476a–76b, 523a–532b.

The following (532b–545a) are in a twelfth-century Beneventan hand:
31) *De taxone liber* 532b–533a.
32) Sextus Placitus, *Liber medicinae ex animalibus*, 533a–545b.
33) (Ps.) Galen *Alphabetum ad Paternum*, 545a–552b.

Bibliography: Langslow 2006, 45–6, 176–80. Wallis 2000. Collins 2000, 179–80. Skinner 1997, 159. Orofino 1994, 58–72. Beccaria 1956, 297–303. Lowe 1935, 19, 133 n.6.

Item 4 ed. De Renzi, 1852, II:73; item 21 ed. Fischer 2003; item 24 ed. Manzanero Cano 1996; item 25 part ed. De Renzi 1852, I:87, and *Bibl. Cas.* II, 371–8. See also Wallis 2000, 275. Item 30, see Riddle 1980, 131.

The manuscript has received a great deal of attention, not least for its illustrations of Ps. Apuleius, for which see the facsimile edition of Hunger 1935, and the commentaries of Orofino 1994 and Collins 2000, both of whom note a ninth-century date for the manuscript proposed by P. Supino-Martini. Many of the smaller texts were edited in Bibliotheca Casinensis (1873–94) and by De Renzi (1852, passim). Various recipes appear in several different hands in the margins (pp. 8, 139, 475, 486, 494, 522). Lowe ruled out the previously suggested hypothesis that this manuscript was one of two (along with Monte Cassino, Archivio della Badia V.69, saec. IX ex) medical codices prepared by abbot Bertarius, though maintained its probable origin at the abbey. The version of the *AG* in *M2* was not copied from *M*: *M2* contains many interpolations from Isidore's *Etymologiae* (book 17), substituted for the *AG* after #8 (*auripigmentum*), thus omitting a large proportion of the *AG* text proper (#9–30, 32). Instead are entries for: *asarum* (*Etym*.17.9.7), *aquorum* (*acorum*) (17.9.10), *acanthus* (17.9.20), *aconei* (*acone*) (17.9.25), *aloe* (17.9.28), *arnoglossos* (17.9.50), *aristolochia* (17.9.52), *absinthium* (17.9.60), *ancusa* (17.9.69), *althea* (17.9.75), *ambrosia* (17.9.80), *agaricum* (17.9.84), *asfodillum* (*asphodelus*) (17.9.84), *apelos* (17.9.90), *apelos melina* (17.9.91), *alca* (*alga*) (17.9.99), *avena* (17.9.106), *aromata* (17.8.1), *aloa* (17.8.9), *amomum* (17.8.11), *buglossos* (17.9.49), *bupthalmos* (17.9.93), *balsami arbor* (17.8.14). Of these, *M* shares only the interpolations of *ancusa, althea, ambrosia, ampelos, ampelos melina, aloa, balsami arbor*, yet it omits none of the *AG* entries for A. Surprisingly, there are no interpolations among the Cs in either *M* or *M2*. As noted, *M* derives from the same tradition as *VF* (see #22 n.2, and above, ch. 5.A). I record here a debt of gratitude to the late Virginia Brown for lending me her microfilm of the manuscript.

B– Vatican City, Biblioteca Apostolica, Barb. lat. 160, fols. 216–235v.
Benevantan minuscule, saec. XI. *Incipit liber Galieni ad Paterninum trans-*

missus de qualitate herbarum et aromatum, lapidum seu animalicularum.
– (Zimirnium) et ad serpentium morsus facit.

Contents:

1) List of plants (*plantago – madgragora*), [fol.] 1v–2.

2) *Breves herbarum*, 2–6v.

3) (Ps.) Hippocrates, *Epistula ad Maecenatem*, 6v–8.

4) Antonius Musa, *De herba vettonica liber*, 8–10.

5) (Ps.) Apuleius, *Herbarius*, 10–27v.

6) Sextus Placitus, *Liber medicinae ex animalibus*, 27v–38.

7) (Ps.) Dioscorides, *Ex herbis feminis*, 38–48v.

8) Galen, *Ad Glauconem de medendi methodo I–II*, 48v–76.

9) (Ps.) Galen *Liber Tertius*, 76v–88.

10) *Liber Aurelii*, 88–94.

11) *Liber Esculapii*, 94–109.

12) *Liber Galieni de podagra* (= Alex. Trall.), 109–112v.

13) *Antidotum Adrianum maiorem*, 113–135v.

14) *De ponderibus signis quae incognita sunt*, 135v–136.

15) Galen, *Liber de urinis*, 136–138v.

16) *De pulsibus et urinis*, 138v–141.

17) *Dogma Yppochratis*, 141–2.

18) Commentary on Hippocrates *Aphorisms I–VII*, 143–198v.

19) Oribasius, *Conspectus ad Eustathium filium I, III, IV*, 199–216.

20) (Ps.) Galen, *Alphabetum ad Paternum*, 216v–235.

21) Theodorus Priscianus, *Euporisticon I–III*, 236–265v.

22) Quintus Serenus, *Liber medicinalis*, 266a–274va.

23) (Ps.) Hippocrates, *Epistula ad Antiochum regem*, 274va–275va.

24) Vindicianus, *Epistula ad Pentadium*, 275va–276.

25) *Dies egyptiaci*, 276a.

26) Recipes, 276b–281vb.

25) *Sapientiae artis medicinae*, 282–282v.

26) Isidore, *Etym. IV*, 282v–286.

27) (Ps.) Galen, *Liber de dinamidiis*, 286–286v.

28) (*Epistula*) *Frusta mortalium genus*, 286v.

29) Hippocrates, *Prognostica*, 286v–287v.

30) (Ps.) Hippocrates, *Indicia valitudinum*, 287v–288.

31) (Ps.) Hippocrates, *Quomodo visitare debeas infirmum*, 288.

32) Vindicianus, *Epistula ad Pentadium*, 288–288v.

33) Vindicianus, *Gynaecia*, 288v–289v.

Marginal additions: Recipes 142, 265v, magic prescription 142, extracts 276a–b.

Bibliography: Langslow 2006, 98–9. Collins 2000, 192. Newton 1999, 389–90, and plate 208 (= fol. 8v). Maggiulli and Buffa Giolito 1996, 112–14. Beccaria 1956, 324–31. Leisinger 1925, 5. Lowe 1914, 19, 152, 365.

Item 9 ed. Fischer 2003; item 11 ed. Manzanero Cano 1996.

Lowe 1914, 19, labelled its Beneventan script as 'Bari-type'; Newton 1999, 245–7, suggests this manuscript was one of the 'cows' (gifts) that Atto of Chieti sent to Alfanus at Monte Cassino, hence dates it to around 1060 (Atto died 1071): he locates its origins in the Abruzzi. Collins notes Carlo Bertelli's proposal that the abbot Theobald had this codex copied at San Liberatore della Maiella (Serramonacesca, Pescara). The manuscript contains a few illustrations accompanying *De herba vettonica* and the first few chapters of Ps. Apuleius. The text of the *AG* is very close to that of *P*, without sharing *P*'s omissions, and has some similarities with *C*. The text contains the addition of an entry for *fillira*: see under *W*.

C – Copenhagen, Det Kongelige Bibliotek, Gamle Kgl Samling cod. 1653. 4., fols. 31v–60. Beneventan minuscule, saec. XI. *Incipit Alphabetum. Haec sunt, frater karissime Paterniane – (Zimirnium) et ad serpentium morsus facit. Explicit liber galieni.*

Contents:

1) Muscio, *Gynaecia*, [fol.] 3–28v.

2) Cleopatra, *Gynaecia*, 28v–31v.

3) (Ps.) Galen, *Alphabetum ad Paternum*, 31v–60.

4) (Ps.) Galen, *Liber de dynamidiis*, 60–60v.

5) *Liber dietarum diversorum*, 61–66v.

6) *Dieta Ypocratis*, 66v–67.

7) *Dieta Theodori*, 67–71v.

8) *Liber artis medicinae*, 72–75v.

9) *Docma Ypocratis*, 76.

10) *In qualis luna debet homo sanguinem minuere*, 76v.

11) Oribasius, *Conspectus ad Eustathium filium I–V*, 77–147v.

12) Recipes, *Ungentum ad padagram, Unguentum albam ad plagam*, 109.

13) *De succedaneis liber*, 148–9.

14) Recipe collection/antidotes, 149–181v.

15) *De ponderibus, signis, quae incognita sunt*, 181v–182.

16) *De ponderibus et mensuris*, 182–183v.

17) *Epitomum Ypocratis de infirmis*, 183v.

18) Galen, *Prognostica*, 183v.

19) *Conservatio flevothomie et dierum canicularum*, 183v–184.

20) *Dies egyptica, quod per totum annum observari debent*, 184.
21) Alphabetical index of medical recipes, 184–5.
22) Recipe collection, 185–188v.
23) *Medicamentarium, quod continet dicta Uribasi doctoris per alfabeta*, 189–215v.

Bibliography: Langslow 2006, 93–4. Fortuna and Raia 2006, 9, 19. Newton 1999 passim. Fischer 1998, 278–9. Beccaria 1956, 120–4. MacKinney 1935–6, 408, and 1952a, 8. Jorgensen 1926, 426–8. Lowe 1914, 19, 338. Rose 1864–70, II:107.

Item 11, bk 5 is the *Liber Byzantiii*, Fischer 1998. Items 8–10 ed. Laux, 1930, 419, 430–2, 432.

The first item, the *Gynaecia* (3–28v), contains interesting illuminations (15 all up) demonstrating the various positions of the foetus in the womb: see Hanson and Green 1994. The manuscript stayed in Italy for several centuries, as revealed by the early Italian recipes, fol. 220; the notes in later fifteenth-century hands on the same folio reveal that by then it was in England, and one entry reveals its possessor to be in Cambridge ('Anno a mundi creacione 4815 Universitas Cantebrigia fuit a Cantebro edificata e a philosophis et scientarum inventoribus multepliciter decorata'). Rose believed the manuscript was used by Conrad Gesner of Zurich in the sixteenth century, but its passage to Denmark is unknown.

C uses the *AG*'s epilogue as a prologue, and like *M2* attaches to it the words of Isidore *Etym.* 17.9.1. *C* also contains many interpolations from Isidore *Etym:* costum (17.9.4), cyperus (17.9.8), c[h]elidonia (17.9.36), camellus (chamaemelos) (17.9.46), citocacia (17.9.35), capillus ueneris (17.9.67), chamaeleon (17.9.70), cicuta (17.9.71), colocasia (17.9.82), centicularis (genicularis) (17.9.83), camepiatus (chamaepitys) (17.9.86), ciclaminos (cyclaminos) (17.9.89), carex (17.9.102), euphorbium (17.9.26), elleborum (17.9.29), [h]eliotropium (17.9.37), epithymum (17.9.13), [h]edera (17.9.22), erpillus (herpyllos) (17.9.51), eriginon (erigeron) (17.9.53), fu (phu) (17.9.7), flominos (phlomos) (17.9.73), flaminos (phlomos) (19.9.94), ferula (17.9.95), fucus (17.9.98), ferum (faenum) (17.9.107), filix (17.9.105) (followed by unidentified recipe 'unguentem ad tumorem seu in quo libero membro ...' etc., then a mangled version of *AG* #76), galbanum (17.9.29), glycyriza (17.9.34), gentiana (17.9.42), gerabone (hierobotane) (17.9.55), gladiolus (17.9.83), gramen (17.9.104), iacantus (hyacinthus) (17.9.15), isopus (hyssopus) (17.9.39), nardus (nardum celticum) (17.9.3). It also interpolates 7.9.26 into *AG* #94. Note the interesting coda to #202.

P – **Paris, Bibliothèque nationale lat. 6837, fols. 1–30v.** Pregothic minuscule, saec. XI/XII. *Incipit praefacio libri Galieni ad Paternianum transmissi de qualitate herbarum aromatum seu animalium – Incipit prologus – Incipit liber.* (No expl.).

Contents:

1) (Ps.) Galen, *Librum ad Paternianum* (= *Alphabetum Galieni*), [fol.] 1–30v.

2) (Ps.) Galen, *Liber de dynamidiis*, 30v–32 .

3) Recipes for specific *emplastra*, 32–37v.

4) (Ps.) Hippocrates, *Epistula fleobotomia*, 38–38v.

5) (Ps.) Hippocrates, *Epistula Ypogratis de incisione fleobotomie*, 39–39v.

6) Apollo, *Epistula de incisione*, 39v–40v.

7) Vindicianus, *Epistula ad Pentadium Gadum nepotem suum*, 40v–42.

8) *Medicinalis sententia de temporibus phlebotomiae*, ('Cum vero colera...') 42.

9) (Ps.) Hippocrates, *Epistula ad Antiochum regem*, 42–4.

10) Various recipes, 44v–45.

11) *Galieni de ephemeris febribus libri tres priores et initium quarti*, 45–110v.

Bibliography: Chandelier, Moulinier-Brogi, and Nicoud 2006, 70, 95, 118. Sconocchia 1998, 123. Halleux-Opsomer 1982a. Kibre 1968, 84. Wickersheimer 1966, 45. Stadler 1902. *Catalogus codicum manuscriptorum Bibliotecae Regiae*. pub. 1774. Paris: Imprimerie royale. Pars tertia, tomus quartus, p. 283.

The manuscript has generally been dated later (Stadler saec. XIII/XIV, Wickersheimer saec. XIII, *Catalogus* [1774] saec. XIV), but as noted by Halleux-Opsomer 1982a, 66n, the manuscript is a composite of different hands from different periods: she suggests an eleventh-century date for the text of the *AG*; I would extend its range into the early twelfth century.

There seems to be no more recent description than the catalogue of 1774 (available on the Bibliothèque nationale website), which states that the manuscript was 'olim Philiberti de la Mare,' and a modern hand wrote 'Dela Mare 511 Reg. 6349' in the top margin of the first folio: this refers to Philibert de la Mare (1615–87), counsellor to the parliament of Burgundy, from whose library the Bibliothèque nationale preserves other scientific and medical manuscripts. The provenance may point to an origin: the script of the *AG* suggests France. The folio numbering from a later hand is behind by one (i.e., '23' is really fol. 24.). My sincere thanks to Klaus-Dietrich Fischer for loaning his microfilm of the manuscript, and to Eliza Glaze and Monica Green for discussions about its date.

The text of the *AG* in *P* is well presented in terms of orthography and grammar, and belongs to the same tradition as *B*, which it parallels verbatim throughout, though the many exceptions show that it was not a direct copy. In particular the scribe mostly omitted the ubiquitous 'omnibus nota' that appears at the start of entries (on which see ch. 4.H). In places the scribe was reluctant to take a chance on Greek or extremely unfamiliar words and phrases, preferring in many cases simply to omit these passages. The marginal notations show active and curious readers at work, though are occasionally unrelated to the text, such as the quotation from John 2.10, 'Omnis homo primum uinum bonum ponit super mensam' (top margin, fol. 24v, [sig. 23v]). The manuscript is defective at #196–206, and although the folio numbering (here 18v–19, though really 19v–20) suggests no awareness of the lacuna, marginal notes in a later hand on these folios indicate a reader's awareness that some plants beginning with O and N were missing. As with *B* and *W*, *P* also contains an entry for *fillira*: see under *W*.

W – Vienna, Österreichische Nationalbibliothek 2425, fols. 125v–137v.
Minuscule, saec. XI. *Alphabetum incipit* – (No expl.).

Contents:
1) Gariopontus, *Passionarius*, [fol.] 1a–123b.
2) (Ps.) Galen, *Alphabetum*, 125v–137v.
3) *De medicina simplicum*, (Inc. 'Adiantus Sparagus'), 137b–148b.
4) *Prognostica de urinis*, 148b–151a.
5) *De pulsibus*, 151a.
6) *Dietarum liber diversorum medicorum* (def.), 151a–155b.
7) *Dieta Galieni de febribus*, 155v–156.
8) *Dieta qua quisque observet anni circulo,* 156v–157.
9) *Diaeta Theodori*, 157–160.

Bibliography: Glaze 2008, 190. Halleux-Opsomer 1982a, 66. Reiche 1973, 120–3. Sudhoff 1915, 379–80. Rose 1864–70, II.178. Catalogue: *Tabulae codicum manuscriptorum praeter Graecos et Orientales in Bibliotheca Palatina Vindobonensi asservatorum*, II (Vienna, 1868, repr. Graz, 1965), 71 no. 2425. My thanks to Ernst Gamillscheg, of the Österreichischen Nationalbibliothek, who kindly arranged a microfilm.

As with *P*, the first few folios of *W* were replaced in a thirteenth- or fourteenth-century hand, causing confusion for dating the manuscript, hence its exclusion by Beccaria and others. Rose, however, dated it to the eleventh century, and noted it as the earliest known copy of Gariopontus. Florence Eliza Glaze (personal communication) dates this manuscript to saec. XI ex.–XII in., and locates its origin in southern Italy (as with Zurich, Zentralbibliothek, MS C 128, another version of Gariopontus). The covers contain notes by Ulric Gräwl, dated 1382, and 'Statuta Colomanni abbatis in S. Cruce, edita in visitatione in Paumgarten a. d. 1371.'

The text of the *AG* in *W* differs considerably from the other manuscripts in that it has been substantially rewritten by the omission of much of the description, any discussion of different types, preparations, methods of adulteration, and so on, in favour of a simple listing of name, type of substance (plant, mineral, animal), the part used (leaves, bones, etc.), properties (*uires*), and one or two therapeutic uses, though these are often omitted also. There are some exceptions, such as the entry for French lavender (#258), which follows verbatim the complete entry found in in *V*, *L*, and *J* (likewise #261). It contains some idiosyncratic additions. In place of line 2 in the entry for soap (#259) ('Quod optimum iudicamus germanicum ... deinde gallicum,' etc.), *W* reads 'alii capitellum alii cursiam gr[ec] i u[er]o putren appellant.' (Michael McVaugh informs me that Guy de Chauliac [fourteenth century] refers to *capitellum* as a type of soap.) For the rose (#232), one of the shortest and most useless of *AG* entries, *W* adds a few recipes with additional ingredients for treating fevers and chills. For spurge (#280), the scribe abbreviates the entry by referring to the library to which the *AG* belongs: 'Titimallum omnibus nota est et species cuius sunt septem quorum nota in herbarii libro repperiuntur. sunt omnium uirtus una esse dignoscitur prouocatur enim uentrem superius riferit.' The *herbarius* in question is undoubtedly Dioscorides, who specifically enumerates and describes seven types of spurge (see ch. 3.D). In this case, as with #281 (*thapsia*) immediately after, the entry is so abbreviated that it is more convenient to include it in full in the apparatus rather than key it into variants. The scribe consistently wrote *acredine/acradine* for *acrimonia*, a phenomenon found in some late antique medical Latin (Palladius, Cassius Felix, Theodorus Priscianus).

As with *M*, *M2*, and *C*, *W* also contains some interpolations from Isidore, concentrated around E–F: *[h]eliotropium* (17.9.37), *elleborum* (17.9.29), *ferula* (17.9.99), *fungi* (17.10.18), *folium* (19.9.2), *galbanum* (17.9.29), *[h] edera* (17.9.22). On other occasions one can detect interpolations either from Pliny or the Latin Dioscorides, such as the note about the beaver's self-castration upon capture ('capituret dum queritur et persequitur ut sibi auferantur testiculi,' #50).

Others defied attempts at identification (e.g., that for *panax* #220). One unidentified entry, which as noted above appears in *B* and *P* also, is for *fillira*:

> Fillira folia stiptica sunt similia oleastri necessario et stipticis rebus admiscentur. Commasticata uero oris uulnera limpidant. Elixata et gargaridiata eundem effectum prestant. Elixatura autem potata urinam prouocant. (fol. 129v)

The version in *B* and *P* differs slightly:

> [Fillira- ... -rebus] admiscitur et maxime vulnera limpidatoris masticatum quoque idem facere nouit. Folia elixantur et gargarisolata similiter praestat effectum et potata elixatura urinam prouocat.

W also contains another unidentified entry that resembles the *AG*'s style, hence could be authentic:

> Eruscus quod est rubus omnibus notus est. Uires habet stypticas. Huius folia tunsa simulque cum suco suo imposita rosulas quae ex rubea colera fiunt aufert et formicationes sedat. (fol. 129)

D. The *Editio princeps*

Galieni Opera (1490) ed. Diomedes Bonardus, Venice: Ph. Pincius, vol. II, [fols. 86–95v], x2 cols. Title, 'Liber Galieni de simplicibus medicinis ad Paternianum.' See Klebs 1938, 4321; Durling 1961, 279, AI. An extremely rare edition, but a digital facsimile is currently accessible on the website for the Bibliothèque interuniversitaire de médicine: http://www.bium. univ-paris5.fr/histmed. This version was reprinted in subsequent editions of Galen by the house of Junta, beginning with that of 1522, *Opera omnia impressio novissima*, ed. Scipio Ferrarius (1522) Venice: L.A. Junta, cc.216va–225vb. See Panzer 1793–1803,VIII:474; Durling 1961, 280, A6. I have used the sixth edition, *Galeni Opera ex Sexta Iuntarum editione*, ed. H. Costaeus (Venice, 1586), fols. 79–92v, where it appears in the tome entitled 'Spuri Galeno ascripti libri qui variam artis medicae farraginem ex variis auctoribus excerptam continentes, optimo, quo fieri potuit ordine sunt depositi et in unum corpus redacti.' Durling 1961, 280, A17. In the following century the text was incorporated (unaltered) into R. Charterius's edition of Galenic and Hippocratic works, *Hippocratis Coi, et Claudi Galeni opera. Renatus Charterius plurima interpretatus, universa emendavit, instauravit notavit, auxit et conjunctum graece et latine primus edidit*, 13.

tom. in 10 vol. (Paris, 1679). Kühn 1821 (K.1, lvi) dismissed it as one of the many *spuria* to be excluded from his edition of Galen's oeuvre.

None of these editions provide any information on the manuscripts used. Here and there the use of alternative titles (such as 'asarum, id est bacchare' #29, 'albucus qui dicitur asphodelus' #33, 'de gramine quae dicitur agrostis' #36, 'armenium uel minium' #37, 'oenanthe id est flos lambruscae' #141), or changes in the text (*squama* for *lepis* #1, cf. tit. #145), demonstrate interpolations from Pliny, or elsewhere the Latin Dioscorides, Isidore, or authors not identifiable (e.g., *tritapta* for *gigarta* in #3), which do not appear in the surviving manuscripts, but we cannot know whether these interpolations were by the editor(s) or were found in other manuscripts now lost. Confusion also led to some odd titles, such as *hasceg* for *indicum* #135, and *hustolex* for *ios scolex* #136.

ALPHABETUM GALIENI (Latin Text) /
THE ALPHABET OF GALEN (English Translation)

Sigla

ᴣ

V	Vatican City, Biblioteca Apostolica, Pal. lat. 187 (saec. VII–VIII)
F	Chartres, Bibliothèque Municipale 62 (115) (saec. IX–X)
L	Lucca, Biblioteca Governativa cod. 296 (saec. IXex–X)
M and *M2*	Monte Cassino, Archivio della Badia, cod. V. 97 (saec. X)
B	Vatican City, Biblioteca Apostolica, Barb. lat. 160 (saec. X–XI)
C	Copenhagen, Det Kongelige Bibliotek, Gamle Kgl Samling cod. 1653. 4 (saec. XI)
P	Paris, Bibliothèque nationale lat. 6837 (saec. XI–XII)
W	Vienna, Österreichische Nationalbibliothek 2425 (saec. XI–XII)
J	*Editio princeps.* 1490, repr. 1522.

Incipit Alphabetum Galieni

Praefatio

Cum mihi proposuisses, carissime Paterniane, omnia smigmata tam metallica quam et aromatica et omnem herbarum nationem describere, optimum duxi, quia doctissimum et peritissimum esse te probaui, ideo tibi exemplum uoluntatis et ingenii mei offerre puto. Igitur, frater, sol-
5 licite atque exquisitissime tota tua peritia nostram hanc scripturam ne calumniareris prospexi. Et ne quis aut herbulam aut aliqua aliam spe- ciem siue aromaticam, siue metallicam requirens diutius erret et totum uolumen euoluat, quapropter ordine primarum litterarum, id est A.B., faciam omnia smigmata nominare quae exemplatio tradit et in usum
10 medicinae cadunt. Incipiamus igitur ab A et sic ad extremam litteram perueniamus.

1 proposuisses] proposuissem *JP* carissime] karissime *P* smigmata] sinigmata *J* 2 her- barum nationem] herbationem *J* 3 doctissimum] studiosissimum *J* te] *om. B* ideo] itaque accipe *J* mei] *om. J* 4 offerre puto] offero *J* 5 atque] *om. J* exquisitissime] exquirire *J* nos- tram] *om. J* ne] ne autem *J* 6 calumniareris] calumniares *BP* quis] ne quis dum *J* herbulam] herbam *P* aliqua] aliquam *JP* aliam] *om. P* 7 requirens] requirit *J* erret] erraret *J* 8 euoluat] euolueret *J* quapropter] *om. J* ordine *P* id est A.B.] A.B.C.D *J* 9 faciam] agam *J* smigmata] sinigmata *J* exemplatio tradit] in simplario tractantur *J* 10 incipiamus] incipimus *B* incip- iemus *J* et sic] et sic inde *B* et sic deinceps *J* 11 perueniamus] peruenimus *B* perueniemus *J*

1) Aes iustum

Aes ustum [quod aliqui chaumenum uocant] sit maxime de clauis cupreis [et] uetustis qui in olla fictili cruda missi in furno incenduntur et aspersi sulfure uel sale, uel utroque, uel alumina et coquuntur. Pro- bamus aes quod est optime ustum utpote rubricae [in cute] colorem

Here begins the Alphabet of Galen

ॐ

Prologue[1]

When you asked me, dearest Paternianus, to describe for you every medicine[2] derived from minerals as well as aromatics[3] and from every species of plant,[3] I thought it an excellent idea. And because I have found you to be most learned and skilled in these matters, I propose to present to you this effort of my will and talents. Therefore, brother, I have anxiously and most searchingly edited this tract of mine, lest, with your complete expertise, you should find fault with it. And if anyone requires information about a particular shrub or some spice or mineral, they need not waste time rummaging through and unfurling the entire roll, for I have arranged in alphabetical order (that is 'A,' 'B,' and so on) all of the medicines which the original copy[4] transmits and which are used in the practice of medicine. Let us therefore commence with 'A' and proceed until we arrive at the final letter.

1 Some vocabulary in the prologue (*prefatio*) suggests that it and the epilogue were late antique additions to the text: see ch. 1.D.
2 *smigmata/smegmata*, usually meaning 'unguents'/'cleansing lotions,' as used below #213. The term is odd here, and may reflect pretensions or simply corruption: see ch. 1.D.
3 On *aromatica* and *natio*, see ch. 1.D.
4 *exemplatio*, see ch. 1.D.

1) Burned copper

Burned copper [which some call *chaumenum*][1] is made mainly from old copper keys, which are placed in an unfired earthenware vessel and heated in a furnace and sprinkled with sulphur or salt or both, or with alum,[2] and then baked. We recommend that the copper is best when burned until it

5 faciet ad ipsa capitula in se quod enim fuscius est combustum. Est ergo
uirtus et efficacia aeris usti similis lepidi, sed tamen fortior est.

1 sit] fit *LC* quod- uocant] *om. VFJLC* chaumenum] chaumenon *BW* caleute caumenum
P 2 cruda] rudi *C* 3 aspersi] sparso *VF* uel- coquuntur] *om. LC* uel sale] ut salent *L* uel
utroque] ueluti *P* probamus] bonum autem aes *J* probamus- combustum] *om.
W* 4 aes] certe *P* utpote] in puce *L* utpote si sectem in cute rubricem *V* strictum in cute *F* et
frictum quam rubrum calore *P* 5 ipsa- in se] *om. C* in se] intera *P* fuscius] fruitius *J*
suscius *L* combustum] combustum dicimus *LFBC* cognitium dicimus *P* ergo] aereo *L*
om. P 6 usti similis] *om. C* utilis lepidum *P* lepidi] squama *J* lippidus *V* laepide *LBC*
lepidus *M* lepidas *W* lepidos *F* sed- est] id est acaciae *L* sicut acacia *BP* id est acere *V*.
plerumque def. M

cf. Diosc. V.76 (κεκαυμένος χαλκός). Pliny 34.107–10. Galen *Simp.* 6.9.36 (K.12, 242).
Isidore 16.20.3

2) Acacia

Acacia fructus est arboris spinosae quae maxime in Aegypto nascitur.
Hic igitur fructus uelut oliua exprimitur et succus illius in sole siccatur
[et in glebam redigitur]. Erit ergo optima acacia quae est penitus laeuis,
leuiter subrubra, suauitate odoris sui subacidas transit et gustu ualde
5 styptica. Haec enim maxime de immaturo fructu premitur, nam nigra
de maturo fructu premitur aut coctura est imbecillior. Omnis autem
acacia uiscide stringere potest.

1 Acacia] agacia *F* acatia *BP* est- igitur] *om. LBC* spinosae] pruni *P* quae- nascitur] *om.
W* 2 igitur] legitur *L* et succus-styptica] *om. W* et - redigitur] *om. VMF* 3 optima] omnia
B penitus] pondere *B* punctus *P* 4 leuiter] *om. VM* subrubra] subrugens *B* subrigens *P*
transit] trahit *CP om. VMF* 5 immaturo] limiaturo *C* maturo *P* fructu] *om. W* premitur]
om. LB nigra] nigra illa ut de immaturo fructu fiat ut coctura *F* 6 aut- imbecillior] *om.
LBCJ* aut conquenda est interciorum *P* omnis-potest] id est sine uirtute *B om. P*

cf. Diosc. I.101 (ἀκακία). Pliny 24.109. Galen *Simp.* 6.1.12. (K.11, 816–17). Isidore
17.7.29.

3) Aerugo

Aerugo circa aliquae metalla in quibusdam pelagis lente et ueluti stil-
licidium manare perhibetur. Verum et conficitur hac ratione. Granula
interiora uuae quasi sunt quaeque graece gigarta dicimus, ualde sub-
struuntur in uase aperto et asperguntur muria et aceto. Deinde super-
5 ponitur lamina aerea et lata et iterum operitur eisdem gigarta similiter
aspersis et lamina alia superponitur et hoc ulterius sit quamdiu uas
capit, deinde oppilatur diligenter ne aliquod habeat spiramentum. Et
sic die septimo aperitur et prolatis laminis aeruginem quam traxerint

is a reddish colour [on the surface] which divides into small sections, and then turns dark when it is fully fired. The effect and efficacy of burned copper is similar to the flake of copper ore[3] but it is more powerful.[4]

1 *chaumen-on, -um*: absent in our earliest manuscripts, it seems to be a neologism from καῦμα, 'burning heat,' 'fever heat': see ch. 4.F n.56.
2 #7.
3 #145. *squama J* is probably an interpolation from Pliny 34.107: see ch. 5.D.
4 'similar to acacia' *LBP*, 'in that it is bitter' *V*.

2) Shittah tree / Acacia (*Acacia* sp. L.)

Acacia is the fruit of a spinous tree which grows abundantly in Egypt. The fruit is pressed just as olives are,[1] and its juice is dried in the sun until it is reduced to a gel. The best acacia is smooth inside and lightly red, and when the sweetness of its fragrance overwhelms its slight acidity, and when it is strongly styptic in taste. It is best when pressed from the unripe fruit, for that which is black and made from the mature fruit or that which is cooked is weaker. All acacia, however, can be used as a powerful astringent.

1 The comparison with pressing olives is unique: cf. Diosc. I.101, Pliny 24.109. Halleux-Opsomer 1982a, 83–4, notes that in the later antique world the word *acacia* began to be used for the fruit or sloe of blackthorn (*Prunus spinosa* L.), which was pressed like olives.

3) Verdigris

Verdigris[1] forms in mines on some sea-coasts where it falls in slow drips for collection. It is also made in the following way: take the interior seeds of grapes, which we call in Greek *gigarta*,[2] and spread them out well in an open container and sprinkle them with brine[3] and vinegar and then place on top of them a flat, copper plate, and just as before sprinkle the granules all over the copper plate, then place another copper plate on top of this, and repeat this process until the container is full, then seal it carefully so that no vapour escapes. Leave it for seven days then open and

leuiter abraditur et rursus intinguntur. Et hoc sit usque quo totae con-
10 sumantur. Alii super acetum acerrimum has ipsas laminas suspendunt
et diebus decem relinquunt, deinde perferunt et aeruginem quantam
inueniunt abradunt. Alii massulas aeris ligant et eis quasi scobis liga-
menta decidunt in acetoque mittunt et quotidie bis aut ter commouent
usque quo aerugo euadat. Est ergo optima aerugo quae est leuis et tactu
15 nullam asperitudinem habet. Vires autem habet quasi aeris usti uel lepi-
dis et maxime acutiores.

1 Aerugo] aerugo dicitur *VLF* aerugo quod alios spondium dicunt *B* aerugo quod alii ios
alii spondium dicunt *P* aerugo quam greci uocant indice *MC* aerugo quod greci uocant
iuspleis *F* arugo quae agricisior dicitur *W* aerugo uel uitreoleus *tit. J* metalla] et ulla *F*
exalbida *BP* pelagis] puru *B* purulenta *P* spleis *L* era *M om. F* stillicidium] fructum *L* **2**
perhibetur] *om. LMCF* emanat *P* Verum - conficitur] *om. B* **3** quasi-quaeque] *om. C*
gigarta] tritapta *J* agigardas *V* gegasta *F* grerata *L* siksarta *B* ΓΗΓΑΡΤΑ *P* gicata *C* gigarta
M giganta *W* dicimus] nuncupant *MCF* ualde] *om. BPC* substruuntur] substernuntur *P*
4 muria] miua *J* moria *V* murra *LB* mirra *P* myrra *W* sale marina *F* sale moria *M* salem
C superponitur] subponitur *FPC* **5** et lata] dilatiata *BP* gigarta] tritaptis *J* gigantus *LB*
gegartis *F* gicaruis *C* gigartis *P* gicartas *M* **6** superponitur] super lamina alia ponatur *F*
ulterius] alterus *LC* alterius *F* **7** oppilatur] operiatur *F* **8** traxerint] contraxerint *L* **10**
acerrimum] acris sunt *F* acrum *W* **11** decem] undecem *B* XI *P* alii- 15 habet] alii limatu-
ram aeris cum acre aceto in aperium uas mittunt et canicularibus diebus in sole ponunt
atque agitare non desinunt donec arugine deponat *W* **12** massulas] maculas *P* ligamenta]
laminata *B* **13** bis aut ter] pisantes *P* **15** aeris] et eaque similiter menta decidunt in aceto *F*
quasi] quae *B* quasi-decidunt] *om. C* **16** acutiores] acutiores uidentur *P*

Aliquanto discrepat textus W sed interpretatio similis.

cf. Diosc.V.79 (ἰὸς ξυστός). Celsus 5.2.3. Pliny 34.110–14. Isidore 16.20.14; 16.21.5

4) Ammoniacum

Ammoniacum est quasi lachryma herbae quae graece nartex, id est
agasylis, appellatur, similis allio sed uiscosior et foliis ueluti pinguio-
ribus quae contrita sunt glutinosa quasi cummen. Nascitur autem in
terris Ammonis quondam regis in ultima Cyrene. Colligitur autem
5 ammonicacum sic. Maximis aestatibus huius herbae radix et thyrsus
inciditur et sic lachrymam fundit et postae in terra colligitur unde et
sordem capit. Verum est optimum ammoniacum quod est mundissimum et recentisissimum et colore candidum et quasi pingue ut thus et
si frangatur spissum et splendidum apparet, odore subsimile castoreo.
Potest igitur efficaciter remollire et relaxare.

1 Ammoniacum] amoniacum *LBP* quasi] *om. M* nartex] aricos *MW om. P* graece nartex]
om. F **2** agasylis] y aus *V* assiaos *F* acyllaic *L* ascilouos *B* asilouis *P* αγυλως *C om. MW*
ferula *J* allio] alium *V* altit id est assae *J* alios *F* aleo *B* ale *P* allio *CW* alio *M* uiscosior]
uiscidiorem *V* uiscosioris *F* uisilioris *L* uiscidior *MC* uiscosior est gustu *B* uiscidior est

take out the copper plates and gently scrape off the verdigris attached to them and then place them back in the container. Continue to do this until the plates are entirely worn away. Some hang the same type of copper plate over extremely sour vinegar and leave for ten days, then at the end of this period they scrape off all the verdigris found upon the plate. Others bind together little lumps of copper in a bandage and then scrape the bandage for the dust left upon it and place this in vinegar, and do this two or three times a day until it ceases to yield verdigris. The best verdigris is that which is smooth to the touch and has no roughness. It has the same properties as burned copper or flake of copper ore,[4] though they are much more severe.

1 Various alternative names are given in the MSS: *spondium BP*, *ios P*, *indice MC*, *iuspleis F*. *Vitreolus*, given as an alternative name in *J* though not attested in any manuscript, is a diminutive of *vitrum* (glass) and designates the sulphate crystals from brass and iron, which have the appearance of glass shards. It is first attested elsewhere in the recipes in MS Lucca 490, ed. Hedfors 1932, 17 (G 20–3), and in those in St Gall and Bamberg (saec. IX), ed. Jörriman 1925, 1. 74. The text throughout betrays considerable corruption, and the reconstruction here is tentative.

2 γίγαρτον can mean grape-seed or olive-pit (or products from such), as can the term στέμφυλα used by Diosc. V.79. The only other attestations in Latin of *gigarton* are Palladius 12.20.4 ('uuae excrementa') and Cassiodorus *In psalmos* 8, *praef.* ('pressae uuae durissimis gigartis'). The reading of *J* (*tritapta*) is obscure.

3 *muria*, despite the corrupted readings of the manuscripts (*moria, myrra, miua, sale marina*, etc.).

4 #1, 145.

4) Gum ammoniac (*Ferula marmarica* L.)

Gum ammoniac is the resin of the plant which the Greeks call *nartex*, that is, *agasylis*,[1] which is similar to garlic, but is more sticky and has fatter leaves which when crushed are glutinous like gum.[2] It grows in the lands of Ammon, who was once king in farthest Cyrene.[3] Ammoniac is collected in the following way: during the hottest period of summer the root and stem of the herb are cut and the resin flows out, which is later picked up from the ground and the dirt removed. The best ammoniac is that which is freshest, extremely pure and bright in colour, greasy like frankincense,[4] and when dried and crumbled it seems to shine and smells a little like castor.[5] It can therefore thoroughly soften and loosen.

1 The manuscripts differ considerably for this opening sentence, tentatively reconstructed here: only *J* qualifies *narthex*, giant fennel (*Ferula communis* L., or possibly *Ferula marmarica* L.), with the Latin gloss 'id est ferula,' though *ferula* appears elsewhere in the text (#55, 64, 119, 147, 281, and in 201 as both the giant fennel and simply as 'stalk,' 'stem'),

gustu *P* et foliis] *om.* *BPC* **3** contrita] est strecata et *B* stricta *P* glutinosa- cummen] tam glutinosa quam gumen *VLF* tam glutinosum sit quod ex ea congritur quasi gummi *W* quasi gluten *J* ut gumma *P* **5** aestatibus] aestiuis temporibus *P* aestatibus temporibus *C* huius] ipsius *LB* thyrsus] obtusus *F om.* *W* **6** fundit et] *om.* *BP* postea-colligitur] *om.* *MW* **7** sordem] fordem *JCB* **8** pingue] porro *B* pingueset opuatus *V* ut thus] prout *F* et per uetus *MW* porro *P om.* *B* **9** splendidum] silendidum *L* splendidum hodorem *V.*

cf. Diosc. III.84 (ἀμμωνιακόν). Pliny 12.107, 13.123. Galen *Simp.* 6.1.37 (K.11, 828).

5) Aloe

Aloe succus est herbae quae est similis scillae, cuius radice incisa hic ipse promanat succus et colligitur, uel tota herba contusa, premitur et siccatur. Aliqui enim dicunt de petris eam colligi in Iudaea et Asia atque in India. Verum optimam iudicamus quae est fragilis, gleba mun-

5   da, pinguis, splendida, subrufa et dum aliter facile exhumidatur odore subuiscosa et gustu amarissima. Effectum autem duplicem habet. Si enim bibatur et deglutiatur uentrem soluit et purgat putredinem et choleram et phlegma deducit. Cum oxymelite datum stomachicis et hepaticis magnum adiutorium est. Alii post coenam bina uel trina gra-

10 na sicut ciceris magnitudine deuorant quam corpus procurat et escas non corrumpit. Et si sitis fuerit extinguitur quoniam triueris cum succo caulium et feceris ad ciceris magnitudinem. Alii miscent colocynthidos interiorem et scamoniam et dant ad typos uel ad uarias infirmitates. Si uero corpus extrinsecus illiniatur efficaciter stringit.

1 Aloe] alues *V* aloen *L* scillae] squillae *P* incisa] ac pressa *BP* hic] *om.* *M* **2** ipse] *om.* *B* promanat] emanat *BP* et lata colligitur *B* ita colligitur *P* uel] ut *BC* premitur] expremitur *FBP* **3** colligi] collectam *J* colligitur *FP* coligere *B* in iudea- atque] *om.* *W* in iudea et asiam quam in indiam *V* in iudeam et in asiam qua india *F* in iudea et asia et india *B* tam in iudea et asia quam in india *LC* [*def*] in india quam in asia *M* **4** Verum- quae] quam optimum uero est quod in india repertus et *W* iudicamus] indicamus *F* **5** pinguis] *om.* *BP* splendida] scilendida et *L* odore] odore pene *LBP* odore penitus *V* exhumidatur] exeximidatur *F* **8** deducit] educit *MP* deponit *W* cum] in *FLBC* datum] dragum *L* **9** est] perstat *W* **10** sicut] *om.* *B* deuorant] languoribus dant *W* quam] quomodo *F* **11** non] *om.* *P* sitis fuerit] *om.* *B* si sitis- magnitudinem] et sitim extinguitur *P* **8** triueris] conteris *FLC* conteritus *W* cum- magnitudinem] *om.* *B* **12** caulium] culiculi *L* **14** illiniatur] inlinitur *B* stringit] destringit *P*

cf. Diosc. III.22 (ἀλόη). Pliny 27.14–20. Galen *Simp.* 6.1.23 (K.11, 821). Isidore 17.8.9, 17.9.28.

despite not receiving an entry of its own, as in Diosc. III.77 (νάρθηξ, *Ferula communis* L.: cf. Pliny 13.123). Diosc. III.84 (ἀμμωνιακόν) is the only author to mention that the whole plant and its root is called 'agasylis,' which is probably the source here of the manuscript variants (*y aus, assiaos, acyllaic, ascilouos,* etc.), and of *J,* 'similis altit id est assae.' This last led Halleux-Opsomer 1982a, 85, to suggests *assa foedita,* as cognate with *assa ferula,* but no manuscript witness supports it. On the other hand, Diosc. makes no comparison with garlic (*allium*), and an alternative reading of 'similis alio [quodam]' may point to some other (now irrecoverable) comparison.

2 #47.

3 Corruption or conjecture: Ammon was the site of a temple of Jupiter in Libya, where 'ammonaic' salt was collected: see #239 n.2. There is no known king of Cyrene with that name. Cf. Diosc. III.84, 'area of Libya where Ammon is.'

4 #279.

5 #50.

5) Aloe (*Aloe vera* L.)

Aloe is the juice of a plant that is similar to squill.[1] Its root when cut produces a juice which is then collected, or the whole plant is crushed, pressed, and dried. Some say it is gathered among rocks in Judea, Asia, and even India. We judge the best aloe to be delicate and of clear gel, plump, shiny, and slightly reddish, and though it is easily pulled from the ground it maintains a sticky fragrance and a very bitter taste. It has two uses. If drunk and swallowed, it loosens the bowels, purges rotting matter, and reduces bile and phlegm. When mixed with oxymel[2] it is a helpful medicine for those suffering from stomach and liver ailments. Some people eat two or three granules the size of a chickpea after a meal, for it maintains a healthy body and does not disturb digestion. It also extinguishes thirst when it is crushed with cabbage juice and is then fashioned to the size of a chickpea. Other people mix it with the inner part of colocynth,[3] and scammony,[4] and use it to treat fevers and a range of illnesses. It acts as an effective astringent when applied externally on the body.

1 #264. The similarity is commonly noted, cf. Diosc. II.22; Pliny 27.14–20. Nutton and Scarborough 1982, 210–12, suggest that Dioscorides and Pliny describe Socotrine aloe (*Aloe perryi* Baker), a type from the island of Socotra, which was imported into the West around the end of the first century BC: see ch. 3.D.

2 Oxymel is a mixture of honey, water, and vinegar (and sometimes spices) boiled down to a syrup: Diosc. V.14.

3 #226.

4 #252.

6) Amylum

Amylum omnibus notum est, enim quasi medulla frumenti quod infunditur [et tunditur donec furfurem suam demittat. Deinde cum aqua mittitur] in sportellam et cum pressura colatur et tunc cum mini-mo ipse nucleus uelut fex desidet aqua effusa in sole siccatur et amylum
5 cognominatur. Cuius est optimum quidem recentissimum et candidum et leue et sine ullo acrore uel lutoso aliquo aut inquinoso odore. Potest autem leniter stringere propter quod collyriis ad lachrymam facienti-bus miscetur et ad profluuium uentris prodesse comprobatur.

1 Amylum] amolum *V* amilos *F* amilum *BPM* amilum conueniens medicaminibus est sit enim ex medulla fructi *W* omnibus- notum] *om. M* notum] nomine *C* quasi] *om. C* 2 infunditur] conditur *LB* infunditur- mittitur] *om. VJ* 3 sportellam] patella *P* pressura *W* pressura] pressa *BP* percolatur *L* cum minimo] in himo *F* et tunc- cognominatur] *om. JC* 4 uelut fex] *om. P* desidet] resederit *PW* resedit *M* effusa] infusa *VMW* in sole] in sale *V* 5 quidem] quod est *LB* 6 acrore] acredine *P* uel- odore] *om. WP* lutoso] ut situ sit *L* sicusa *F* ut lutoso *B* autem amulum *C* 8 miscetur- comprobatur] *om. LBC* miscetur] optime prodest *B* prodesse combrobatur] optime prodest *P* comprobatur] *om. M*

cf. Diosc. II.101 (ἄμυλον). Pliny 18.76. Isidore 20.2.19.

7) Alumen

Alumen dicitur rotundum et scissum et liquidum. Inuenitur autem omne alumen in insula quae Melos dicitur, sed et in Aegypto et in Macedonia et pluribus locis. Verum est optimum scissum alumen, quod graece schiston dicitur, quodque illius rotundi ueluti flos est.
5 Hoc igitur simile est albis capillis uexatis et quodammodo carminatis. Liquidum deinde, quod est leue, perlucidum et aliquid in summo in modum floris quasi lactis habet, quod ipsum probamus siccum ad dul-corem. Sincerum autem si sit, non inficitur ab illo rotundum, eo quod est subalbidum et aliquo pallore uelut pinguamen, fragile, mundum et
10 colorem mutat asperso succo quem diximus. Potest autem omne alu-men uehementer stringere.

1 scissum] sistum dicitur *L* scisto dicitur *FBP* et liquidum] *om. B* quod stipterea dicitur *P* 2 melos] mellus *V* nylos *M* 3 scissum] sistium *L om. P* 4 quod- quem diximus] *om. W* schiston] tricitis *V* trikips *L* erigitens *F* strcidis *M* illius] est *B* 5 igitur] ergo *FB* capillis] *om. P* quodammodo] quod ad modum *LB* carminatis] carpiriatis *L* carpinamis *B* carpi-natus *FPM* liquidum et candidum *B* 6 deinde] candidum *P* 7 siccum-dulcorem] *om. B* sucus alucore *F* malecori *L* suco mali odori sincera ulla diuersa pigmentorum adhibimus et omnibus ipsum probabmus. In cera autem si non inficitur *P* 8 quod est] quam et *BP* si sit] siscit *L om. B* eo] *om. BC* 9 pinguamen] pinguinatum *M* 10 asperso] scisso *B* succo] *om. C*

cf. Diosc. V.106 (στυπτηρία). Pliny 31.79. Galen *Simp.* 9.3.30 (K.12, 236–7). Isidore 16.2.2.

6) Starch

Starch is known to everyone, for it is the kernel of wheat[1] [which is pressed until it sheds its skin and is then mixed with water and] poured into a little basket which with applied pressure acts as a filter. After a short while a little kernel remains as the dregs and this is rinsed in water and dried out in the sun, from which it takes the name *amylum*.[2] Starch is best when it is freshest, white and smooth, without any coarseness or debris in it and without a foul smell. It can be used as a mild astringent, and is therefore mixed with eye-salves for teary eyes, and is a proven antidiarrhoeal.

1 *frumentum* usually means corn or grain, but can be used for wheat also, the usual source of starch: Diosc. II.101, spring-wheat (πυρὸς σητανίος), and Pliny 18.17 'ex omni tritico ac siligine, sed optimum e trimestri.'
2 The phrase is probably referring to the etymology of 'without a mill' (ἀ + μύλος), mentioned by Diosc. II.101 and Pliny 18.76. *amocillum J* is likely a corruption.

7) Alum

Alum is said to come in rounded, pressed, and liquid forms. All types of alum are found on the island called Melos, but it can also be found in Egypt and Macedonia and many other places. The best alum is pressed, which in Greek is called *schiston*,[1] and which has a rounded shape like a flower. It also has white frizzy hairs, as though they have been disentangled. Then there is the liquid type, which is smooth, transparent, and has a pattern on its surface that looks like a flower traced in milk, which we recommend should be dry though sweet in taste. Make sure that it is pure and not corrupted by that rounded type, which is whitish in colour, of the same pallor as grease, and is brittle, clear, and changes colour when sprinkled with the juice we mentioned.[2] All types of alum can be powerful astringents.

1 *schiston*, otherwise known as 'split,' 'feathery,' or 'cloven' (trans. Beck, Diosc. V.106, σχιστή) alum, or *trichitis* (as in *VL*, that is, 'hairy,' from 'θρίξ, τριχός', hair), which, according to Diosc. ibid. is a type of *schiston* found in Egypt. It was well known for its capacity to subdue inflammation: Pliny 28.126, 31.39, 33.84; Celsus 5.2.
2 No juice is mentioned: the text here seems corrupt or unfinished.

8) Auripigmentum et psilotrum

Auripigmentum, quod aliqui arsenicum dicunt et alii lamna dicunt, naturale est. Intrinsecum colorem habet aureum, quod comminutum in laminas tenues comminutas funditur, quod in modum auri refulget. Inuenitur in montibus Asiae et Armeniae, Ponticae et Cappadociae.
5 Verum optimum est quod de Amista et Hellesponto affertur, hoc enim laminas habet latiores et est sincerissimum et aurigeneo colore limpidissimum. Vires quoque habet sicut ignis, quod urere posit et alienationes consumere. Si cum calce uiua mixtum fuerit et corpori illinitum pilos tollit. Vnde graece haec ipsa mixtura psilotrum dicitur, latine acilea nuncupatur.

1 Auripigmentum] auripimento *V* alii lamna dicunt] *om. JMFC* 2 naturale] natura talis est *V* naturaliter *P* intrinsecum] extrinsecus *LBPMF om. C* habet] *om. LBCM* 3 laminas] iacolas *VFP* lacculas *LB* lanulis *MC* tenues] *om. C* comminutas] et minutas *VLC om. M* in mutis *F* funditur- modum] *om. C* quod- limpidissumum] *om. W* 4 ponticae] *om. BP* et verum est optimum *B* 5 amista et hellesponto] in amista et ponto *V* amista et disponto *J* dinamista et eclisponto *L* ab asia et helisponto *B* ellesponto *P* amista et cydisponto *M* 6 laminas] lamnolas *V* lannulis *LBC* lamolas *P* 7 sicut] *om. CM* quod urere] quae similiter ut *LBF* deuorare *P* alienationes] algones *B* hanheltium *C* 8 mixtum] commistum *BC* fuerit] *om. L* 9 unde- *fin.*] *om. BP* graece- mistura] *om. F* psilotrum] silitron *V* ΠΣΥΛΛΩΤΗΡΟΜ *L* acilea] acileea *F*

cf. Diosc V.104 (ἀρσενικόν). Pliny 33.79. Celsus 5.5. Vitruvius 7.7. *DTh.* 5. Isidore 19.17.2.

9) Aphronitrum

Aphronitrum optimum est quod est leuissimum, candidissimum, latum et fragile. Potest enim et calefacere et cum aliqua mordacitate relaxare et omnes sordes de corpore abstergere.

1 Aphronitrum] afronitrum *WPF* candidissimum] candidum *VF om. B* 2 calefacere] excalfacere *C* mordacitate] mordicatione *F* omnem sordem *V* 3 sordes] sudorem *C* abstergere] extergere *M*

cf. Diosc. V.113 (ἀφρὸς νίτρου). Pliny 31.113. Isidore 16.2.8.

10) Alosanthos

Alosanthos est ueluti pinguamen quod aquis supernatat stantibus prout stant alia, colore croceato. Cuius optimum est quod est in Nilo flumine colligitur, hoc enim coloratissimum est et cum aliqua pinguedine, aquosum et paene putridum. Potest igitur acriter uiscide et calefacere et quasi sal aliena consumere. Item collyria et malagmata colorare.

8) Yellow orpiment

Yellow orpiment is a natural product which some call arsenic and others call *lamna*.[1] It is golden in colour on the inside and is stripped into small, slender reeds which are beaten until they sparkle like gold. It is found in the mountains of Asia, Armenia, Pontus, and Cappadocia. The best is that which is imported from Amista and the Hellespont,[2] which has wider sheaves for its leaves, is thoroughly purified, and has a transparent, golden hue to it. It contains properties similar to fire in that it burns and hence can consume abnormal growths. When mixed with quicklime[3] and smeared on the body it removes hair. In Greek this mixture is called *psilotron*, but is known in Latin as *acilea*.[4]

1 This meaning of *lamna* ('plate,' 'blade,' 'coin') is unattested elsewhere. Cf. Diosc. V.104 (ἀρσενικόν), 'found in the same mines as red sulphide of arsenic' (σανδαράκη, below #242).
2 Amista: Amisos, today Samsun, on the Black Sea coast. The *AG* is our only source for this type: *ThLL* I,1198–9. It is possibly a corruption of 'Mysia,' for Diosc. V.104 records that this type is imported from 'Mysia, at the Hellespont.' On the latter location, see Nairn 1899.
3 #67.
4 The term *acilea* is unattested elsewhere, and this particular depilatory reflects sources now lost: see ch. 3.E.

9) Foam of soda

Foam of soda is best when it is very smooth, bright, flat, and brittle. It can be used to both warm and loosen with a degree of stinging, and to remove all filth from the body.

10) Salt efflorescence

Salt efflorescence[1] is like a grease which floats upon still waters just like other debris,[2] and is saffron-coloured. The best salt efflorescence is that which is harvested from the Nile river, for this is the most vibrant in colour, watery yet with a degree of oiliness, and has begun to decay. It can heat piercingly and powerfully, and consume abnormal growths just like salt. It is also used to colour eye-salves and emollients.

1 Alosanthos] alcionkium *V* alosardium *J* alosantos *F* alosanteum *LB* alosanitum *P* alo-
sanctum *W* alosardis *M* pinguamen] liquarum *M* prout stant alia] porro existant *B* porro
extat *P* porro dystant a *V* porro exstat *W* perstat *M* prout stat ac *F* 2 in nilo flumine] nullo
fallimine *J* flumine *C* 3 colligitur] et ibi colligitur *BP* in hoc *V* 4 et paene putridum] pene
putitur *V* puridum *LC om.* *BP* putidum *M* pene potandum *F* uiscide] incidere *J* uiride *L*
inscide *CP* et calefacere] excalefacere *V* 5 quasi sal aliena] salmigena *L* quasi aliena *C om.*
PB item- colorare] *om.* *W* colorare] curare *L* colure *V*

cf. Diosc. V.112 (ἁλὸς ἄνθος). Galen *Simp.* 11.2.7 (K.12, 374).

11) Alcimonium

Alcimonium species habet duas, uidetur autem siccum et liquidum.
Aliqui aiunt utrumque lachrymam esse arboris quae in Iudaea et aliis
locis nascitur. Alii quidem certius veluti pinguamen stagnorum esse
dicunt, ualde glutinosum et in se cohaerens et aquis palustribus super-
5 natare. Hoc igitur in scapulis pusillis attrahi et imponi et impletis scapulis non aliter praecidi et sic in terra expositum siccari et hoc ipsum
maxime in Iudaea et aliis locis fieri. Liquidum autem quod in Babylo-
ne et Iacintho colligitur circa quosdam lacus. Est igitur siccum [sicut]
asphaltum, quod est optimum fragilissimum et colore rufo ueluti pur-
10 pura splendens. Quod autem nigrescit pice adulteratur liquida, deinde
quod est lucidum et glutinosum. Vtrunque odore grauissimo. Vires
autem habet uiscidas et efficaciter relaxantes.

1 Alcimonium] algimonium *V* alcimonium *M* aloouium *P* alcionium *BW* autem] hic et
V et liquidum] et gluminosum *B* et liquidum et glutinosum *P* et glutinosum *W* 2 aiunt]
dicunt *FW om.* *C* iudea] india *MCW* 3 alii quidem] aliquod est *VLC* quod est *B* certius]
est cerosum *M* pinguamen] pingues *VFM* stagnorum] tenuosum *M* 4 cohaerens] coerris
L sincerum *BP* palustribus] paludestris *VF* paludastris *LB* paludis *M* 5 pusillis] post illos
F scapulis] si sit *C* attrahi] trahit *V* subtrahi *L* imponi] imponitur *V om.* *BPC* impletis]
repleti *LB* replentissimi *M om.* *C* impletis- praecidi] pletis scapolis muscariis precidi
quam si panno tamen gruescaneatur *V* repleti sumus carus alia poidi quam ci panno
tamen corruscanatur *F* repletis scasis in terra deponent et siccant *P* praecidi] quasi panno
tam iscritis cantatur *L* que spannum tamen cruis[..]re[.]tur *M om.* *C* scapulis] scaphis *J* 6
et sic in terra expositum] inter rade ponunt et *B* 7 liquidum] quod *F* hoc- fieri *om.* *C* fieri]
solet *B* autem] hic *V* 8 iacintho] cyngo *V* cyncito *M* iam quinto *F* lacus] loca *VFPM* est-
relaxantes] *om.* *V* Est igitur siccum] *om.* *CP* siccum] optimum *B* 9 asphaltum] *nov. cap.*
VFPCW aspolatum *C* fragilissimum] frigilissimum *V* frigidissimum *J* fragile *P* colore] et
siccum colore *BP* rufo] optimo *C om.* *F* 12 uiscidas] inscidas *P* relaxantes] *om.* *C*

5–11 *plerumque om.* *W*

cf. Diosc. V.118 (ἀλκυόνειον). Pliny 32.86–7. Celsus 5.6.18. Galen *Simp.* 11.2.3 (K.12,
370–1). Garg. *Med.* 27.

1 The name is a Latinized version of ἄνθος ἁλός (Diosc. V.122, cf. V.66: cf. Diosc. Lat. V.87, 137, *alosanthos*: also in Lat. Oribasius 2.1.A14, Alex. Tral. I.27 *aloxanti*). The reading of *alosardium* (J) baffled previous attempts to identify this entry (e.g., Halleux-Opsomer 1982a, 95).

2 Difficult passage: given the variants of *VBPW*, it perhaps should read something like 'floats upon still waters and stands apart from other floating things by its saffron colour.' Cf. Diosc. ibid., 'Salt scum flows down the river Nile and settles on top of certain marshes.'

11) Alcimonium

Alcimonium[1] has two types, one that seems dry and the other liquid. Some say that both are derived from the sap of a tree which grows in Judea and other places. Yet others affirm that it is a greasy substance found in lagoons and which is very gluey and coagulates to float upon the surface of swamp waters. This is then harvested by placing it on small plates which when full are placed on the ground to dry, without the alcimonium being cut,[2] and this is certainly the method mostly used in Judea and other places. The liquid version is collected in Babylonia and Iacinthus,[3] and around certain lakes. The dry type is like camelthorn,[4] and is best when very delicate and has a reddish colour with a purple shine to it. It blackens when it is adulterated with liquid tar, and becomes gluey and translucent. Both types have a heavy fragrance. It has powerful and thoroughly loosening properties.

1 The term *alcionion* (or [h]*alcimoneum*, [h]*alcymonium*, [h]*alcionium*, [h]*alcyonium* in ancient and medieval writings) was applied variously in Greek medical writings (as early as Hippocrates' *Gynaeceia* I.106 as a depilatory) to several species of marine zoophytes, including sponges and holothurians, which resist modern identification: its modern taxonomic application to types of coral (e.g., *Alcionium digitatum* L., 'dead man's fingers') should not mislead. Diosc. V.118, Pliny 32.86–7, and Galen *Simp.* 11.2.3 (K.12, 370) (ex Diosc., on which, Berendes 1902, 541: and cf. Puschmann 1879, I:442n, on Alexander of Tralles' attempts at identification) all describe several types of alcionium as a floating sponge ('bastard sponge' in LSJ) that originates from the sea: Pliny recorded a putative origin from the nest of the halcyon bird, which may explain the name. The description given here is unique (and odd) enough among the few early sources mentioning alcimonium to suggest that the text has been corrupted, or is confused with another subject, which seems to be also reflected in *VFPWC*, all of which start a new entry for camelthorn (n.4 below). I thank Professor Eleni Voultsiadou, University of Thessonaliki, for correspondence on ancient Greek knowledge of sea sponges, on which see Voultsiadou 2007.

2 Tentative: the variants *VFBPM* all point to an additional phrase here ('not cut until it glimmers [*corruscat*?] upon a cloth'?), but the text is too corrupted.

3 Presumably the island of *Zakynthos* (mod. Zante) in the Ionian sea.

4 Corrupted (n.1 above): *asphaltum*, camelthorn (*Alhagi maurorum* L.), is not listed in the *AG*, cf. Diosc. I.20, Pliny, 12.110, 15.30, and note André 1985, 28 (a type of Astragalus).

12) Aphroselinum

Aphroselinum in Aegypto tantummodo inuenitur. Quod ita creatur, ros coelestis positus ad lunae claritatem in specie lapidis, quem specularem uocamus, coagulatus stringitur, unde et nomen sumpsisse uidetur. Cuius eligimus quod est colore coeruleo et lucido. Potest autem
5 omnia uitia capitis potatum emendare et cephalalgicis et epilepticis utiliter prodesse.

1 Aphroselinum] afrosilino *F* afroselinum *P* erosilinum *C* in aegypto] *om. V* inuenitur] nascitur *W* creatur ros] creatura *M* 2 ad lunae] alunt *L* specularem] specialem *P* 3 coagulatus] qua gulatum *L* coagulatus est *B* stringitur] constringitur *VLW* constrigimus *F* sumpsisse uidetur] sic appellatur *M* sumpsisse- colore] sumpsit cui est colore *V* lucido optimum est *V* 4 potest- emendare] *om. W* 5 cephalalgicis] cefelargias *V* cephalargicis *M* 6 utiliter] unici *V* uniuersis *BPC om. F* prodesse] prodesse creditur *M*

cf. Diosc. V.141 (λίθος σεληνίτης). Galen *Simp.* 9.2.21 (K.12, 208). Ps. Apul. 80.45 (rosmarinum).

13) Argemone

Argemone herba est pluribus nota cuius uires faciunt contra uenena ad diuersas compositiones. Huius igitur radicem Graeci eupatorium dicunt et est diuretica ualde.

1 Argemone] argemonia *V* argimonia *WF* agrimonium *P* agrimonia *M* argemone et eupatorium *J* pluribus nota] pluribus locis nata *V* floribus micata *F* 2 diuersas compositiones] *om. W* compositiones recipiatur *C* radicem] atque *M* graeci] *om. F* eupatorium] HOY-PATOPYO *L*

cf. Diosc. II.177 (ἀργεμώνη). Pliny 24.176, 26.92. Celsus 5.27.10. Galen *Simp.* 6.1.54 (K.11, 835).

14) Apium

Apium herba est quae in hortis nascitur omnibus nota et cuius semen facile nouerunt magis tamen ad medicinam aptum est. Mittitur autem in purgatoriis et in antidotis. Est autem uiribus acrius et excalfactorium et diureticum certum.

1 Apium] appiu *M* nascitur] inuenitur *VWF* omnibus nota et] *om. BW* cuius] eius *VF* 2 facile] aptum est creticum *etc.* *#15 infra P* facile- magis] *om. C* medicinam aptum] usibus medicinae *VLFC* usibus medicinis *M* ipsum semen *LFCM* aptum est] *B* ipsum semen apii *V om.* autem] enim *FBC* 3 acrius] agris *V* acris *F* 4 certum] *om. B*

1–3 *plerumque om. W*: semen eius uiribus acre et calfactorium et dioreticum est.

12) Moon foam / Selenite

Moon foam is found only in Egypt. It is fashioned by placing heavenly dew[1] under moonlight on the type of stone which we call *specularis*,[2] upon which it coagulates and is then pressed together, from which its name seems to derive.[3] We select that which is azure in colour and clear. When drunk it can cure all ailments of the head, and is helpful to those suffering from chronic headaches[4] and epilepsy.

1 *ros coelestis*, also mentioned below as the constituent ingredient for honey, #179; see ch.3B. Though there are few attestations of the term in classical sources (e.g., Ovid, *Fast.* 1.312: cf. Vergil, Georg. 4.1, *caelestia dona*), it was picked up by Christian writers as a synonym for *manna*. Cf. Isidore 13.10.9, *ros* derived from 'fine' (*rarus*), 'because it is not thick like rain': also 12.6.49, a type of shellfish which conceives pearls 'ex caelesti rore.'
2 *lapis specularis*: a kind of transparent stone (muscovy-glass, isinglass-stone, mica) often used for windows. Pliny 36.150, 9.113.
3 *aphros + selene*, lit. 'foam [of the] moon.' Diosc. V.141 notes this as an alternative name, 'because it is found at night when the moon waxes.'
4 *cephalalgicus*: Langslow 2000, 481.

13) Wind rose (*Papaver argemone* L.)

Wind rose is a plant known to many, and its properties are used in different compound medicines to fight against poisons. The Greeks call the root of this herb 'agrimony,'[1] and it is a strong diuretic.

1 Incorrect: *eupatorium* is common agrimony (*Agrimonia eupatoria* L.), not listed in the *AG*. Diosc. IV.41 (εὐπατόριος, and II.176 ἀνεμώνη, 177 ἀργμώνη) records the the common confusion of agrimony with wild poppy anemone (*Anemone coronaria* L.) and the wind rose: ch. 3.E, and below #235 n.1. Cf. Pliny 21.165, 26.23. Celsus 5.27.10.

14) Celery (*Apium graveolens* L., var. *sativum*)

Celery is a plant which grows in the garden and is known to everyone, and its seed is even better known to be useful for medicine. It is put in purgatives and antidotes. It is a heating agent with quite sharp properties, and a reliable diuretic.

cf. Diosc. III.64 (σέλινον κηπαῖον). Pliny 19.124, 20.113. Galen *Simp.* 8.18.6 (K.12, 118–19) Garg. *Med.* 2. Ps. Apul. 119. *Dyn. Vat.* 21/*SGall* 25. Isidore 17.11.1 (syluaticum), 17.9.80.

15) Anisum

Anisum optimum est Creticum, deinde Aegypticum. Gallicum incertissimum est. Verum tamen habet uires relaxantes et excalfactorias et diureticas ad usum idoneas et est semen eius simile cicutae. De nutrientibus si bibatur lac prouocat et auget, odore eius suaui.

1 Anisum] anissu *M* anisum- est *om. cont. #14 supra P* gallicum] galbeum *P* incertissimum] inertissimum *FMP* 2 uerum- idoneas] *om. W* relaxantes] acriter *FLC* acriores *BP* acram *M* 3 ad usum] adsume *and sscr.* mulgiapata *B* ad summe *P* nutrientibus] *om. LCB* 4 lac prouocat] lacae probatur *M* auget] ad gustum *M* eius] *om. FLBC* nameque *P*

cf. Diosc. III.56 (ἄνησσον). Pliny 20.73, 25.151. Galen *Simp.* 6.1.48 (K.11, 833). Isidore 17.11.6.

16) Agaricum

Agaricum ab arboribus excrescit prout tubera uel boleti et est fractu leue et quasi uanum. Est fragile ualde et aspectu subrubrum. Penitus subalbidum non dissimile furfuri. Gustu initio uidetur dulce sed ex lento euadit amarissimum. Quod autem est optimum in Ponto maxime inuenitur, quod autem in aliis locis nascitur multum est ineptius. Vires autem habet acriter relaxantes et uenenosis morsibus et potibus resistentes. Sic et uentrem purgat aliquatenus et choleram et phlegma deponit. Ideo nec dolorem nec uomitum facit.

1 prout tubera] pro tuuera *F* in tubera *C* ubera *BW* perisum uero *M* uel boleti] *om. LB* uolitas uidemus crescere *F* ut uelutos uidimus crescere *VLC* uolens tumescere *P* fractu] tactu *LBPC* 2 uanum] libanum *BP* fragile] pallidum *J* aspectu-ineptius] *om. W* penitus subalbidum] ponitur albidinis *F* uolens tumescere *B* penitus albedinis *P* 3 furfuri] solfuri *V* sulfuri *F* ilperi *L* gustu *P om. B* 5 nascitur] iuenitur *V* ineptius] inhortius *L* incrotius *B* inertius *VFCM* inercius *P* 6 acriter relaxantes] acres et laxatiues *P* potibus] punctibus *FB* punctis *P* moribus *W* 8 nec dolorem] *om. V*

cf. Diosc. III.1 (ἀγαρικόν). Pliny 25.57. Galen *Simp.* 6.1.5 (K.11, 813). Isidore 17.9.4.

17) Absinthium

Absinthium omnibus notum est, et Santonicum tam Gallicum et marinum quam Ponticum, quod ipsum ex omnibus notum est. Hoc enim et in succum redigitur contusa tota herba, et succus eius expressus siccatur, uel in aqua usque ad tertias decoquitur, et sic ipsa aqua recocta in

15) Anise (*Pimpinella anisum* L.)

The best anise is Cretan, then the Egyptian.[1] The Gallic type is really unreliable. Anise is used for its excellent softening, warming, and diuretic properties. Its seed is similar to hemlock.[2] When drunk it stimulates and increases lactation for infants, and its fragrance is sweet.

1 Similarly Diosc. III.56, though mentions nothing of Gaul: noted by Meyer 1856, IV:492.
2 *Cicuta*: hemlock (*Conium maculatum* L.) given below #49 under the Greek name of *conium* (Gr. κώνειον), though also referred to again as *cicuta* in #42, where it is likened to coriander (#68) and celery (#14). It is also given as an alternative name for fleabane (#79) in some manuscripts.

16) Agaric (*Polysporus* sp.)

Agaric grows near trees in the form of tubes or *boleti*[1] which are easy to break, and almost hollow inside. It is very delicate, with a shade of redness to it. The inside is almost white, not unlike bran. Initially its taste seems sweet, but slowly this is replaced by an extremely bitter taste.[2] The best agaric is found mainly in Pontus, and though it grows in other places it is much weaker. It has severely loosening properties and helps resist against venomous bites or poisonous drinks. It also purges the stomach a little, and reduces bile and phlegm. It causes no pain or vomiting.

1 *boletus*, an expensive mushroom valued by the Romans, probably that which we call *porcini* (*Boletus edulus* sp. Bull., Fr.), but the term was also applied to other types of mushroom: André 1985, 37.
2 Likewise Diosc. III.1, and Pliny 25.57 'initio gustus dulcis mox in amaritudinem transit.'

17) Absinth (*Artemisia absinthium* L.)

Absinth is known to everyone, whether it be the Santonic,[1] the Gallic, the marine,[2] or indeed the Pontic type which is known to everyone. The whole plant is crushed and reduced to a juice which is then pressed and dried, or is reduced by cooking to one third, then this is cooked again with

5 mellis redigitur spissitudinem. Vsus quoque et uires absinthii et quod
lumbricos expellit nemo ignorat.

1 Absinthium] absentio *V* apsintium *L* omnibus] pluribus *W* santonicum] *om. W* tam
gallicum] *om. C* tam] tamen *V* gutat *M om. C* marinum- ponticum] *om. F* **2** ponticum]
quam ponticum *VL* quam et ponticum *B* quod *J* expellit *C* ex omnibus] et hominibus
C notum] optimum *FLBC* melius est *W* **3** redigitur] egeritur *M* redigeretur *F* siccatur]
sic inoxo *B* sic inoxio *P* **4** in aqua] aliqua *C* aceto aut aqua *W* ad- decoquitur] *om. P*
decoquitur] et exponitur *C* recoctum *M* sic- spissitudinem] *om. W* recocta] sit decocta *V*
decocta *FBP* **5** redigitur] *om. FBCM* uires] uires abensti coleram deducit *V* uires absinthii calidesunt *C* uis *P* **6** nemo ignorat] nemo ignorare debit *V* nemo ignoranti *F* ualde
P om. C

cf. Diosc. III.23 (ἀψίνθιον). Pliny 27.45–53. Celsus 3.21.6. Galen *Simp.* 6.1.1 (K.11, 798–
807). Ps. Apul. 101. Isidore 17.9.60.

18) Amurca

Amurca faex est olei quando premitur nigra et aquosa, quae coqui solet
in mellis spissitudinem. Cuius est optima quae de recenti coquitur et
sic uetustatem accipit. Vires habet stypticas, ualde propinquas ei quod
in emplastris inuenitur et in aliis multis compositionibus.

1 Amurca] murco *C* amurcha vel faece olei *tit. J* quando] cum *VBCMW* nigra] aqua *L*
nigra et aquosa] *om. W* enim quae *V* coqui solet] coquitur *V* soluitur quoque *M* quae-
cumque sola *F* **2** cuius- accipit] *om. W* quae] aqua *VF* **3** sic] siue *B* si *C* uetustatem]
uetustitatem *V* uenustatem *J* accipit] accipitur *V* propinquas] propter *MPW* pro *F* **4**
emplastris] in plastra *V* inuenitur] mittitur *BPCM* compositionibus] passionis *V* causis *C*

cf. Diosc. I.102 (ἀμόργη). Pliny 15.9–34. Galen *Simp.* 6.1.30 (K.11, 824–5). Isidore
17.7.69.

19) Aristolochia

Aristolochia est rotunda et oblonga et magis tamen ad usus medicinae
elegitur rotunda et hoc ipsum quod non est uetustum et graue, siccum et quodammodo spissum et gustu amarum. In emplastris mittitur
et caua uulnera comprendit et replet. Ad uiperarum morsus et ad alia
uenena intrinsecus facit.

1 Aristolochia] aristologium *V* aristologia *FPM* et- rotunda] *om. J* **2** hoc- amarum] *om.*
W ipsum quod *om. P* siccum] sucus ipsius *P* **3** gustu amarum] *om. C* amarum] *om. M*
gustu- replet] quod ad modum replit *V* hoc igitur in emplastris *FBC* mittitur] mittitur et
anagdotis *C* **4** comprendit] comprehendit *F* **5** intrinsecus] extrinsecus *P*

cf. Diosc. III.4 (ἀριστολοχεία). Pliny 25.95–8. Galen *Simp.* 6.1.56 (K.11, 835–6). Ps. Apul.
19. *Ex herb. fem.* 12. Isidore 17.9.52.

water until it obtains the consistency of honey. No one is unaware of the uses and properties[3] of absinth, and that it expels intestinal parasites

1 Central western Gaul.
2 From Taposiris, Egypt, according to Pliny 27.45–53; likewise Diosc. III.23, although also from Mount Taurus in Cappadocia.
3 'which reduce cholera' *V*, 'which are warming' *C*.

18) Amorge / Olive lees

Amorge is the black and watery lees of olives when they are pressed, and is usually cooked until it reaches the consistency of honey. It is best when cooked from fresh lees and then left to age. It contains styptic properties, especially when placed near anything to which a plaster has been applied, and is used in many other compound medicines.

19) Birthwort (*Aristolochia rotunda* L., *A. longa* L., *A. clematitis* L.)

Birthwort can be either spherical or long, but the spherical type is better for medical uses,[1] and this type should not be used when it is old, pungent, dry, slightly thickened, or has a bitter taste. It is added to plasters and used to fill and seal hollow wounds. It is an antidote for bites and some types of internal poisoning.

1 According to Diosc. III.4 (ἀριστολοχεία), the round type (στρογγύλη) is female, the long (μακρά) is male.

20) Atramentum sutorium

Atramentum sutorium est quod foditur in metallis et quasi agglutinatur circa foueas in quibus aes inuenitur. Est quod in summa terra
metallorum colligitur aliquid et pari ad ipsum dicunt de ferro rugino-
sissimo in aceto acrissimo multo tempore maceratur et sic feculentum
5 quod subsiderit siccatur. Vero optimum enim quod est colore quasi
sulfureo et aliquatenus subuiridi. Optime denigrat et uicem denigratur
si aqua spargatur. Vires enim habit acriter et uiscide stringentes et refrigerantes et inficientes unde lanas et alia cito denigrat.

2 aes] ares *F* sepe *BP* est] et *F* est- siccatur] *om. W* 3 aliquid] alii *P* pari] parasti *F* parari
BC deparari *M* pastaria *P* ruginosissimo] rubiginosi suprascripti qui *F* rubiginissimo
L 4 acrissimo] acerrimo *P* feculentem] ferculentem *P* 5 subsiderit] subrederit *M* 6 sul-
fureo] sulphureo *FP* sulfura *M* sulpburine *B* aliquatenus- spargatur] *om. W* subuiridi]
subuiridicans *VFP* subuiridicatur *BC* optime] optimus *VBC* denigratur] nigratum *M*
uicem] uitae *F* 8 inficientes] refitaentes *F om. B* et alia] *om. L* alia cito] cito et diu *P*
deest J

cf. Diosc. V.101 (μελαντηρία). Pliny 34.123, 35.41. Celsus 5.1.1, 5.8.1. Galen *Simp*. 9.3.20
(K.12, 226). Isidore 19.17.17.

21) Aconitum

Aconitum herbula est quae folia habet a sua radice duo uel tria similia
cucumeri et thyrsulum non altum, radicem oblongam, ad unum magis
tenuiorem et minutiorem, speciem et colores ad similitudinem scor-
pionis pusilli sine pedibus. Vires autem habet quasi thapsia et ualde
uenenosa esse dicitur.

1 Aconitum] agantus *M* herbula] herba *FMW* 2 non altum] *om. B* eodem modo *W* altum]
alium *LP* ad unum] ad immo *V* imum *L* et adimo *F om. M* ad unum- minutiorem] magis
tam tenuiorem *B* magisque *L om. WP* 3 minutiorem] mitiorem *V* 4 pusilli] postillu *VF*
om. W sine pedibus] in pedibus *VF om. W* 5 esse dicitur] *om.W* dicitur *P*

cf. Diosc. IV.76 (ἀκόνιτον). Pliny 20.50, 27.4, 27.10. Scrib. 187. Galen *Simp*. 6.1.19–20
(K.11, 820). Isidore 17.9.25.

22) Artemisia

Artemisia herba est subsimilis absinthio, sed per omnia uastior, id est
foliis latior, ramis altior et fortior, sed aspectu et colore talis fere et
odore et gustu quoque et ipsa amara. Nascitur locis humidis. Potest
autem relaxare, menstrua et urinam prouocare, et partum expellere [et
adnotationis uero quas super uersus posuimus per capita sunt].

20) Shoemaker's black

Shoemaker's black[1] is dug out from mines where it sticks to sides of pits in which copper is found. There is another type collected from the topsoil above mines, and they say that a type equal to this is found upon well-rusted iron that is then soaked in extremely sour vinegar for a long time, and then the tiny dregs that sink to the bottom are dried out. The best is that which has a sulphurous colour with hues of green to it. It is excellent for blackening, and itself turns black when sprinkled with water. It has severe and powerful astringent and cooling properties, as well as staining properties, and therefore quickly blackens wool and other items.

1 Cf. Diosc. V.101 (μελαντηρία), which parallels the above but differs in several respects: see ch. 3.B. Both Pliny 34.123 and Celsus 5.1.1 state that the Greek name for this substance is *chalcanthon* (below #51, 52).

21) Leopard's bane (*Doronicum pardalianches* Jacq.)

Leopard's bane is a shrub which has two or three leaves like those of the cucumber[1] sprouting from near its root, and a stalk that is not very tall. Its root is oblong in shape, much more slender and smaller at one end, and which in appearance and colour resembles a small scorpion without legs. It has the same properties as deadly carrot[2] and it is reputed to be extremely poisonous.

1 #78. Similarly Diosc. IV.76.
2 #281.

22) Wormwood (*Artemisia campestris* L., *A. arborescens* L.)

Wormwood is a plant very similar to absinth,[1] but is altogether much larger, in that its leaves are wider and its branches are much thicker and higher, but in appearance and colour it is generally the same and is likewise bitter in both fragrance and taste. It grows in moist places. It can loosen, promote menstruation or urination, and expels the foetus [And we made a note about this on the lines above among the headings[2]].

1 herba] *om. FLB* per omnia] *om. MW* foliis] solus *L* 2 ramis altior] *om. BPW* aspectu]
hispida *P* talis- locis] *om. JC* fere] proerit *V* refert *M* 4 autem] acritem *LB* acriter *VMP*
agiliter *F* relaxare] reuixare *V* urinam] omnia *F* et adnotationis- sunt] *om. JBLPCW* sunt]
sunt notata *M* sunt adnotandas

cf. Diosc. III.113 (ἀρτεμισία). Pliny 25.73–5. Galen *Simp.* 6.1.62 (K.11, 839–40). Ps. Apul.
·12 (10, 11?). Isidore 17.9.45.

23) Asparagus

Asparagus et semen et radix eius urinam prouocat. Potest duritiem
relaxare. Vires habet aliquatenus calefactories et diureticas.

1 Aparagus] aspargo *F* aparagi semen *P* semen] herba *W* duritiem] omnem duritiem *LBP* 2
relaxare] relaxare menstrua et urinam et partum expellere *B* menstrua prouocare et partum
expellere *P* aliquatenus] *om. P* calefactorias] calefactiones *F* diureticas] relaxantes *W om. M*

cf. Diosc. II.125 (ἀσπάραγος πετραῖος). Pliny 19.145, 20.108. Galen *Simp.* 6.1.66 (K.11,
841). Garg. *Med.* 31. Ps. Apul. 85. Isidore 17.10.19.

24) Amygdala

Amygdala sunt et suaues et amara et utraque uiribus styptica. Magis
tamen sunt in usu medicinae et aptiores amarae eo quod stringunt
magis.

1 Amygdala] amigdolas *V* amigoalam *L* amydalas *M* amigdolae *P* suaues] dulces *VW*
utraque uiribus] uires habent *B* in usu] *om. BP* 2 stringunt] subtiles *JC* magis] *om. BP*

cf. Diosc. I.123 (ἀμυγδάλη). Pliny 15.26, 15.42, 15.89, 16.117, 17.252, 23.85, 23.144.
Galen *Simp.* 6.1.36 (K.11, 827–8). Garg. *Med.* 53. *DynSGall.* 89–91 (dulcia, amara). Isidore 17.7.23.

25) Anagallis

Anagallis est herba quae nascitur in locis humidis et ipsa species habet
duas et enim una coeruleo flore, altera quasi roseo, uerum utraeque
sunt in reliquo similes. Sunt autem thyrsulo humili, foliis mollibus et
pusilli et quasi rotundis, habentes circa ramulos suos radiculas minutas
5 et subdulces, uiribus stringentes. Efficacior quidem est illa caerulea,
quae in emplastris mittitur et succus illarum cum melle attico ad oculorum facit claritatem.

1 est herba quae] *om. VFLBP* quae- 5 caerula] quam alii simphicum dicunt *W* 2 et enim]
est enim *VM* una] *om. FB* roseo] rubeo *P* 3 mollibus et pusillis] aspera *M* utraeque herbae
VFLB in reliquo] et reliquo *V* 4 habentes] *om. BP* 5 efficacior] efficacia *V* efficaciem *F*
est] eam *L* illa caerulea] in lato oleo *F* 7 facit] prastat *W*

cf. Diosc II.178 (ἀναγαλλίς). Pliny 25.144. Galen *Simp.* 6.1.39 (K.11, 829). *Ex herb.fem.*
62.

1 #17. Similarity also noted by Diosc. III.113, and Pliny 25.73–4.
2 A conjectured translation of this obscure and intriguing note found only in *VFM*, though nothing appears among the chapter headings for the 'A's in *V* (fols. 8–9), or in *F* (fol. 54v), and *M* does not use chapter headings. See ch. 5.C.

23) **Asparagus** (*Asparagus officinalis* L.)

Asparagus, as well as both its seed and root, stimulates urination. It can loosen indurations. It has some warming and diuretic properties.

24) **Almond** (*Prunus amygdalus* Stokes)

There are both sweet and bitter almonds, and both have styptic properties. The bitter type is better and used more frequently for medicinal purposes because it is more astringent.

25) **Pimpernel** (*Anagallis arvensis* L., *A. caerula* Schreb., *A. phoenicea* Scop.)

Pimpernel is a plant which grows in moist places, and is of two different types: one with a dark blue flower, the other with rose-coloured flower,[1] but otherwise they are similar in everything else. They have a small stalk, tiny leaves that are soft and almost round, and their boughs have tiny little roots which are quite sweet and have astringent properties. The blue-flowered type is the most effective, and is used in plasters. The juice of both types mixed with Attic honey clears up dim-sightedness.

1 Both Diosc. II.178, and Pliny 25.144, correctly note that the blue-flowered pimpernel is male, the red-flowered is female.

26) Ami

Ami semen est aethiopicum, cymino minutius quidem et exalbidum et quodammodo spissum. Cuius optimum est quod est recens et non furforosum. Vires habet acres et excalfactorias et relaxantes. Vnde prorsus inflationem sedat et urinam et menstrua prouocat.

1 Ami] aei *M* aethiopicum] poene *FLM* cymino] sanam *W* cymino minutius] qui minutis *B* quod minutis *P om. LC* exalbidum] est ex lapidibus *P* et- optimum] *om. B* **2** quodammodo] quod ad modum *VF* furfurosum] uaporosum *BWP* **4** sedat] facit *M* reddat *F* urinam] *om. W*

cf. Diosc. III.62 (ἄμι). Pliny 20.58. Galen *Simp.* 6.1.28 (K.11, 824).

27) Adarcha

Adarcha nascitur in Gallia, locis humidis, circa calamos quibus adhaeret et est spuma quae est in specie spongiae, consimilis halycyonio, sed asperior et siccior et leuis ualde. Vires habet acres et in modum ignis excalfacientes et si corpus de ea fricemus cum magno calore punctionis densat et sanitatem facit.

1 Adarcha] adarces *M* adarcis *F* adartes *P* adhaeret] adderit *V om. P* **2** spuma] spumosa albida *W* consimilis] adsimilis *VF* similis *B* similat *P om. W* halycyonio] alcionis *M om. W* halycyonio alboris *B* albores *P* halycyonio- acres] altionoi redas agris *F* sed asperior] *om. BP* **4** punctionis] mordicationes *W sscr.* i ubipungias *B* **5** densat] *sscr.* spissat cocquat *B* sanitatem] anathema *JW* natas *F* ardorem *C om. M*

cf. Diosc. V.119 (ἀδάρκης). Pliny 32.140, 16.167, 20.241. Galen *Simp.* 10.2.2 (K.12, 370).

28) Amomum

Amomum genus est herbae odoratissimae, quae ramulos habet duros et siccos, uirgulis assimiles, colore subrufo, circaque folia habet iuncta et copiosa et in speciem racemorum implicata. Flos eius quasi uiolae, tota gustu acriter excalfaciens. Cuius etiam optimum dicitur Armenium,
5 sequitur Ponticum quod ipsum asperius est et durius. Nam de Media omnium est inertissimum eo quod locis humidis nascitur. Vires autem habet omne amomum calefacientes et stringentes. Eligere tamen debemus quod ab una radicula omnes suos ramulos excrescentes et integros habet foliis et floribus qualiter diximus. Item reprobare debemus quod
10 gummi conglutinatur coniunctum cum amomo, aut amomide herba quae est assimilis, sed caret odore et flores habet origano similes.

1 odoratissimae] optimi odoris *W* **2** uirgulis assimiles] sed durus uirgolis *V om. W* circa-
6 nascitur] *om. W* folia] foliola *P* **3** speciem racemorum] *om. LBC* ragimarum *V* ipsis

26) Ajowan (*Carum copticum* B., H.)

Ajowan is a seed from Ethiopia, smaller than cumin, whitish in colour, and a little dry. It is best when fresh and not chaffed.[1] Its properties are sharp, heating, and loosening, therefore it is very good for reducing swelling, and stimulates both urination and menstruation.

1 Similarly Diosc. III.62. Cf. Pliny 20.58.

27) Adarces (Salt inflorescence)

Adarces grows in the marshy places of Gaul[1] among reeds to which it attaches itself, and is an incrustation like a sponge, very similar to alcimonium,[2] but rougher, drier, and very light. Its properties are sharp and heat in a manner like fire, and when it is rubbed on the body it seals punctures with a great deal of warmth and sanitizes them.[3]

1 Cf. Diosc. V.119, 'occurs in Galatia' (difference noted by Meyer 1856, IV:492) but likewise similar to alcimonium. Pliny 32.140 states that *adarces* was the Latin name for the Greek *calamochnus*, a name seemingly unattested elsewhere in Greek or Latin writings.
2 #11.
3 Or 'and causes pain.' *JW*; 'and burns.' *C*.

28) Nepal cardamom (*Amomum subulatum*, Roxb.)

Nepal cardamom belongs to the family of the most fragrant plants. It has small, hard, and dry branches that are more like twigs, reddish in colour, and upon these its large, rounded, and full leaves grow in entangled clusters. Its flower is like the violet,[1] with a fully bitter taste that heats the mouth. The best is said to be Armenian, followed by that from Pontus, which is coarser and hardier. That from Media is the weakest of all, because it grows in wet places. All Nepal cardamom has warming and astringent properties. We ought to select that which has all of its branches still intact and growing out from a singular small root, and which bears leaves and flowers as we have described. Likewise we ought to reject the mixing of Nepal cardamom with a gluey concoction containing gum, or with a similar plant known as '*amomis*,'[2] which has no fragrance and has flowers similar to oregano.[3]

ramis per implicata *BP* implicata] per impletas *M* eius] -cellum *F* - cella *BP* tota- 4 optimum] *om.* *F* 5 sequitur] releuans *F* quod ipsum- inertissimum] *om.* *P* media] armenia *M* 7 stringentes] siccantes *JC* eligere- *fin.*] *om.* *W* 9 habet foliis] cum foliis *C* reprobare] probare *P* debemus] uidemus *F* quod gummi] quod gumen *V* cum augmento *BP* cum gummi *C* 10 coniunctum cum] coniunctum invicta *C* naturam est iuncta *F* cum amomo] *om.* *P* amomide] *om* *BP* amomi *C* de herba *C* herba- assimilis] simile uirgulis siccis *P* 11 sed caret] *om.* *LC* caret] siccat et odore *B* caret dolorem *P*

cf. Diosc. I.15 (ἄμωμον). Pliny 12.48. Galen *Simp.* 6.1.38 (K.11, 828). Isidore 17.8.11.

29) Asarum

Asarum radicula est minuta et nodosa et transuersa, suauiter olens, quae et in Gallia et in Phrygia inuenitur. Verum optimum asarum Ponticum et huius quod est recentissimum. Vires autem habet excalfacientes et cum acrimonia relaxantes, unde et urinam et menstrua prouocat et totum corpus calefacit cum oleo et uino iunctum atque laeuigatum.

1 Asarum] asara id est bacchare *tit.* *J* asaron *F* transuersa] quasi transuersa *VF* euersa *B* quasi euersa *P* suauiter] uitis *V* 2 quae- 4 relaxantes] *om.* *W* quae- asarum] in est asarum *F* 4 cum acrimonia] acritem *W* 5 oleo et] *om* *C.* iunctum atque laeuigatum] immixtum uires habet similes nardo *W* in unguito *M* atque] alique *F*

cf. Diosc. I.10 (ἄσαρον). Pliny 12.47. Galen *Simp.* 6.1.63 (K.11, 840). Isidore 17.9.7.

30) Acorus

Acorus herba est quae et thyrsum emittit et folia iridi similia, lata et oblonga et exalbida et in summo extensa et acuta, quasi gladius. Florem aureum habet, radicem quoque implicatam et nigriorem quam iris habet nodosam et gustu acrem et aliquatenus ad odoratam. Semen plenum, spissum, integrum et ipsum odore suaui perstringitur quo et meu et per omnia est efficacius et magis excalfacit et extrahere splenem uidetur consumptum et potum et succus radicis illius claritatem oculis praestat mirifice.

1 Acorus] acorion *F* acoro *P* emittit] eius *F* iridi] faidis *V* addissimilata *F* fridis *P* gladiolis *C* folia- exalbida] *om.* *W* folia] foliola *P* 2 extensa] densa *V* tensa *FP* gladius] gladiolus *M* florem] *om.* *M* 4 iris] yreos *W* mires *F* hyris *P* uires *C* nodosam] calidas *C* gustu- 6 excalfacit] uires habet calfaciendi *W* acrem] amarum *M* anuras *C* semen] semper *M* 5 suaui] *om.* *CM* suauiter *P* 6 meu] meon *BP* mou *M* extrahere] intra *JC* stringat *B* stringit *P* unde *W* 7 consumptum et potum] consumere potum *J* sumptus et putus *V* sumptus et potatus *F* sucus eius potatur *M* si sumatur in potum *BP* *om.* *C* 8 praestat] facit *JB* mirifice] *om* *P*

cf. Diosc. I.2 (ἄκορον). Pliny 25.157 (25.144). Galen *Simp.* 6.1.18 (K.11, 819–20). Ps. Apul. 6. Isidore 17.9.10.

1 #287.
2 Unidentified: André 1985, 14. Similarly referred to by Diosc. I.15, who adds that it is from Armenia; and Pliny 12.49.
3 #206.

29) Hazelwort (*Asarum europaeum* L.)

Hazelwort[1] has small, knotty, and crosswise roots, smells sweetly, and it is found in Gaul and Phrygia. The best hazelwort is from Pontus, and especially when this is freshest. It has heating properties, and loosening properties with vigour. It therefore causes urination and menstruation, and warms the whole body when it is pulverized and mixed with oil and wine.

1 The chapter heading in *J* 'id est bacchare' probably derives from Pliny 12.45 (and 21.29), who claims *baccar* is a type of wild nard (*nardum rusticum*): similarly Diosc. I.10, 'ἄσαρον ... οἱ δὲ νάρδον ἀγρίαν καλοῦσι', but does not employ the term *bacchare,* which he uses for sowbread (III.44 βάκχαρις = *Gnaphalium sanguineum* L.)

30) Yellow flag (*Iris pseudoacorus* L.)

Yellow flag is a plant which shoots up a stalk and has leaves similar to iris[1] in that they are wide, long, and pale, and extend all the way to a sharp tip, like a sword. It has a golden flower, and tangled roots that are blackish in colour and knotty like the iris, which have a sharp taste and a subtle smell. The seed is plump, thick and whole, and is permeated with a pleasant fragrance like that of spignel,[2] but it is more powerful than spignel in all its uses, it heats more, and reduces the spleen when eaten or drunk. The juice from its root miraculously clears up vision.[3]

1 #137. Likewise Diosc. I.2, Pliny 25.157. Isidore 16.9.10.
2 #189.
3 Similarly Diosc. I.2, Pliny 25.157. Beck 2005, 7 n.18 notes the etymology of ἀ- (privative) + κόρον, 'eye-pupil.'

31) Auena

Auena crescit in segetibus et est tota similis foliis et culmo frumento, sed fructu differt. Haec igitur uires habet efficaciter relaxantes et ad omnem tumorem facientes.

1 Auena] auina *V* abina *F* segetibus] segetibus egyptum *B* egypto *P* tota similis] *om. W* 2 fructu] *om. FMM2* habet] habet leniam *BC* habet leniter *FPWMM2* relaxantes] proprie tamen *etc. ex #32 infra M2* 3 facientes] efficaciter facientes *WM*

cf. Diosc. II.94 (βρόμος). Pliny 18.143, 149. Galen *Simp.* 6.2.17 (K.11, 855). Isidore 17.9. 105.

32) Andrachne

Andrachne omnibus nota est et uiscide stringit. Proprie tamen masticata dentium offensam quae ab immaturis pomis fit quam graece haemodiam dicimus sanat.

1 Andrachne] andraunia *V* portulaca *J* andragine *F* andragne *BC* andrabene *P* andragmis *M* andraginis *W* et] ea *J* et- tamen] *om. W* uiscide] uiscu destringit *P* masticata] massata *VF* hiuat *B* uiuat *P* 2 immaturis] matis primis *F* pomisat *B* immaturis spumi *M* examinaturis pomis *W* quam- sanat] *om. P* haemodiam] limonia *V* limodia *F* lymodiam *M* himoina *C om. W* 3 sanat] sumta *C* fortiter reprimit *W*

cf. Diosc. II. 124 (ἀνδράχνη). Pliny 13.120, 25.162. Celsus 2.33. Galen *Simp.* 6.1.43 (K.11, 830–1). Ps. Apul. 104. *DynVat* 39/*SGall* 43.

33) Asphodelus

Asphodelus nascitur in molli terra et habet folia similia porro capitato et circa terram diffusa, thyrsum lenem, rotundum et oblongum, in summo ramulos excrescentes habentes florem exalbidum, fructiculum nigrum, radices quoque oblongas et rotundas, quae sunt potentiores.
5 Combustae enim et in cinerem redactae alopeciam emendant et albidas maculas. Item coctae aqua ipsa bibita renibus dolentibus prodest. Vrinam et menstrua prouocat et ad tussim ueterem facit. Nam tunsa et imposita duritiem emollit.

1 Asphodelus] asfodillus *FWM2* asfodilu *M* albucus qui dicitur asphodelus *JC* albutium *P* capitato] a capite *P om. W* 2 lenem] uero *V* leuem *B* lene molle *F* et oblongum] *om. CW* in- 3 exalbidum] *om. W* 4 rotundas] ramindas *C* radices- potentiores] *om. W* 7 prouocat] *om. L* tussim] missem *C* nam] inde non est *C* 8 emollit] remollit *VBCMM2* mollet *F* mollit *P* repellit *W*

cf. Diosc. II.169 (ἀσφόδελος). Pliny 21.108. Galen *Simp.* 6.1.71 (K.11, 842). Ps. Apul. 32 [52?]. Isidore 17.9.85.

31) Oats (*Avena sativa* L.)

Oats grow in fields, and their leaf and stem are utterly similar to wheat, but the seed is different. Oats have effective loosening properties, and are good for all swellings.

32) Purslane (*Portulaca oleracea* L.)

Purslane[1] is known to everyone, and is a powerful astringent. It is chewed to relieve the type of toothache which we call in Greek *haemodia*,[2] that results from eating unripe fruit.

1 All manuscript witnesses use a variation of the Greek term (ἀνδράχνη, or ἀνδράχλη), but *J* employed the Latin equivalent *portulaca* (though still listed among the As), as found in other sources: e.g., Diosc. Lat. 'de andracla, id est portulaca'; *Dyn.SGall.* 43. The different terms are noted by Pliny 13.120, 25.162, specifying that Italians call it *inlecbra*.
2 *haemodia*: defined by LSJ as the 'sensation of having the teeth set on edge [caused by acid food or vomit].' Diosc. II.124 also refers to αἰμωδία, but does not qualify it or mention fruit: the only other author to do so is Galen, *De alimentorum facultatibus libri* III (K.6, 689), 'from eating unripe mulberry fruit,' thus constituting the only (and still tenuous) link between Galen and our Ps. Galenic text: see ch. 4.F.

33) Asphodel (*Asphodelus* sp. L)

Asphodel[1] grows in soft soils, spreads along the ground, and has leaves similar to the head of the leek. Its stem is smooth, rounded and lengthy, and at the very top it sprouts small branches that have a white flower and bear a small black seed. Its roots are also oblong and rounded, and are the most powerful part of the plant. These when burned down and reduced to ashes cure alopecia and pale blemishes. The water in which asphodel is cooked is drunk to relieve pain in the kidneys. Asphodel stimulates urination and menstruation and helps with chronic coughs. When crushed and applied topically it mollifies indurations.

1 The Latin equivalent of *albucus* given in *JPC* is probably an interpolation from Pliny 21.109.

34) Abrotanum

Abrotanum omnibus notum est et amaritudine et acrimonia sua similiter ut absinthium proficit, sic et lumbricos occidit.

1 Abrotanum] aprotanum *FM2* aprotanus *M* aparnum *C* omnibus] multis *W* acrimonia sua] acrimoniosos *F* 2 ut] *om. VF* sic] similiter *V om. FLP* occidit] expellit *J* educit et occidit *MM2W*

cf. Diosc. III.24 (ἀβρότονον). Pliny 21.60. *Ex herb. fem.* 69. *DynVat* 41/*SGall* 45. Isidore 17.11.7.

35) Anethum

Anethum uires habet acres et excalfactorias et diureticas aptas et idoneas rursus ad inflationes omnes intestinorum.

1 Anethum] anetum *BP* anemi herba est omnibus nascitur *C* herba omnibus nota *MM2* 2 rursus] prorsus ad *BLP* inflationes] infrigidationes *J* inflammationes *FM2* intestinorum] intestinorum mirifice facit et sanat *MM2* intestinorum uiuat *F* intestinorum facit *C* intestinorum mirifice facit *W*

cf. Diosc. III.58 (ἄνηθον). Pliny 20.196. Galen *Simp.* 6.1.45 (K.11, 832). Garg. *Med.* 28. Ps. Apul. 122. *DynVat* 34/*SGall* 38. Isidore 17.11.6.

36) Agrostis

Agrostis herba est quae in campis nascitur et habet folia quasi auena pusilla et ueluti uinculis in terra repentia, nodosa, radiculas quoque longas, spissas, nodosas, quae faucium uulnera glutinat. Potata urinam prouocat. Canes etiam quando uolunt purgari hanc herbam manducant.

1 Agrostis] acrotes *F* agrustis *P* de gramine quae dicitur agrostis *tit. J* 2 uinculis] uiticulas *V* iuticulas *F* lugiculas *B* in iaculas *L* lenticulos *C* pusilla- repentia] *om. W* 3 nodosas] nudos *F* faucium] fauciis *F* facem *B* facit *P* faciant *M2 om. W* 4 canes] canes cum his *B* cum his *P* canes- manducant] *om. W* quando] qui *B* cum *VFLCM* hanc herbam] acer acerua *B* acerba *P* 5 manducant] comedunt *C*

cf. Diosc. IV.29 (ἄγροστις). Pliny 24.178. Galen *Simp.* 6.1.3 (K.11, 810–11). Ps. Apul. 78. Isidore 17.9.104.

37) Armenium

Armenium optimum dicimus quod est leue, aequale, coeruleum, fragile, sine lapillis. Vires habet chrysocollae et tamen inferiores. Proprietate genas siccat et nutrit palpebras.

34) Southernwood (*Artemesia abrotonon* L.)

Southernwood is known to everyone. Both its bitterness and corrosive-
ness act like that of absinth,[1] and hence it also kills intestinal parasites.

1 #17, #22.

35) Dill (*Anethum graveolens* L.)

Dill[1] has sharp, heating, and diuretic properties which are especially good
for easing all types of gas in the intestines.

1 On its different uses in medieval medicine, see Stannard 1982b.

36) Dog's-tooth grass (*Cynodon dactylon* Pers.)

Dog's-tooth grass[1] is a plant which grows in fields and has leaves like little
oats,[2] which are knotty and creep along the ground like chains, and have
small roots that are long, thick and knotty, which agglutinate wounds in
the throat, and when drunk stimulate urination. Dogs eat this plant when
they wish to purge their stomachs.

1 The heading in *J*, 'De gramine, quae dicitur agrostis' (still listed among the As), seems
owed to Pliny 24.178–83, a general discussion of grasses that specifies types of *dactylon*,
but does not use the term *agrostis*. See ch. 2.J and plate 5.
2 #31.

37) Azurite

We say that the best azurite[1] is that which is smooth, flat, dark blue,
delicate, and without pebbles. It has the same properties as chrysocolla,[2]
though of lesser strength. It is particularly used for drying up watery eyes
and nourishing the eyelids.[3]

1 Armenium] armenio *V* armeniu *M* arminimum *P* armeonium uel minium *J* arminum *W* optimum] optimum est *B* leue] lenem *VM2* aequale] quasi *BPM2W* **2** uires- palpebras] *om. W* chrysocollae] criscole *V* ocellae *BP* inferiores] si enertiores *V P* inestiores *F* si incertiores *B* proprietate] propriam *V* proprie tamen *F* tamen proprie *P* **3** genas] genus *F* agenes *B* agenes oris *P* gignuntur *MM2* siccat] *om. F* palpebras] palestras *VFBPLC M2*

cf. Diosc. V.90 (Ἀρμένιον). Pliny 34.47. Varro 3.2.4. Vitruvius 7.5.8.

38) Alosachne

Alosachne est ueluti spuma maris quae in petris insidet et remanet. Vires habet quasi sal et omne nitrum.

1 Alosachne] alosaggno *V* halosachne *J* alosagine *F* alosacine *MM2* alosagne *P* alosanaus *C* alosachne *W* maris] *om. BP* quae in petris] *om. C* remanet] permanet *W* **2** omne] *om. WP* nitrum] aniatum *B*

cf. Diosc. V.110 (ἁλὸς ἄχνη). Pliny 31.74 (ex aquis maris … spuma). Galen *Simp.* 11.2.8 (K.12, 374–5).

39) Bdellium

Bdellium lachryma est arboris quae in Arabia nascitur quod in modum resinae [manat]. Optimum est quod perlucet cum aliqua pinguedine prout chrysocolla et gustu est amarum, mundum et nihil in se asclusum aut alienum habet. Molle et ab igne odorem suauem emittit. Potest
5   tamen penitus et in instanti excalefacere, durities neruorum remollire et stricturas relaxare et ea quae sinus collectionem habet euaporare. Sed et adulteratur gummi ipsum quod gustu et sapore deprehendimus, amittit enim suam mordicationem.

1 Bdellium] bdelli *V* bidellium *F* adellium *P* arabia] armenia *BP* rabia *C* optimum- perlucet] *def. L* manat] *om. VFMW* **2** optimum] *om. P* quod perlucet] *om. V* perlucidum *FMW* **3** prout-emittit] *om. W* chrysocolla] crasso colla *P* nihil- habet] idem animae adsilusum et nihil in se stercorusum aut alienum habit *V* id est animae asilosum inter se corosum *F* id est anima asylosum et *L* id est nimium clusum in eodem dextrosum *B* nihil in eo est tupe aut alienum molle habet et lene tactu odore *P* nihil in se sordidum aut alienum *J* sordidum idem anima ad filosum *M* asclusum] asilosum *C* **4** ab igne] lenem accatum *B* emittit] sufficiens *M def. L* potest tamen] potens hoc eum *V* potens autem eo quod *LFBCM* potest quidem quae multum calefacit *P* **5** et in instanti] instanter *V* instar *M om. P* neruorum] suboris *B* ruboris *P def. L* remollire] emollire *J* mollire *P* **6** sinus] nus *V* sine collectione *J* syrus *B* sirus *P* senus *MW* **7** gummi] cummen qua *L* ipsum] ipsius *VJL* illud *P om. F* quod gustu] *om. P* deprehendimus] deprehendere possum *B* **8** suam mordicationem] autem ab amartitudine *L* autem suam morditudine *B* mordica […]ant *C*

cf. Diosc. I.67 (βδέλλιον). Pliny. 12.35. Galen *Simp.* 6.2.6 (K.11, 849–50). Isidore 17.8.6.

1 Similarly Diosc. V.90, almost verbatim. According to Pliny 35.47, so-named because of its Armenian origin.

2 #66.

3 Despite the common manuscript readings of *palestras* ('it encourages exercise'), *palpebras* (*J*) matches Diosc. (τριχῶν τῶν ἐν βλεφάροις) and Pliny (*in palpebris*).

38) Salt froth

Salt froth[1] is like a sea-foam which settles on rocks and remains there. It has properties like those of salt and all the alkalis.

1 *alosachne*, from the Greek ἁλὸς ἄχνη, transposed likewise in Lat. Diosc.V.135 (*alosagne*, and V.144).

39) Bdellium (*Commiphora mukul* Engl.)

Bdellium is the sap of a tree that grows in Arabia and which is fashioned into a type of resin. It is best when it is transparent and a little oily like chrysocolla,[1] has a bitter taste, is pure, containing no splinters or anything extraneous,[2] is pliable, and releases its fragrance when heated by fire. It can heat deeply in an instant, soften indurations of the ligaments, relieve cramps, and disperse congestion in the chest.[3] We can recognize that which has been adulterated with gum from its taste and flavour, for it has a particular bite to it.

1 #66. On bdellium, see Innes Miller 1969, 69–81.

2 *asclusum (asclusum,* or *assulosus, astulosus),* from *astula*, 'splinter,' 'chip' (of wood or stone). *ThLL* records its use with *resinae (osclosae/asclusae)* among medical writers (Pelagonius, Chiron, Marcellus, *Antidotarium Bruxellensis*; cf. Pliny 12.105, *assulose* adv.), but especially the *AG*, noting manuscript variants (*VLMC*). Used also in #53 and #119. Cf. Diosc. I.67 'ἀμιγὲς ξύλων καὶ ῥυπαρίας'.

3 A difficult passage: I follow *FLCMW* (poss. *V*) in reading *sinus/senus,* which seems more likely than *sine collectione J*, 'disperse without causing abscesses' (?).

40) Balaustium

Balaustium flos est syluastici maligranati cuius floris aliquae sunt species. Inuenitur autem et album et purpureum et simile rosae et quidem ipsum uires easdem cytinis habet.

1 Balaustium] balaustia *WMM2* balaustion *F* syluastici] *om. CW* floris aliquae] aliquod *V* tres *W* duae *P* species] stipticis *V* 2 inuenitur] *om. P* rosae] roris *MM2* quidem] quod *V* 3 ipsum] usum *M2* cytinis] quid in his *V* ordinis *F* cum ceteris *BP* id est stipticas *C* caetinis *W* cidinis *MM2*

cf. Diosc. I.110 (ῥόα), I.111 (βαλαύστιον). Pliny 13.113, 23.110–14. Galen *Simp.* 6.2.3 (K.11, 847).

41) Butyrum

Butyrum fit ex lacte bubulo et ouillo. Est autem optimum ouillum quod colore subrufo et amplius est pingue. Quod ita conficitur et mittitur enim in cupam longam et cum uirga quae habet in capite rotulam, quae intra cupam capere possit, agitatur diutissime hoc ipsum lac. Et fit
5 ut quod est illius pinguissimum supernatet quod colligitur et in uasculo reponitur. Optimum est ergo butyrum quod est pinguissimum et ex ouillo lacte aestate factum. Vires autem habet leuiter relaxantes unde ad uuluas dolores et tumores et loca neruosa et delicata et siccitatem pulmonis proprie uidetur facere.

1 Butyrum] bytirum *V* butyris *F* butirum *P* bubulo] uaccino *P* ouillo] ouium *MM2* ouino *PW* 2 amplius] magis *W* pingue] pinguoribus *V* pingiores *F* quod- 7 factum] *om. W* et mittitur] emittitur *JLCB* 3 cupam] capillam *JLCB* cupulam longulam *P* quae habet- et fit] *om. BPC* 4 quae intra] quasi intra *V* et fit ut] sicut *VFM* 5 uasculo] uase *BP* 7 lacte] *om. M* aestate] statim *V def. M2* leuiter] leniter *VMM2PW* 8 dolores] *om. MM2* tumores] tumores facit *BP* loca] uocat *P* neruosa] neruosissima *VFMW* dolorosa *C* 9 facere] iubare *V* innare *F* adiuuare *CMW*

cf. Diosc. II.72 (βούτυρον). Pliny 28.133. Celsus 3.22.14. Scrib. 43. Galen *Simp.* 10.2.10 (K.12, 272–3).

42) Batrachium

Batrachium herba est in modum cicutae similis coriandro uel apio crescens ueluti ferula et in summo florem habet aureum et splendidum ualde et radices subalbidas et longas nascitur in locis humidis. Vires habet causticas et exulcerantes.

1 Batrachium] butracion *VW* butratium *F* botracium *MM2P* crescens] criscit *VWM* acrescens *J* 2 aureum] aurusum *VM* auro simile *P* 3 ualde] *om. V* radices- longas] *om. W* subalbidas] subexalbidas *VP* 4 causticas] chasticas *P* exulcerantes] efficaciam *B* exulcerantes efficaciter *P* cuius atriculas sanare cermini est *C*

cf. Diosc. II.175 (βατράχιον). Pliny 25.172. Galen *Simp.* 6.2.5 (K.11, 849). Ps. Apul. 8–9.

40) Flower of wild pomegranate (*Punica granatum* L.)

Balaustium is the flower of the wild pomegranate, and there are different types of flower. There is the white flower, the purple, and another that is pinkish, and they have the same properties as *cytini*.[1]

1 The flower of the cultivated pomegranate: see #183. Almost verbatim description in Diosc. I.111, although the βαλαύστιον varieties merely resemble (ἔοικε) the *cytini* (I.110): see ch. 3.B. Pliny 13.113, 23.110 seems to think *balaustium* is the cultivated pomegranate.

41) Butter

Butter is made from the milk of cows and sheep. The best, however, is that made from sheep's milk, when it has a slightly reddish hue and is very fatty. It is made by placing the milk in a deep cask, at the mouth of which is a small wheel with a stick attached that is submerged into the cask and used to stir the milk for a long time.[1] The fattiest elements then rise to the surface and are skimmed off and stored in a vase. The best butter is therefore that which contains the most fat and is made from the sheep's milk in summer. It has gently loosening properties and therefore seems to help specifically with vaginal pains and swellings, and similar complaints in other sensitive and delicate places, and also with dryness in the lungs.

1 This description of a butter churn with wheel device is unique: cf. Auberger 1999. Diosc. II.72 mentions only sheep's and goat's milk; cf. Pliny 28.133, 'the richest butter is made from sheep's milk – it is also made from goat's milk – but in winter the milk is warmed, while in summer the butter is extracted merely by shaking it rapidly in a tall vessel.' All three texts are correct in identifying sheep's milk with the highest fat content (7.50%, compared to goat 4.40%, cow 3.20%, horse 0.55%: Auberger 1999, 18 n.16).

42) Ranunculus (*Ranunculus* sp. L)

Ranunculus is a plant which like hemlock[1] is similar to coriander[2] or celery[3] and grows like giant fennel.[4] It is crowned by a golden flower which is truly resplendent, has roots that are long and whitish, and grows in moist places. It has caustic and ulcerative properties.

1 #15 n.2, and #49.
2 #68.
3 #14.
4 #4 n.1.

43) Bryonia

Bryonia herbae radix est plerumque habens uastitudinem rapae et longitudinem pastinacae. Viribus acris et calefactoria, propter quod oleum in quo est cocta ad acopa et malagmata facit.

1 Bryonia] brionia *VFPM2W* bryonia ut uiticella alba *C* (cf. Pliny 23.21) habens] *om.* *FLBCM* reperitur *W* uastitudinem] magnitudinem *P* rapae] propie *VFM* rapae inuenitur *P* pastinacae *M2* longitudinem] *om.* *M2* 2 pastinacae] pastinace soauem *V om.* *M2* oleum in quo] *om.* *LP* 3 est cocta] *om.* *BP* acopa] aguma *F* caput *MM2* facit] facit naturaliter nascens in plurimis locis *J* sit *M2*
cf. Diosc. IV.182. (ἄμπελος λευκή). Pliny 23.21–8. Galen *Simp.* 6.1.34 (K.11, 826–7). Ps. Apul. 67. Isidore 17.9.90.

44) Crocus

Croci [naturaliter nascitur pluribus locis] flos est rufus et odoratus, cuius est optimus quod longitudinem et aequalem habet uastitudinem et nimium in summo ueluti apicem albicantem. Recens et aspectus prout succinum perlucens et ad tactum non humidus et ad gustum
5 linguam submordens et saliuam diu multumque inficiens. Huiusmodi enim erit optimus licet cornetum, licet sium uel aegaeum uel selgium, uel syriacum et cyrenaicum et centuripinum siculumque appellamus, uel si quid est aliud a suo tactu aliter cognominatur. Vires autem habet omne crocum leuiter stypticantes. Mitigat enim omnes dolores com-
10 mistum collyriis, antidotis, et aliis quam pluribus compositionibus. Verum adulteratur oleo rosaceo perunctum atque de fructu laeuigata cum ea spuma argenti aut molybdaena aut lapide parto. Sed dignoscitur aliquid enim tactu puluerastrum inuenitur, odor sitosus et terrenus et de frustra austerus occurrit et est inertissimus.

1 Croci] crocum *VFMP* croco *M2* bulbi erratici *J* naturaliter- aequalem] conatur alibus siui nascens pluribus locis flus habit ruffu ut sit de hudoratu *V* naturalium sine nascitur pluribus locis errati pului flos est rustum et ut sit de auratum *F* natura orientalem siue nascens pluribus locus herratici bulbi flos est rufum et ut si *LBC* orientale nascitur pluribus locis cuius flos est rufus et odiferus. est optimum quod est bonae longitudinis et uastitatis *P* naturalibus notum est nascitur in pluribus *M id...* plurimis locis et radices plurimas *M2* pluribus notum nascitur in multis locis *W* flos- 8 cognominatur] *om.* *W* 2 longitudinem] cumbina longitudine *VLC* benum longitudinem *F* bona longitudine *B* 3 nimium] nimiam uastitudinem *J* nimium in somno *F om.* *P* albicantem] abiicientem *JL* recens- aspectus] *om.* *P* 4 non humidus] homini *V* ut humidus *B* 5 linguam] longum *B* saliuam- inficiens] aliquandiu multumque insitiens *F* diu] *om.* *BC* 6 licet- selgium] *om.* *M* cornetum] sicut rectum *V* corinetum *F* coryctum *C* coruptum *BP* corp(..)etum *M2* sium- selgium] *om.* *P* sium uel aegaeum] iscium uel egium *V* scium uel igneum *F* serum uel egenum *L* sicut uellerum *B* sicut aureum *C* scium uiraeum *M2* uel selgium] uel sifyacum *V* uel segium *F* ut selgum *C* scilicet *B* uel religeo *M2* 7 cyrenaicum] quirinaicum *VP*

43) Bryony (*Bryonia dioica* Jacq.)

Bryony is the root of a plant which is around the same width as the turnip,[1] and the same length as the parsnip.[2] In terms of properties it is sharp and warming. For this reason the oil in which it is cooked is commonly used in salves and emollients.

1 #233. See plate 6.
2 #222.

44) Saffron (*Crocus sativus* L.)

The saffron flower[1] [which grows naturally in many places] is red and fragrant, and is best when it is equally proportioned in length and width, and turns very white at the tips. When fresh its appearance is transparent like amber, it is dry to the touch, and its taste bites into the tongue and causes a considerable amount of salivation that lasts for a while. Among the best types are the 'horned' [or 'Corycian'], the 'sium' [or 'Lycian'], or 'Aegean' [Aigain], the 'selgian,' the Syrian, the Cyrenaic, and the Sicilian type we call the 'Centuripan,'[2] and other types are usually named after the way they feel to the touch. All types of saffron have mildly styptic properties, and they ease all types of pain when mixed with eye-salves, antidotes and in many other compound medicines. It is adulterated with rose-oil and applied as an unguent, and its flower is crushed together with litharge, or galena, or with Parthinian stone.[3] But the adulteration is revealed by its dusty feel and by its mouldy and earthy fragrance, and is pointless anyway because it turns sour and becomes extremely weak.[4]

1 The text for this entry is quite corrupt, owed to confusion with meadow saffron (*colchicon* #80), evident in the references to *bulbus erraticus*, esp. in *J*, where it served as the title term (hence listed among Bs).
2 The types here ('licet cornetum, licet sium') appear to be a corruption of the information found in Diosc. I.126, who ranks the 'Corycian' first, followed by that from 'Corycos which faces Lycia, and the one that comes from Lycian Olympus, and then that from Aigai in Aitolia. But the Cyrenaic and that from Centuripae in Sicily have no strength.' The 'Syrian' is not mentioned, nor any other types. Similarly Pliny 21.31–4. On medicinal uses in the ancient world, Goubeau 1993.
3 litharge #256, galena #182. Parthinian stone (following *VL*, *lapide parto*) is obscure: *Part[h]us* most likely refers to the town in Illyricum near Dyrrachium (mod. Durrës, Albania), though it may also be a corruption of *parthicus*, that is, 'Parthian stone,' from Parthia. Cf. *J* 'Spartan stone.' The parallel passage in Diosc. ibid. has 'daubed with concentrated must' (ἐψήματί τε ἀλείφεται).

cimiriacum *C* sirinaicum *B* centuripinum- cognominatur] *om. M* **8** tactu] tractu *VFPM2* cognominatur] cui nominatur *V* cum nominatur *M2* **9** leuiter] leniter *VFPC* mitigat] dolores sedat *BP* **10** compositionibus] compassionibus *P* **11** Verum- 14 inertissimus] *om.* *W* fructu] fractu *F* fricat *M* fluctu *P* **12** cum ea] commeos *V* cum eo *F* idoneo *M* cum oleo *P* parto] partos *F* partho *BP* pasto *M* piriat *C* sparto *J* dignoscitur] agnoscitur *M* **13** puluerastrum *L*] berastrum *V* puluerastum *F* puluis statim *BPC* lactata puluerastum *M* bulluastrum *J* sitosus] spissus *BP* infusus *C* sit usus *M* terrenus] *om. C* **14** et de frustra- inertissimus] de frustra martis occurit et est quoniam ementissimum *V* de frustra asterito occurit et est inistis *F* sed cum mixtum et alienum occurrit et est inertissimus *BC* defrustatum occurrit et est inertissimum *P* de frustra austere occurrit et ertissimum *M* de fructu austerus occurrit et est entissimus *J*

cf. Diosc. I.26 (κρόκος), IV.83 (κολχικόν). Pliny 21.31–3 et passim. Galen *Simp.* 7.10.57 (K.12, 48). Isidore 17.9.5.

45) Cadmia

Cadmia fit in metallis ubi aes, cyprum et argentum coquitur et conflatur quomodo prima terrea refossa est. Nam quod est terreum coctura ipsa separatur et quasi fuligo lateribus et cameris ex ipsa plenitate furnorum agglutinatur. Et sic tempore crassatum et in speciem lati lapi-
5 dis induratum auellitur. Est ergo optima cadmia et quasi prima quam Graeci botrytin appellant, quae extrinsecus quasi racemulos habet et intrinsecus subuiridicat. Secunda quam onychiten appellamus. Tertia ostracitis, quae est tenuissima et aspectu testae assimilis. Omnis tamen cadmia styptica est et uehementer stringere potest. Propter quod col-
10 lyriis miscetur et uulnera sordida expurgat et mundat et ad cicatricem perducit.

1 Cadmia] catmia *VFM2* cathmia *P* aes] *om. VM* est *FP* **2** quomodo- refossa est] quoniam prima terrea pressa sunt *J* refossa] refulsa *F* effodiendo inueniunt *W om. C* terreum] terreneum *F* arenum *C* separatur] reparatur *V* terrenum contra ipsa separat *L* **3** fuligo lateribus] foligola terribus *V* metallis lateribus *W* et cameris] cornaris *C om. P* ex ipsa- et in speciem] formonsum adinglutinatur *V* formosum *F* **4** agglutinatur] conglutinator *P* lati] luti *V* uelut *F om. BP* **5** auellitur] euellitur *VMP* **6** graeci] crescenatem *B* crescentem *P* botrytin] butrin *V* bustrin *F* BOTPYITEN *L* batroidem *B* botroidem *P* BWTRIZEN *C* botros *W* botrite *M* extrinsecus] *om.W* racemulos] circa cimulos *M* quasi- subuiridicat] *om. L* **7** subuiridicat] *om. F* onychiten] moniticin *V* AECUNIKITEN *L* monicites *B* monieites *P* AEVHIKYTEN *C* monicitis *W* unitrici *M* **8** ostracitis] osiradicis apellamusque *V* oerraticis *F* WCERATIKIC *L* articis *B* straticis *P* WCTPATITEN *C* strautes *W* honeraticis *M* aspectu testae] aspecto est *V* apectu tere *F* colore terrestreo *W* **9** quod] hoc *VF* **10** et mundat] *om. J* mundat- perducit] *om. W*

cf. Diosc.V.74 (καδμεία). Pliny 34.100–5. Galen *Simp.* 9.3.11 (K.12, 219–21). Isidore 16.20.2, 16.20.11–12.

4 Corrupt, here following the inferences of *BCPM*.

45) Calamine / Zinc Ore

Calamine is made when the metals of bronze, copper,[1] or silver are heated and forged as soon as they are dug out from the ground, for it is separated from the dirt when heated, and then it sticks like a thick soot upon the walls and ceiling of the furnace. It is left there to thicken until it hardens like a type of flat rock before it is chipped off. The best calamine and ranked first is that which the Greeks call *botrytis*, and which on the outside has small clusters, and inside turns a pale green colour. The second best is that we call *onychitis*. The third best is *ostracitis*, which is very delicate, and looks like earthenware.[2] All calamine is styptic and can be extremely astringent. Because of this it is mixed with eye-salves, cleans and sanitizes infectious wounds, and cicatrizes them.

1 Possibly 'Cyprian copper': cf. Diosc. V.74; Pliny, 34.100–5.
2 These three types are similarly recorded by Diosc. and Pliny, ibid.

46) Cerusa

Cerusa sic sit. Acetum acerrimum funditur in amphoram ut faciat quasi dimidiam, deinde plumbum dilatatum et extenuatum super acetum
suspenditur. Et sic ipsa ui aceti remissa in solutu plumbo in amphora,
ueluti faex subsidet, quam faecem de limpidato aceto laeuigamus et in
5 sole siccatam terimus. Et ex aqua tamdiu lauamus ut nihil habeat asperum et laeuigatam ex aqua in sole siccamus. Cuius est optima quae est
candidissima et non grauissima et quae lingua non denigratur. Vires
autem habet stypticas, sed ex aqua et uino potata uenenosa esse dicitur.

1 Cerusa] cerussa *V* cerosa *FP* cerossa *WM2* acerrimum] agrissimum *V* acris *F om. C*
faciat] sit *V* 2 dimidiam] media *V* 3 solutu plumbo in] ima *L om. FBMW* ui] uiam *V om.*
FP 4 subsidet] residet *FW* faecem] confecit *V* 5 ex- 7 denigratur] *om. W* tamdiu lauamus]
dilutuamus *M* 6 ex aqua] ex qua *J* ex aqua] aquam *JVFBL* 7 lingua] lingua incidat *V*
lingua inudatur *F* inudata *L* lingua inundata *M* liniata *B* linita *P* 8 ex] *om. V* et] uel *V*
et lingua submordit dentes quoque et saliua et manus inficit grauem uino *M2* et uino
potata] uel humo portata *F* dicitur] meditur *C*

cf. Diosc V. 88 (ψιμύθιον). Pliny 35.175. Isidore 19.17.11, 19.17.23.

47) Cummen

Cummen lacrymum est quam plurimarum arborum omnibus notus
est, cuius quasi optimum eligimus quod est candidum et mundum et
glutinosum. Praestat autem quod et tragacantham facere dicimus.

1 Cummen] cumen *VF* chitran *J* gummi *LBCW* lacrymum] lacticinium *J* 2 est] *om. VW*
3 tragacantham] tracantum *VFBM* dragacantum *M2* dracantum *C* dragantum *LJ* draga-
gantum *W*

deest P

cf. Diosc. (κόμμι) I. 97, 101, 113, 121, 123. Pliny 16.108, 13.66, 24.3. Galen *Simp.* 7.10.40
(K.12, 34). Isidore 17.7.70 (gummi).

48) Chalcitis

Chalcitis gleba est naturalis quae est in Cypro insula in metallis inuenitur, colore subaureo. Intus uenas habens diffusas prout alumen scissum
et in modum stellarum fulgentes. Cuius est optima quae est recentissima et mundissima et aequaliter fragilis. Potest igitur sine mora uehe-
5 menter stringere. Vnde ad profluuium sanguinis, humoris et uentris
habetur utilissima.

1 Chalcitis] calcetes *V* calcites *FP* chalcidis *M2* naturalis] *om. W* cypro] ebron *M2* 2
subaureo] subauroso *P* habens] *om. FP* scissum] iscissum *V* sciston *BP* 4 aequaliter]

46) Cerussite (White lead ore)

Cerussite is made in the following way. Extremely sour vinegar is poured into a pitcher until about half-full, then lead is thinned out and placed over the top so that it hovers over the vinegar. That which falls off the lead and into the pitcher from the force of the vinegar then settles like dregs, which are washed until free of vinegar and then are pulverized and spread out in the sun to dry. We then wash them again in water for a long while until they lose all coarseness. We then remove them from the water, and again pulverize them and dry them in the sun. The best is that which is brightest and not extremely heavy, and which does not blacken the tongue. It has styptic properties, but is said to be poisonous when drunk with water or wine.

47) Gum (from trees)

Gum[1] is a sap obtained from many trees and is well known. We select the best, which is when it is white, pure, and sticky. It can be used for the same things we state for tragacanth.[2]

1 The reading of *chitran* in *J* puzzled Meyer 1856, IV: 492, who raised the possibility that the word derives from Arabic, but it is not attested in any manuscripts. Diosc. refers to gum in several places (e.g., I.97, 101, 113, 121, 123) but accords it no entry of its own. Cf. Pliny 16.108, 13.66, 24.3.
2 #278.

48) Rock alum

Rock alum[1] is a natural deposit the colour of bronze which is found in mines on the island of Cyprus. It has extensive veins inside just like cut alum and which sparkle like stars. Rock alum is best when it is freshest, purest, and evenly brittle. It can be a fast-acting astringent, and is considered most helpful to staunch discharges of blood [and humour], and stopping diarrhoea.

1 The description here (*Chalcitis- stellarum fulgentes*) is found verbatim among the artisan recipes in Lucca, Biblioteca Capitolare 490 (ed. Hedfors 1932, 15; context, Everett 2003,

qualiter *V om. W* potest igitur] *om. F* unde] *om. V* **5** humoris] *om. VFMM2* humoris et tumores *B* tumorem *P* **6** habetur utilissima] utilis est *W*

cf. Diosc. V.99 (χαλκῖτις). Pliny 34.117. Galen *Simp.* 9.3.35 (K.12, 241–2). Isidore 16.15.9, 16.20.11.

49) Conium

Conium herba est omnibus nota, cuius semen et simillimum aniso et ad usum medicinalem colligitur, refrigerat autem et stringit. Sed et succus illius expressus seruatur nam et ipsum eadem praestat. Verum optimum conium habetur Creticum.

1 Conium] comum *J* conobium *BP* simillimum aniso] silis mannis sic *V* similis numanersi *B* simile mumanersi *P* similis aniso *FMW* **3** expressus] *om. M* uerum- creticum] *om. WM* **4** habetur] est *P*

cf. Diosc. IV.78 (κώνειον). Pliny 25.151, 20.132 (cicuta). Galen *Simp.* 7.10.68 (K.12, 55).

50) Castoreum

Castoreum animalculum quadrupes et qui castor appellatur testiculus eius est quod partim in aqua partim in terra uiuit et circa aliqua flumina capitur. Verum optimum castoreum est ponticum, hoc enim plenissimum inuenitur et cohaerentibus suis testiculis positum, siccum 5 colore cereo, odore grauissimo. Potest igitur firmiter excalfacere et relaxare et sic elimare longinqua uitia neruorum quae nimiis doloribus contrahuntur.

1 Castoreum] castorium *JM2* animalculum] animal *V* culines sunt animalis *P* castor] fiber *W* appellatur] uocatur *P* **2** eius- uiuit] *om. P* eius- aqua] est pars quoque in qua parte terra *V* est quoque parte aqua est in terra *F* quod in qua parte in terra *L* in terra aliquando *W* in terra uiuit et] ut uite *VMM2* aliqua] *om. P* **3** capitur] capituret dum queritur et persequitur ut sibi auferantur testiculi *W* hoc- odore] nigrum *P* **4** plenissimum] hoc enim optimum est et plurissimum *V* plenis *F* plurissimum *MM2 om. W* cohaerentibus] quaerentibus *B om. W* positum] penitus *VFLBC* penitus abscidit *W* **5** cereo] renio *V* rineo *FL* nigro *BW* sine *M* subrufum *C* grauissimo] gratissimo *F* **6** longinqua] diuturna *P*

cf. Diosc. II.24 (κάστορος ὄρχις). Pliny 8.109, 32.101. Celsus 6.7.8. Scrib. 3. Galen *Simp.* 11.1.15 (K.12, 337–41).

51) Chalcanthum

Chalcanthum in metallis inuenitur cuius est quod naturaliter foditur. Cuius quoque optimum dicimus quod est colore coeruleo potius quam pallido. Et aliud quod ex hoc quidem fit admixta quoque aqua copiosa coquitur in plumbeo uase satis amplo. Et sic ex partibus copiosis

155–9, 309), the date of which (saec. VIII ex) is close to that of *L*, suggesting that an earlier copy of the *AG* existed at Lucca.

49) Hemlock (*Conium maculatum* L.)

Hemlock is a plant known to everyone. Its seed is very similar to anise,[1] and is harvested for medicinal use, for it cools and is an astringent. But also the extracted juice can be used, and achieves the same effects as the plant itself. The hemlock from Crete is considered the best.

1 #15 n. 2.

50) Castoreum

Castoreum is the testicle[1] of a little four-legged animal called a beaver which lives both in water and on land, and is captured next to rivers in one way or another. The best castoreum is that from Pontus, for this is found to be the richest, and when it is from two testicles that are joined together[2] and is dry, pale yellow in colour, and extremely pungent in fragrance. It can therefore thoroughly heat and loosen, and thus can help eliminate chronic ailments of the tendons which cause excessive pain.

1 As believed until the sixteenth century, but castoreum is obtained from an internal gland located near the testicles: Berendes 1902, 161.

2 Similarly Diosc. II.24, 'Always choose testicles that are joined together at one end – for it is not possible to find two sacs joined together in one membrane.'

51) Vitriol

Vitriol occurs naturally in mines where it is dug out. We declare the best to be that which is bluish in colour rather than the pale type. There is another which is made from this last by cooking it in a lot of water in a spacious leaden pot with plenty of room. On the larger parts that are submerged a

5 instanter compositis ueluti laminae insequuntur. Post quod ipsum
chalcanthum inuenitur in glebulas dispartitum particulas agglutinat.
Verum est optimum, quod est colore prout diximus coeruleo, graue,
mundum, perlucidum et uiscide stringere potest.

1 Chalcanthum] calcantum *VFWM2* chalchantum *P* cuius- foditur] *om. C* est- est] *om.*
M quod] optimum est quod naturaliter *etc. P* **2** potius quam] potiusque *J* **3** et aliud] si
est graue mundum perlucidum. est aliud *P* aliud- uerum] *om. W* copiosa] quo pixa *V*
om. M **4** plumbeo] plumbeo modum *B* in modum uarosum *P* satis amplo] amplius *P* sic
ex partibus] siccat partica *L* sic ex particulis *P* copiosis- *fin.*] coponitur et conglutinatur.
Omne calcanthum uiscide stringere et efficaciter potest *P* **5** laminae] *om. VFLBP* formu-
lis *C* insequuntur] inseruntur *FM* **6** inuenitur] *om. J* particulas] asparti cotilis *F* sparti-
toris *J* sparticulis *L* ex partis cuius *B* ex parte utile *M* utile *C* agglutinat] adglutinatur *VF*
7 graue- perlucidum] *om. W* **8** potest] potest efficacia *B*

cf. Diosc. V.98 (χαλκανθές). Pliny 34.123. Galen *Simp.* 9.3.34 (K.12, 238–41). Isidore
16.2.9.

52) Chalcanthos

Chalcanthos in metallis fit quando enim aes remissum est et purificatur, aqua frigida aspergitur subito autem in summo cocit. Ipsum flos
est in specie milii, et optimum est quod est rotundum, graue, et aliquatenus splendidum, siccum, fragile si laeuigetur. Adulteratur tamen sco-
5 bae aeris quae frangitur. Verum enim uero aeris flos uires habet uiscidas
et stringentes, sic enim cicatrices deducit et factas in oculis extenuat.

1 Chalcanthos] calcantum *V* calcantes *F* item de chalcantho *tit. J* calcanthos *B* chalanthos
L quando] cum *F* enim aes] minime *BL* enim est *F* est *C* purificatur] potatur *V* putatur
L portatus *F* comminutum *B om. C* **2** frigida] stringida *B* in summo cocit] subito autem
in sume cocet *V* aut in fimo conficitur *J* sub informatum coquitur *B* **4** adulteratur- fran-
gitur] *om. B* tamen scobae aeris] enim uel scopea eris *V* enim uel ut scabia et res *F* scobae]
spinae *C* **6** oculis] in gulis *J* in culis *L* maculas *F* in collyris *B* extenuat] *om. V*

deest MM2PW

cf. Diosc. V.77 (χαλκοῦ ἄνθος). Pliny 34.107. Galen *Simp.* 9.3.37 (K.12, 242). Isidore
16.2.9.

53) Crocomagma

Crocomagma fit expressis aromaticis crocini unguenti et est quasi gleba lata et oblonga. Cuius optimum dicimus quod croci habet colorem
et similiter redolet ut myrrha et gustu linguam submordet, dentes quoque et saliuam et manus inficit, graue et minime asclusum. Vires quo-
5 que et ipsum croco repraesentat, id est stypticas, nam et hoc leniter

thin layer quickly develops, and the vitriol is found stuck to this layer in little separate lumps of particles.[1] The best type is that which almost blue in colour, as we have said, but also heavy, pure, translucent, and it can be used as a powerful astringent.

1 Tentative, following *V*, and the comments of Diosc. V.98, concerning the type of *chalcanthes* (*epthon*, 'boiled') made in Spain which 'divid[es] into cubical shapes that touch each other like grape-clusters' ('διαιρούμενον κυβοειδῆ σχήματα βοτρυδὸν ἀλλήλοις προσεχόμενα'); similarly Pliny 34.123 'limus vitreis acinis imaginem quandam uvae reddit.'

52) Flower of copper

Flower of copper[1] is formed in mines when copper has been smelted, purified, and then immediately sprinkled with cold water, causing the flower to condense on its surface.[2] The flower looks like a type of millet, and it is best when it is rounded, heavy, somewhat glittery, dry, and brittle so that it is easily pulverized. It is adulterated with copper filings, which weaken it. For indeed flower of copper has powerful and astringent properties, and so it reduces scar tissue and corneal abrasions.[3]

1 Transliteration of the Greek (χαλκοῦ ἄνθος) seems to have caused confusion with the previous entry, perhaps explaining its absence in *MWP*. The text here has been reconstructed with a view to that of Diosc. V.77, which it follows closely in places.
2 Difficult passage, but clearly following the sense of Diosc. ibid. 'the attendants, after removing the dirt, pour on it very fresh water, their purpose being to cool it. From the sudden condensation and contracting the flower forms on the surface, as if it were spat out.' Cf. *J*, 'sprinkled with water ... or covered in dung.'
3 Difficult passage: see ch. 4.H.

53) Saffron residuum

Saffron residuum is made by pressing the aromatics used to make unguent of saffron[1] to form a lump that is flat and oblong. We declare the best type to be that which has the colour of saffron, emits a fragrance similar to myrrh, has a taste that gently bites the tongue, but also stains the teeth, saliva, and hands, is heavy, and has no splinters.[2] It has the same properties

stringit. Plures autem in compositione propter precii distantiam pro crocomagmate mittunt crocum.

1 Crocomagma] crocomagna *FM2* expressis] ex tritico *P* crocini unguenti] *om. P* gleba] gliuola *VF* 2 lata] *om. F* 3 redolet ut] *om. F* olet *P* submordet dentes] submordentes *V* 4 manus] nimis *WP* graue- asclosum] *om.W* asclusum] usclosum *L* arnoglosum *C* asolosum *P* asclosum *J* 5 repraesentat] ut crocum *P* id est] *om. VFMM2WP* lenitur] *om. BPC* 6 plures] flores *F* compositione] profetiones *B* distantiam] substantiam *M* 7 crocum] *om. FCM*

cf. Diosc. I.27 (κροκόμαγμα). Pliny 21.139. Isidore 17.9.6.

54) Cucurbita syluatica

Cucurbita syluatica nascitur maxime in Aegypto, magnitudine similis infantili pilae. Gustu amarissima nota quam pluribus. Haec igitur amaritudine et acrimonia sua uiscide et cum uexatione totum corpus commouet, sic et uentrem purgat et in antidotis commiscetur.

1 Cucurbita] coloquintida est cucurbita siluatica *CWMM2* maxime] maxime coloquintidas *F* aegypto] eos ponti *BP* magnitudine] *om. P* magnitudine-pilae] *om. CW* 2 nota- pluribus] *om.W* 3 acrimonia] acredine *P* sua] *om. F* sua- uexatione] *om.W* in- commiscetur] unde catarticis adhibetur *W* commiscetur] miscetur *P om. L*

deest V

cf. Diosc. IV.176 (κολοκύνθα ἀγρία). Pliny 19.69–75, 20.8, 20.14–17. Garg. *Med.* 6. *Dyn-Vat* 27/*SGall* 31. Isidore 17.9.32

55) Centaurea

Centaurea herba est origano assimilis, aliqua maior et aliqua minor. Verum folia habet magis oblonga quam rotunda, incisa quasi ferula. Ramulos ab ipsa radice plures et longos in quibus summis floscellum est purpurastrum. Vires quoque habet cum aliqua exasperatione laxan-
5 tes. Vnde totum uentrem purgat et nigram bilem prouocat et menstrua amouet et partum expellit. Radix autem eius est uasta, candens uelut ignis, mollis, fragilis, succosa, acris cum magna amaritudine. Cuius succus est colore subrufo et facit suspiriosis et tussientibus.

1 Centaurea] centauria *LFP* est] est omnibus nota et *BP* est medicis omnibus nota *M* origano assimilis] origano foliis similis *W om. M* aliqua maior] *om. F* 2 uerum- purpurastrum] *om. W* magis oblonga] magis sublonga *F* sublonga *M* incisa] in cera *L* incisaferula] *om. C* ferula] sertulas *P* 3 summis] ramis *M* summitate *P* floscellum] flos illius *J* 4 pupurastrum *FLBCMM2*] purpuratus *PJ* exasperatione laxantes] asperitat relaxantes *B* relaxantes *FMW* cum amaritudine relaxantes *P* exasperatione non relaxantes *M2* 5 uentrem] corpus *C* purgat] diriuant *P* bilem] coleram *BP* prouocat] *om. C* 6 amouet] ammonet *BP* prouocat *M* mouit *M2* mouet *W* radix- subrufo] *om. W* radix- succosa]

as saffron itself, that is, styptic, from which it is gently astringent. Many use saffron in place of saffron residuum in making compound medicines because of the difference in price.

1 Similarly Diosc. I.27, who informs us elsewhere (I.54 κρόκινον) that these are saffron and myrrh. On 'aromatics,' see ch. 1.D.
2 *asclusum*, see above #38 n.2.

54) Wild gourd (*Citrullus colocynthis* Schrader)

Wild gourd[1] grows chiefly in Egypt,[2] to about the size of a child's playing-ball. It has an exceedingly bitter taste which is well known to many. Its bitterness and pungency forcefully agitate the entire body with shaking, hence it purges the bowels, and is mixed with antidotes.

1 Given again below as colocynth #226 (and note *CWMM2*, idem Scrib. 20, Marc. *Med.* 32).
2 Despite their considerably lengthy descriptions, neither Diosc. nor Pliny mentions Egypt, but the *Mulomedicina Chironis* (201, 304), a veterinary treatise of the fourth century AD (and subsequently Vegetius's *Mulomedicina* [1.17.2], which borrowed from heavily from it), calls colocynth 'cucurbita aegyptia.' André 1985, 80.

55) Centaury (*Centaurea centaurium* L.)

Centaury is a plant that is similar to oregano,[1] though there is one type that is larger, and another type that is smaller than oregano. The leaves are more oblong than rounded, yet have jagged edges like giant fennel.[2] From right down near the roots it sprouts many small, long branches, at the tips of which appears a small purplish flower. Centaury has properties that loosen with a degree of irritation. It therefore purges the bowels, forces out black bile, brings on menstruation, and expels the foetus. Its root is large, gleams like fire, yet is soft, delicate, and juicy, and has a sharp taste with a good deal of bitterness to it. Its juice is a tawny colour, and it helps those suffering from respiratory problems and coughing.

1 #206. See plate 7.
2 #4 n.1. 'more oblong' is probably corrupt: *magis* was more likely intended to distinguish the type.

radix ep(…) ipsa herba et semen et flos uehementer sunt *C* uasta] ista *F* iuxta *M om. P* candens] ardens *JC* cadens *B* 7 succosa- cuius] *om. FP* 8 facit] prodest *W*

deest ex lacuna V

cf. Diosc. III.6 (κενταύριον τὸ μέγα). Pliny 25.66, 25.142. Galen *Simp.* 7.10.17 (K.12, 19–20). Ps.Apul. 34. Isidore 17.9.33.

56) Chelidonia

Chelidonia herba est omnibus nota quae succum quasi croceum emittit, qui in ictericis citrinis et in nigris bibitur et collyriis mista ad caliginem oculorum facit.

1 Chelidonia] celidonia *LBPCMM2* celedonia *F* omnibus] medicis omnibus *W* succum] succum expressa *W* signum *M* croceum] crocatum *LBPC* 2 in ictericis citrinis] et sibi sinceris *FLBM* sincerem *P* sincerum *CW* bibitur] missum *FM om. BPW* sed post [eius] inigratur quod 3 *C* facit] prodest *W om. M2*

deest ex lacuna V

cf. Diosc. II. 180 (χελιδόνιον μέγα). Pliny 25.50, 37.36. Galen *Simp.* 8.12.9 (K.12, 156). Ps. Apul. 74. Isidore 16.9.6, 17.9.36.

57) Cedrum

Cedrum lachryma est arboris eiusdem nominis, solidum, uiscidum quoque et ipsum efficaciter et cum uexatione refrigerat et stringit, propter quod maxime causticis admiscetur.

1 Cedrum] cetrum *M2* solidum] alidum *LC* uiscidum] uiscide *P* 2 cum] omni *J* con *M* cito *W* 3 maxime] *om. W* causticis] catusilicis *F* causaricis *P* cum sincis *C*

deest ex lacuna V

cf. Diosc. I.77 (κέδρος). Pliny 24.17, 13.53–4. Galen *Simp.* 7.10.16 (K.12, 16–19). Isidore 17.7.33.

58) Cinnabar

Cinnabar non est sanguis draconis, quod aliqui putant, sed fit de lapide quae dicitur argyritis arena, et est optimum quod in Hispania fit hoc nam splendidissimum est et coloratissimum est, colore sanguineo ueluti coccineo. Potest igitur leniter stringere, quapropter quidem collyriis
5 additur. Sed et coloris gratia inficit enim colorat adhuc magis quam haematites. Aliqui dicunt ipsum esse minium cinnabar.

1 Cinnabar] cennuar *V* cinobar *F* cinadar *J* cynnaber *L* cennaber *M* cennabar *M2* cuinabar *C* cinnabaros *W* cimnabar *P* 2 dicitur argyritis] cum mariayntides *BP* cum artentide sablaita *F* argentidis subablata *M* arcintide *C om. W* arena] alleta *L* abletat *B* abiectat *P*

56) Celadine (*Chelidonium majus* L.)

Celadine is a plant known to everyone and which discharges a juice the colour of saffron that is taken as a drink for yellow and black jaundice. Mixed with eye-salves it clears up dim-sightedness.

57) Cedar (*Juniperus* L.)

Cedar is the sap from the tree of the same name,[1] and is solid, sticky, and thoroughly cools and contracts with some discomfort, and therefore is often used in caustics.

1 Cf. Diosc. I.77, and Pliny 24.17, noting that the sap is called *cedria*.

58) Cinnabar

Cinnabar is not dragon's blood, as some think,[1] but is made from a stone which is said to be a sand containing silver-ore,[2] the best of which is found in Spain, for it sparkles the most and is the richest in colour, the colour of blood or scarlet. It can be used as a gentle astringent, and for this reason is added to eye-salves. But its rich colour also stains and colours more than hematite.[3] Some claim that minium is actually cinnabar.

1 Likewise Diosc. V.94, and also Pliny 33.111–22, who blames Greek authors for confusing it with 'Indian cinnabar,' which he believed was the 'gore of a snake crushed by the weight of dying elephants, when the blood of each animal gets mixed together' (33.116). Cinnabar

om. C hispania] spania *FM* ispania *P* vim spalua *L* hoc- igitur] *om.W* 3 coloratissimum]
odoratissimum *L* coloratis colore *F* ueluti] id est *F* 4 coccineo] coctineo *P* leniter] leuiter
P 5 additur] adhibitur *V* adhibetur *F* adhiuiatur *M* sed- gratia] *om. BP* gratia] grasta *V*
graue *M om. W* 6 haematites] emmatiraitis *V* aematicis *F* emeor *M* aliqui- minium] aliqui
dicunt ioierrenmium mela et senebar *V* alii putant quod ipsum sit cinnabar *W* minium]
nimium *FLBPM om.CW*

partim def. V

cf. Diosc. V.94 (κιννάβαρι). Pliny 33.111–22. Galen *Simp.* 9.3.12 (K.12, 221). Isidore
19.17.7.

59) Cera

Cera omnibus locis inuenitur, sed optimam dicimus quae aliquatenus
pinguis est et munda et odore suaui et quasi melleo. Verum omnis cera
remollit et laxat durities. Propter quod in emplastris et acopis et malagmatibus mittitur.

1 omnibus locis] ubique gentium *W* dicimus-pinguis] *om. P* **2** pinguis] adhuc pinguis *F*
melleo] mel *BP* uerum] uires habet *MW* **3** remollit et laxat] remollitur relaxat *V* molliter
et relaxat *FM* remollire et relaxare possunt *B def. L* emplastris] inest plastris *C*

cf. Diosc. II.83 (κηρός). Pliny 21.83–5, 22.116–18. Galen *Simp.* 7.10.23 (K.12, 25–6).

60) Careum

Careum est semen omnibus notum, cuius et efficacia fere talis est qualis et apio hoc tamen aliquatenus uirosius uidetur esse et magis ad compositionem idoneum. Verum in antidotis et in purgatoriis mittitur.

1 Careum] carium *V* careum id est carnate *F* careo *M* **2** omnibus notum] optimum est *CW*
cuius] cuius et usus *JBPC* fere] fertur *V* fere talis est] tenetur *C* tenetur talis *P* **2** uirosius]
uirius *V* maioris *W* uariosuis *C* uiriosius *P* uidetur] inuenitur *F* **3** uerum- mittitur] *om.*
W mittitur] *om. F*

cf. Diosc. III.57 (καρώ). Pliny 19.164.

61) Cyperus

Cyperus radiculae sunt pusillae non dissimiles oblongis oliuis, quarum
eligimus quae sunt grauiores et pleniores et duriores et ueluti lanugi-
nem habentes ad similitudinem spicae et ipse colore fusciores, redolentes suauiter et gustu acres et subamare. Viribus excalefacientes et
aliquatenus stringentes.

1 Cyperus] cypirus *V* ciperus *FPM2* oblongis] *om. C* oliuis] sublimes *BP* quarum- sub-
amare] *om. W* **2** pleniores- duriores] *om. V* pleniores] sed splendidiores *J* lanuginem] lan-
agines *V* **3** ad- fusciores] *om. P* similitudinem] similem *VLM* assimiles *B* spicae] specie

is an exudation that can be obtained from several different plants, most commonly of the Draceana genus (esp. *Dracaena draco* L. and *Dracaena cinnibari* L.), but also the Croton, Daemonorops, and Pterocarpus genera. Cf. Riddle 1985, 154–5.

2 *argritis*: a type of silver dross, or litharge of silver: Pliny 33.106. According to Diosc. ibid., this is called *minion* and is erroneously confused with cinnabar.

3 #161.

59) Wax

Wax is found everywhere, but we declare the best to be that which is somewhat greasy, pure, and has a sweet, honey-like fragrance.[1] Certainly all wax softens and weakens indurations. Because of this it is used in plasters, salves, and emollients.

1 Similarly Diosc. II.83.

60) Caraway (*Carum carvi* L.)

Caraway is a seed that is known to everyone. Its effectiveness is almost the same as celery[1] but it seems to be a little more fetid, and makes a better additive to compound medicines. It is also mixed with antidotes and purgatives.

1 #14. Similarly Diosc. III.57, but likens its properties to anise seed.

61) Galingale (*Cyperus rotundus* L.)

Galingale consists of its small roots which look like oblong olives. We select those roots which are heavier, more plump, and also a little hardier, which have a woolly down similar to spikenard;[1] their colour is darker, they emit a sweet fragrance, and have a sharp and slightly sour taste. They have heating properties, and some astringent properties also.

1 #236.

V picces *B* fusciores] fortius *B* fortiorem *M* redolentes suauiter] habentem sed *F* fortius redolentes *P* **4** gustu] sucus gustu *P* excalefacientes] excalefacientes et dioreticas *W* **5** aliquantenus] aliquatorius *L*

cf. Diosc. I.4 (κύπερος), I.5 (ἕτερον εἶδος κυπέρου). Pliny 21.115–17. Galen *Simp.* 7.10.65 (K.12, 54). Isidore 17.9.8.

62) Coccus cnidius

Coccus cnidius fructus est herbae quae in montuosis locis crescit, autumno colligitur et in sole siccatur. Duos habet ueluti cortices, unum extrinsecus rubrum, alium intra se, quasi nucleum album habentem. Verum purgat uehementer, [et si sincera agglutinatur] et nimio suo calore prout ignis [orem et fauces] excalefacit.

1 Coccus cnidius] coco nidius *VF* cocco nidium *BP* cocco gnidium *CMM2W* locis] *om.* *J* crescit] nascitur *P* **2** autumno] aut omnino *F* aut anno *L* locis auerno *C* tamen omnino *M* et uocatur alliptados uel lauriola *W* in sole- habentem] *om.* *W* **3** rubrum] rubeum *VMP* alium] altram *P* alatrum *C* intra] unde *L* se] intus *B* intra se] *om.* *VFCM* **4** Verum] *om.* *VFM* uehementer] uehementer et si sincera(s) aggluttinatur *VLPCM* si integra glutiatur *W* **5** prout] probatur *F* ignis] ignis orem et fauces *FL* ignis ures et faucium *M* ignis exurit et fauces *W* ignis est et fauces *BPC*

cf. Diosc. I.36 (κνιδίου κόκκου), IV.171 (χαμελαία). Pliny 27.70. Galen *Simp.* 7.10.33 (K.12, 32). Ps. Apul. 112. *Ex herb. fem.* 28 (?). Isidore 19.28.1.

63) Cyclaminus

Cyclaminus herba est quae radicem habet nigro cortice clausam in speciem rapae et thyrsulum pusillum, circa quam folia similia hederae, sed nigriora tria uel quatuor et flosculum roseum in speciem violae. Haec igitur radix quo tempore tota herba florescit conciditur et in pila tundi-
5 tur et sic in pressorio succus eius exprimitur et in patella fictili siccatur uel etiam coquitur usque dum mellis habeat spissitudinem. Vires autem habet acres et excalfactorias, unde et uentrem purgat. Hic ipse succus potatus et commixtus cum melle aut cum lacte aut aceto per nares infusus caput purgat et suffusiones super oculos inunctus dissipat.

1 Cyclaminus] cyclmenos *F* ciclamus *P* herba] *om.* *F* cortice] colore *BP* nigro- clausam] *om.* *W* clausam] inclusa *P* **2** pusillum] postillum *F* **3** tria vel quatuor et] itaque *BP* uiolae] *om.* *J* haec- 6 spissitudinem] *om.* *W* **4** quo- conciditur] uel succus si cum uino fuerit commixtus ebrum facit *M* tunditur] mittitis *V* mittitur *FM* **5** pressorio] pressuriolum *V* pressorio oliaro *M2 om.* *P* fictili] sic cale *V* coquitur] quoque *V om.* *P* spissitudinem] ispissitudinem *V* **7** unde] ualde *V* hic- commixtus] *om.* *W* **8** commixtus] caput comixtur *V* **9** caput purgat] *om.* *VFP* purgat] *om.* *BCM* et suffusiones] ad infusionem *BP* ad fasiones *C* dissipat] *om.* *BP* dissoluit *M* uitium dissoluit *M2* dissoluit uicia *W* uitium collitur *C*

62) Berries of spurge flax (*Daphne cnidium* L.)

Spurge flax is the fruit of a plant which grows in mountainous places. It is collected in autumn and dried in the sun. It has two husks, that on the outside which is reddish in colour, and the other on the inside, as though encasing a white kernel. It is a very powerful purgative, and [when pure it coats the mouth and throat and] its excessive warmth heats [them] like fire.

63) Cyclamen (*Cyclamen graecum* Link.)

Cyclamen is a plant whose root is encased in black bark as found on the turnip,[1] and which has a small stalk, around which its leaves grow like ivy, though they are darker and clumped in groups of three or four, and it produces a small pink flower like the violet. When the whole plant is in full bloom the root of this plant is cut out and crushed in a mortar and then placed in a press to extract its juice, which is then dried on a clay plate, or even cooked, until it obtains the consistency of honey. It has sharp and heating properties, and therefore purges the bowels. If the juice is drunk, or mixed with honey, milk, or vinegar and infused through the nostrils, it clears the head, and it reduces cataracts on the eyes when applied topically as a lotion.[2]

1 #233. Similarly Diosc. II.164, who describes the same species, whereas the description of Pliny 25.114 seems to be of *Cyclamen hederifolium* L. or *Cyclamen europaeum* L., found in Italy.
2 Similarly Diosc. ibid., and see ch. 4.H.

cf. Diosc. II. 164 (κυκλάμινος). Pliny 21.51, 25.114. Galen *Simp.* 7.10.60 (K.12, 50–2). Ps. Apul. 17. Isidore 17.9.89.

64) Costum

Costum radiculae sunt non ignotae et plenulae et spissae et siccae, non uermiculosae, quod ducimus ex Creta, eo quod sit recens tale est enim optimum costum. Sed et gustu submordet et excalfacit. Haec omnia in se habet, quod ab Aegypto affertur, sequens est Arabicum. Hoc enim
5 leuius quidem et odore suauius. Deinde Indicum, quod est uastum et nigrum et leue prout ferula. Infirmius Syrium est, strictum est enim colore buxeo et graue ualde et odore acre. Verum omne costum uires habet acriter stringentes et excalfacientes.

1 ignotae] ignotae subalbide *FBPMC* ignorantem *M2* subalbida et splendida *W* plenulae] *om. W* plenae *P* siccae] *om. W* 2 uermiculosae] per uermiculares *P* quod- 7 costum] *om. W* quod- Creta] *om. BPC* 3 submordet] subamarum *BP* et excalfacit] dentes calefacit *BP* 4 affertur] infertur *VF* inferatur *M* 5 leuius] leuioris *VFM* de indicum *V* 6 infirmius] infirmissimum *BCMM2* inferius *F* syrium] *om. FLBPCM* strictum est] om. *LBCM* tricumero *P* 7 acre] agru *V* graui *J* agre *F* acero *BC* acro *P* om. *M*

cf. Diosc. I.16 (κόστος). Pliny 12.41. Galen *Simp.* 7.10.45 (K.12, 40–1). Isidore 17.9.4.

65) Cantharides

Cantharides animalicula sunt, quorum sunt efficacissimae quae in spicis frumenti adhuc florescentis inueniuntur, uariolae et quasi districtae circa pinnulas suas, uiolaceo colore, oblongae et uastae, inertiores enim sunt et minores et latiores quae circa rosas colliguntur et adhuc infir-
5 miores quae uno colore sunt. Haec enim animalicula momento a uapore aceti suffocantur et in uas non picatum reponuntur. Faciunt enim ad aliquas antidotos et in aliqua caustica, sed uenenosa esse dicuntur.

1 Cantharides] canteridas *FM* canterides *P* cantaredas *M2* animalicula] animalia *P* efficacissimae] efficaces *F* in spicis] ispissis *V* in species *F* 2 florescentis] florentes *F* florentibus *W* uariolae] uirides *J* uariae *W* districtae] restricte *V* stricte *FM om. W* 3 pinnulas] papillas *J* oblongae] oblongula *P* uastae] castae *J* uastae- colore sunt] *om. W* enim- minores] *om. F* minores- reponuntur] *om. M2* 4 circa- quae] *om. M* rosas] sunt *F* 5 quae uno] quoque ut *J* momento] *om. VFCPW* animalicula] animalia *J om. W* 6 aceti] angusto *M* picatum] non fracto *F* faciunt] intrant *P* mittuntur *C* 7 aliqua] alia *F* caustica] calasticis *CM* chaustia *P* dicuntur] creduntur *W* sunt *C*

partim def. L

cf. Diosc. II.61 (κανθαρίδες). Pliny 11.118, 23.29, 23.62, 23.80, 29.93, 29.96, 29.110, et passim. Galen *Simp.* 11.1.44 (K.12, 363–4). Isidore 12.5.5.

64) Costus root (*Saussurea lappa* Clarke)

Costus root is formed from its well-known small roots that are full, thick, and dry, not worm-like.[1] We choose first that which comes from Crete, and this when fresh is the best costus of all. It has a slightly biting taste and warms the mouth. All of this also applies to that which is imported from Egypt. Ranked after this is the costus from Arabia, which is much lighter and has a sweeter fragrance. After this is ranked that from India, which is large and black yet light in weight, like giant fennel.[1] Very unreliable is the Syrian type, which is rigid, the colour of boxwood,[1] and quite heavy, with a sharp odour. All costus has severely astringent and heating properties.

1 Similarly Diosc. I.16. Despite these similarities, Diosc. does not mention the Cretan type (noted by Meyer 1856, IV:492), or the Egyptian, nor is the Syrian type criticized. On giant fennel, see #4 n.1.

65) Blister beetles

Blister beetles are insects, and the most effective are those found in ears of grain when they begin to flower. These are mottled, as though they have stripes across their wings, and are violet-coloured, oblong, and quite large, while the slower-moving ones which are collected from roses are smaller and fatter and those which are a single colour are less potent. These insects are quickly stifled by the steam off vinegar and then stored in a container that has not been coated with pitch. They are used in a number of antidotes and several caustics, but are said to be poisonous.[1]

1 On the medical use of blister beetles in Antiquity, see Scarborough 1979, 73–80; Beavis 1988, 168–75.

66) Chrysocolla

Chrysocolla est quasi puluis uiridis ualde, cuius optimam dicimus de
Armenia est enim leuis et sine ulla asperitate et omnium uiridissima.
Sequens est Macedonica, postque Cypria, aerugini similis. Verum
omnis chrysocolla uires habet acriter stypticas. Vnde et collyriis ad
5 lachrymam facientibus miscetur et in cerotis posita carnes luxuriantes
stringit.

1 Chrysocolla] crysocola *V* chrisocolla *F* crisocolla *PW* crissocola *M2* puluis] aerra *C* de
armenia] arminiaco *VM* 2 omnium uiridissima] *om. W* 3 macedonica- unde] alta macedonia effectum habet ut aerugo cupri aliter styptica *W* postque cypria] opus eius est
quod et apri *P* 5 lachrymam facientibus] ad lacrimas efficiuntur stringendas *W* cerotis]
citiris *V* ceteras *FMP* sinceris *C om. W* posita] *om. VFWM* luxuriantes] humidas *W* 6
stringit] restringit et exsiccat *V* restringit et desiccat *F* stringit et siccat *WM* stringit et
siccat efficaciter *BPC*.

partim def. B

cf. Diosc. V. 89 (χρυσοκόλλα). Pliny 33.86–94. Galen *Simp.* 9.3.38 (K.12, 242–3). Isidore
16.15.7, 19.17.10.

67) Calx uiua

Calx uiua fit de lapide qui affectatim uritur et est uiscidissima quae est
recens et non mundata. Mittitur autem in emplastris et in causticis, est
enim et ipsa caustica et uires habet acerrimas.

1 Calx uiua] calces uiua *VFM* calce uiua *M2* calisiua *J* calx- quae] calem uiuam omnis
sciunt *W* qui affectatim] quia defecta est *F* qui affatim *P om. C* uritur] aduritur *P* 2 non]
om. C mundata] inundata *VBM* inudata *F* huddum *C* uda sed infusa *W* causticis] calasticis *MC* est- ipsa] *om. BPMW* 3 caustica] *om. MW* acerrimas] *om. C*

partim def. F

cf. Diosc. V.115 (ἄσβεστος). Pliny 36.180. Galen *Simp.* 9.3.31 (K.12, 237, τίτανος). Isidore
16.4.4 (asbestos).

68) Coriandrum

Coriandrum notissimum est omnibus. Huius ergo succus uiscide uirtutis refrigerat et stringit, propter quod ad ignem sacrum proprie facit
sed et ipsa folia trita et imposita. Item semen eius tritum et potatum
cum oleo lumbricos occidit.

1 omnibus] *om. JBPL* multis *W* uirtutis] *om. JB* 2 sacrum] agri *V* acrum *FM* acrem *W* 3 sed-
imposita] *om. W* imposita] uiris habit uiscida *V* incomposita *F* potatum] utatum *F om. W*

cf. Diosc. III.63 (κόριον). Pliny 20.82. Galen *Simp.* 7.10.43 (K.12, 36–40). Garg. *Med.* 4.
Ps. Apul. 103. *DynVat* 31/*SGall* 35. Isidore 17.11.7.

66) Chrysocolla

Chrysocolla is like a richly green dust, and we declare the best to be that from Armenia, for it is smooth, without any coarseness, and greener than all the other types. Following this is that from Macedonia, then that from Cyprus, which looks like verdigris.[1] All chrysocolla has severely styptic properties. It is therefore mixed with eye-salves for healing watery eyes, and when applied with a wax-salve it strips away overgrown flesh.

1 #3.

67) Quicklime

Quicklime is made from a stone which is thoroughly burned and is most effective when fresh and not yet purified. It is placed in plasters and in caustics, for it is a caustic itself and has severely sharp properties.

68) Coriander (*Coriandrum sativum* L.)

Coriander is very well known to everyone. Its juice has powerful cooling and astringent properties, hence it is extremely good for erysipelas[1] when you crush its leaves and place them on the area affected. Also its seed when crushed and drunk with oil kills intestinal parasites.

1 *ignis sacer*: I follow here the conventional translation of erysipelas (an acute streptococcus bacterial infection of the dermis), but *ignis sacer* ('sacred fire') was used by Latin writers to describe a range of different conditions, including ergot poisoning or ergotoxicosis (commonly known as St Anthony's fire, usually caused by the ingestion of alkaloids produced

69) Cyminum

Cyminum uiribus praecipuis praeditum uidetur Aethiopicum, sequens
Afrum. Verum omne cyminum leuiter laxat et calefacere potest, et sic
inflationibus et tortionibus prodest et uiscide dolorem placat et si bibatur pallorem facit.

1 Cyminum] cymino *V* ciminum *FW* ciminum thebaicum *P* uiribus- uerum] *om.* *W* uiri-
bus praecipuis praeditum] uiridissimum uidetur *VM*2 uiriosissimum *B* uerissimum *CM*
aethiopicum] idropicum *C om. P* 2 leuiter] *om. V* leniter *M* laxat] relaxare *VB* relaxat
CM 3 tortionibus] rosionibus *BP* uiscide] quiescit *B om. W* uiscide- *fin.*] mittigat dolores
P placat] facit *V* sedat *W* placat- facit] *om. B* sed potius thebaicum utile deinde afrum
W.

partim def. F

cf. Diosc. III.59 (κύμινον ἥμερον). Pliny 20.2. Galen *Simp.* 7.10.61 (K.12, 52). *Ex herb.*
fem. 5.

70) Chamaeleon

Chamaeleon herbae est non una specie, inuenitur autem nigra et albida.
Verum albida est quae folia habet circa suam radicem in terra diffusa,
uaria, densa, aspera, quasi spinosa et in medio uelut surculum acutum.
Florem purpureum, densum et minutum, quid et albescit, semen quasi
5   anethum, radicem albam et uastam mollem grauiter redolentem, gustu
subdulcem, quaecum passione urinam mouet et hydropicos siccat et
lumbricos occidit. Nigra deinde quae est folia habet tenuiora et paucio-
ra et subrubicunda, thyrsulum oblongum, crassitudine digiti et ipsum
rufum. In quo summo cumulum et florem spinosum, hyacinthino
10   colore. Radicem uastam et nigram uel uariam, spissam, subrufam et
submordentem si masticetur, quae facit ad maligna uulnera quae graece
cacoethe dicuntur. Item maculas et lentigines cum aliis rebus commista
tollit. Vtrumque tamen quibusdam locis circa folia sua dicunt uiscum
emittere.

1 Chamaeleon] camillion *V* emeleon *C* chameleon *W* cameleon *P* camelleon *M*2 non
una] duas *M om. F* albida] alba *VM* 2 diffusa] *om. VFM* 3 densa] unde *M* uelut] stat quasi
P surculum] sorcellum *V* surcellum *LB* acutum] acutum statem *VJ* acutum stantem *F*

by the fungus Claviceps purpurea, which infects cereals), shingles (a neurological disease caused by reactivation of varicella zoster virus), lupus (systemic lupus erythematosus, a chronic autoimmune disease), or more simply putrefaction or mortification of tissues on or near the skin. Galen (*Simp.* VII.43, K.12, 36–40) used the example of coriander's effect on erysipelas to demonstrate the superiority of his drug theories over that of Dioscorides: Riddle 1985, 72–3.

69) Cumin (*Cuminon cyminon* L.)

The cumin with the strongest properties appears to be that procured in Ethiopia, followed by that in Africa. All cumin gently softens and can be warming, hence it is good for flatulence and stomach cramps, is a powerful pain-reliever, and when drunk it turns the skin pale.[1]

1 Similarly Diosc. III.59 (drunk or applied topically).

70) Chameleon thistle (*Cardopatium corymbosum* L.)

Chameleon thistle is not a single type of plant, for there is both black and white chameleon thistle. The white type is that which has its leaves sprouting from near its roots along the ground and which are extensive, variegated, thick, coarse, a little prickly, and have a sharp nettle in the middle. It bears a thick, small purple flower which whitens with age, and seed like anise.[1] Its root is large, white, soft, and heavily fragrant, with a slightly sweet taste, and it helps with painful urination, dries out those suffering from dropsy, and kills intestinal parasites. Then there is the black chameleon thistle, which has thinner and smaller leaves that are ruddy in colour, and a small stalk which is oblong, about the width of a finger, and red in colour. At the top of the stalk there is a crown with a spiky flower the colour of the hyacinth.[2] Its root is large and black, or varied in colour, thick and reddish, and when chewed has a slightly biting taste. The root is good for the malignant wounds which in Greek are called *cacoethes*.[3] Likewise it removes blemishes and freckles when mixed with other substances. It is also said that in some locations its leaves discharge a resin.

1 #35.

4 florem- densum] *om. M* **5** anethum] nigrum *B* enicum *W* anicum *P* albam] ut maluae *P om. W* uastam] astam *M* mollem] mollem ut malita *W* **6** passione] passo *P* hydropicos] hidropem *V* **7** deinde] nempe *P* pauciora] faciliora *F* **8** crassitudine] magnitudine *M* grossitudine *FPW* ipsum rufum] *om. M* **9** cumulum] cimula *P* hyacinthino colore] yantineno calore *P* spissam subrufam] *om. W* **11** submordentem] morsu submordentem *C* mordatur *B* si masticetur] sicut afferitur *J* si mascitur *LC om. W* si mordeatur *P* si- **14** emittere] *om. M* uulnera] uicera *J* **12** cacoethe] cacoitica *V* cacoethace *F* KAKWCTHHC *LC* kake *P* cacimaculas *W* cecacoentici *M2* item- emittere] lentigines aufert *W* **13** tollit] tollit et deborat *M2* locis] legis *L* uiscum] uirgam *BP om. M2* **14** emittere] emitterre et stanat florem purpureum densum et minutim *C similiter sed def. M2*

cf. Diosc. III.9 (χαμαιλέων ὁ μέλας). Pliny 22.45–7. Galen *Simp*. 8.22.6 (K.12, 154). Ps. Apul. 110. Alex. Tral. 2.259. *Ex herb.fem*. 5. Isidore 17.9.70.

71) Cardamomum

Cardamomum semen est quod nascitur in Arabia et in aliis locis. Est optimum quod est flauum, plenum, graue, durum et quasi lentum et non rugosum et totum conclusum, et si frangatur puluerosum, gustum subamarum excalfaciens, odore graue. Vires habet acres et excalfacto-
5 rias, relaxantes. Vnde tortiones et inflationes uentris sedat et urinam et menstrua prouocat et partum expellit et ad tussem ueterem facit.

1 Cardamomum] cardamomu *V* locis] *om. V* est optimum- **4** graue] *om. W* **2** durum] indurum *M2* **3** rugosum] sucosum *M2* **5** relaxantes- urinam] *om. F* relaxantes] *om. W* Vnde-sedat] doloris quidem inflationire restit *B id.* inflatione resistit *P* sedat] sanat *VWM2* **6** prouocat] prouocat et ad [....]inarem uentris *C* expellit- facit] *om. L* tussem ueterem] dolorem uentris *V* tumorem uentris *F* tumorem uentris ac tussim *W* tumorem ueterem *M2*

cf. Diosc. I.6 (καρδάμωμον). Pliny 12.50. Galen *Simp*. 7.10.9 (K.12, 12). Isidore 17.9.11.

72) Chamaemelus

Chamaemelus herba est omnibus nota, cuius virtus et efficacia est cum suauitate calefactoria et cum omni dulcedine temperata et sudorem prouocans.

1 Chamaemelus] camimolon *V* camilon *F* camaleon *L* camemilla *W* chamomilus *P* herba] *om. F* omnibus nota] *om.W* efficacia] *om. P* **2** temperata] temperatiua *FP* sudorem] hodoris *V*

desinit M

cf. Diosc. III.137 (ἀνθεμίς). Pliny 22.53. *Ex. herb. fem*. 19. Galen *Simp*. 6.1.47 (K.11, 833). Isidore 17.9.46.

73) Calamus aromaticus

Calamus aromaticus [optimus] est qui in Indian crescit, et subrufus et

2 #300.

3 The term could be used for a number of ailments. In Hippocratic writings κακοήθης is used to mean 'malignant' or 'malignacy,' as in Diosc. I.43, I.54, I.68 (κακοήθεις σκληρίας), I.128, II.104, II.126, etc., though the term is not used for the entry on chameleon thistle, III.9 (instead 'τὰ φαγεδαινικὰ ἕλκη καὶ τὰ τεθηριωμένα'). Celsus 5.28.2 refers to it as the first stage of a carcinoma ('quod cacoethes a Graecis nominatur'), which can be safely removed by surgery. Cf. Pliny 22.132, a sore or ulcer (*ulcus*), 24.7, a type of induration (*duritia*).

71) Cardamom (*Elettaria cardamomum* White and Maton)

Cardamom is a seed which grows in Arabia and other places. The best type is that which is of flaxen colour, is plump, heavy, tough yet pliant, not wrinkled, is entirely enclosed, and when broken becomes powdery, with a slightly bitter, warming taste and a pungent fragrance. It has sharp, heating, and loosening properties, and therefore relieves cramps and gas in the bowels. It causes urination and menstruation, expels the foetus, and also cures chronic coughs.[1]

1 Or 'cures pain/tumours in the stomach,' *VFCW*. See also plate 8.

72) Camomile (*Anthemis* sp.)

Camomile is a plant known to everyone for its property and effectiveness as a soothing calefacient, and that is blended with every type of sweetener, and causes perspiration.

73) Sweet flag (*Acorus calamus* L.)

[The best] sweet flag grows in India, and is tawny in colour, is very knotty

nodosus maxime iuxta radicem, intus exalbidus et quasi araneas habet, et qui frangitur in plura uel fusticia sicca fiunt et minuta et odore suaui et gustu uiribus stypticis et glutinosis, acris et excalfaciens. Reproba-
5　mus autem qui aut exalbidus est, aut nigricans, et nodos raros et paucos habens.

1 est] est optimus *VLWPM2* 2 iuxta- exalbidus] *om. J* laxata radices *B* de intus autem exalbidum *W* quasi- suaui] *om. W* 3 qui frangitur] confrangitur *V* constringitur *P* plura] flora *P* pura *M2* uel fusticia] in frustra *CJ* flores ut frustu cas siccas frun[....] *B* 4 gustu] fructu *JC* acris] *om. W* reprobamus- habens] *om. W* aut 5 nigricans] *om. BP* nigro *M2*

plerumque def. F

cf. Diosc. I.18 (κάλαμος ἀρωματικός). Pliny 12.104 (odoratus), 25.46. Scrib. 269. Galen *Simp.* 7.10.2 (K.12, 6–7). Marc. *Med.* 35.7. Alex. Tral. 1.111. Isidore 17.8.13.

74) Chamaedrys

Chamaedrys herba est oblonga, cuius folia sunt simillima quercus sed minora, gustu amaro, thyrsulum et florem ex parte rubicundum habens et pusillum. Nascitur in locis asperis. Vires habet acres et relaxantes, propter quod ad tussim ueterem facit, et menstrua et urinam prouocat. Sed et partum expellit et hydropicis prodest.

1 Chamaedrys] cametreus *V* camidreus *F* chamedris *P* camedreos *M2* oblonga] *om. W* quercus] minora ut querco *V* 2 thyrsulum- asperis] tyrsulum et florem sub rubicundum et pusillum *W* florem] folia *P* habens-pusillum] *om. VM2* 3 acres] acriter *W* 4 propter quod] *om. W.*

plerumque def. F

cf. Diosc. III.98 (χαμαίρωψ). Pliny 24.131. Galen *Simp.* 8.22.2 (K.12, 153). Ps. Apul. 24. *Ex herb. fem.* 8, 65 (zamalention). Isidore 17.9.47.

75) Casia

Casia fistula cum plures habeat species et aliquibus locis colligatur, optima est quae in Arabia inuenitur. Est enim aspectu grauissima, colore rufo, tenuis, leuis, longa fistula maiori, ad tactum leuissima. Haec igitur ipsa cortex est cum uirgulis quae sibi crescunt tenues et oblon-
5　gae, quarum cortex tempore certo a uirgulis suis recedit et quasi tribu-
lus adducitur prout dracones squamis suis se exuere dicuntur. Verum omnis casia fistula uires habet acriter calfacientes et stringentes. Vnde ad omnia uitia thoracis facit et in antidotis mittitur, aliquando pro se, aliquando pro cinnamo, duplicatis ponderibus eius.

1 Casia] cassia *VFBLP* fistula cum] *om. CJ* cum- enim] *om. W* aliquibus] in diuersis *F om. P* 3 longa- 7 fistula] *om. W* fistula maiori] fistulata *F* 4 cortex] certis *LCJFM2* uir-

around its roots, and on the inside is white and a little like spider-webs, which when well crushed by hammering become small and dry splinters that have a sweet fragrance and taste, and are sharp and warming, containing styptic and glutinous properties.[1] We reject the types that turn white, or blacken, and those that have few nodes.

1 A singular reference to 'glutinous' properties: cf. *Liber de dynamidiis* (no. 7) 'Colleta,' ch. 1.E.

74) Wall germander (*Teucrium chamaedrys* L.; *Teucrium lucidum* L.)

Wall germander is a plant of oblong shape, with leaves that are just like those of the oak but smaller and have a bitter taste.[1] Its stem and flower are small and reddish in parts. It grows in harsh places. It has sharp and loosening properties, and is therefore good for chronic coughs and stimulates menstruation and urination. It expels the foetus, and is good for dropsy.

1 Similarly Diosc. III.98.

75) Cassia / Chinese cinnamon (*Cinnamomum cassia* Bl., syn. C. *aromaticum Nees*)

Although the reed cassia[1] has many species and is harvested in a number of places, the best is that found in Arabia. It has the appearance of being most heavy, red in colour, and slender, with long tubes that widen, and are very light to the touch. Cassia is in fact the bark upon the slender and oblong twigs which grow out from the trunk, and at some point it breaks off from the twigs, so that the tree looks like a caltrop,[2] in the same way that reptiles are said to remove their scales. Every type of cassia has sharply warming and astringent properties, hence it is used for every ailment of the chest, and blended with antidotes, using double the amount in weight that you would for cinnamon.[3]

gulis] uirgularum *LP* sibi] ibi *B* **5** certo] suo *V* tribulus] turbulosa *B* turbida *P* stipula *F*
tribulas *M2* **6** squamis] ramis *B* squamis missu suis *V* quam *M2* exuere] *om.* *V* excurgere
B **7** calfacientes] excalfacientes *V* **8** aliquando pro se] *om.* *BFM2* **9** eius] *om.* *W* eius mittitur *V*

cf. Diosc I.13 (κασσία). Pliny 12.95. Galen *Simp.* 7.10.11 (K.12, 13). Isidore 17.8.12.

76) Cyphi

Cyphi est quasi suffimenti genus, quod ex his rebus conficitur, calamo
aromatico [et] cinere aspalthi libras singulas, myrrha libra una, primo
cypero, iuniperis ueteribus et uua elapidata puluerizari quae possunt,
tunduntur et cribrantur et sic cum reliquis in uino maceratur una die.
5   Deinde aspalathum comburitur et laeuigatum aspergitur et tunc mel
decoctum admiscetur. Et sic omnia in unum diligenter commouentur
et in uase fictili reponuntur. Hoc antiqui ad deos utebantur. Medici
quoque illud in aliquas compositiones mittunt. Potest autem acriter et
efficaciter relaxare.

1 Cyphi] cyfy *V* cifi *P* cyfi *FC* chifien *J* suffimenti] fumenti *V* frumenti *F* pigmenti *P* sub
pigmenti *B* rebus] *om.* *P* **2** cinere] cymas *V* cimas *F* amaro *L* amas *P* aspalthi] aspalthi
libras singulas *VFBLP* myrrha] ismirnis *V* *om.* *F* smurminis *C* myrrha- primo] *om.* *L*
3 iuniperis] uini *VFBCP* ueteribus] uniuersis *L* puluerizari] puluere *VF* puluerisque *P*
4 cribrantur] cribellantur *VC* sic cum reliquis] siccum reliquum *J* **7** ad deos] qui eis *J*
a deus *V* ad eos *F* ad deos *B* a deos *L* adeo *C* medici- potest] qui et pistelentius iscaribus laborabant meditur (*scil.* medici?) *V* quia pestilentiosis casibus laborans medici *C* **9**
efficaciter] *om.* *C*

deest WM2, intra 'F' in C

cf. Diosc. I.25 (κῦφι). Scrib. 70.

77) Cinnamum

Cinnamum sunt quasi ramuli et uirgulae quae ab una radice plures
excrescunt rectae, et minutae et non adeo longae et colore subrufae
et extenso cortice non rugoso. Qui cortex ungue non facile abraditur,
nodulos plures et inuerualla habent et in medio leues et si frangan-
5   tur, quasi puluerem emittunt odore omnia uincentes. Fere enim haec
maxima probatio cinnami est quae est omnium rerum odoratissimum.
Gustu uero et uiribus est acre et excalfactorium et extensos stringit et
submordet, propter quod in multas compositiones mittitur, tamen in
antidotis preciosum est ualde.

1 Cassia and cinnamon are species of *Cinnamomum aromaticum*, both commonly referred to as cinnamon (though cf. Exod. 30.23–4). Cassia is name of the Chinese species, cinnamon of the Malayan species. Innes Miller 1969, 42–7.

2 *tribulus*, a unique description among our sources, which may mean here the plant caltrop (*Tribulus terrestris* L.), also known as star thistle or puncture vine, among many other names (not recorded in *AG*: see Diosc. IV.15, Pliny 21.91, 98), or the spiky military device of the same name used against enemy cavalry, as seems to be the case in #256 below.

3 Similarly Diosc. I.13, Isidore 16.9.12.

76) Cyphi

Cyphi[1] is a type of incense which is made from these ingredients: a pound each of sweet flag[2] and ashes of camelthorn,[3] a pound of myrrh, young galingale, ripe juniper berries,[4] and crushed grapes,[5] all of which can then be pulverized together, pulped, sifted, and then that which remains is marinated in wine for a day. Camelthorn is then burned, crushed, and sprinkled on top, then this is mixed with honey that has been reduced by cooking. All of this is carefully and thoroughly blended together then stored in an earthenware vessel. The ancients used to burn it for their gods. Doctors place it in other compound medicines also.[6] It can sharply and effectively loosen.

1 This ancient Egyptian recipe appears in the Ebers Papyri (c. 1553–1550 B.C.) and was well known to Greek authors, including Diosc. (I.125): see Scarborough 1984, 229–32, Riddle 1985, 60, 90.

2 #73.

3 #11 n.4.

4 Or 'old wine' *VFBCP*.

5 Or 'crushed nuts' *JLC*.

6 Note *V*, '[ancients ...], and whose doctors unwholesomely prepared rites' (see chs. 2.I and 5.C.*V*). Cf. Diosc. '[it] pleases the gods; the priests of Egypt use it lavishly. It is also mixed with antidotes, and is given to asthmatics in drinks.'

77) Cinnamon (*Cinnamomum cassia* Blume, ~ *Cinnamomum zeylanicum* Blume)

Cinnamon[1] is the branches and twigs which grow abundantly straight up from the near root, and are small, not very long, slightly reddish in colour, and with loose bark that is not wrinkled. The bark is not easily removed with your finger-nails, as it has many nodules on the inside which are thinner toward the middle, and when these nodules are pounded they emit a substance like dust and a fragrance that overpowers any other aromatic. And this is generally the best test of cinnamon, whether its fragrance is more powerful than anything. It is sharp in taste and properties, it is a

1 Cinnamum] cinnamomu *V* cinnamomum *FM2* cinama *P* ab una] habent *V* habet *FM2* ab ima *L* 2 non adeo] om. *M2* 3 non rugoso qui cortex] *om. BLP* non rugoso] id est sine rugore *FPB* id est non rugo sequi cortex *L* suum sucusus est *M2* ungue] unde *BP om. M2* 6 probatio cinnami] *om. F* cinnami] *om. V (BP begin second entry on same)* 7 extensos] extendens *F om. P* 8 multas] malagmas *V* maculas *F* magnas *M2* antidotis 9 preciosum] antidota preciosa *FP* est ualde] *om. VFBLP*

cf. Diosc. I.14 (κιννάμωμον). Pliny 12.86, [12.30?]. Galen *Simp.* 7.10.25 (K.12, 26). Isidore 17.8.10.

78) Cucumer erraticus

Cucumer erraticus nascitur maxime in locis sabulosis, et est per omnia dulci cucumeri similis, sed pilosior, id est asperior et multo breuior et minor radicem habet quidem uastam quae laeuigata et pro fomento imposita ad dolorem uel durities neruorum et ad podagras facit. Et siccata et rasa et cribrata, omnes maculas de corpore tollit.

1 Cucumer] cocumere *V* erraticus] agrestis *F* siluaticus *P* seluaticus *B* nascitur- sabulosis] *om. P* sabulosis] plurimis *B* sabulosis ad duritias neruorum et ad omnem maculam de corpore tollenda *F* sablonosis *M2* maxime] *om. W* et est- dulci] domestico *W* 2 pilosior] pilosus *VWM2* 3 habet] *om. VPM2* uastam] subrufam et nigram *BP* quae- tollit] intrin- secus molli dulci lacte plena quod dicitur generaliter strangulationi occurrere et ferro precisos sanare *P* quidem] *om. W* quae] atque *VW* pro- ad] *om. W* ad dolorem] ad multas duritias *V* 5 siccata et rasa] item facit tunsa et cribillat cum *V* rasa] tunsa *LW* omnes maculas] omnia mala *J*

cf. Diosc. II.135 (σίκυς ἥμερος). Pliny 20.9. Garg. *Med.* 16. Marc. *Med.* 36.67. Ps. Apul. 114. *DynVat* 46/*SGall* 49. Isidore 17.10.16.

79) Conyza

Conyza est omnibus nota, cuius folia prae acrimonia sua uiperas et serpentes fugant, et ad morsus earum faciunt, sed et uuluae longinqua uicia remediant.

1 Conyza] coniza *V* coniza id est cicuta *F* cotita *L* choinizaie codanarum *C* coniza uel cicuta *W* 2 fugant] effugare faciunt *M2* uuluae- uicia] uulnera uitia *V* bulla et longinqua *F* uuluae necnon uicia emendat *W* longinqua- *fin.*] *om. M2*

deest P

cf. Diosc. III.121 (κόνυζα). Pliny 21.58. Galen *Simp.* 7.10.42 (K.12, 35–6). *Ex herb. fem.* 28.

80) Colchicon

Colchicon uel ephemerum hanc ipsam herbam aliqui bulbum erra- ticum dicunt. Effert autem florem album, deinde emittit quasi bulbi

calefacient, it binds things that are strained,[2] and it has an under-bite. For this reason it is put in many compound medicines, but it is particularly valuable for antidotes.

1 See above #75 n.1.
2 *extensos stringit*: an odd phrase, probably corrupt.

78) Wild cucumber (~ *Cucumis sativus* L.)

Wild cucumber grows abundantly in sandy places, and it is altogether similar to the sweet cucumber except that it is more hairy, that is, more coarse, and much shorter, and although small it has a very large root, which is pulverized and then placed in a compress to relieve pain, indurations of the ligaments, and gout. When dried, shaved, and then sifted, it removes all blemishes from the body.

79) Fleabane (*Inula graveolens* Desf.; *Inula viscosa* Aiton; *Inula britannica* L.)

Fleabane is known to everyone. The pungency of its leaves puts snakes and vipers to flight, and is good for their bites. It also cures chronic wounds on the vulva.

80) Meadow saffron / Autumn crocus (*Colchicum autumnale* L.)

Meadow saffron or ephemeron is the same herb called 'wild onion' by others.[1] It blooms a white flower, and produces an onion-like red fruit on

fructum in thyrsulo rubrum, radicem subrufam et nigricantem, intrinsecus albam et mollem, dulcem, lacte plenam, quae dicitur [fungi] generis strangulationis occurrere, et ferro praecisos sanare.

1 Colchicon] colciton *V* calciones *F* 2 effert- *fin.*] afferre *M2* deinde] deinde folia *LFC* 3 in thyrsulo rubrum] uoluitur solum rubeum *V om.* C intrinsecus albam] liam *B* bam *L* 4 lacte] *om.* V dicitur] dicitur uicem fungi *J* 5 strangulationis] strangulatoris *J* stranquilationis *C* occurrere] facere *J* sanare] sanare linita *J*

deest WP, desinit M2

cf. Diosc. IV. 83 (κολχικόν). Pliny 25.170 (ephemeron), 28.129, 28.160–1.

81) Clematis

Clematis hanc aliqui uitem albam dicunt, flexibilem, subrufam et quasi circumflexam. Folia habet acria et exulcerantia, fructum uuae similem. Phlegma deorsum deducit et purgat.

1 Clematis] climatis *VP* climates *F* flexibilem] uitis autem habit flexibele *V* uitem autem habet flexibilem *F* 2 circumflexam] circumplexam *P* circumflicta *F* uuae similem] quasi uilem *V* quoque similem *F* quasi bilem *BL* exulcerantem fructum habet quassibilem *P* que similem et *C*

deest W

cf. Diosc. IV.180 (κληματίς). Pliny 24.84. Galen *Simp.* 7.10.31 (K.12, 31).

82) Cnicum

Cnicum thyrsos habet bene longos, spinosos, capitella et folia habentes oblonga, aspera et incisa et ipsa spinosa. Florem quasi croceum, semen oblongum et quasi angulosum, rubrum et album. Quod totum uentrem purgat et phlegma et choleram deponit. Sed moderate tunditur
5　semen eius et expresso succo eius datoque in ptisana, aut in iure pulli, alii mel, uel anisum uel amygdalas [amaras] miscent et faciunt dysentericis magnum adiutorium drachmas VI.

1 Cnicium] cinicum *V* cynicum *F* cardon *C* gnicum *W* bene] *om.* W longos] *om.* F capitella] capitiolas *P* capita *W* 2 aspera] aspersa *J* et ipsa spinosa *V* incisa] uitiosa *J* uitiosa incisa *V* croceum] cucumeris *BP* 3 oblongum] album *W* et quasi- totum] *om.* W angulosum] arnoglossum *VFLBCP* uentrem purgat] *om.* P 4 et phlegma] *om.* L moderate] moritur *C* tunditur- ptisana] hoc tunsum et expressum datur cum ptissana *W* 5 ptisana] tesacius *V* tysana *F* iure] ius *VFP* uire *C* 6 alii mel] alumen uel *J* alii cum melle et aniso *W* et odie in mel et anism *V* mel *FL* alii meti *C* amygdalas] amygdalas amaras *BP* miscent- drachma VI] amaris mixtum hictericis dantes *W* dysentericis] hictericis *P* 7 drachmas] et dabes hoc *VF* dabis hoc dragmas fex *B* dabis hoc drag. vi *L* dabis vo[...] de hoc z. vii *C* dabis drachmas VI *P*

cf. Diosc. IV.188 (κνῆκος). Pliny 21.90, 21.94, 21.184. Scrib. 135. Galen *Simp.* 7.10.32 (K.12, 32).

its stem. It has a tawny-coloured root, which darkens in places, and on the inside is soft, white, sweet, and full of milk, which is said to induce a type of choking [like a mushroom], and sanitizes sword cuts.[2]

1 Similarly Diosc. IV.83. Note above #44 n.1.
2 Text probably corrupt: Diosc. ibid. records that 'when eaten it kills by choking like mushrooms,' and the necessary antidote is cow's milk, as it is for mushrooms (antidote also in Pliny 28.129, 160). 'Mushroom' appears in *J* ('uicem fungi generis') but in no manuscript reading, though 'generis' does, hence 'choking like a type of mushroom does' may be the intended meaning.

81) Traveller's joy (*Clematis vitalba* L.; *Clematis cirrosa* L.; *Clematis flammula* L.)

Traveller's joy, which others call white vine, is pliant, reddish in colour, and winds around things.[1] It has bitter and ulcerating leaves, and a fruit similar to grapes. Its reduces phlegm downward[1] and purges it.

1 Similarly Diosc. IV.180: see ch. 2.G.

82) Safflower (*Catharmus tinctorius* L.)

Safflower has very long, prickly stalks crowned by little heads and which have oblong leaves that are rough with serrated edges and are also quite prickly. Its flower is saffron-coloured, its seed oblong and angular,[1] and red and white. The seed entirely purges the bowels, and reduces both phlegm and bile. But when the seed is carefully ground and its juice is extracted, this is diluted in barley-water, or blended with chicken broth – others mix it with anise or [bitter] almonds – and six drachmas of this provides great relief for those suffering dysentery.

1 *angulosum*: though all manuscripts read 'like plantain' (*quasi arnoglossum*), plantain (*Plantago major* sp.L.) appears under the Latin name of *plantago* below #217, and the very similar entry on safflower in Diosc. (IV.188) also notes that its seed is 'white and red, longish and angular' (γεγωνιωμένον).

83) Corallus

Corallus est colore rubro quasi arbustum. Vires autem habet stypticas, unde sanguinem excreantibus mire proficit.

1 Corallus] coralium *F* curallium *LB* corallium *CP* corallus- arbustum] corallium quasi subnigrum extrinsecus intus rubicum est *W* rubro] nigrum *BLC* rubeum *F* album et rufum sed rufum est melius *P* arbustum] arbor rustus *V* ´arborustum *F* rubor astrum *B* arbor astrum *L* arborosum *C om. P* 2 excreantibus] reicientibus *P* mire proficit] con mire facet *V* mirefice proficit *W* facit *LF*

cf. Diosc. V.121 (τὸ κουράλιον). Pliny 32.21–4. Isidore 16.8.1 (corallius).

84) Cuculus

Cuculus graece cinenon dicitur, alii strychnon uocant, et est herba facile nota. Nascitur in hortis, ramos habet plures et diffusos, folia autem maiora et latiora. Quae facit grana rotunda et uiridia primum, similia quasi granis uuae, deinde quando maturauerint nigra efficiuntur. Et dicitur uua lupina. Vires igitur habet stringentes et bene refrigerantes.

1 Cuculus] cuculus id est uua lupina *tit. J* id est solatium *C* cuculis *P* cinenon] crochonon *V* croconum *F* CYNONAN *B* CYNKNWN *L* cingeon *C om. P* strychnon] strignum *V* stignum *P* strignum siue uita lupina *B* uocant- hortis] dicitur *P* facile] difficile *F* 2 ramoslupina] *om. W* ramos] ramulos *V* 3 facit] cymeni *V* ucimum *B* ucimi *L* uouum *P* habet *C* 4 granis] *om. VFCP* 5 bene] ualde *C*

cf. Diosc. IV.70–3 (στρύχνον κηπαῖον, ἕτερον –, – ὑπνωτικόν). Pliny 21.89, 21.180, 27.68 (alio nomine strumum). Celsus 2.33.2. Galen *Simp.* 8.19.15 (K.12, 145–6) (τρύχνον ἢ στρύχνον). Marc. *Med.* 36.63. Ps. Apul. 22 (apollinaris), 75 (solata). *Ex herb. fem.* 29 (manicos, cucabulum). Isidore 17.9.78.

85) Cyanum

Cyanum optimum habetur quod ualidissimum est et coeruleum. Inuenitur in metallis, in quibus aes conflatur et quibusdam littoribus maris. Vires habet verum acriter stringentes et quodammodo causticas.

1 Cyanum] canum *J* cianum *PC* clanum *W* habetur] *om. WP* ualidissimum] utilissimum *C om. P* 2 aes] *om. JFLBC* maris] *om. F.*

cf. Diosc. V.91 (κύανος), V.139 (λίθος σάπφειρος). Pliny 37.119. Galen *Simp.* 9.3.15 (K.12, 223).

86) Cornu cerui

Cornu cerui uires habet stringentes. Vnde ad omne profluuium facit, sed et limatur et raditur et in ouo in potum datur et lumbricos expellit. Item si suffumigatur, uiperas et serpentes fugat.

83) Coral

Coral is red in colour, like a lobster. It has styptic properties, hence is it wonderful for healing those spitting blood.[1]

1 Likewise Diosc. V.121 and Pliny 32.4. No other source compares its colour to lobster.

84) Hound's berry (*Solanum nigrum* L.)

Hound's berry[1] is called cinenon in Greek, and others call it strychnon, and it is a plant easily known about. It grows in gardens, has many dense branches, and leaves that are quite large and quite broad. It produces rounded seeds which are at first green like grape seeds, and then turn black when they mature. It is also called 'wolf grapes.' It has astringent and thoroughly cooling properties.

1 The number of names used reflects the common confusion surrounding the term *strychnon*, applied to various toxic plants, especially those with soporific properties: see André 1985, 251. Diosc. IV.70–3 was rare in distinguishing three different types, including winter cherry (ἕτερον στρύχνον = *Physalis alkekengi* L.) and sleepy nightshade (στρύχνον ὑπνωτικόν = *Withania somnifera* L.). While *uva lupina* was commonly used for hound's berry (André, ibid., 277, to which we should add *De observ.* 59, 'herba lupina'), the term *cinenon* (*J*) and its variants are obscure.

85) Lapis lazuli

Lapis lazuli is considered to be best when it is very hard and deep blue in colour. It is found in mines where copper is forged, and along certain seacoasts. It has sharply astringent properties, and some caustic properties also.

86) Stag horn

Stag horn has astringent properties. It therefore staunches all types of haemorrhage, and is filed down and shaved, and when given to drink with an egg it drives out intestinal parasites. Also, if it is fumigated it puts vipers and serpents to flight.

1 cerui] ceruinum *FP* 2 ouo] ouo mittit *V* ouo potata *W* 3 uiperas] uespere *P* fugat] expellit *C* effugat *W*

cf. Diosc. II.59 (ἐλάφου κέρας). Pliny 8.118, 10.195, 28.178 (cinis etc. passim). Scrib. 60, 141. Galen *Simp*. 11.1.8 (K.12, 334–5).

87) Coagulum

Coagulum omne stringere et glutinare potest. Vnde ad omne profluuium aptissimum et utilissimum habetur.

1 Coagulum] coculum *V* coaculum *F* omne] omne masculinis *F* stringere] constringit *W* glutinare] *om*. *VBF* glutinat *W* 2 utilissimum] uale *W* uiribus *F* imussimum *L*

cf. Diosc. II.75. (πιτύα). Pliny 11.239 et passim. Galen *Simp*. 10.2.11 (K.12, 274)

88) Dictamnum

Dictamnum herba est mollis quae quasi ramos emittit rotundos et exalbidos et oblongos, nodulos copiosos habentes circa quos foliola excrescunt similia menthae, sed exalbidiora et angustiora et in summis ramulis flosculum roseum. Et in hoc ipso ueluti semen nigrum et
5   similastrum odore pene artemisiae sed suauiore et gustu lento excalfacit cum amaritudine. Vires habet excalfacientes et leniter stringentes et proprie tamen urinam et menstrua prouocat, sed et partum mortuum expellit et odore suo uiperam et serpentes interfecit.

1 Dictamnum] diptamnum *F* dipacomus *C* 2 exalbidos] *om*. *J* oblongos] *om*. *W* foliola] folia *VF om*. *B* 3 summis ramulis] sumo ramulo *LB* 4 roseum] rubeum *BP* et hoc- uires] *om*. *W* 5 similastrum *FJ*] semlastrum *V* similiastrum *L* simile ad extrema quae et artemisia *B* simile asaro *C* simile artemisae *P* pene] *om*. *V* lento] exlento *J* lenis *C om*. *BP* 6 cum amaritudine- habet] *om*. *V* uires- excalfacientes] *om*. *CP* leniter] leuiter *JLC* 7 urinam] urinam mouit *VFW* partum] f(o)etum *JBP* 8 interfecit] occidit *JC*

cf. Diosc. III. 32 (δίκταμον). Pliny 25.92. Celsus 5.23.16. Galen *Simp*. 6.4.6 (K.11, 863). Ps. Apul. 62. Isidore 12.1.18, 17.9.29.

89) Diphriges

Diphriges est ueluti faex quae a cupro uero crescit et calefacitur et in uino uel aqua extinguitur, aut de lapide pyrite conficitur qui similiter ut calx uiua uritur. Virtus est quoque robustus et quod naturaliter in metallis lutosis et liquidum inuenitur, quodque primum in sole siccatur
5   deinde coquitur. Potest igitur uiscidissime stringere et in modum niuis refrigerare. Vnde optimum dicimus quod gustu ualidissime stypticum est et similiter ut alumen exiccat.

87) Rennet

All types of rennet can contract and agglutinate, and it is therefore most suitable for all types of haemorrhaging, and is considered to be very useful.

88) Dittany of Crete (*Origanum dictamnus* L.)

Dittany of Crete is a soft plant that shoots branches that are round, pale, and oblong, with lots of nodules upon them, around which sprout little leaves that are similar to mint,[1] but are paler and more narrow, and on the small branches near the top there appears a small, pink flower. Inside this flower is a black seed that has full fragrance a little similar to that of wormwood[2] but more pleasant, with a bitter taste that slowly begins to warm the mouth. Dittany has heating and gently astringent properties, and is particularly useful for stimulating urination or menstruation. It also expels a dead foetus, and its odour kills off vipers and snakes.

1 #185.
2 #17, #22. 'a little similar': *similastrum* is the reading of our three earliest manuscripts (*VFL*), as well as *J*, hence another adjective with the *-astrum* suffix (see ch. 4.E). Another possible reading is *simile astro*, 'seed a little like a star/star-shaped,' or *astrum* as a corruption of *asterem* (*aster, -eris*, Gr. *asterion*), blue daisy/Italian aster (*Aster amellus* L.), mentioned in #125 but it seems to be a corruption. The inaccurate descriptions of Diosc. and Pliny, who both say that dittany of Crete has no flower or fruit, provide no solution (see ch. 1.I and plate 3).

89) Copper pyrites

Copper pyrites[1] is the slag that forms upon copper when it is heated and then cooled in wine or water, or it is made from mineral pyrites burned in the same manner as quicklime.[2] Equally powerful is the liquid form found as a natural deposit among the mud inside mines, which is first sun-dried and then smelted. Copper pyrites can be an extremely powerful astringent, and can also cool just like snow. We declare the best to be that which is most strongly styptic in taste, and which dries like alum.[3]

1 Diphriges] defrigis *V* diffriges *F* dyfriges *LB* difriges *W* cupro- crescit] *om. VCP* peru- ide erepitur *B* in uino- extinguitur] quod subinde calefacit *J* 2 lapide- naturaliter] per inde eripit et coquitur ut calix est de lapide pene lubrica *P* est autem de lapide quod pene lumbricum *LB* 3 uritur] *om. VB* est *C* uirtus- naturaliter] *om. W* 4 lutosis] lutusum *VC* luturum *F* liquidum et lutosi *P* liquidum lutosum *B* liquidus et luteus *W* 5 uiscidissime] uiscide namque constringit *W* 6 unde– quod] *om. W* 7 exiccat] desiccat *F* desiccatium *W*

cf. Diosc V.125 (πυρίτης λίθος). Pliny 34.135. Celsus 5.7.22. Scrib. 227, 247. Galen *Simp.* 9.2.6 (K.12, 199), 9.3.8 (K.12, 214–17). Isidore 16.4.5 (pyrites).

90) Dryopteris

Dryopteris filix est quae in arbore crescit, quae Graeci polypodium uocant. Per omnia illi terrestri et specie similis et efficacia.

1 Dryopteris] dioterix *F* diopestri *P* filix- crescit quae] id est *L* id est filicida filie *C* crescit] nascitur *PL* nascitur et crescit *W* graeci] *om. C* 2 illi] iridi *J om. W* similis] *om. F* efficacia] non ineffectu *W*

deest V, etsi titulus fol. 32v inter capitula.

cf. Diosc. IV.187 (δρυοπτερίς). Pliny 27.72 [26.58?]. Galen *Simp.* 6.4.10 (K.11, 865).

91) Daucum

Daucum optimum est creticum. Hoc igitur radiculam habet oblongam crassitudine digiti, thyrsulum eadem longitudine, in quo summo flo- sculum exalbidum et in medio semen quasi pilosum, magnitudine milii. Est et aliud cuius semen oblongum uidetur et cymino simile. Tertium
5 uero quod cymulam habet quasi coriandri. Verum omne daucum uires habet acres et excalfactorias, sic protinus ad gustum linguam remordet et excalfacit. Vrinam quoque et menstrua prouocat et partum deiicit et ad tussem ueterem et ad omnia uitia thoracis quae cumque fuerint idonee facit.

1 optimum est creticum] herba est folis fenuculi similis *C* 2 crassitudine- longitudine] grossam *P om. W* digiti thyrsulum] dicitur solum *F* longitudine] *om. V* 3 exalbidum] roseum *W* et in- daucium] *om. W* milii] melo *V* milii optimum est creticum *C* 4 cuius] quasi *V* 5 cymulam] cumolum *V* comulam *L* coniulam *C* culmum *P* 6 sic] ne *L* protinus] sicut aprotanum *P* 7 urinam- *fin.*] *om. C* deiicit] deducit *VFW* 8 ueterem] *om. L* et ad- facit] *om. W* thoracis] stomachi *F* 9 idonee] *om. P*

cf. Diosc. III.72 (δαῦκος). Pliny 25.110–12. Celsus 5.23.3B. Scrib. 177. Galen *Simp.* 6.4.2 (K.11, 862). Isidore 17.9.65.

92) Dorycnium

Dorycnium, et alicacabum aliqui dicunt, folia habet colore similia

1 Text quite corrupt. Diosc. V.125 describes only that formed from the 'stone from which copper is mined,' and mentions that the name *diphryges* refers to its 'twice-roasted' preparation, as does Pliny 34.135, but Pliny describes three types similar to the above: a mineral pyrites heated in a furnace; mud from a certain cave in Cyprus which is dried then heated by burning brushwood around it; and slag from the bottom of copper furnaces.

2 #67.

3 #7.

90) Black oak fern (~*Asplenium onopteris* L.; ~*Dryopteris* sp. L.)

Black oak fern is a fern which grows on trees, and which the Greeks call polypody, and is similar to the terrestrial type of polypody in all aspects of appearance and effectiveness.[1]

1 Text probably corrupt, the absence in *V* unfortunate. Though André 1985, 91, cites this (Gal. *alf.* 90) as *Polypodium vulgare* L., the tentative identification given here is from Beck 2005, 327, for despite the stated homonym, polypody is given below (#221), and Diosc. IV.186, 187, recognizes two different plants, with δρυοπτερίς growing specifically in old oaks: likewise Pliny 27.72 (*in arboribus nascitur*), cf. 26.58.

91) Daucos (~ *Athamenta cretensis* L.)

The best daucos is from Crete, which has a small, oblong root as thick as a finger, and a stalk about the length of a finger,[1] at the top of which is a small white flower containing seeds in the middle that are a little hairy and are about the size of millet grains. There is another type with a seed that is oblong and similar to cumin.[2] A third type has a tender sprout like coriander.[3] All varieties of daucos have sharp and heating properties, so that when tasted they immediately bite and heat the tongue. It stimulates urination and menstruation, and ejects the foetus. It is also excellent for chronic coughs and all chest complaints which might occur.

1 Similarly Diosc. III.72, and also gives three types, a number conceded by Pliny 25.110. See Andrews 1949b.

2 #69.

3 #68.

92) Dorycnion (~ *Convulvulus oleaefolius* Desr.).

Dorycnion, which some say is alicacabum,[1] has leaves the same colour as

oliuae sed longiora et aspera. Florem habet album et semen quasi orobi et in folliculos, quod aliqui quasi amatorium describunt. Verum tota herba uenenosa est.

1 Dorycnium] dorignum *VBP* doregnium *F* dongnium *W* orignum *C* alicacabum] alicadabum *L* alicaclauum *C* calclabum *B* cacabum *W* cadauum *P om. F* 2 florem] floscellum *VBWP* 3 quod- est] *om. BWP* 4 uenenosa] uenusta *C*

cf. Diosc. IV.74 (δορύκνιον). Pliny 21.179–82. Scrib. 191 Galen *Simp.* 6.4.8 (K.11, 864). Ps. Apul. 22.7 (?).

93) Eridrium

Eridrium radicula est plerumque pusilli digiti, colore humaceo, maxime penitus. Cuius eligimus quae est sicca, et spissa, non uermiculosa, gustu excalfaciens. Vires habet acriter relaxantes, et maxime ad oculorum claritatem facit.

1 Eridrium *V*] elidrium *F* eludrium *L* eliudrium *W* lidrium *C* eudrium *B* dyrium *P* cridrium *J* plerumque pusilli] quod potitionem *VFC* in modum pusilli *W* humaceo] humatium *V* ut appium *W* umatio *L* apium *P om. B* humaceo id est terreo *J* maxime penitus] *om. CB* 2 sicca] *om. V* 3 gustu] *om. P* oculorum] *om. W* 4 claritatem] lacrimantes *C*

94) Euphorbium

Euphorbium est lachryma arboris quae in Mauritania nascitur et ut aliqui uolunt et in partibus Italiae quae Galliae adiacent. Optimum est ergo euphorbium, quod est recens, subalbidum et glutinosum assimile minuto thuri, gustu ex lento linguam uehementer mordens et excalfaciens. Potest autem purgando humores stringere et calefacere aquaticos humores, prout sinapi Alexandrinum, quod est nimis acutum et igneum. Hydropicis maxime datur et colicis omnibus qui frigidi sunt. Ventres autem cholericos turbat et sitim grauem facit, quia nimis exiccat, commisces autem ei petroselinum, aut anisum, aut cyminum.

1 Euphorbium] euforuio *V* euforbium *FPW* et ut- adiacent] *om. W* et ut] nam hoc *BC* 2 quae- adiacent] *om. C* adiacent] adiungit *V* iungit *F* 3 recens] *P* glutinosum] guttulis suis *JP* glutinusius *V* 4 minuto] minutim *V om. W* thuri] *om. V* gustu- excalfaciens] *om. W* 5 potest- calefacere] *om. V* aquaticos- sitim] *om. W* 6 prout] uelut *C* nimis] minus *J* minus gustus *F* 7 maxime] ualde *V* 8 grauem] grandem *C* exiccat] desiccat *LP* 9 commisces- cyminum] *om. W* commisces] commiscebis *JLBC* miscetur *P*

adicit C Isidorum

cf. Diosc. III.82 (εὐφόρβιον). Pliny 25.77–9, 5.16, 27.2. Scrib. 38, 67. Galen *Simp.* 6.5.24 (K.11, 879). Isidore 17.9.26.

the olive tree, but longer and coarser. It has a [small] white flower and its seed in little pods like bitter vetch,[2] which some claim are used in aphrodisiacs. The whole plant, however, is poisonous.

1 Diosc. IV.74 mentions that Crateuas called *dorycnion* '*hallicaccabon* or *calleas*,' but elsewhere he also noted that these were alternate names for varieties of hound's berry (IV.71, 72): see above #84 n.1, and André 1985, 115. Scarborough, 'Thornapple' (forthcoming), argues that *dorycnion* (in Diosc. IV.74, also Nicander and Galen) is another kind of thornapple, *Datura stramonium* L. (Diosc. IV.73). On Crateuas, Wellmann 1897; Riddle 1985, 20–1, 185–91.

2 #133 (*herbum*), #207 (*orobum*). Aphrodisiacs: likewise Diosc. ibid., φίλτρα. See ch. 2.J.

93) Eridrium

Eridrium[1] is a small root commonly the size of a little finger, and is an earthy colour, especially inside. We select that which is dry, thick, not worm-like, and which warms the tongue. It has severely relaxant properties, and in particular is used to clear up vision.

1 Unidentified: privileging here the name given in *V*. André 1985, 78, cites it (*cridrium*, as in *J*) as 'forme obscure: plante indéterminée,' and suggests *eridrium*, 'tout aussi obscure,' because of its location in *J* (end of Ds), which is where it appears in most MSS also. The variants show it was no less puzzling to medieval scribes.

94) Spurge (*Euphorbia resinifera* Berg.)

Spurge is the sap from a tree which grows in Mauritania and, as some would have it, in the regions of Italy that border Gaul.[1] The best spurge is that which is fresh, a little pale, and sticky like crumbled frankincense, with a slowly emerging taste that bites hard into the tongue and warms it. Spurge can contract humours as it purges, and can heat watery humours just like Alexandrian mustard (seed),[2] which is very intense and quite fiery. It is often given to those suffering from dropsy and all those suffering colic who have the chills. It upsets the stomachs of those with cholera, and creates tremendous thirst which really dries you out unless you mix it with parsley,[3] or anise[4] or cumin.[5]

1 Cf. Diosc. III.82, grows 'in Autololia, next to Mauretania.' Pliny 25.7 mentions an inferior Gallic type, and claims to be using a (now lost) treatise on the plant by King Juba (II) of Numidia (c. 52–23 BC): see ch. 1.F, and ch. 3.D.

2 #261.

3 #223.

4 #15.

5 #69.

95) Ebenum

Ebenum genus ligni quod in Aethiopia nascitur, cuius optimum est quod est nigrum et aequaliter leue, simile sua lenitudine cornu. Est autem aliud in India quod aliqua in se abiectitia habet subrufa. Verum melius est ad usum medicinalem quod prius diximus. Vires autem
5　habet ad gustum acres et aliquatenus stypticas. Vnde scobes illius in collyriis miscentur et ad uulnera olentia facit.

1 aethiopia] egypto et aethiopia *BWP* 2 quod- aequaliter] *om. C* aequaliter] *om. W* simile sua lenitudine] ualidum ut cornu *W* lenitudine] ualetudine *BP* cornu] coruo *J* 3 abiectitia] abgicientia *V* habicienti *F* eotricitentiam *B* habicitentiam *L* adicentia *C om. P* 5 scobes] scopis *VLC om. BPW* 6 olentia facit] polythema *J* uulneribus utilitatem facit *C*

cf. Diosc. I.98 (ἔβενος). Pliny 12.20. Celsus 3.21.7, 5.7, 5.12. Galen *Simp.* 6.5.2 (K.11, 867–8). Isidore 17.7.36.

96) Eleas dacrua

Eleas dacrua oliuae lachryma est et gummi assimilatur et similiter ab ipsa matrice oliuae manat et maxime in Aethiopia inuenitur. Cuius est optimum quod est recens, mundum, pingue, rufum, gustu submordens. Vires autem habet non leniter excalfactorias et ad oculorum caliginem facit.

1 Eleas dacrua] eleas draci *VBP* elias drachi *F* dracu *L* eleais dracu *C* ebeas drachi *J* oliuiae lachryma seu elea *tit J.* gummi] gumen *VF* 2 matrice] *om. W* 4 non leniter] uehementer *J* non leuiter *P* acriter *W* 5 facit] *om. F* est aptum *W*

cf. Diosc. I.105 (ἀγριελαία). Pliny 15.24–5.

97) Elaterium

Elaterium succus est [radicis] erratici cucumeris, quod hac ratione conficitur. Collectis quam pluribus cucumeribus priusquam in plenam maturitatem ueniant, hos ipsos omnes incidimus et in colum mittimus, supposito alio uase quod succum excipiat illorum et sic maxime illos
5　supprimimus, ut non solum humorem exprimamus sed etiam ipsius cucumeris quicquid est carnosum. Deinde si fieri potest siccamus in patella fictili, nimisque probatum delimpidamus. Si quid aquosum est supernatat et quod est admodum quasi saeculentum in mortario laeu-

95) Ebony (*Diospyrus ebenum* Koenig; *Diospyrus melanoxylon* Roxb.)

Ebony is a type of wood that grows in Ethiopia, and the best is black and smooth all over, similar in its smoothness to wrought horn.[1] But there is another type in India which has a somewhat dishevelled appearance[2] and is slightly reddish in colour. The first type we mentioned, however, is better for medical uses.[1] It has properties that are sharp to the taste and a little styptic. Shavings of it are mixed with eye-salves and are placed on fetid wounds.

1 Similarly Diosc. I.98.
2 Cf. Diosc. ibid. 'having white and yellow divisions and equally frequent knots.'

96) Sap of wild olive (*Olea europea* L.: var. *Olea silvestris* Miller)

Sap of wild olive[1] is the sap of an olive tree that is like a type of gum.[2] It oozes out of the fully grown tree like the regular olive,[3] and is found chiefly in Ethiopia. It is best when it is fresh, pure, thick, reddish in colour, and has a slightly biting taste. It has quite effective heating properties, and is good for dim eyesight.

1 Oddly using a Greek name (ἐλαίας δάκρυα, largely bungled by our medieval scribes) rather than any Latin equivalent (*lacryma oleastri*). Diosc. I.105 calls it wild olive (ἀγριελαία) though notes that it is also called Ethiopian olive, and switches to this when discussing its sap (I.105.6, 'δάκρυον τῆς Αἰθιοπικῆς ἐλαίας'), suggesting similar or the same sources here (see n.3, and ch. 3.B). Pliny 15.25 calls this plant *cici* (or *croton*, an alternative also given by Diosc.), which he distinguishes from the wild olive (15.24).
2 #47.
3 Difficult but similar to Diosc. I.105, 'both the olive that grows in our land and the wild olive tree produce that sort of sap.'

97) (Juice of) Squirting cucumber (*Ecballium elaterium* Rich.)

Elaterium is the juice [from the root] of a wild cucumber which is prepared in the following way. We harvest many cucumbers before they are fully ripe, then we slice them all up and put them in a colander, under which we place another container to collect their juice. We then vigorously crush them so that not only all the moisture is squeezed out, but even the fleshy pulp of the cucumber is forced through. If possible we leave it to dry in an earthenware container, then carefully rinse that which remains.[1] The watery elements float to the top, and this is siphoned off to leave the

igamus et in pastillos redigimus. Aliqui etiam aquam miscent aut plu-
10 uialem aut marinam et rursus delimpidant. Deinceps caetera similiter
faciunt. Verum optimum est elaterium cyrenaicum, hoc enim est candidissimum et leuissimum et maximam habet amaritudinem et gustu aliquatenus excalfacit et saliuam euocat in modum ueratri. Quod autem
est quasi pallidum uel fuscum et asperum et graue peius est. Vires quo-
15 que habet similes elleboro uel etiam uastiores. Totum enim corpus promouet sic et uentrem et uiscera depurgat et uomitum suscitat et si plus
modo sumatur uenenum esse dicitur. Sed et adulteratur amylo propter
candorem et leuitatem, item cinere pollinis, item farina orobi, uel succo
ipsius radicis. Haec tamen intelliguntur eo quod amaritudinis et can-
20 doris, quod diximus, aliquid amittat et aliquid grauitudinis assumat,
nec ad gustum sic excalfacit.

1 erratici] de radicibus *V* radicis *BP* quod] eius *V* quod- 14 uires] *om.
W* 2 confici-
tur] colligitur et conficitur *VP* quam] cum *V* 3 incidimus] admordemus *J* adincidimus
L homines adindicimus *C* in- mittimus] *om. C* colum] sacculum *BP* 5 supprimimus]
subpraemimus *L* supponimus *C* 6 si fieri potest] *om. BP* 7 nimisque probatum] minus
prolatem *V* minus prolacte *F* nimis perlates *B* minus neperlatim *L* nimis tam prolatos
P deinde *C* delimpidamus] lipidamus *C* aquosum] *om. C* 8 admodum] quid exit mundum *V* mundum *F* in medium *L* inmundum *C om. BP* saeculentum] feculentum *P* 10
rursus] tyrsos *P* 11 cyrenaicum] ciruzaicu *J* cirinaicum *VBC* cirumaicum *F* cirinaku *L*
quiriniacum *P* 13 euocat] reuocat *V* in modum] in mundum *V* ueratri] ueretri *F* uelatri
BP 14 pallidum] placidum *BP* fuscum] fosius *L* graue] grandem *V* 15 similes- uiscera]
om. W totum] tantum *L* totum pro uiscera purgat *V* promouet] mouet *P* 16 uiscera]
uiscide *FLBPC* plus modo sumatur] plusquam oporta bibatur multum officit quia *W*
17 adulteratur] ulceratur *F* 18 cinere] cineris pullinae *V* cineris pollionem *F* ciceris *B*
cinere pollines *P* cinis pollinis mixatur *C* 20 aliquid] aliud *F* aliqui *P* grauitudinis] aliqua
grauitudine assumendo

cf. Diosc. IV.150 (σίκυς ἄγριος). Pliny 20.3. Celsus 5.12.6. Scrib.70, 224, 237. Galen *Simp.*
8.18.15 (K.12, 122–3). Marc. *Med.* 30, 14. Ps.Apul. 114.12.

98) Erificium

Erificium herba est quae in summis montibus inuenitur, quae folia
habet similia apio et thyrsulum oblongum in cuius summo flosculum
quasi uiolaceum habet et semen in medio. Radicem quoque ad magnitudinem et ualetudinem in specie cepae oblongae et ueluti ad imum
5 extenuatae habetque radices alias, quae sic radiculas minutas e lateribus
emittunt nigro cortice clausas. Haec ipsa uiribus est aconito similis et
thapsiae. Proprio gustu omnes illas probare debetur.

1 Erificium] erificum *P* quae- inuenitur] *om. P* 2 in cuius- radicem] habens in summitate
sita uiolatium florem intrinsecus retinentem semen *W* cuius] in quo *VC* 3 quoque- ual-

sediment, which we then pound in a mortar and shape into lozenges. Some mix it with rain water or sea water to rinse it yet again, or continue with other, similar methods of preparation. The best *elaterium* is the Cyrene when it is very bright in colour and very smooth, and is extremely bitter, with a taste that somewhat warms and causes the mouth to salivate in the same way as hellebore.[2] The varieties that are pale, swarthy, coarse, or heavy are inferior. Squirting cucumber has properties similar to black hellebore,[2] but much more powerful. It agitates the whole body in completely purging the bowels and internal organs, causes vomiting, and is said to be poisonous if more than the correct dose is taken. It is adulterated with starch[3] for brightness and levity, or similarly it is mixed with extremely fine flour,[4] or with meal of bitter vetch, or the juice from its own roots. The inclusion of these ingredients is revealed by the reduction in bitterness and the change in colour we mentioned, as well as by the change in weight and that it ceases to heat the mouth upon tasting.

1 Difficult passage. Diosc. IV.150 describes similar (and equally puzzling) methods of preparation: the corresponding passage says, 'Heap up in the sieve the pieces that have been squeezed, douse them with fresh water, press them and discard them.'

2 #285, where *helleborum* is used for black hellebore (*Helleborus niger* L.). Cf. *F* (!)

3 #6. Cf. Diosc. ibid., 'some add starch to imitate elaterion that is white and light,' and its juice is also used.

4 *cinere pollinis*: lit. 'with the ash [fine dust] of flour,' though *pollinis* may be a corruption of *polii*, hence 'ashes of hulwort' (*polion/polium* = *Teucrium polium* L.), not cited in *AG*, but see Diosc. III.110, Pliny 21.44, 21.145.

98) **Wolfsbane** (*Aconitum napellus* L.)

Wolfsbane[1] is a plant which is found on top of mountains. It has leaves similar to garden celery[2] and an oblong stalk, at the top of which is a small, violet-coloured flower, with seed in the middle. Its root is about the same size and width as the type of onion that is oblong, and has at the bottom more roots that thin out, and which themselves sprout from the sides tiny roots that are encased in a black bark. These have properties that are similar to those of leopard's bane[3] and deadly carrot.[4] These should all be tested by taste.

1 Identification uncertain, description unique to the *AG*. Both André 1985, 96, and Halleux-

etudinem] *om.* W **4** ualetudinem] uastitudinem *FLC om.* *BP* imum] unum *JB* ima *B* **5** extenuatae] extuatae *VJ* tenuem circum se radiculas minutas emittens nigra cortice clausa *W* lateribus] interibus *P* **6** nigro] in *C* similis] *om. VLBCP* **7** proprio] propter *LB* propter quod *VFP* debetur] deuita *J* debet *F* debemus *BPW*

cf. Diosc. IV.77 (ἀκόνιτον ἕτερον). Pliny 27.4–11 (?).

99) (H)Elenion

Helenion herba est omnibus nota, nam et in hortis nascitur et est assimilis lapathio siluatico. Folia habet asperiora et latiora, habet thyrsulum alium in cuius summo flos est aureus. Aliqui sine thyrsulo describunt radicem bene olentem, uastam, subrufam, amaram. Huius 5 igitur folia ex uino cocta et imposita renibus coxiosos emundant. Item succus ipsius cum ruta tussientibus et ruptis prodest, urinam quoque et menstrua [prouocat] et partum expellit.

1 Helenion] elenion *VFLBPC* elenium *W* omnibus- et] *om.* WP **2** siluatico] *om.* JL habet- igitur] quia languinosa et albidiora *W* **3** alium] altum *P* cuius] in quo *VC* aureus] aurosum *VCP* **4** describunt] scribunt *V om. BP* radicem] radicem habet *JP* huius] huiusmodi *J* **5** coxiosos] toxiosos *F* uitia *BP* dolores *C* dolentibus *W* emundant] emendat *VBC* meduntur *W* tundat *P* **6** ipsius] florum *J* eorum *BP* illarum *L* cum ruta] potatus *BP* mixtus cum ruta *W* tussientibus] satientibus *B* sitientibus *P* ruptis] *om. C* **7** prouocat] *om. VFLCP*

cf. Diosc. I.28 (ἐλένιον), [29 (ἐλένιον ἄλλο)?]. Pliny 14.108, 25.159. Celsus 5.11. Galen *Simp.* 6.5.7 (K.11, 873). Marc. *Med.* 22.5. Ps. Apul. 96.

100) Eruca

Eruca omnibus est nota, et habet vires excalfactorias et diureticas et coitum per suam uirtutem ad uenerias inspirat.

1 Eruca] eruga *V* ruba *P* nota] nota hominibus *C om.* WP **2** coitum] ubiter suam *V* ubi cerpes *F* clities *L* probabiliter *BP* obiter *C* uenerias] uenas *J* ueria *L* ueneriam *B* uenerios *C* ueterem *P* inspirat] portitat *J* speritur *V* pruritat *BF* prodest *C* excitat actus *W* prouocat *P*

cf. Diosc. II.140 (εὔζωμον). Pliny 19.123, 20.125. Garg. *Med.* 14. *DynVat* 52/*SGall* 64. Isidore 17.10.21.

101) (H)Edera

Hedera nascitur in diuersis arboribus, parietibus vel petris, et terra, est

Opsomer 1982a, 96, suggest *Aconitum napellus* L. Diosc. IV.77 lists 'another kind of leopard's bane' (*aconiton*), which grows on the Vestini mountains in Italy, has 'elongated pods and roots like black legs of a shrimp,' and is used to hunt wolves: his description differs considerably. The conjecture of Meyer 1856, IV:493–4, 'aus der *Peristerion hyptios* des Diosc, und der Name entstellt aus *Erigenion*, einem Synonym derselben Planze,' cannot stand.

2 #14.
3 #21.
4 #281.

99) Elecampane (*Inula helenium* L.)

Elecampane is a plant known to everyone, for it grows in gardens and is similar to wild sorrel.[1] It has leaves that are quite wide and coarse, and one type has a stalk which bears a golden flower at the top. Others describe another type without a stalk and which has a root that is very fragrant, large, tawny-coloured, and bitter. The leaves of this type of elecampane are cooked in wine and then applied topically near the kidneys to relieve hip ailments. The juice from the leaves mixed with rue[2] is good for healing coughs and ruptures, stimulates urination and menstruation, and expels the foetus.

1 #154.
2 #227.

100) Rocket (*Eruca sativa* Lam.)

Rocket is known to everyone. It has heating and diuretic properties, and it strengthens the desire for sexual intercourse.[1]

1 Following *V*. Cf. Diosc. II.140, 'is an aphrodisiac when eaten in large quantities.' Similarly Garg. *Med*. 14, *DynVat*.52/*SGall* 64.

101) Ivy (*Hedera helix* L.)

Ivy grows on different kinds of trees, or on walls, rocks, and along the

omnibus nota. Huius tota folia dysenterias remediant. Item trita et pro fomento imposita ad combusta et uulnera cancerosa faciunt.

1 Hedera] edera *VLBCJW* ederanas *F* in] per *LB* pro *F* parietibus] *om. V* terra] cetera terra *F* terram serpit *W* est- nota] *om. P* **2** omnibus] omnibus hominibus *C* tota] trita *J* rarita *C om. VW* dysenterias] remedia haberit *P* item] bene *VF* nam *P* item- combusta] et frumento mixta combusturae *W* **3** faciunt] posunt *W* faciunt ualde *C*

bis W, secundum ex Isidoro 17.9.22

cf. Diosc. II.179 (κισσός). Pliny 16.114–16, 16.243–4, 24.75–81 et passim. Galen *Simp.* 7.10.29 (K.12, 29–30). Ps. Apul. 98. Isidore 17.9.22.

102) Elelisphacos

Elelisphacos herba exalbida, uasta et oblonga, uirgam habet uastam et quadratam. Folia aliquatenus aspera, odore non insuaui. Fructus in summis ramis quasi ormini syluani, nascitur locis asperis. Potest autem acriter stringere, propter quod ad dysentericos facit et succus foliorum eius capillos denigrat et urinam prouocat.

1 Elelisphacos] elilisfacum *V* elisfagus *F* elilisfagus *W* elelisfagos *P* elissacum *C* uasta] *om. FC* oblonga] *om. W* uirgam] *om. VF* **2** odore- asperis] *om. W* insuaui] soaui *V* insucciu *C* **3** ramis- syluani] *om. C* quae in sylua *BP* **4** stringere] restringere *VW* **5** capillos] *om. J* denigrat] denigrat generatque *C*

cf. Diosc. III.33 (ἐλελίσφακον). Pliny 22.146–7, 26.31. Galen *Simp.* 6.5.8 (K.11, 873). Ps. Apul. 102. *Ex herb. fem.* 4.

103) Epithymum

Epithymum est flos herbae assimilis setae, subuiridis, et nauseosum uentrem temperat et uomitum facit. Nam uero epithymi flos exalbidus est et asper et plurimum in Pamphilia inuenitur. Ambo igitur aequalis uirtute per os purgat phlegma et per aluum fortiter. Choleram nigram 5 soluit tritum [drachmas VI] aut cum uino dulci, aut cum melicrato, miscens et modicum salis et his dabis quibus praecordia ab inflatione uexantur et hepaticis et qui suspirium patiuntur

1 Epithymum] epytimum *VL* epithimum *FBPWC* setae] sitani *V* sit *F* subuiridis-qui] *om. W* **2** temperat] purgat *W* nam- *fin.*] *om. W* **3** plurimum] om. *P* pamphilia] phamolia *V* ambo- fortiter] *om. JC* aluum] oclutio *V* album *F* arium *W* alueum *P* aluunt *LB* fortiter] *om. P* **5** drachmas VI] *om. V* melicrato] mellegratium *V* mellicuato *P* **6** dabis] *om. P* **7** patiuntur] pariuntur *JLB* medetur *W* parientibus *C* pacientibus *P*

cf. Diosc. IV.177 (ἐπίθυμον). Pliny 26.55–6. Galen *Simp.* 6.5.14 (K.11, 875). Marc. *Med* 20.74. Isidore 17.9.13.

ground, and is known to everyone. Any of its leaves when crushed alleviate dysenteries, or when ground down and applied in a poultice they heal burns and cancerous wounds.

102) Sage (*Salvia* sp. L.)

Sage is a plant that is pale, quite large and oblong in shape, and has a large, squared stem. Its leaves are fairly coarse and have a mildly pleasant fragrance. Its fruit appears on the highest branches like wild asparagus[1] and it grows in rough terrains. Sage can act as a severe astringent, and is therefore good for treating dysenteries. The juice of its leaves is used as a black hair-dye, and stimulates urination.

1 *ormini*: Pliny 19.151, 20.110, 26.94; Diosc. II.125. *Dyn. Vat.* 9, *Dyn. SGall.* 13. On various types see Andrews 1956a; André 1985, 126.

103) Epithymon / Dodder (*Cuscuta epithymum* L.)

Epithymon is the flower of a plant that is brush-like[1] and slightly green in colour. It calms a nauseous stomach and causes vomiting. The flower of epithymon is whitish in colour and coarse, and is found mostly in Pamphylia.[2] It powerfully purges phlegm with equal force through the mouth and through the bowels. It dissolves black bile when [six drachmas' worth is] pounded and mixed with sweet wine, or water-mead,[3] and when mixed with salt you give it to those who suffer from gas in the diaphragm, those with liver problems, and those who suffer from respiratory problems.

1 *seta*: 'brush-like' is cognate, and an accurate description. Cf. Diosc. IV.177 'flower of the tougher type of thyme that resembles savoury. The little heads ... having hair-like stems.' V suggests *saetania*, common medlar (*Mespilus germanica* L.), on which Pliny 15.84, 23.141, and Diosc. I.118 (*setanion*, a type that grows in Italy). Also a type of onion or bulb, Pliny 19.95, 19.102.
2 South Asia Minor, mod. Antalya province, Turkey.
3 *melicratus*: a type of hydromel consisting of a large amount of water: Diosc. V.9; Vegetius, *Mulomed.* 3.15.22.

104) Erpillum

Erpillum herba est quae in hortis seritur et in campis [sudibus] et lapidosis crescit, iacens ut serpens ramulos et oblongos foliolis densis pusillis odore suaui in modum samsuci. Floscellum huius in summo rubicundum inuenitur aliquando. Potest igitur urinam et menstrua
5 prouocare et in fomentatione capitis freneticorum et lethargicorum aliqui illud utuntur.

1 Erpillum] erpullum *B* epillum *C* seritur] *om. V* inuenitur *F* campis- menstrua] *def. F* sudibus] *om. VW* saudis *C* 2 crescit-potest] *om.W* ut] uel *V* ramulos] ramulos tenues et oblongos habens *P* 3 pusillis] postillis *V* huius] *om. LP* habet *B* 5 prouocare- fomentatione] prouocant et ancusa id est semen iunci et ad ignem *BP* fomentatione] adsumatione *LC* lethargicorum] litargiorum *F* 6 aliqui- utunter] facit *BP* adhibita prodest *W*

deest J, partim def F

cf. Diosc. III.38 (ἔρπυλλος). Galen *Simp.* 6.5.20 (K.11, 877). Ps. Apul. 100.

105) Erythrodanum

Erythrodanum ramulos habet tenues et quasi nodosos, folia ex interuallo et iuxta singula ueluti stellulas. Fructum rotundum, uiride, deinde rubicundum et ad maturitatem nigrum. Radiculam tenuem, pusillam, rubicundam, diureticam. Cuius usus ad eadem proficit ad quae alias diureticas herbas facere diximus.

1 Erythrodanum] eritodanum *V* eritrodamum *FLP* eridrodanum *B* eritrodanum *C* erirodaprium *W* folia] folia extensa ualde *WC* iuxta- *fin.*] uirtus eius est dioretica *seq.* Isid. iuxta- stellulas] *om. FP* 4 pusillam rubicundam] *om. F* usus] sucus *C*

ex Isidoro 17.9.13 W

cf. Diosc. III.143 (ἐρυθρόδανον). Pliny 1.24, 1.56, 24.94. Galen *Simp.* 6.5.22 (K.11, 878). Ps. Apul. 28. Alex. Tral. 2.145.

106) Elatine

Elatine ramulos habet quinque uel sex tenues et oblongos, folia quasi elxine pilosa et styptica, unde pro fomento imposita lachrymam oculorum stringit, et cocta deinde et manducata ad dysentericos facit.

1 Elatine] elatin *V* latine *P* quinque ut sex] VI uel VII *J* uel VI *C* V.L.VI *P* 2 elxine] ex se *J* elexine *V* quasi ectine *F* quasi *B* in eo *C om. J* fomento] fronti *J* fumigatione *L* fermento *W* perfuma *C* 3 manducata] comesta *W* facit] optime facit *BP* sistit *W*

cf. Diosc. IV.40 (ἐλατίνη). Pliny 1.27, 1.50, 27.74. Galen *Simp.* 6.5.5 (K.11, 873).

104) Tufted thyme (*Thymus serpyllum* L.)

Tufted thyme[1] is a plant that is cultivated in gardens and also grows in thorny and stony fields, where it spreads along the ground like a snake,[2] with slender long branches that bear thick foliage of small leaves that have a pleasant fragrance like that of marjoram, and occasionally it produces a small, ruddy flower on top. It can stimulate urination and menstruation, and some use it in a warm compress applied to the head of those who are delirious or lethargic.

1 Identification from André 1985, 122: cf. Beck 2005, 197, *Thymus Sibth.* for Diosc. III.38, which is similar (marjoram [#260], emmenagogue, diuretic, lethargy, delirium), but otherwise his detailed description differs considerably. On types of thyme: Andrews 1958b, 1961a, and see #282 (Cretan thyme).
2 The name *erpyllum* derives from the verb 'to creep' (ἕρπειν), pointed out by Diosc. III.38.

105) Madder (*Rubia tinctorum* L.)

Madder has small, slender branches that are quite noduled, and leaves that are spaced out, each leaf the shape of a star.[1] Its fruit is round, green, but turns reddish and then blackens upon ripening.[1] Its little root is slender, small, ruddy and is a diuretic. It can be used for the same purposes we have mentioned for other diuretic herbs.

1 Similarly Diosc. III.143.

106) Cankerwort (*Linaria spuria* Miller)

Cankerwort[1] has five or six small, slender branches, and leaves which are hairy like those of bindweed,[2] and very styptic, hence they are applied in a compress to dry teary eyes, and are good for dysenteries when cooked and eaten.

1 Entry was conjoined to *erythrodanum* in *JP.*
2 Similarly Diosc. IV.40; entire entry parallels almost verbatim. Bindweed (*Convolvulus arvensis* L.) not in *AG*: see Diosc. IV.39 (ἐλξίνη), Pliny 22.41.

107) Ebulus et sabucus

Ebulus et sabucus uidentur quasi magnitudinem differe sed similitudinem et folia et florem et grana similia habere et fere easdem habere uires. Proprie tamen ebuli radix hydropicos exinanit et purgat, sed et ipsa folia cocta et manducata uentris sunt purgatiua.

1 Ebulus et sabucus] euolum et sambugum *V* ebolum et sabucus *F* ebulus uidetur a sambuco magnitudine distare *P* magnitudinem] magnitudinem differe sed similitudinem et folia *V* diferre foliis autem flore et semine similia sunt *W* sed est sibib foliis *P* 2 folia] *om.* *F* similia] *om.* *V* habere] *om.* *FP* 3 uires] uirtutes *F* hypropicos] *om.* *BP* exinanit] et inanit *J* et- purgat] *om.* *B Addit #108* multo- fugat *P* 4 uentris sunt purgatiua] uentrem purgat *VFC* uentrem prouocat *W*

cf. Diosc. IV.173 (ἀκτῆ, ἔλειος ἀκτῆ). Pliny 26.120, 24.51, 16.179–80. Galen *Simp.* 6.1.21 (K.11, 820–1). Ps. Apul. 92. *DynVat* 54, 64.

108) Erice

Erice adsimilis myricae sed multo breuior est et acris ualde, propter quod et comula illius serpentes et uiperas fugat.

1 adsimilis] adsumit *C* myricae] hiricae *J* ericis *V* erige *F* irice *C* crucae *W* eructe *P* propter- illius] huius uis floris et thursi *W* 2 comula] cimula *L* cunula *P* et] et flos *LB* fugat] expellit *P* effugere facit *C*

cf. Diosc. I.88 (ἐρείκη). Pliny 13.114, 24.64. Galen *Simp.* 6.5.19 (K.11, 877).

109) Ficus

Ficus fructus est arboris eiusdem uocabuli, quae nimium comeste stomachum conturbant, uentrem deponunt, feruores nutriunt, sudorem prouocant, sitim prohibent. Siccae uero, quae caricae uocantur, stomacho utiles sunt, sitim operantur et uentrem stringunt. Faucibus,
5 arteriis, renibus et uesicae utiliter prosunt et eos qui spiritum tarde recipiunt iuuant. Illos etiam qui de longa aegritudine colore pallidi uidentur meliorant. Hydropicis prosunt, thoracis sorditiem purgant. Decoctae uero cum hyssopo et careno, potui ieiuno datae, tussibus antiquis prosunt et pulmonum uitia emendant. Cum aqua uero mul-
10 sa decoctae arteriis et tonsillis tumentibus optimum gargarismum est. Duricies et parotidas et feruores soluunt si cum iride et nitro decoctae imponantur. Crudae quoque similiter, nitro et calce permisto duricies et uulnerum rheumatismum emendant. Si cum foenegreco in aceto coquantur et cum eodem succo permistae et imponantur impetigini et maculis utiles sunt.

107) Danewort / Dwarf elder (*Sambucus ebulus* L.) and Elder (*Sambucus niger* L.)

Danewort and elder may differ in size but they are similar in their leaves, flower, and seed, and they have almost the same properties. The root of danewort is especially good for clearing up and purging dropsy, and its leaves when cooked and eaten purge the bowels.

108) Tree heath (*Erica arborea* L.)

Tree heath is similar to tamarisk[1] but is much shorter, and very sharp, and therefore its foliage puts snakes and vipers to flight.

1 #193. *myricae*: surmised from Diosc. I.88.

109) Figs (*Ficus caria* L.)

Figs are the fruit of the tree of the same name, and when too many are eaten they upset the stomach, empty the bowels, nurture fevers, cause perspiration, and prevent thirst. When dried, which are called '*caricae*,'[1] figs are good for the stomach, quench thirst, and fortify the bowels. They are also highly beneficial for ailments of the throat, trachea, kidneys, and bladder, and they help those who inhale slowly.[2] Dried figs also improve the condition of those made pale from suffering a chronic illness. They help with dropsy, and remove filth in the chest. When stewed with hyssop[3] and reduced sweet wine and given as a drink while the patient is fasting, they clear up old coughs and help with ailments of the lungs. When stewed with honeyed water they make an excellent gargle for treating the trachea and swollen tonsils. They dissolve indurations and reduce fevers and parotidal swelling when stewed with iris[4] and soda and applied topically. Likewise when unripe and mixed with soda and quicklime[5] they dissolve indurations and heal rheumatic wounds. If cooked in vinegar with fenugreek[6]

1 uocabuli] *om. W* comeste] comixtium *V* comestit *F* comettus *C* 2 conturbant] comburunt *J* 4 utiles] utilissimum *VP* uentrem stringit] *om. W* uentrem alaxat *V* uentrem laxant *F* 6 colore- hydropcicis] debilitantur *W* 7 uidentur meliorant] *om. VF* habent optime facit *BP* thoracis- datae] *om. LP* 8 decoctae] de nocte *L* hyssopo] ysuphu *V* ieiuno] ieiunu sputiu *V om. FC* 9 prosunt] produnt *V* emendant] mundat *J* mulsa] multa *VJW* 10 tonsillis] tussilis *FLC om. W* 11 si cum- emendant] *om. W* si cum iride et nitro] sicut iris et intra *J* miseus et nitro *V* sicum ysopo et nitrum *F* sicum yreos et nitro *BP* sucum iris et nigrum *L* 12 calce] sale *BC* calde ·*P* 13 si cum fenugreco] *om. JLC* sua fenigreco *B* 14 permistae] permistae conquantur] *VJC* 15 utiles sunt] curat *BP*

cf. Diosc. I.128 (σῦκα). Pliny 13.51, 23.117–24. Galen *Simp.* 8.18.43 (K.12, 132–3). Garg. *Med.* 49. *DynVat* 76. Isidore 17.7.17.

110) Foeniculum

Foeniculum est omnibus notum, de cuius radicibus contusis in pila lignea et per linteum succus exprimitur. Idoneum est ad multa medicinalia et maxime ad uitia oculorum premitur. Similiter et de semine et de ramis illius adhuc uiridibus. Verum optimus est succus qui de radicibus fit, leniter enim calefacit et remollit.

1 Foeniculum] fenucolus *V* feniculum *FP* fenuculi *C* omnibus notum] herba *P* 2 per linteum] ut bete *J* cum lenteo *V* lenteo *F* uete *L* per linteolum *WC om. BP* medicinalia] medicina *P* 3 ad- *fin.*] *om. W* 4 uiridibus] uiribus *VFC* qui] dicunt *B* dicimus *P* qui de] quia *L*

cf. Diosc. III.70 (μάραθον). Pliny 20. 245–6. Galen *Simp.* 7.12.5 (K.12, 131). Garg. *Med.* 25. Ps. Apul. 125. *DynVat* 43/*SGall* 47. Isidore 17.11.4.

111) Folium

Folium folia sunt herbulae quae in India nascitur locis humidissimis et paludibus colliguntur autem haec ipsa folia in aqua supernatantia. Verum est optimum quod pilosum et hastile est minutissimum et integrum leue, exalbidum uiscide. Suauiter redolens in modum spicae et eundem odorem diu multumque seruans, gustu subamarum. Potest autem uires habere similes quas spicam habere diximus.

2 autem- 6 autem] *om. W* 3 hastile est notissimum] pastillis est minutissimum *VLBP* pastillis et minutis *F* 4 redolens] per olentes *V* praeolentes *F* 5 eundem] tendens *B* extendens *P* 6 similes] *om. WB* quas] quasi *L* quales *BC* spicam] spicam indicam *W*

cf. Diosc. I.12 (μαλάβαθρον). Pliny 12.42–5, 12.129. Galen *Simp.* 7.12.2 (K.12, 66). Isidore 17.9.1.

and then diluted with their own juice and applied topically they are useful for treating freckles and blemishes.

1 While much of the above directly parallels Diosc. I.128, he does not mention the distinctly Latinate term *caricae* (dried figs from Caria, Asia Minor), a word unknown in Greek, but used in Latin from the time of Cicero up to the Visigothic law code (c. 650), though the Latin Diosc. uses it on two occasions (IV.177, 183) where the Greek text (IV.182, 188) simply used 'figs' (σύκα).
2 Cf. Diosc. 'ἀσθματικοί.'
3 #297.
4 #137. Likewise, Diosc.
5 #67.
6 #117.

110) Fennel (*Foeniculum vulgare* Gaertn.)

Fennel is known to everyone. Its root is pounded with a wooden mortar and the extracted juice is then strained through a linen cloth. It is useful for many medicines, but is especially applied to eye complaints. Its seed and branches are used for the same when still green, but the juice from its roots works best, for it gently warms and softens.

111) Folium (Leaf of malabar?) (~ *Cinnamomum* sp., *Pogostemon patchouli* Pell.)

Folium[1] is the leaves of a shrub that grows in India and is harvested in swamps and very damp places where the leaves are found floating upon water. The best is that which is hairy and which has a very small spear-like stalk that is entirely smooth, and is very pale in colour. *Folium* is fragrant like spikenard,[2] producing the same, strong odour for a long time, and has a slightly bitter taste. It has properties that are similar to that which we described for spikenard.

1 Identification uncertain: the description is very similar to that for malabar #190, but the *AG* seems to recognize two different plants. The name *folium* was applied to different aromatic plants by Latin writers, including patchouli (*Pogostemon patchouli* Pell.), different types of spikenard (#236), or types of malabar: André 1985, 105, 151; Innes Miller 1969, 74–7. Interestingly, Isidore 17.9.2 (*folium*) seems to describe the same plant as above. Cf. Pliny 12.129 (*malabathrum indicum*).
2 # 236.

112) Faecula

Faecula faex est uini uel aceti quae coquitur et reponitur. Cuius est efficacissima quae est recentissima. Mittitur autem in caustica.

1 Faecula] fecola *V* fecula *FLCPW* uini] uni *C* cuius- recentissima] *om. W* 2 quae- recentissima] *om. F* caustica] caustica medicamenta *W*

cf. Diosc. V.114 (τρύξ). Pliny 14.130–1, 23.63–7. Scrib. 226, 228. Marc. *Med.* 19.12, 20.75, 34.76. Theod. *Eup.* 15H. Isidore 20.3.13.

113) Filix

Filix herba est sine flore, sine fructu, ramulos habens longos, uastos, patulos, diffusos et in specie alarum et pinnarum adapertos uastos. Extrinsecus nigra, penitus uiridis, nascitur maxime in locis petrosis. Vires habet acres et relaxantes et aliquantulum uentrem mollientes et proprie tineas et lumbricos expellit ex mulsa aqua potata.

1 Filix] felex *F* felix *LBC* filex *P* ramulos- 3 petrosis] *om. W* 2 alarum] malorum *LBP* adapertos- nigra] *om. P* uastos] radicem uastam *C* 4 aliquantulum] *om. W* uentrem] uenerem *F* 5 tineas] et si *W* densas *V* aqua] *om. FBP*

cf. Diosc. IV.184 (πτερίς). Pliny 27.78. Galen *Simp.* 8.16.39 (K.12, 109–10). Ps. Apul. 77. Isidore 17.9.105, 20.14.4.

114) Phu

Phu herbae genus est herbae quae in Ponto nascitur, quaeque habet folia similia apii syluatici, thyrsum rufum, mollem, cauum, leuem, ex interuallo nodos habentem et flores maiores quam nascissi cum aliquo pallore purpurastrum. Radiculam subrufam, uastitudine pusilli digiti, ex qua alii excrescunt minores in se implicatae. Et quae tota radix uires habet excalfactorias et diureticas.

1 Phu] fu *VCP* fuhu *F* genus- nascitur] *om. W* quae- quaeque] *om. P* 2 thyrsum- tota] *om. W* 2 cauum] album *J* 4 purpurastrum *F*] porporastrum *V* purpurascentes *J* purpurastros *LC* purpure *B* purpureo *P* pusilli] postilles *V*

cf. Diosc. I.11 (φοῦ). Pliny 12.45, 21.136. Scrib. 170. Galen *Simp.* 8.21.8 (K.12, 152). Marc. *Med.* 20.19. Isidore 16.9.7 (fu).

115) Fimus

Fimus omnis efficacissime tumorem lenit et ueteres durities remollit. Item podagricos in accessionibus mitigat. Caninus fimus albus, tritus cum melle impositus et tumores et uulnera delicate mundat. Asininus

112) Wine lees

Wine lees are the dregs of wine or vinegar[1] which is cooked and then stored. It is most effective when it is freshest. It is mixed with caustics.

1 Diosc. V.114 recommends specifically old Italian wine, and that lees from vinegar are too strong.

113) Male fern (*Polystichium filix mas* Roth.)

Male fern is a plant without flowers or fruit, and has twigs that are long, thick, wide-spreading, and extensive, and look like extended wings or feathers. It is black on the outside, green inside, and grows chiefly in rocky places. It has sharp and loosening properties, and properties which slightly soften the stomach. When drunk with honeyed wine and water it drives out tapeworms[1] and intestinal parasites.

1 *tineas*: possibly 'itch,' Niermeyer 1997. On honeyed wine, Balandier 1993.

114) Cretan spikenard (*Valeriana phu* L.)

Cretan spikenard is a type of plant which is grown in Pontus. It has leaves similar to wild celery,[1] with a red, soft, and smooth stalk that is hollow and sports nodules in the middle section. Its flowers are larger than those of narcissus,[2] and have a slightly purple shade to them. It has a small, reddish root about the size of a little finger, from which other smaller roots grow and entangle themselves. The entire root of Cretan spikenard has heating and diuretic properties.

1 #14.
2 #197.

115) Dung

All dung thoroughly abates swelling, and softens chronic indurations. Likewise it soothes the onset of pain from gout. Whitened dog's dung, when crushed with honey and applied topically, gently cleanses swellings

autem cum aceto tritus, uel succus eiusdem, dentium dolorem tollit.

1 Fimus] fymus *V* fimum *FL* lenit] repremit *VFL* ad tumorem illinitus *WC* ueteres] uentris *W* **3** tumores] *om. W* septimia *C* mundat] permundat *V* promutat *W* asininus- tollit] *om. W* eiusdem] eius *V*

deest BP

cf. Diosc. II. 80 (ἀπόπατος). Pliny 28. 155 (caprae), 156 (uituli), 165 (felis), 174 (asini), 198 (bubuli) etc. passim. Isidore 17.2.3.

116) Fel

Fel omne uires habet acres et thermanticas. Vnde potest omnem sordem de corpore abluere et abstergere. Item caligines et cicatrices oculorum extenuat.

1 Fel] fex *VF* felosne *W* el *C* thermanticas] genitigats *V* lenitica sunt *F* instacas *L* sine antica *C om. WBP* **3** extenuat] et sanat *V* eximit *JC*

cf. Diosc. II.78 (χολή). Pliny 28.146. Galen *Simp.* 10.2.13 (K.12, 275–81). Isidore 11.1. 127–8.

117) Foenumgraecum

Foenumgraecum est omnibus notum. Vires autem habet leniter et efficaciter relaxantes. Vnde ad omnem rigorem et tumorem idonee facit.

1 Foenumgraecum] grecumfenum *VWC* grecofenum *FL* fenugrecum *BP (bis)* *C* estnotum] *om. P* efficaciter] *om. JW* **2** rigorem- facit] *om. W* et tumorem] *om. VF* idonee] *om. C*

intra 'G' in JVFLCW, bis C intra 'F'

cf. Diosc. II.102 (τήλεως ἄλευρον), IV.111(λωτὸς ἄγριος).

118) Glaucium

Glaucium succus est herbae eiusdem nominis, humilis, huius folia similia sunt papaueri. Nascitur autem locis planis et humidis. Hic ipsa folia siccant et calido cinere opprimunt ut sic succum suum remittant. Alii hanc herbam tantummodo in olla noua addunt et sic ollam luto
5 undique circumliniunt et in furnum feruentem mittunt et sic foliis maceratis succum eorum exprimunt et siccatum in sole in pastillos redigunt. Alii ex aqua dulci coquunt quo usque aqua croci colorem habeat et tunc colant et similiter siccant. Est autem optimum glaucium quod recentissimum et intus crocatum, quasi similiter inficiens croco
10 mundo, odore graui et defrustrato, gustu amaro. Vires autem habet

and sores. Donkey dung, when crushed and mixed with vinegar, or when reduced to a liquid form,[1] relieves toothaches.

1 Diosc. II.80 discusses the dung of cow, goat, sheep, horse, pig, pigeon, chicken, stork, vulture, mouse, lizard, and human faeces, as well as dog and donkey dung, to which two he attributes different uses (binding bowels, scorpion stings, respectively). The only other author to refer to a 'juice' of donkey dung is Theodorus Priscianus, *Eup.* 39.4 ('fimi asini recentis sucus'), to help cure those who suffer from day blindness (*nyctalopes*, on which, Langslow 2000, 488).

116) Bile

All bile has sharp and heating properties, and therefore can wash and remove all types of filth from the body. It also reduces dim-sightedness and corneal abrasions.

117) Fenugreek (*Trigonella foenum-graecum* L.)

Fenugreek is known to everyone. It has mild but thoroughly loosening properties, and therefore is excellent for every kind of induration and swelling.

118) Glaucion (*Glaucium cornuculatum* Curtis)

Glaucion is the juice of the plant of the same name, which is quite small, and has leaves similar to the poppy.[1] It grows on wet plains, and there they dry the leaves by placing hot embers on top of them, which causes them to excrete their juice. Others simply place the plant in a new pot and then seal the pot with clay and place it in an extremely hot furnace, and in this manner the leaves are steamed and release their juice, which is then dried in the sun and made into lozenges. Others cook it with sweetened water until the water turns the colour of saffron, and then they strain it and similarly dry it. The best glaucion is that which is freshest and has a hue of saffron within it, as though it had been dyed in pure saffron, and has a deceptively

stypticas. Vnde resolutum in aqua et inunctum prius, impetus lippitudinis et lachrymas stringit.

1 succus] cuius sucus simile optinet nomen *W* huius] foliofae quae *J om. V* est foliis *B* foliore *C* 2 humidis] *om. W* 3 siccant] sunt *F* ficus *L om. B* opprimunt] conpremuntis *V* operiuntur *W* suum] *om. JW* 4 addunt] mittunt *WBP* 5 circumlinunt] liniunt *W* linium *C* 7 quo usque] cuiusque *F* donec *W* 8 tunc- siccant] et sic colatum desiccant *W* 9 intus crocatum] tinctum croci colore *J* quasi- mundo] *om.* `W` croco] cum *V* 10 defrustrato] *om. FC* gusto amaro] *om. F* uires- stypticas] *om. W* 11 prius] primus *VLBPC* lippitudinis] lippitudines mitigat *W* 12 stringit] restringit *VP*

cf. Diosc. III.86 (γλαύκιον). Pliny 27.83. Scrib. 22. Galen *Simp.* 6.3.5 (K.11, 857). Ps. Apul. 74.26

119) Galbanum

Galbanum est lachryma naturaliter manans ex arbuscula assimili ferulae quae in Nicomedia nascitur. Cuius lachrymam eligimus quae est mundissima et similis ammoniaco non uda, nec dura et quam minime astulosa. Cuius vires hoc praestant, quod et ammoniacum.

1 Galbanum] galuanum *F* galbanon *P* nicomedia] Media *J* manans] mamatus *F* 2 lachrymam- *fin.*] uires hoc idem praestant quod dicitur praestare amoniacum *W* 3 uda] ut dum *V* nudus *P om. C* minime] nimis *F* 4 astulosa] asclusum *VF* asclosum *LBCP* uires] uires habent et *V*

cf. Diosc. III.83 (χαλβάνη). Pliny 12.126. Celsus 3.20.3. Galen *Simp.* 8.22.1 (K.12, 153). Isidore 17.9.28, 17.9.95.

120) Git

Git herba est quae in segetibus crescit, habens semen nigrum et minutum et odore non insuaui pluribus notum. Haec igitur relaxare et calefacere potest. Si potetur menstrua et vrinam prouocat et lumbricos expellit. Imposita tumores et durities soluit.

1 Git] gitter *V* gitti *FL* gyt *C* segetibus] segrae *C* habens- igitur] *om.W* 2 non] *om. V* nomen *F* insuaui] suaui *FP* pluribus notum] *om. P* 3 si potetur] *om. W* sic potum *L* sic potatum *BP* sic *tamque per nudatam C* 4 imposita] uice cataplasmatis impositum *W*

cf. Diosc. III.79 (μελάνθιον). Pliny 20.182. Celsus 2.33. Scrib. 131. *Ex herb.fem.* 3.63

121) Galla

Galla est quasi nux parua uel malum arboris assimilis quercus. Quarum sunt aliquae pertusae, aliquae integrae et extrinsecus asperae, quae ipsae sunt uiscidiores. Verum omnis galla uehementer stringit, elegimus tamenque sunt nigriores, integrae, grauiores et quodammodo pleniores, quas Graeci cicidas appellant.

pungent fragrance and a bitter taste. Glaucion has styptic properties, and therefore when dissolved in water and applied topically early it prevents ophthalmia[2] and staunches tears.

1 #199. Cf. Diosc. III.86 like the 'horned poppy' (*Glaucum flavum L*) = IV.65 (μήκων κερατῖτις), commonly mistaken as source for glaucion.

2 *lippitudo* (also #272, and ch. 4.H): see Celsus 1.9.5, and esp. 6.5.6, Scrib. 135. Pliny 28.29. Cf. Diosc. III.186, 'it cools and is therefore good for incipient eye problems.'

119) Galbanum (*Ferula galbaniflua* Boiss. and Buhse.) More comp.

Galbanum is the sap which flows naturally from a shrub that is similar to giant fennel[1] and which grows in Nicomedia. We prefer the sap that is well purified and similar to gum ammoniac,[2] and is neither watery nor hard and has no splinters.[3] Its properties are the same as those of gum ammoniac.

1 #4 n.1. Cf. Diosc. III.83, 'juice of the giant fennel giant that grows in Syria.'

2 #4.

3 *astulosa*: see #39 n.2.

120) Black cumin (*Nigella sativa* L.)

Black cumin[1] is a plant that grows in grain fields and has a small, black seed with a slightly sweet fragrance which is known to many. It can warm and loosen, and when drunk it stimulates urination and menstruation, and drives out intestinal parasites. When applied topically it breaks down swellings and indurations.

1 *git* (*gitti, giddi, gitter*), a word of Semitic origin, is the preferred name among Latin writers for *melanthion*, Diosc. III.79. Halleux-Opsomer 1982a, 96, suggested corncockle (*Agrostemma githago* L.); I follow André 1985, 110.

121) Nutgall

Nutgall is like the small nut or the fruit of a tree similar to the oak. There are two types, one which is perforated, and the other without apertures and rough on the outside, which is more effective. All nutgalls contract tissue, but we prefer those that are darker, heavier, without apertures, and which are a little plump, which the Greeks call cicidas.[1]

1 Galla] gille *P* nux parua] ncercole *V* nuciculas *F* nucula *W* nucelle *BP* nucicula *C* malum] malule *L* mamule *C* sunt *BP* assimils] *om.* *F* quercus] querere *W* quarum sunt] *om.* *F* 2 pertusae] procusse *J* portus *V* protuse *FBPC* asperae] aspre *V* aspae *J* 4 integrae grauiores] *om.* *VFW* 5 quas- appellant] *om.* *P* cicidas] cidos *F* cicidos *W* KIKIDWS *B*

cf. Diosc. I.107 (κηκίς). Pliny 24.9. Isidore 17.7.38, 19.19.5.

122) Glycyrhiza

Glycyrhiza herba est quae folia habet tactu pinguia et glutinosa circa uirgulas oblongas, et flosculum quasi hyacinthi, fructiculum magnitudine similem pilulis. Plantae radices quasi buxeo colore oblongas, et in omnem partem diffusas in modum uiticularum. Nascitur autem locis 5 glareosis et in campis. Ad gustum dulces, ex quibus succus coquitur similis glaucio. Nam habet uires suauiter relaxantes et cum dulcedine exhumidantes. Vnde et sitim sedat, et renum et uesicae dolorem mitigant. Prouocat enim urinam et menstrua et omnem tumorem tunsa et imposita placat.

1 Glycyrhiza] gliciriza *V* gliriza *F* gliqiricia *W* gliciritia *B* glyutciriza *C* gliciricia *P* folia- 4 relaxantes] *om.* *W* glutinosa] gummosa *J* glutinosa lentisco simila *VF* 3 plantae] platanis *B* plantani *P* 4 uiticularum] uitis *P* nascitur- campis] *om.* *LBP* 6 glaucio] glantino *P* glaucio- uires] *om.* *F* suauiter] similiter *V* 7 exhumidantes] exhumedantes *V* exhumidas *F* humidantes *LBCP* humidantis *W* exhumectantes *J* et renum] *om.* *B* 8 tunsa] tusa *BP* contusa *C om.* *W* 9 placat] pagat *V* sanat *W* sedat *C*

cf. Diosc. III.5 (γλυκύρριζα). Pliny 21.91, 22.24. Scrib. 75. Galen *Simp.* 6.3.9 (K.11, 858). *Ex herb. fem.* 42. Isidore 17.9.34.

123) Gentiana

Gentiana herba est assimilis lactucae, folia quidem in gyro habet incisa in specie serrae, et thyrsum oblongum crassitudine digiti, cauum, mollem, ab imo rubicundum, semen in foliolis, radicem longam, similem aristolochiae, sed nigrorem et lentiorem corticem. Haec igitur gustu 5 est amarissima et habet uires acres et excalfactorias et relaxantes et acrimonia et amaritudine sua potest partum expellere et omnem serpentem fugare et aduersus uenenosa prodesse.

1 lactucae] lactucae coctae *J* lactucae tota *LBP* folia] habens folia lactucae similia *W* quidem- imo] *om.* *W* gyro] nigro *B* 2 serrae] ferre *J* ferri *VLBCW* crassitudine] magnitudine *VF* grossitudine *CP* 3 imo] *om.* *BP* uno *C* 4 aristolochiae] gladio *B* aristolochiae gladio *P* sed- corticem] *om.* *W* lentiorem] leniorem *F* 5 et relaxantes] *om.* *BPW* 7 uenenosa] uenena *VW* uena *L* prodesse] expellere *W*

cf. Diosc. III. 3 (γεντιανή). Pliny 25.71. Celsus 5.23.3A. Scrib. 170. Galen *Simp.* 6.3.2 (K.11, 856). Ps. Apul. 16. *DynVat* II.5. Isidore 17.9.42.

1 *cicidas*: cf. Gr. κηκίς, Diosc. I.107 (cf. Lat. Diosc. 'De cecidos, id est galla'); similarly Marc. *Med.* 7.17, *Phys. Plin.* 2.28. Nutgall is not the the fruit or seed of the oak (likewise Diosc. καρπός), but an excrescence produced by the presence of parasitic organisms, both vegetable (fungi, mould) and insects: Beck 2005, 78 n.133.

122) Liquorice (*Glycyrrhiza glabra* L.)

Liquorice is a plant with leaves that are greasy and gummy to the touch clustered upon its oblong twigs. It has a little flower like the hyacinth,[1] and a small fruit the same size as pellets. The roots of the young liquorice are the colour of boxwood and oblong in shape, and spread out far in different directions like vine tendrils. It grows in gravelly places and in fields. It has a sweet taste, and its juice is cooked in the same manner as glaucion.[2] It has pleasantly relaxant and agreeably moistening properties. It therefore quenches thirst, and relieves pain in the kidneys and bladder. It stimulates urination and menstruation, and when crushed and applied topically it eases all types of swelling.

1 #300. Similarly Diosc. III.5.
2 #118.

123) Gentian (*Gentiana lutea* L; *Gentiana purpurea* L.)

Gentian is a plant that is very similar to lettuce, for it has leaves that are circular with jagged edges like a saw, and an oblong stalk about the thickness of a finger that is hollow, soft, and reddish in colour near the bottom. Its seed is surrounded by little leaves, and it has a long root like that of birthwort[1] but with bark that is darker and more pliant. The root has an extremely bitter taste, and has properties that are sharp, heating, and loosening. Its pungency and bitterness can be used to expel the foetus, put all snakes to flight, and it is useful against poisons.

1 #19. Above description very similar to Diosc. III.3.

124) Gladiolum

Gladiolum omnes nouerunt. Huius igitur radix potata ex uino et ad uenerea percitat et menstrua prouocat. Item trita et in modum fomenti posita surculos et spinas de corpore extrahit et omnes durities dissoluit et euaporat.

1 omnes nouerunt] omnis pene medici noscunt *W om. P* potata ex uino] puta ex humo *F* 2 uenerea] uenerium actum *W* uenena *L* uienarum *C* uenerit *F* percitat] proprietatem *V* proficit *W* peritat *LFP* perritet *B* idritat *C* menstrua- trita] *om. B* item- euporat] *om. W* 3 extrahit] abstrait *V* omnes] enim *J*

cf. Diosc. IV.20 (ξίφιον). Pliny 21.65, 21.107–8. Galen *Simp.* 8.24.3 (K.12, 87). Marc. *Med.* 15.13. Ps. Apul. 46, 79 (?). Isidore 17.9.83.

125) Terra samia

Terra samia terra est naturalis et genus cretae, cuius optimam iudica-mus quae est leuissima, candidissima, mollis et fragilis et quae bolaria oblonga habet et linguae applicata uehementer adhaeret, huiusmodi quoque est quam florem asterem appellamus. Facit autem terra samia 5   ad ea quae instantius stringere uolumus, sic et coeliacis et dysentericis et ad omne profluuium fortiter prodesse obseruatur.

1 Terra samia] gessamia *VFLBW* gesamia *P* naturalis- cuius] naturaliter apud graecum *W* genus] *om. VF* 2 quae] optimum *F* candissima- *fin.*] *om. FW* mollis- *fin.*] *om. VBP* bolaria] solaria *J* 3 applicata] ampliciae *C* huiusmodi] huius *LBC* 4 quam] quam quasi *LC* asterem] austere *C*

cf. Diosc. V.153 (σαμία γῆ). Pliny 34.194. Galen *Simp.* 9.1.4 (K.12, 178–92). Isidore 16.1.7.

126) Gypsum

Gypsum lapidis genus est simile uitro, cuius optimum est quod uenas habet latas, lucentesque et hac ratione paratur. In clibano missum ut calx candidum fit. Vires habet quas et terra samia et amplius stypticas et refrigerantes fortiter. Vnde illinitus fronti sanguinis cursum restrin-git, potatus uero et dysentericis et coeliacis omne profluuium statuit.

1 Gypsum] gipsum *VW* gipsus *P* cuius- fit] *om. W* uenas] *om. VF* 3 terra samnia] gessamia *VFLBC* 4 et refrigerantes fortiter] *om. W* cursum] fluxum *J* restringit] sistit *W* extinguit *BP* 5 potatus] potus *LC* statuit] statuit mollis et fragilis...*etc. ex* #125 - obse-

124) Corn flag (*Gladiolus segetum* Gawler)

Everyone knows corn flag, the root of which is drunk in wine as an aphrodisiac, and it stimulates menstruation.[1] It is also ground up and applied in a compress to extract splinters and thorns from the body, and it also reduces and disperses all types of induration.

1 Similarly Diosc. IV.20, though distinguishing the effect of the upper and lower root, the latter causing sterility. He also notes (IV.22, ξυρίς) that Romans call gladwyn (*Iris sp*.L.) *gladiolus*. André 1985, 111, cites this entry (Gal. *alf*. 117) as an unidentified species of wild iris, or possibly stinking iris (*Iris foetidissima* L.).

125) Samian earth

Samian earth is a type of natural soil from Crete. We judge the best to be that which is the lightest in weight and brightest in colour, yet soft and delicate, and which has little oblong lumps[1] which when placed on the tongue stick to it fast, in the same manner as the flower we call *aster*.[2] Samian earth is used when we want to contract things quickly, hence it is observed to be powerfully effective against intestinal pain, dysentery, and all types of haemorrhages.

1 *bolarium*, -a, from Gr. βωλάριον (a diminutive of βῶλος, lump or clod), an extremely rare word, seemingly attested in Latin only in *AG* and the second-century poet Septimus Serenus, from whose work only fragments remain. It is not used by Diosc. for his entry on same (V.153) or anywhere else, but he does use the adverb βωληδόν in V.106 (alum). In any case, lines 2–6, absent in *VFBPW*, may well be an interpolation.
2 This is probably confusion of the information found in Diosc. V.153, who says there are two kinds of Samian earth, *collourion* and *aster*, though it is the former that 'sticks to the tongue like glue.' See also above #88 n.1. *aster* would be Italian *aster* (*Aster amellus* L.), not accorded an entry in *AG*.

126) Gypsum

Gypsum is a type of stone that is similar to glass, the best of which has extensive, conspicuous veins for which it is sought. It is placed in an iron furnace[1] until it turns white like limestone.[2] It has the same properties as Samian earth, though they are even more styptic, and vigorously cooling. Therefore when smeared on the forehead it restricts the flow of blood, and when drunk it staunches all discharges from those suffering dysentery and bowel ailments.

1 *clibanum*: the only reference in *AG* to this particular type of oven, broad at the bottom,

ruatur *VFBPC* ualde proficere creditur flos uero eius austeris…*etc. ex* #125 - uolumus *W*

cf. Diosc. V.116 (γύψος). Pliny 36.182. Galen *Simp*. 9.3.6 (K.12, 213-14). Isidore 16.3.9, 19.10.20.

127) Terra ampelitis

Terra ampelitis terra est genus cretae colore nigra. Cuius optimam dicimus quae est nigerrima, splendida, leuis et quae ab humore cito resoluitur. Stringens ualde et denigrans propter quod rubeos et canities inficit. Sed et in calliblephara mittitur.

1 Terra ampelitis] gy ampilodis *V* gy ampelotis *F* gemapelitus *W* giampilotis *C* gemapelitis et ut aspaltum *P* cretae] getenus *F* rea *L* colore] ut aspaltum colore nigro *W* huius quoque quasi flus est asticus terrae nigra *V* **2** ab humore] humore attacta *W* **3** rubeos et canities] capillos *J* rubeum et canas *V* rubeum sit canas *F* robeas et canas *L* rubeas cannas *C* rubos et carras *P* rubos et calamos *W* **4** sed- calliblephara] in oculorum ideo blefara *W* in oculorum blefar *P* calliblephara] culurbrefar *V* calibrefor *F* calliblem mar *L* calibe id·est ferro *C*

deest B

cf. Diosc.V.160 (ἀμπελῖτις γῆ). Pliny 35.194.

128) Terra erethria

Terra erethria gleba est rubicunda. Cuius est optima quae sine sabulo est et sine lapillis. Vires habet ualde stringentes et efficacius quam terra samia.

1 Terra erethria] ge eritria *V* gi chia *V* gis erithreas *W* ge eretrias *L* ge eritheas *B* gi erithas *C* ge erithreas *P* cuius] illa uero *W* **2** efficacius] efficatiores *W* terra samia] ges samia *VFWLBPC*

cf. Diosc. V.152 (ἐρετριάδος γῆ). Pliny 35.192.

129) Terra chia

Terra chia gleba est alba et lata non ualde candida, aufertur aliquando et in patellis rotundis aut quadratis. Et ipsa efficaciter stringens, sed et cutem purgat et extendit et splendidam facit.

1 Terra chia] gy cia *V* ge chia *LBP* gi chia *FW* gi cia *C* gleba] *om. F* candida] candula *L* **2** patellis] pastellis *F om. W* **3** extendit] extenuat *J*

cf. Diosc.V.155 (χία γῆ). Pliny 35.194.

usually associated with baking bread: Pliny 18.105, 20.99; Celsus, 2.17, 3.21. The bare
notice for gypsum in Diosc. V.116 demonstrates the independence of the *AG*: see ch. 3.B.
2 Cf. #67.

127) Ampelitis earth

Ampelitis earth is a type of Cretan soil that is black in colour. We affirm
that the best is that which is blackest, shiny, smooth, and which dissolves
quickly if it encounters moisture. It is a strong astringent and black dye,
hence is used to dye red and white hair, and is used as eye make-up.

128) Eretrian earth

Eretrian earth takes the form of a reddish-coloured clod. The best is that
without any sand or gravel in it. It has strongly astringent properties that
are more effective than those of Samian earth.[1]

1 #125.

129) Chian earth

Chian earth is a clod that is thick and white, but not brightly so, and
is obtained in small round or square plates. It is an effective astringent,
cleanses the skin, tightens it, and makes it shine.

130) Terra selinusa

Terra selinusia uires habet quas et terra chia et est optima quae est lucens, candidissima, splendens et quae ab humore cito remittitur.

1 Terra selinusia] ge selinufia *B* giyeselinosa *F* gi selitiusia *W* ges salama *C* gelinusia *P* terra chia] egifya *V* egi chia *F* ge chia *LB* egialia *C* chechia *P* 2 lucens] lactes *VF* lactens *BC* latens *P om. W* candidissima] splendidissima solint dinsit *V* ab- *fin.*] ad tactu humore liquescit *W* cito] cita *J om. F* remittitur] remadet *J*

cf. Diosc. V.155.2 (χία γῆ, σελινουσία γῆ).

131) Terra melina

Terra melina est colore spodii et dura ualde, ad gustum sapore aluminis et linguam exiccat. Eligimus autem quae est mollior et fragilior et recens et quae ab humore cito resoluitur, nec in se quicquam lapidosum habet. Hanc autem probationem in omni creta desideramus et haec quoque prioribus illis similiter proficit.

1 Terra melina] gi mellia *F* gy melia *V* ge melia *L* gimolia terra *WC* gamolina *P* spodii] splendidi *V* spondii *FLP om. W* 2 exiccat] adsiccant *V* desiccans *BP* desiccat *W* mollior et fragilior] melior *J* mollior *W* 3 humore] ore *F* resoluitur] remittitur *P* nec- *fin.*] *om. W* lapidosum] lapidositatis *J* habitosum *F* 4 creta] certa *P* desideramus] habemus *LP* et haec- proficit] *om. BP*

cf. Diosc. V.159 (μήλια γῆ).

132) Terra cimolia

Terra cimolia est aliqua candida et spissa et purpurea uelut pinguis. Et melior uidetur et ad tactum frigida. Valde quoque illinitur ad ignem sacrum et combusta. Pustulas fieri prohibet.

1 Terra cimolia] gy chimolia *V* gyptimolia *F* ge chimolia *L* ge chimolea *BP* gi cimolia *WC* 2 melior] mollior *C* melior- tactum] om. *W* 3 sacrum] acrum *VFLBCP* pustulas] stupulas *C* fieri] facere *W*

cf. Diosc.V.156 (κιμωλία γῆ). Pliny 35.195–7.

133) Herbum

Herbum nascitur in lenticula [quod et feritur] et est aliud pallidum, aliud subrufum. Vires habet acres et diureticas ualde. Ita et si comedatur extra rationem caput grauat et uentrem soluit et per urinam sanguinem educit. Boues saginat si decoctum detur. Fit autem farina de illis
5   quae albiora sunt, ita ut aqua parum aspergantur et in patellis assentur

130) Selinus earth

Selinus earth has the same properties as Chian earth, and is best when it is clear, very bright and shiny, and crumbles apart quickly upon contact with moisture.

131) Melian earth

Melian earth is of the colour of ash and is very hard, with a taste like alum,[1] and makes the tongue dry. We recommend that which is fresh, quite soft and delicate, and which crumbles apart quickly when wet and contains no rubble. We require that all clays be tested this way, and likewise for all those which we have mentioned above.[2]

1 #7.
2 Similarly Diosc. V.159: see ch. 3.B.

132) Cimolian earth

Cimolian earth has two types, one white and dense, the other purplish and like grease, the best of which is cold to the touch.[1] It is applied topically for erysipelas,[2] and upon burns. It also prevents blisters.

1 Similarly Diosc. V.156.
2 #68 n.1.

133) Bitter vetch (*Vicia ervilia* L., Willd. = *Ervum ervilia* L.)

Bitter vetch[1] grows in the form of small pods and is [said to be] of two types, one pale in colour, the other reddish. It has sharp and powerful diuretic properties. If too much is eaten it causes headaches and diarrhoea, and draws blood through the urine. If boiled and offered to cattle it fattens them. Meal is made from the bitter vetch seeds that are whitish by sprin-

ut corium dimittant, et sic tundantur et tenuissime cribellentur, et uentrem utiliter malaxant et urinam mouent et colores malos emendant ad mensuram trium cochleariorum cum melle uel uino data. Vulnera quoque similiter cum melle optime purgat et implet. Lentigines uel maculas tollit, duritiem mamillarum soluit. Carbunculos malignos sanat et contra canis et uiperae morsum utiliter imponitur. Cum aceto uero acceptam urinam mouet, tortionibus prodest. Aqua uero ubi cocti fuerint pruritum prohibet si totum corpus inde lauetur.

10

1 Herbum] horobum *BP* eruum *C* quod- feritur] seritur *FP om. VW* est] *om. J* 2 ita] *om. J* 3 rationem] ronnem *J* et per- *fin.*] *om. W* 4 boues saginat] uobices sauat *C* 5 albiora] albi *VLBP* ut aqua] ut si aqua *J* parum] partim *P* 7 malaxant] laxant *JFC* urinam] *om. BP* colores malos] dolores oculorum *F* 8 trium] *om. BP* melle] oximelle *P* data] *om. F* 10 carbuncolus] forunculos *V* furunculos *F* farunculos *C* 11 canis] canis rabide *C* 12 tortionibus] tortionibus uentris *C*

cf. Diosc. II.108 (ὄροβος). Pliny 18.57, 139, 22.151. *Dyn. Vat.* 15/*SGall* 19.

134) Herba sauina

Herba sauina omnibus nota est. Habet uires acres et excalfactorias. Propter quod et in uino potata partum expellit et per urinam sanguinem prouocat.

1 sauina] sabina *LC* omnibus nota] *om. P* habet] inter *JL om. W* 2 propter- in] quae cum *W* per] super *JL om. W* 3 prouocat] inducit *J*

cf. Diosc. I.76 (βράθυ). Pliny 24.102. Scrib. 154. Ps. Apul. 86. *Dyn. Vat.* 58/*SGall* 61.

135) (h)Indicum

Indicum, id est minium, species habet duas. Est enim unum leue et puluerastrum et ualde caeruleum, quod inueniri solet circa primitia et quasi radiculas foliorum calamorum qui in India nascitur. Est et aliud quod maxime in Italia fit, cum purpura tingitur nam quod est spumosum et quasi supernatat et lateribus uasorum illorum adhaeret colligitur. Et contusa argyritide terra mixtum in similitudine digitorum formatur uel in pilulas siccatur. Cuius est optimum quod uehementissime caeruleo colore fulget leue aequaliter et fragile. Potest igitur siccare et refrigerare. Vnde locis turgentibus et distensis impositum cito emollit et quasi rugulas facit.

5

1 Indicum] hindicum *VFB* hasceg *J* hindictis *P* id -minium] *om. FWP* leue] lene *J* 2 puluerastrum *VLBP*] puluerastris *F* puluereum *W* plus uenustum *J* quod-radiculas] *om. W* 4 cum] cum quo *LB* quando *J* 5 lateribus- contusa] *om. W* uasorum] illorum uastum *P* 6 contusa] cum ipsa *J* argyritide] sieretide *V* argille *WP* mixtum] mistum *J* 7 uel- pilulas]

kling them with a little water and then baking them on plates until their skins crack, and then they are crushed and finely sifted. This meal softens stools, promotes urination, and a measure of three spoonfuls given with honey or wine helps those who are off-colour. The same amount with honey is excellent for filling and cleansing wounds. It removes blemishes and freckles, and softens induration of the nipples. It heals malignant carbuncles and is effective against dog and viper bites when applied topically. Taken with vinegar it unblocks difficult urination, and is good for cramps. The water in which bitter vetch seeds are cooked stops itching if applied to the whole body as a rinse.

1 Also given below under Greek name of *orobum* #207. Text is corrupt: liberties taken with translation here are owed to the similarity of this entry, in places verbatim, to Diosc. II.108.

134) Savin (*Juniperus sabina* L.)

Savin is known to everyone. It has sharp and heating properties. Therefore when drunk with wine it expels the foetus and draws down blood through the urine.

135) Indigo (*Indigofera tinctoria* L.)

True indigo, that is, minium, has two different types. One is smooth, dusty, and a deep blue in colour, and is usually taken from among the first shoots, which look like little roots, of the leaves of reeds which grow in India. The other type is found chiefly in Italy and is the scum collected from using purple dye[1] which floats upon the surface and sticks to the sides of the cauldrons. It is ground, mixed with earth containing silver-ore,[2] and fashioned into the shape of fingers, or is dried in the form of lozenges. The best is that which gleams the deepest blue in colour and is evenly smooth and delicate. It can dry and cool, and therefore quickly assuages swellings and inflammations when applied, and creates folds like wrinkles.

1 Language difficult, but similarly Diosc.V.92, *baphicon*, 'scum from purple murex,' and Pliny 35.46, though neither mentions Italy: see ch. 1.F.
2 #58 n.2.

om. W **8** leue] *om.* J **9** unde- distensis] ualde turgentia et distincte P
deest C

cf. Diosc. V.92 (ἰνδικόν). Pliny 33.163, 35.46. Isidore 19.17.16 (indicum).

136) Ios scolex

Ios scolex est quasi collyrii genus subcaeruleastrum quod confringitur
aliquatenus. Huius ipsius sunt genere duo, est enim quod foditur et
in terra inuenitur, est quod potius reparatur ratione tali. Mortarium
fit cupreum cum pistillo et sic laeuigatur alumen liquidum et sal uel
5 nitrum eisdem ponderibus ex aceto acerrimo. Alii et aceti lutei uetu-
stissimi duas partes commiscent. Et haec laeuigant aestibus maximis
et in sole calidissimo usque quo spissitudinem aliquam accipiat et
colorem caeruleum. Deinde fingitur uelut collyrium. Aliqui ex urina
infantis masculi hoc faciunt. Et fit aurificibus idoneum ualde ad aurum
10 glutinandum. Vires autem habet ios scolex stringentes ualde et nimis
cum magna exasperatione.

1 Ios scolex] iuscolex *VFL* iusem *W* huscolex *C* hus *BP* hustolex *J* subcaeruleastrum
VBLP] subrufi oleastrum *F* subrubeastrum *C* fragile et subceruleum *W* subcaerulem *J* **3**
potius] *om.* B **4** fit] est *J* **5** acerrimo] fortissime *F* alii- commiscent] *om.* W lutei] *om.* VFC
latei *L* **6** aestibus] erit ibi *F* hictibus *LP* extimo *C* **7** calidissimo] feruente *W* **9** aurificibus]
artificibus *VFLBP* **10** ios scolex] uiscide *J* ualde-nimis] *om.* FC nimis- exasperatione]
asperas stringentes *W* nimis] *om.* P **11** magna] maxime *C* exasperatione] asperitate *P*

cf. Diosc. V.79.6 (ἰὸς σκώληκος). Pliny 34.116.

137) Iris

Iris ad similitudinem iridis quam in coelo uidemus, dicitur et haec
cognominatam. Quomodo enim illa plures et dissimiles habet colores,
sic et haec uarios et differentes emittit flores, aliqui albi enim purpu-
rei sunt, aliqui caeruli, aliqui uiolacei. Verum folia sunt similia gladio-
5 li, sed maiora et latiora et pinguiora, ramulos quoque ipsos. Similiter
radiculus nodosas et duras suauiter olentes in modum uiolae. Quarum
optima est iris illyrica et macedonica, est enim spississima et breuis et
non fragilis et subrufa, odore suauissima et gustu linguam uiscide excalefacit et dum tunditur sternutare facit. Ab istis Africana laudatur quae
10 est colore candidior et gustu amara. Verum si tempore relinquantur et
quasi senescant fiunt quidem uermiculosae sed odore suauiores. Possunt igitur excalfacere leniter et uulnera concaua replere.

1 Iris] ireus *F* ireos *L* iridis] yris *V* ireis *F* aeris *C* yreos *P* dicitur- cognominatam] ita et
cum nominatur *V* **2** plures] flores *F* **3** uarios] *om.* B albi] *om.* VJ **4** caeruli] *om.* J **5** sed-
pinguiora] *om.* F latiora] altiora *P* similiter] *om.* V

136) Worm-like verdigris

Worm-like verdigris[1] is like a blue-green variety of eye-salve that has been mashed up a little. There are two kinds: one is found in the ground and mined, the other is prepared in the following way. Place it in a copper mortar and grind it with the pestle along with liquid alum[2] and an equal weight of salt, or soda, and add very sour vinegar. Some also mix in two parts of well-aged yellow vinegar. They beat these in the hottest part of summer under a burning sun until the mixture achieves a certain thickness and blue colour. It is then fashioned into an eye-salve. Others do the same but mix it with the urine of an infant boy, and goldsmiths find this excellent for soldering gold together. Worm-eaten verdigris has powerful astringent properties, but not without causing a great deal of irritation.

1 *Ios scolex*, following *VFL*: the name is a corrupted version of the Greek ἰὸς σκώληξ which Diosc. V.79 records as a type of verdigris (above #3) and about which he provides information very similar to that given here, as does Pliny 34.116, who calls it *scolex*. No other Latin texts used the name except the translations of Oribasius and Alexander of Tralles: Opsomer 1989, I:358.
2 #7.

137) Iris (*Iris florentia* L.: *Iris germanica* L.: *Iris pallida* Lam.)

Iris is so named because of its resemblance to the rainbow we see in the sky. For just as a rainbow has many different types of colours in it, so the iris produces varied and different coloured flowers, some white, some purple, some blue, some violet. The leaves are similar to corn flag[1] though are larger, wider, and fatter, but its twigs are very alike. Its small roots are knotty and tough, and pleasantly fragrant like the violet. The best iris is from Illyricum and Macedonia, for it is very thick, short, and not so delicate, tawny in colour, with the most pleasant fragrance and a taste that vigorously heats the tongue, and when pulverized it causes sneezing. Recommended is the African iris, which is brighter in colour and has a bitter taste. When left for some time to age these become worm-eaten, yet they smell even more pleasant. They can be used for gentle heating, and to fill the cavities of wounds.

1 #124.

deest W.

cf. Diosc. I.1 (ἶρις). Pliny 21.40, 21.143. Ps. Apul. 79. Isidore 17.9.9 (illyrica).

138) Ibiscus

Ibiscus genus est herbae similis erraticae maluae, omnibus notum. Cuius radix et tota quidem herba leniter et efficaciter omnem duritiem et rigorem mollire et relaxare potest.

1 Ibiscus] iuiscum *VFB* hibiscum *LC* ibicus *P* genus] *om.P* erraticae] siluatica *VF om.P* omnibus notum] *om.P* **2** leniter] lenit *JLC om.* B **3** rigorem] *om.* V

deest W.

cf. Diosc. III.146 (ἀλθαία). Pliny 19.89, 20.29. Celsus 4.31.4.

139) Iuniper

Iuniperi fructus arbusculae rubicundus et dulcis omnibus notus. Cuius vires sunt excalefactoriae et diureticae et ad digestionem facientes.

1 arbusculae] parue arboris *W* dulcis] gustu dulcis *W* **2** sunt] habet *F* digestionem] indigestionem *C*

cf. Diosc. I. 75 (ἄρκευθος). Pliny 24.54. Galen *Simp.* 6.1.57 (K.11, 836–7). Isidore 17.7.85.

140) Ichthyocolla

Ichthyocolla est aliqua quae conficitur coctis ex aqua corticibus quorundam piscium. Est aliqua quae uentriculus uidetur esse piscis quae in Ponto nascitur. Verum omnis ichthyocolla stringere potest. Vnde et cutem extendit et splendidam facit.

1 Ichthyocolla] igiocolla *VBP* iciocolo *F* ithiocolla *W* aliqua] *bis om.* W quorundam] aliquorum *WP* **3** uerum] uerum aluta piscium *BPW* **4** extendit- facit] *om.* C

cf. Diosc. III. 88 (ἰχθυοκόλλα). Pliny 32.73.

141) Inanthe (Oenanthe)

Inanthe flores sunt uitis syluaticae qui ipso solstitio colliguntur Idus VII. Nam qui ante hunc diem uel post eum flos colligitur infirmus est. Vires autem habet oenanthe bene refrigerantes et stomacho praestantes.

1 Inanthe] inantes *VFLBCWP* oenanthe id est flos labruscae *tit. J* uitis syluaticae] agristis uastis *W* ipso] *om. J* **2** Idus] infirmior *V* idus viii kaend. iul. *F* idus iulias *B* ido octavus kalendas *C* octauo calendarum iunii *W* VII] VI *L* VIII *J* post eum] postea *FB* infirmus]

138) Marsh mallow (*Althaia officinalis* L.)

Marsh mallow is a type of plant similar to wild mallow, and is known to everyone. Its root and indeed the whole plant can mildly yet effectively loosen and soften every type of induration and stiffness.

139) Juniper (*Juniperus* L.)

The red, sweet fruit of the juniper bush is known to everyone. Its properties are heating, diuretic, and aid digestion.

140) Fish glue

Fish glue has two varieties: one kind is made from the boiled skins of certain types of fish, but another kind is made from the gizzards of a fish native to the Black Sea. All types of fish glue can be used as an astringent. It also stretches the skin tight and makes it shine.

141) Dropwort (*Spiraea filipendula* L.)

Dropwort are the flowers of a wild vine which are harvested right at the time of the summer solstice, seventh day after the Ides.[1] The flowers harvested before or after this day are untrustworthy. Dropwort has good cooling properties, and others excellent for the stomach.

1 Text seems corrupt: the summer solstice usually fell around 21 June, hence should properly be termed 'eleven days before the kalends of July' ('a.d. XI Kal. Iulii'), though the date above is approximate (Ides = 13 June). Neither Diosc. (III.20) nor any other source men-

infirmiores *F* debiliores *P* **3** et] et uires *LBC* **4** praestantes] praestantes ut flore bullis similis *C*

cf. Diosc. III.120 (οἰνάνθη). Pliny 10.87 (auis), 12.132, 21.65, 21.167, 23.8. Ps. Apul. 54.

142) Kamaepitys (Chamaepitys)

Kamaepitys species habet tres. Est autem quae ramulos emittit oblongos et curuos in similitudinem anchorae et fructiculum tenuem comulam, flores tenues et aureo colore, semen nigrum et odore plenum. Est et alia quae masculus dicitur, quae folia habet tenuiora et quasi pilosa, thyrsulum asperum, tenuem, molliculum et subaureo colore semen et ipsum pinum redolens. Est et tertium quod quasi in terra serpit, folia habens similia cardis, sed quasi pinguiora, uelut pilosa, flosculum aurei coloris. Et ipsa similiter olens prioribus. Verum omnes uires habent acres et diureticas.

5

1 Kamaepitys] chamaepitys *J* kamipedem *V* kamipiteus *F* khameputis *L* khamepiteas *B* kamepitheos *WP* species- emittit] enim quae ramos *F* **2** anchorae] annuose *F* comulam] cumulum *W* cumolem *L* **3** tenues] aurosos *VW* odore plenum] modo sepius *F* odoriferum *W* plenum] pinus *VL* **4** et alia] *om. F* aliquoties *P* quae] quinque *F* pilosa] *om. B* **5** molliculum] mollem *WP* subaureo] subrufum *C* semen- *fin.*] uires diuertica habent *W* **6** pinum redolens] plenum odorifero *P* serpit] repit *F om. B* **7** cardis] cardi *V* cardus *P*

cf. Diosc. III.158 (χαμαίπιτυς, ἕτερα –, τρίτη –). Pliny 19.54, 19.152–4, 20.262, 24.29. Galen *Simp*. 8.22.7 (K.12, 155). Ps. Apul. 26. Isidore 17.9.86.

143) Kamaedaphne (Chamaedaphne)

Chamaedaphne folia habet quasi laurus, sed leuiora, tenuiora et in extremo spinosa, fructum in ipsis foliis rotundum et subrufum et diureticum.

1 Chamaedaphne] kamidapne *VF* kamedapne *W* khamedaphne *L* kamedafne *CP* tenuiora] *om. F* **2** spinosa] speciosa *BP* **3** diureticum] diureticum ualde *WC*

cf. Diosc. IV.147 (χαμαιδάφνη). Pliny 15.131, 24.132. Ps. Apul. 27. Pelagonius 184.

144) Kamelea (Chamelea)

Chamelea ramulos habet oblongos, folia fortiora quam lapathi, in speciem quasi foliis oliuae, densa et tenuiora et spinosa et gustu mordentia. Tota herba uentrem purgat.

1 Chamelea] kamimella *V* kamimula *F* kamomilla *W* kamellea *W bis* khamillea *L* kamomillie *C* kamelea *P* in speciem] similia *W* **2** tenuiora] subtiliora *W* gustu- purgat] *om. C* **3** purgat] soluit *F* purgat tepida attractione *J*

tions a specific harvest time for dropwort: Pliny (12.132) reports that it is picked when it flowers; elsewhere (10.87) he refers to the bird also called *oenanthe* (the wheatear) which hides at the rise of Sirius ('the dogstar,' early July) and re-emerges at its setting (mid-August), which may be connected to the tradition reported above.

142) Mountain germander (*Ajuga chia* Schreb.) Herb ivy (*Ajuga iva* Schreb.) Ground pine (*Ajuga chamaepitys* Schreb. L.)

The ground pine has three different varieties.[1] There is that (mountain germander) which sprouts oblong, curved twigs that look like anchors, bears a small fruit covered with fine hairs, and has delicate flowers that are golden in colour, and a black seed with a full fragrance. There is another type (herb ivy), which is called the male, which has thinner leaves that are a little hairy, a coarse stem that is thin, rather tender, slightly golden in colour, and its seed smells like pine. There is a third type (ground pine) which creeps along the ground, has leaves like a thistle,[2] but a little fatter and slightly hairy, and a small, golden-coloured flower. It also smells like the other types mentioned, all of which have sharp and diuretic properties.

1 Similarly Diosc. III.158 (though order inverted), cf. Pliny 24.29.
2 *cardus*: commonly used for thistle, most likely artichoke thistle (*Cynara cardunculus* L.), but there was great variation: André 1985, 50 lists twelve different species of plant for *cardus*. The equivalent comparison in Diosc., ibid., is ἀείζωον τὸ μικρόν, common house-leek (*Sempervivum tectorum* L.: below #267), which if intended makes this use of *cardus* unique to the *AG*.

143) Chamaidaphne (*Ruscus ramosus* L.)

Chamaidaphne has leaves like the bay laurel,[1] but lighter, thinner, and thorny at the ends. Its fruit, which grows upon the leaves themselves, is round, orange, and is a diuretic.

1 #149.

144) Spurge olive (*Daphne oleides* L.; *Daphne oleafolia* L.)

Spurge olive has small, oblong branches and leaves which are tougher than those of sorrel[1] and which bear a resemblance to olive leaves, but are hardy, quite thin, thorny, and biting in taste. The entire plant purges the stomach.

1 #154. Despite the name, the description also matches that given by Diosc. IV.146 (δαφνοειδές), spurge laurel: see ch. 3.D.

cf. Diosc. IV.171 (χαμελαία), IV. 146 (δαφνοειδές). Pliny 15.24, 24.133. Scrib. 133, 166. Ps. Apul. 112.6 (?). *Ex herb. fem.* 38.

145) Lepis

Lepis hac ratione colligitur, quando massae cupreae diu multumque in igne calefacto et in aceto luteo extinctae super incudem atteruntur, ut lamulae et ueluti spumellae decidant, has igitur colligimus. Et Graeco nomine lepidas appellamus, quae et ipse uiscide stringere possunt. Eligimus autem quae sunt uastissimae et colore subrufae.

1 Lepis] lippidum *V* lepida *LC* lepula *P* lepidus *W* lepis id est squama *tit. J* hac ratione] *om. J* cupreae diu] praetium *V* massa aeris ignita *W* 2 luteo] uelut *VF* extinctae] iustinctis *V* extenta *P* atteruntur] attuntur *V* battuntur *FLP* baptuntur *B* intunditur *W* ut lamulae] iam nullae *V* 3 spumellae] is cum multe *F* squamae *P* umulas *W* et graeco- *fin.*] et usui medicamentorum usui reseruamus. possit enim uiscide et acrtier stringit *W* 4 lepidas] lippida *V* lepida *F* lipidem *L* lipida *B* lempida *C* 5 subrufae] sobria *P*

cf Diosc. V.78 (λεπίς). Pliny 34.107.

146) Lycium

Lycium succus est arbusculae pusillae et spinosae. Cuius grana et folia et radices in unum ex aqua dulci coquuntur et diebus admodum quinque relinquuntur. Et iterum coquuntur donec faecis habeant crassitudinem et sic in uascula defunditur. Maxime Ponticum tale est aliqui
5   diutius coquunt sic ut in glebam redigant. Verum est optimum lycium radicum. Haec enim gleba est extrinsecus nigricans et intrinsecus subrufa et assimilis aloe, gustum habens amarum et stypticum. Vnde adulteratur amurca et absinthii succo.

1 Lycium] licium *FW* litium *P* 2 unum] unum combattuntur *VFLB* unum conturbatuntur *C* unum tunduntur *P* admodum] *om. LC* 3 faecis] haecis *J* crassitudinem] spissitudinem *P* 4 uascula] uascello *LC* uase *WP* 6 radicum] aliut *V* indicum *BPCW* intrinsecus] *om. F* 7 habens] *om. LBC* 8 amurca] amaricat *VLCP* mauricat *FL* absinthi succo] malocori et absenthum *V* malo colori *FW* mali colores *BP* mallo cori *L*

cf. Diosc. I.100 (λύκιον). Pliny 12.30–2, 24.125–7. Scrib.142. Marc. *Med.* 28.3.

147) Laser

Laser lachryma est herbae quae silphium appellatur, quae in multis locis in speciem et magnitudinem ferulae crescit. Folia circa se habens maiora quam apii, semen simile. Optimum est ergo laser quod Cyrenaicum dicitur, hoc enim subrufum est et perlucidum et intus exal-

145) Flake of copper ore

Flake of copper ore is prepared in this way: molten copper is placed on a hot fire for a long time and then is dipped in yellow vinegar and placed over an anvil and hammered until the residue falls off like slag, which we then collect. In Greek we call these 'flakes,' and they can be used as a powerful astringent. We select those which are by far the largest and are orange in colour.

146) Dyer's buckthorn *(Rhamnus petiolaris, Rhamnus lycoides, Rhamnus punctata* Boiss.)

Dyer's buckthorn is the juice of a small, spiny shrub. Its seeds, leaves, and roots are cooked together with sweetened water, and then left alone for five days. This is then cooked again until it obtains the consistency of lees,[1] and then is poured into a small vase. Some often cook the boxthorn from Pontus[2] this way for longer until they reduce it to a mass. But definitely the best part of the buckthorn is its roots. These are reduced to a mass which is dark on the outside, and orange on the inside, similar in appearance to aloe,[3] and with a bitter and styptic taste. It is adulterated with amorge[4] and the juice of wormwood.[5]

1 #112, though cf. *J, haecis,* 'even consistency': Diosc. I.100, 'honey-like consistency.'
2 This Pontic variety is unique to *AG*: André 1985, 149.
3 #5.
4 #18.
5 #22. *VFBP* add 'and pomegranate rind' (*malicorium*); see #183 n.1.

147) Sap of laserwort (~ *Ferula tingitana* L.)

Laserwort is the sap from the plant which is called silphium,[1] which grows in many places in the same manner and to the same size as giant fennel.[2] It bears leaves all over that are larger than those of celery,[3] but their seed is similar. The best laserwort is said to be the Cyrenaic, which is reddish

5 bidum, odore uiscidum et ab humore cito resoluitur et ad gustum
comfestim humorem prouocat. Vires autem habet acerrimas unde cum
magnis uexationibus excalfacit et relaxat.

1 Laser] lasar *VB* lasser *F* silphium] silper *VL* insilper *F* silfer *BWP* 6 acerrimas] agris-
simas *V* acres fortiter *BP* nimium acres *W* 7 et relaxat] *om. J*

cf. Diosc. III.80 (σίλφιον). Pliny 19.38–40, 20.100–1. Scrib. 67. Chiron 421, 455. Isidore
17.9.27.

148) Libysticum

Libysticum semen est herbae eiusdem uocabuli et acre excalefactorium
utile stomacho et ad digestionem ualde utile. Propter quod ad inflatio-
nis potum facit et in antidotis quae colicis prosunt admiscetur.

1 Libysticum] liuisticum *VFC* leuisticum *B* libisticum *WP* acre] adre *V om. F* 3 potum]
potatum *FPW* admiscetur] *om. W*

cf. Diosc. III.51 (λιγυστικόν). Pliny 19.165 Galen *Simp.* 7.11.16 (K.12, 62). *DynSGall* 63.

149) Laurus

Laurus omnibus nota arbor est, fructus illius est quem baccam lauri
appellamus. Quae bacca et tota arbor uires habet acriter calefacientes,
euaporantes et relaxantes. Lixiuium eius in caustica mittitur et est ad
usum medicinae efficacissimum, quod est ex cinere [quae de] quer-
5 cu fit, non solum hominibus prodest sed et pecoribus. Multis enim
causis necessarium est in uulneribus, sed et interioribus partibus ut
potio datur. Habet enim propriam rationem quibus uero causis detur
demonstrabo. Pecora quibus pulmo tensione uexatur et quibus uene-
nata aliqua res per cibum obrepserit ex hac potione liberantur. Hac
ratione multum et uulneribus praestat.

1 baccam] baga *V* bacca *bis*] eliaca *F* 3 eius] *om. VFLBPC* eius- efficacissimum] *om.W*
4 quae de] *om. VFLBCP* 5 fit] *om. F* 6 necessarium] prodest *V* in uulneribus- *fin.*] item
fit lexiuium de oliua siue aperfico uel de brassica quae acrior amonibus nota est. Licium
frigidum et siccum est inscido gradu *W* sed] quam *BC* 8 pulmo] bulto *F* 9 obrepserit]
oppresserit *VFP* potione] ratione *J* potati *P* 10 praestat] praestat remedium *C*

alterum ex Lat. Diosc. in W

cf. Diosc. I.78 (δάφνη) Pliny 15.127–31, 23.158. Ps. Apul. 58.10. Isidore 17.7.2.

in colour and transparent, whitish in the middle, has a strong fragrance, quickly dissolves from contact with moisture, and also immediately causes salivation when tasted. It has severely sharp properties and therefore heats and loosens, but with great discomfort.

1 Also given below #250. Cyrenaic silphium was renowned for medicine: images of it appeared on Cyrenian coins. See Gemmill 1966; Tameanko 1992; Wright 2001. This species of the Ferula genus is extinct: Pliny 19.38–40 reports that the plant was no longer grown in Cyrenaica by his day, though Diosc., a contemporary, says nothing similar, and Soranus (fl. c. 130 AD) recommended it as a contraceptive. Riddle 1992a, 28, suggests that its efficacy and popularity as a contraceptive 'drove it to extinction probably soon after Soranus' time.'

2 #4 n.1.

3 #14.

148) Lovage (*Levisticum officinale* Koch)

Lovage[1] is the seed of the herb of the same name that serves as a severe calefacient useful for stomach complaints, and is especially useful for digestion. It is therefore drunk to ease gas and mixed with antidotes which help with colic.

1 See Andrews 1941.

149) Bay laurel / Sweet bay / Bay tree (*Laurus nobilis* L.)

Bay laurel is a tree known to everyone, the fruit of which we call 'bay-berries.' The berries and indeed the whole tree have sharply warming, dispersive, and loosening properties. The lye[1] made from it and the ashes of oakwood is placed in caustics, and is most effective for medical uses, and not only for humans but also for sheep. In many instances it is essential for treating wounds, but is consumed as a drink for internal problems. There is a test one can apply to determine whether this form should be used for these purposes, which I shall now describe: any sheep which are suffering from contraction of the lungs, or are suddenly sick from something poisonous that they ate, should be cured by drinking the lye. And if it achieves this it is even better for healing wounds.

1 At this point all manuscripts (*VFLBPCW*) begin a new entry (for *lixivium*, lye) attached to the beginning of the text for #150 below: 'superius ... dixi' may support this, but I follow *J* in separating at 'lixiuia uires habent': see #150 n.1.

150) Lixiuia

Lixiuia uires habent stypticas et acres [quae] inurunt enim exuulnerando instaurant. Humectantia uulnera exstaurando et exuulnerando
siccant. Ideo uix coherentia uulnera et fistule et his similia quae fuerint,
lixiuia calida clysterizantur sepe. Facillime curam medicamenti sentien-
5 tia ad sanitatem perducit, difficilia ad sanandum uulnera callosa longo
tempore sordida, quae nullam medicamentorum uirtutem pro longo
tempore senserunt. Vtile est lixiuuia quotidie lauari, ex cuius uirtute
omnem limum et quaecunque pro uetustate stupida facta sunt auferun-
tur. Si quotidie fiat et post hoc deinde curari oportet sicut ipsa uulnera
10 dictauerunt medicamentis aptis. Virtus enim lixiuiae non sanat sed ad
sanandum uulnera disponit. Hoc idem superius in interioribus fieri
partibus dixi. Ex quo cinere lixiuia facta, cui rei necessaria fit, et qui
cinis et quomodo temperatus, impositus quod praestet uide. Etenim ex
sarmentis cineribus lixiuia facta danda enim est, res arbusta et styptica
15 cum tam solide curet sed magis stringat. Sed et ipse cinis in recentia
uulnera in profluuio positus sanguinem statuit a cursu. Ideoque omnibus pecoribus castratis cinis impositus prodest omnibus abscisis. De
sarmentis tamen oliuae cinis acerrimus est et ex nucleo solo acrior est.
Et ex hoc lixiuia uulnerum cancerata aufert: [etenim acriore] lixiuio
20 sordida uulnera lauanda sunt. Caprifici cinis [nam brassice] omnibus
superioribus acrior est

1 Lixiuia] lexibus *L* lixiua *B* liciua *C* lexiua *P* uires] uirtutes *F* inurunt] inure *V* inurit *J*
inurum *P* exuulnerando] *om. V* uulnerando *J* ulcerando *C* exstringendo desiccat *P* 2 ins-
taurant- 2 exuulnerando] *om. F* exstaurando] excurando *L om. B* exuulnerando] exstringendo *B* 4 clysterizantur] clisteriantur *V* disterantur *P* sepe- sententia] *om. BP* 5 callosa]
calorosa *F* 7 utile] ualde *VF* optimum ualde *BP* utile ualde *C* 8 stupida] studia *F* auferun-
tur] *om. V* 9 deinde] *om. VB* 10 aptis] apertis *LC* sanat] satis *B* 11 hoc idem- sed] *om. P*
interioribus] inferioribus *F* 12 dixi] dixit *V* rei] ue *J* 13 cinis] scinis *B* praestet] prostet *J*
praestituit *V* praeest *L* uide] uitae *V om. B* unde *C* etenim] *om. VL* de quibus *B* 14 facta
danda] fit dicenda enim *J* arbusta] combusta *B* arbute *C* acris *J* 15 cum] quae *LBC* sed]
om.LBC 16 profluuio] profundum *LJ* statuit] stringit *P* 17 abscisis] incisis *VF* aliis *BP*
18 solo] puro *VF* ipso *P* 19 uulnerum- lixiuio] *om. B* uulnerum- aufert] *om. P* cancerata]
carcinomata omnia *VLC* etenim acriore] *om. VLC* ex hac enim *F* acrior isto est caprifici cinis succus omnium est acerrimus superiorum *fin. P* 20 nam brassice] *om. JF* cinis
VLBC acrior est] *om. J* caprifici- superioribus] *om. F*

deest W (*uide* #149, 8)

cf. Diosc. V.75 (ἀντίσποδα).

150) Lye

Lyes[1] have styptic and sharp properties which cauterize and ulcerate tissue in order to restore it. They dry out weeping wounds by ulcerating and repairing them.[2] Therefore warm lyes are often administered by syringe into those wounds which are not closing well, or into fistulas, and other similar sores. The healing action of lye as a medicament that leads to sanitizing the wound is very easily felt, especially in regard to cleansing problematic, indurated wounds that have festered for a long period, and which have not been treated by any drugs for a long period. It profits to bathe in lye every day, for its properties cause it to remove all encrustations and whatever has become insensitive over time. Also, if it is applied daily, any wounds later received can be treated and cured by the appropriate medicaments required for those particular wounds. For the property of lyes does not actually cleanse the wound, but prepares it to be cleansed. This is also true for the internal uses I mentioned above. Depending on what ashes the lye is made from, what it might be used for, and what it is mixed with, it can be applied topically, then monitor whether it is helping.[3] Lye is made from the ashes of brushwood, and that from arbutus wood is styptic and heals steadfastly but is even more astringent.[4] This ash is applied deep inside recent wounds to staunch the discharge of blood while the injured is still moving. Likewise the brushwood ash is always applied topically after castrating sheep and is good for healing cuts in general. Ash from olive tree twigs is extremely sharp, though that made from the olive drupe alone is even sharper, and this ash removes all tumorous growths upon wounds, and dirty wounds should be washed with (an even more bitter) lye. Ashes from the wild fig tree [and from cabbages] are sharper than all those mentioned above.

1 This entry is extremely corrupt, and was confused with that above (#149) in the manuscripts: the reconstruction here is very tentative. I note the most opaque passages below, and urge readers to consult the variants to construct their own meanings for these, and for that matter the entire entry, which has no parallel in other sources. On the use of ash in medicines, Gaillard-Seux 2003.

2 *exuulnerando*: found only in the Lat. Diosc. to translate terms such as ἑλκωτικός (ἑλκωματικός), 'ulcerating' (e.g., II.175–Lat. II.161) and ἐποιδεῖ γὰρ ... φλυχταινοῦται, 'causes swelling ... and blistering' (IV.153, Lat. IV.148). *Exstaurando* is unattested elsewhere, and should perhaps be *restaurando/instaurando*.

3 Difficult passage: *quod praestet uide*, an odd expression, is the best reading in agreement with the manuscripts, though *praestuit uitae V* points to another option.

4 *arbutus* (*res arbusta*), following *VFC* (and *bombusta B*), but *acris J* is also possible. The difficult phrase *cum tam solide curet sed magis stringat* may reflect a garbled translation from Greek.

151) Lentiscus

Lentiscus arbuscula est similis myrta, corticem in cyma subrubrum habet, foliis pusillioribus. Virtutis est stypticae. Huius folia aut extremitatem cymatum uel radicis corticem in aqua uehementer decoques quo usque mellis habeat spissitudinem, et sic uteris. Facit autem eis qui
5 sanguinem reiiciunt. Ex posca frigida dato scripulas quinque uel sex et ad fluxum uentris et ad dysentericos similiter oblatum et ad ueretri uitia tritum imponito. Fit autem succus de foliis eius uel cymis contusis et expressis, similem in omnibus habens uirtutem. Aqua in qua folia eius fuerint decocta si fracturas frequenter fomentaueris aiunt
10 celerius solidari. Facit et ad dentes se agitantes si enim frequentius in ore teneatur stringuntur. Fit quoque de semine eius oleum similibus causis necessarium. Capillis cadentibus inunctum prodest, eo quod calidam et stypticam habet uirtutem. Nascitur uero in Chio insula. Vtile sanguinem reiicientibus et tussibus antiquis in potatione sum-
15 ptum. Prodest haemoptoicis cum aqua frigida seu posca, tussientibus in mulsa calida. Est autem stomacho optimum et ructuosis et foetori oris prodest masticatum. Pilos etiam qui in oculis nascuntur contrarii acu uel graphio calefacto optime colligat. Gingiuas tumentes impositum emendat et sanat.

1 Lentiscus] lentisco *V* cyma] cimas *VLB* cimis *C* 2 pusillioribus] minora *P* 3 cymatum] amaram *J* cymorum *VFP* cimerum *LB* cimare *C* 4 sic uteris] ueteris *L om. BP* 5 scripulas] *om. VJ* scrupuli *P* quinque uel sex] tur uel que *V* V uel VI *JP* ut VI *C* 7 uitia] uitia idem *LB* tritum] timeritum *V* contusis] cocturas *L* 9 aiunt] dicunt *VB* 10 solidari] solidabus *B* si enim] *om. LBC* frequentius] frequetius sanguinem reiicentes ad dentes laxos *B* agitantes] laxas *P* 12 inunctum] unctum *VBL* inuinctum *C* 13 nascitur] pesina *B* in Chio insula] quae in ea nascitur *B* in carosinum *L* in ea resina *VFCP* 14 utile] utilissima *V* tussibus] *om. B* potatione] potione *VP* 15 haemoptoicis] inmotuicis *V* emotoicis *C* 16 ructuosis] ruptuosis *VF* foetori] fetorusis *V* fetorosis *F* faturiosis *BP* fetoriosis *L* 17 oris] *om. FLBCP* masticatum] manducatum *V* qui] qui iam *J* contrarii- graphio] contractum ut gladio *BP* graffio *V* 18 acu] hac *F*

multum in breve W

cf. Diosc. I.70 (σχῖνος). Pliny 24.42. Galen *Simp*. 8.18.48 (K.12, 135). Isidore 17.7.51.

152) Lenticula

Lenticula herbae semen est quod seritur in locis sabulosis, lapillosis et tostis. Quae si saepius fuerit comesta non cocta caliginem turbando operatur et tarde digeritur inflationem uentris. Excoriata uero et decocta in aceto uentrem stringit. Podagricis cum polline mista pra-

151) Mastic tree (*Pistacia lentiscus* L.)

The mastic tree[1] is a shrub similar to myrtle,[2] but has tawny-coloured bark around its spring shoots, and tiny leaves. Its property is styptic, and you stew thoroughly in water the leaves at the end of the spring shoots or the bark from the roots until it obtains the consistency of honey, which you then use. It is good for those coughing up blood. When five or six scruples'[3] worth are drunk from a cold cup it helps with diarrhoea, and likewise for dysentery, and when crushed it is applied topically to genital sores. The juice from its leaves or its spring shoots crushed and pressed has the same property in every respect. They say that if you continually dress fractures with the water in which the leaves are stewed it quickly strengthens the bones. It is also used for treating loose teeth; for if held in the mouth fairly frequently it firms them up. An oil made from its seed is also essential for treating the same conditions. This oil is applied as an unguent to stop hair loss, for it has a warm and styptic property. It grows on the island of Chios. It is helpful for those coughing up blood and chronic coughs when drunk. With cold water, or served in a cold cup, it is good for those coughing up blood, and is mixed with warm honeyed wine for coughs in general. It is excellent for the stomach, hiccups, and alleviates bad breath when chewed. Applied with a hot needle or stylus it also destroys ingrown hair around the eyes. It reduces and cures swollen gums when applied.

1 See also #188, 224.
2 #188.
3 *scripulus*: one twenty-fourth of an ounce. On measures in the *AG*, see ch. 4.F.

152) Lentil (*Ervum lens* L.)

The lentil is the seed of a plant that is cultivated in sandy, rocky, and extremely dry places. If lentils are eaten too often without being cooked they cause an irritating dimness of vision and eventually cause gas in the bowels. Shelled and cooked in vinegar they firm up the bowels. Mixed

5 estat medelam, cum melle autem uulnera purgat et glutinat. In aceto
cocta et imposita durities et scrofas malaxat. Oculorum doloribus uel
tumoribus optimum adiutorium est. Si cum cydonio in aqua coelesti
fuerit decocta et oleo rosaceo admista ueretri uitia emendat. Et si melle
adiecto ponatur [in succo maligranati] sanat et ad vesicas et ad ignem
sacrum et perniones, simili modo imponito.

1 Lenticula] lenticola *V* lentiscula *F* seritur] fertur *J* lapillosis- uero] *om.*
W lapillosis] *om. LBP* lapidosis *F* 2 tostis] itus *L* costis *CP* comesta] cum menta *P* caliginem] inducit
caliginem *J* turbando] *om. V* superat *P* contrita *F* 3 operatur- uentris] *om. J* 4 podagri-
cis- medelam] *om. W* polline] pulenta *B* 5 aceto] acero postia *J* 6 imposita] *om. F* mal-
axat] emollit *J* 7 tumoribus] *om. F* si cum- *fin.*] *om. W* cydonio] cidonia *FLBCP* 8 melle
adiecto] *om. BP* in succo maligranati] malignos *VFLBP* uulnus malignum *C* 10 sacrum]
sacrum et ad duritias *C* perniones] ad didonas *V* modo] modo decoctum *JL*

cf. Diosc. II.107 (φακός). Celsus 2.18.5. Pliny 18.123. *Dyn.Vat.* 13/*SGall* 17. Isidore
20.7.4.

153) Lupinus

Lupinus herbae fructus est quae seritur omnibus prouinciis. Virtus
eorum est calida. Nutriunt corpus. Sunt digestibiles et uentrem tem-
perant si cocti manducentur. Aqua uero ubi decocti fuerint cum ruta et
pipere, modice potui data spleniticis prodest. Scabiei maculis uel feruo-
5 ribus apta, optima fomentatio est. Cum aceto uero in modum cataplasmatis impositus scrofas, ustiones et malignos bene accipit. Farina
uero eorum cum marrubio pari pondere in umbilico imposita lumbricos expellit. Radix eorum in aqua decocta et potui data urinam mouet.
Ipsis autem comesti lupini stomaci uitia compescunt. Sunt autem et
10 siluatici lupini minores et uirtute similis, tamen optimi sunt qui maiores sunt. Nam possunt acriter relaxare et sua amaritudine non solum
lumbricos occidere sed et partum expellere.

1 Lupinus] lupinum *V* herbae- prouinciis] omnes nouerunt *W* quae- prouinciis] *om. P* 2
digestibiles] indigestibiles *J* 4 potui data] datur potui *V* potata *BP* scabiei] cum aceto scabiei *VF* 5 apta] *om. FLBCPW* 6 ustiones] *om. BP* bene] humores *J* benefacit *BP* optime
facit *C* uulnera sedat *W* 7 imposita] *om. V* lumbricos- data] *om. W* 8 eorum- decocta] in
umbilicum impositum *F* 9 ipsis- compescunt] lipsiani *(novum capitulum)* comesti lupi-
nis minores sed uirtute similes *J* comesti] comixti *V* sunt autem- 11 sunt] eandem uim
habent lupini siluatici *W* 11 sua- expellere] lumbricos necant *W* non solum] solent *B om.*
P 12 expellere] expellere dicuntur *BP*

cf. Diosc. II.109 (θέρμος.). Pliny 18.133–5, 22.154–6. Ps. Apul. 111. *Dyn.Vat.* 18/*SGall*
22. Isidore 17.4.7.

with fine flour they constitute a cure for gout, while mixed with honey they clean and seal wounds. When cooked in vinegar and applied topically they soften indurations and scrofulous swellings.[1] They are an excellent aid for curing tumours and aches in the eyes. When stewed with quince in rain-water and then mixed with rose-oil they heal genital sores. When cut with honey they heal malignant wounds[2] and bladder problems, and this can be applied topically to treat both erysipelas[3] and chilblain.

1 *scrofa* (*scrofulae*): the term was loosely applied any type of swelling, but often meant in particular tuberculous swellings in the lymphatic glands of the neck, axilla, and groin. Cf. Celsus 5.28.7 (*struma*).
2 Or 'with honey and juice of pomegranate rind,' *J*, perhaps an interpolation from Diosc. II.107.
3 'sacred fire': see # 68 n.1.

153) Lupine (*Lupinus* L.)

Lupine is the fruit of a plant which is cultivated in every province. Its property is warming, and it nourishes the body. They are good digestives, for when cooked and eaten they calm the bowels. The water in which they have been stewed with rue and pepper[1] is drunk in moderate amounts for problems of the spleen. They are a suitable treatment for scabby blemishes and fevers, and are best when applied by fomentation. Lupine is mixed with vinegar to make a type of plaster which when applied treats scrofulous swellings, burns, and malignant sores. Flour made from it is mixed with an equal amount in weight of horehound[2] and placed in the navel to expel intestinal parasites. Water in which lupine roots are stewed is drunk to stimulate urination. When[3] the roots are eaten together with the lupines themselves they ease stomach complaints. There is also a wild lupine which is smaller but contains the same properties, though the larger type is still the best. These wild lupines can severely loosen, and their bitterness not only kills intestinal parasites but also expels the foetus.

1 #227, #209.
2 #187.
3 Here *J* creates an entirely separate entry for 'lipsiani,' which baffled previous interpreters (Meyer 1856, IV:494: Halleux-Opsomer 1982a, 92), but it is likely a corruption of the following on *lupini minores/siluatici*. Smaller, wild lupine is also mentioned as a coda in Diosc. II.109: cf. Diosc. Lat. II.92, 'est etiam et agrestis similis nostro, amarissimus satis, tanta habet efficacia quantum noster.'

154) Lapathium

Lapathium herba est omnibus nota [et inuiscatiuum esse dicitur]. Huius igitur radix et radicis cortex trita et aceto uel alumine liquido omnem maculam a sudore factam infricata in balneo tollit.

1 Lapathium] lapatum *V* lapatium *BP* lapacium *W* inuisciatuum- dicitur] uisgatium esse mittitur *L om.VFBCPW* 2 radix] *om. LBC* radicis] *om. FP* radicis cortex] *om. W* trita] intus *JLC* 3 sudore factam] sordem fractam *C* tollit] mirifice tollit *W om.C*

cf. Diosc. II.114 (λάπαθον). Pliny 19.184, 20.231. Garg. *Med.* 8. Ps. Apul. 13 (29). *DynVat* 30/*SGall* 34. *Ex herb. fem.* 48. Isidore 17.10.20.

155) Lilium

Lilium omnes nouerunt, cuius uires possunt durities neruorum remollire et ipsa folia cocta et imposita ad combusta proficere. Sed et potata aut ipsum semen menstrua prouocat. Partum deiicit et ad serpentis morsum facit.

1 nouerunt] notum *V* herba est *PW* 2 ad combusta] *om. BP* 3 ipsum] ipsum remollire *V* menstrua- *fin.*] menstrua et partum deiicit *W* 4 facit] optime facit *BP*

bis in W

cf. Diosc. III.102 (κρίνον). Pliny 21.22–5 et passim. Galen *Simp.* 7.10.55 (K.12, 45–7). Ps. Apul. 108. Isidore 17.9.18.

156) Ladanum

Ladanum hac ratione colligitur, [funiculi ligati] circa hederam distrahunter, et sic quicquid illis glutinosum in modum sordis adhaeret abraditur et in glebulas expansas redigitur. Aliquando etiam geniculis et femoribus caprarum quae edera uescuntur adherens hoc ipsum lada-
5 num inuenitur et similiter abraditur hoc totum. Verum est optimum quod olet et quod est subuiride et pingue et resinosum sine ullo sabulo. Maxime tamen in Cypro colligitur. Quod enim a Cilicia uel Arabia affertur infirmum est. Vires autem habet ladanum acres et excalfactorias et relaxantes, propter quod in malagmata et in emplastra mittitur
10 et ad tussem ueterem facit. Item capillos fluentes cum oleo myrtino continet.

1 funiculi ligati] sit pontehilice ut *V* sponte ilicatice *F* sepuncta et elimata ultra *BP* sepuncta et elimata eradistrahitur *W* seponataque late *C* saepo[....] *L* circa ederam] *om. F* 2 sic- adhaerens] et quod coagulat *W* sordis adhaeret] sordis hedere *V* cardis ederis *BP* 3 expansas] ex passo *LBP om. VC* redigitur] pargitur *V* reditur *L* geniculis] genibus *J* 4 femoribus] foemenibus *VJ* pedibus *BPW* femioribus *C* hoc- inuenitur] *om. W* 5 hoc

154) Sorrel / Monk's rhubarb (*Rumex* sp. L.)

Sorrel is a plant known to everyone [and is said to be quite sticky]. Its root and the bark upon it, when ground up with vinegar or liquid alum[1] and rubbed on while in the bath, removes all blemishes caused by sweating.[2]

1 #7.
2 Cf. Diosc. II.114, 'decoction [of the root] assuages also itching, either poured all over or when mixed with the bath water.'

155) Lily (*Lilium candidum* L.)

Everyone knows the lily, whose properties can soften indurations of the ligaments, and its leaves when cooked are applied topically to burns. When drunk it stimulates menstruation, as does its seed. It ejects the foetus and is good for snake bites.

156) Rockrose (*Cistus* sp. L.)

Rockrose is collected in this way: the small vines entwined with ivy are disentangled and the glutinous substance that sticks to them like a sort of scum is scraped off and laid out in small globules. Sometimes that which sticks to the beards and thighs of she-goats which feed on the vines is used as though rockrose itself, and all of this is scraped off and prepared likewise.[1] The best is that which emits fragrance, and which is a little green, greasy, and resinous, without any sand in it. It is chiefly harvested in Cyprus. That which is imported from Cilicia and Arabia is unreliable. Rockrose has sharp, heating, and loosening properties, and is therefore placed in emollients and in plasters, and is good for chronic coughs. It also mixed with oil of myrtle to tame frizzy hair.[2]

1 Similarly Diosc. I.97, who notes this as a subtype of rockrose, 'called by some *ledon*'; likewise Pliny 12.75, 26.47. On their types and medical uses in antiquity, see Mouget 1993.
2 Cf. Diosc. ibid. 'stops hair from dropping off.'

totum] et reponitur *W* **6** olet] odiferum *W* sabulo] sablune *V* sablone *FP* ablonem *L*
7 quod enim] *om.* *V* cilicia] alicutia *V* licia *F* alicia *LC* uel arabia] et asia *B* **9** malagmata]
malagmata miscit *F* mittitur- facit] *om.* *F* **10** oleo] oleclio *V* **11** continet] potentes retinet
W continet efficaciter *P*

cf. Diosc. I.97 (κίσθος). Pliny 12.74–5, 26.47–8. Galen *Simp.* 7.10.28 (K.12, 28–9).

157) Lonchitis

Lonchitis folia habet rubicunda, florem nigrum, assimilem personae
comoedorum quorum orificia patent. Semen album, oblongum in specie lancearam, radicem quasi dauci, similiter diureticam.

1 Lonchitis] loncidis *V* longites *F* longichis *P* lonchigis *W* rubicunda] subrubicunda *C*
2 comoedorum] herbea *C* personae comorum *L om.* *W* orificia] ora *W* semen] florem *W*
3 lancearam] patiarum *F* diurecticam] diureticam id est urinam prouocantem *P* urinam
prouocantem *W*

deest B

cf. Diosc. III.144. (λογχῖτις). Pliny 25.137. Galen *Simp.* 7.11.19 (K.12, 63).

158) Lauer

Lauer nascitur circa aquam et in ipsa aqua foliola habentur pusilla
rotunda, flosculum exalbidum. Tota herba est acris diuretica.

1 Lauer] labar *VFLCP* laber *W* in- aqua] ipsa a qua *J om.* *W* foliola] folia *VCWP* **2**
rotunda] rotundula *JPW* exalbidum- diuretica] album est acra *W* acris et diuretica *C*

deest B

cf Diosc. II.127 (σίον). Pliny 26.50, 26.80, 26.87. *Ex herb. fem.* 69.

159) Lathyris

Lathyris thyrsum [aridum] habet oblongum, uastitudine digiti, cauum
et in summo quasi ramulos diffusos folia longa. Fructiculum in summis
ramis triangularem et pene subrotundum quasi capparis, in quo sunt
tria granula rotunda et interstrata tenui ueluti membranula. Quorum
5   septena uel nouena uentrem uiscide purgant, aqua frigida superbibita.
Quae choleram rubram et phlegma deponunt.

1 Lathyris] lactura *V* lacturia *F* lactarea *L* lacterides *C* letura *W* acura / actura *JP* (*fine
#158*) aridum] *om.* *JFCP* **2** longa] oblonga *LCP* **3** triangularem] trigonum *P* quasi cap-
paris] et sic appare *V* **4** tria] terna *FL* interstrata- membrulana] inuenitur *C om.* *W* mem-
branula] membrano *V* **5** uiscide] inscide *P* **6** quae] atque *V om.* *C* rubram] rubea *V*

deest B

cf. Diosc. IV.166 (λάθυρις). Pliny 27.95. Galen *Simp.* 7.11.2 (K.12, 56). Marc. *Med.* 30.26.
Ps. Apul. 112. Theod. Prisc. *Log.* 99.

157) Lonchitis (*Serapias lingua* L.)

Lonchitis has ruddy leaves and a dark flower that looks like one of those masks with the gaping mouth used in comedies.[1] It has a white seed that is oblong and shaped like a spear-head, and a root that is like that of daucos,[2] and it is likewise a diuretic.

1 Likewise Diosc. III.144, '[masks ...] and resembling little felt hats.'
2 #91.

158) Water parsnip (*Sium angustifolium* L., *Sium erectum* Huds.)

Water parsnip[1] grows next to water, and upon the water itself it dangles tiny, rounded leaves and produces a small white flower. The entire plant is used as a strong diuretic.

1 The entry was attached to #159 in *JP*. Identification, André 1985, 140. Cf. #277. See also ch. 1.I and plate 3.

159) Caper spurge (*Euphorbia lathryis* L.)

Caper spurge has an oblong [and dry] stalk, the width of a finger and hollow, and long leaves at the top that form a thatch like little branches. At the end of the branches it bears a small, triangular fruit, a little rounded like a caper, in which there are three small, rounded seedlets kept apart from one another by a thin membrane.[1] Around seven to nine of these, followed by a drink of cold water, thoroughly purge the bowels. They also reduce red bile and phlegm.

1 Similarly Diosc. IV.166.

160) Linozostis

Linozostis folia habet quasi ocimum, thyrsulos binis nodulis et densis et pluribus ramulis. Fructum foemina quae dicitur habet copiosum et quasi racemosum. [Quae deinde masculus] pusillum et rotundum, combinatum quasi testiculos foliis ipsis adhaerentem. Verum utru-
5 mque uentrem soluit, similiter et phlegma et humores calefaciendo deponit. Et quasi olus decoctu manducatur et succus ipsius bibitur.

1 Linozostis] linozotes *F* linoiosies *L* linogotis *P* linozotis *W* linozostis uel mercurialis herba *tit. J* ocimum] ozimum *V* acrinum *F* nodulis] nodosis *L* 2 ramulis] cauos *J* foemina] semina *F om. P* quae- habet] *om. LC* copiosum] folia copiosa *PW* 3 quae- masculus] *om. VFLCPW* pusillum] *om. V* 4 utrumque] *om. F* 5 et- olus] si solum *C* humores] alios humores *P* 6 ipsius] idem et ius *JC* eius *P*

deest B, aliquanto def. W

cf. Diosc. IV.189 (λινόζωστις). Pliny 25.38. Cael. *Chron.* 5.10. Ps. Apul. 83. *Ex herb. fem.* 22.

161) Lapis haematites

Lapis haematites est naturalis qui ab Aegypto et India et Hispania affertur. Durus est extrinsecus leuis et ueluti colore liuidus intus deinde sanguineus. Cuius maxime idoneum iudicamus qui tritus in cote ex aliquo humore in modum sanguinis inficitur. Verum potest uiscide stringere et tamen sine magna acrimonia.

1 haematites] ematitis *VLBW* hematitis *F* aegypto] *om. W* hispania] spania *VL* 2 extrinsecus- colore] *om. V* intrinsecus liuidus et niger *W* leuis] lenis *J* 3 sanguineus] quasi sanguinem emittit *W* iudicamus] indicamus *F om. B* cote] cute *VFLBW* 4 inficitur] effecitur *V* fricetur *W* 5 et] non *B* tamen] *om. V* cum magna acredine *W*

deest P

cf. Diosc. V.126 (αἱματίτης λίθος). Pliny 36.129, 37.169. Celsus 5.5.2. Scrib. 26. Galen *Simp.*9.2.2 (K.12, 195–6). Isidore 16.8.5.

162) Lapis schistus

Lapis schistus colore fusco est, aliqua in se ostendens subrufa, oblongus et spinas habens directas prout alumen scissum. Hunc quoque optimum iudicamus qui et ipse uiscide stringere potest.

1 schistus] excistis *V* citos *F* scistus *L* scistusem *B* chistus *C* fusco] rufus *C* aliqua] aliquando *C* oblongus] longus *F* 2 spinas] pinnas *L* pennas *BC* directas] erectas *B*

deest PW

cf. Diosc. V.127 (σχιστὸς λίθος). Galen *Simp.*9.2.3 (K.12, 196–8). Isidore 16.4.18.

160) Mercury (*Mercurialis annua* L.)

Mercury has leaves like basil.[1] Its sprays have two joints and sport many closely packed branches. That which is called the female plant bears much fruit in the form of bunches. [That which is called male] has a small, rounded fruit that clings to the leaves in pairs like testicles. Both types loosen the bowels, and also reduce phlegm and humours by a warming action. They are stewed like pot of vegetables and eaten, or their juice is drunk.

1 #205. Similarly Diosc. IV.189.

161) Hematite

Hematite stone is produced naturally in Egypt, India, and Spain, whence it is exported. Its surface is hard and smooth, and inside it is slate-coloured, and then blood red. We consider it to be most effective when it is crushed by a grinding stone until it turns the colour of blood but contains no moisture. It can act as a powerful astringent, without any major discomfort.

162) Talc

Talc [stone] is a swarthy colour with a light shade of red to it, oblong in shape, and has spikes that stick up like pressed alum.[1] We consider this type the best and it can be used as a powerful astringent.

1 #7.

163) Lapis asius

Lapis asius in Alexandria tantummodo inuenitur. Cuius est optimus ad usus medicinae qui est colore candidus, fragilis, leuis in modum pumicis et ueluti puluis in manibus insidens. Effectu autem uiscide relaxare potest. Aliquid enim salsuginis et nitrosis saporis ostendit. Vnde et
5 corpora quae in eo sepeliuntur, tempore consumit. Quapropter Graeci sarcophagon appellant.

1 ad usus] et summe *B* 2 usus medicinae] *om. W* fragilis] *om. VF* 3 manibus] manibus latis cum tangitur *W* 4 ostendit unde] sentiens deinde *B* unde- appellant] *om. W* 5 quae in] ex *B* sepeliuntur] repleantur *F* sepeliuntur- consumit] sepe lauamus et abluimus *B* sepe lauamus corpora *W* tempore] breui tempore *C* quapropter- appellant] *om. B* 6 sarcophagon] sarcofa *V* sartoragrum *F* sarcofagum *LC*

deest P

cf. Diosc.V.124 (ἄσσιος λίθος). Pliny 2.211, 36.131. Celsus 4.31.7. Galen *Simp.* 9.2.9 (K.12, 202).

164) Lapis pyrites

Lapis pyrites est durissimus et grauissimus, colore aeris. Et si ferro percutiatur scintillam et ignem mittit. Vires habet acres et ad similitudinem cadmiae et proprie ad oculorum claritatem facit.

1 pyrites] piritis *VF* durissimus] purissumus *J* aeris] eris *J* acris *BC om. W* 2 scintillam] stillat *B* mittit] omittit *L* emittit *VB* 3 facit] *om. J*

deest P

cf. Diosc.V.125 (πυρίτης λίθος). Pliny 34.135, 36.137–8. Galen *Simp.* 9.2.6 (K.12, 199–201). Isidore 16.4.5.

165) Lapis magnes

Lapis magnes inuenitur circa littus oceani, colore coeruleo qui probatur. Hic quoque ferrum ad se trahit et tenet. Vires autem et purgatorias habet, propter quod hydropicis in potu datur et omnem humorem per uentrem eministrat.

1 magnes] magnitis *VB* magni *F* magnetis *W* 2 tenet] *om. W* autem] *om. FLB* 4 eministrat] ministrat *V* amministrat *B om. W*

deest P

cf. Diosc.V.130 (μαγνήτης λίθος). Pliny 34.147–8, 36.126. Galen *Simp.* 9.2.10 (K.12, 204). Isidore 16.4.1.

163) Assian Stone

Assian stone is found only in Alexandria. The best type for medical uses is bright, delicate, and smooth like pumice stone,[1] and sinks into your hands like powder. In terms of action it can vigorously loosen. It has something of a salinated and nitrous flavour. Over time it consumes the bodies which are buried in it, which is why the Greeks call it sarcophagon.[2]

1 #213.
2 Likewise Celsus 4.31.7, and Pliny 2.211, 36.131, noting its origin from the town of Assos. Cf. Diosc. V.124.

164) Pyrite

Pyrite[1] is extremely hard and heavy, and the colour of copper. If struck with iron it emits a spark and can start fire. It has sharp properties that are much like those of calamine,[2] and in particular helps clear up vision.

1 Mineral pyrite, or 'fool's gold,' is an iron sulphide (FeS_2) that was easily confused with chalcopyrite ($CuFeS_2$) or 'copper pyrite,' often found alongside pyrite, and probably what is meant here, as with Diosc. V.125, 'a type of stone from which copper is mined' (cf. *AG* #89). Pliny 36.137–8 rightly distinguishes between different types of *molaris pyrites*, one of which 'resembles copper.'
2 #45.

165) Lodestone

Lodestone is found on ocean shores, and is identified by its dark blue colour. It also attracts iron to itself and holds it. It has detersive properties, and is therefore taken as a drink by those suffering from dropsy, and draws out all humours through the bowels.

166) Lapis gagates

Lapis gagates colligitur in Lycia circa ripam fluminis quod Gage appellatur et est colore niger et aridus et quasi lactuca, leuis ualde. Cuius est melior qui ad lucernam accenditur facile et odorem asphalti habet. Potest autem remissus ex aqua dentes qui agitantur restringere et stabilire.

1 gagates] gagatis *F* gagathis *W* lycia] litia *V* lidda *B* gage] gantes *VL* gangis *F* gagates *BW* appellatur] *om. V* 2 et est- habet] *om. C* lactuca] latus *J* lectula *VF* latulas *B om.* *W* 3 facile] fragile *B* odorem] colore *W* 4 potest autem] potestatem *C* dentes- restringere] *om. V* deest *P*

cf. Diosc. V.128 (γαγάτης). Galen *Simp.* 9.2.10 (K.12, 203–4). Isidore 16.4.3.

167) Lapis thracias

Lapis thracias est niger quasi calculus. Nascitur in Scythia in amne quodam qui Pontus dicitur. Vires easdem habet quas gagates et dicitur accendi ab aqua et extingui ab oleo.

1 thracias] tracius *V* stratius *F* trachius *W* calculus] calchus *V* calclus *L* scythia- dicitur] in ponto amne qui est in yciria *V* nec est in ifia *F* in ponto qui est in yciphia *L* enim in egyptum *B* in ponto amne qui est in egyptum *W* 2 easdem] *om. V* autem *B* 3 ab aqua] aliquid *B* parum *W*

deest *P*

cf. Diosc. V.129 (Θρᾳκίας). Pliny 33.94, 37.183. Galen *Simp.* 9.2.10 (K.12, 203–4). Isidore 16.4.8.

168) Lapis quadratus

Lapis quadratus in Aegypto inuenitur et est exalbidus. Vires habet uiscidas et stringentes. Vnde conceptum uetare dicitur. Inuenitur et in Aethiopia.

1 et] cuius melior *F* 2 uetare] uitare *J*
deest *PC*

169) Lapis phrygius

Lapis phrygius in Phrygia inuenitur. Cuius melior est qui est pallidaster et aliquatenus grauis. Hic eo quod nimium stringit et exulcerare solet.

1 phrygius] frigius *VFW* pallidaster *BL*] paledaster *VF* pallidus *W* pallidior *J* 2 aliquatenus] *om. W* nimium] aliquatenus *W* 3 solet] *om. V* resoluit *F*

166) Lignite

Lignite is obtained in Lycia,[1] from the banks of the river which is called Gage, and is black in colour, dry, very light, and a little like lettuce.[2] The better type is what which ignites easily from the flame of an oil lamp and gives off an odour like camelthorn.[3] Dissolved in water it can firm up and stabilize loose teeth.

1 A Roman province located on the southern coast of Turkey (mod. provinces of Antalya and Mugla).
2 *lactuca*: following *L*, and Diosc. V.128, 'πλακώδης' 'flaky.'
3 #76 n.2.

167) Thracian stone

Thracian stone is black and looks like an accounting pebble. It grows in a certain river called the Pontus in Sintia.[1] It has exactly the same properties as lignite,[2] and is said to be ignited by water and extinguished by oil.

1 *Sintia*: likewise Diosc. V.129, which clarifies the muddle in our manuscripts (yciria *V*, ifia *F*, yciphia *L*, egyptum *BW*, scythia *CJ*), though the location remains uncertain: 'the modern Radovitz [river],' according to Fortenbaugh and Sharples 1991, 185 n.533; on the Macedonian border of Thracia, according to Dana and Brush 1868, 760.
2 #166.

168) Squared stone

Squared stone[1] is found in Egypt and is whitish in colour. It has powerful and astringent properties. It is said to prevent conception. It is also found in Ethiopia.

1 Unattested elsewhere, but cf. #175, which may be the same. *Quadratus* may also mean simply 'quarry' stone.

169) Phrygian stone

Phrygian stone is found in Phrygia. The better type is somewhat pale and moderately heavy. This type is an excessively powerful astringent, and usually causes suppuration.

deest PC

cf. Diosc. V.123 (λίθος φρύγιος). Pliny 36.143–4. Galen *Simp*. 9.2.7 (K.12, 201). Isidore 16.4.9.

170) Lapis galactites

Lapis galactites est quasi cineraster et dulcis ad gustum, et humorem facit quasi lacteum. Leniter stringens, propter quod ad lachrymam oculorum facit.

1 galactites] galactitis *V* galactis *F* glaterius *W* cineraster *FB*] ceraster *V* cinerastes *L* cinereus *JW* humorem] tumorem *F* humectat *W* **2** lacteum] lactens *LB* propter quod] propterea *V* quod] *om. LB*

deest PC

cf. Diosc. V.132 (γαλακτίτης). Pliny 37.162. Galen *Simp*. 9.2.2 (K.12, 196). Isidore 16.4.20, 16.10.4.

171) Lapis melitites

Lapis melitites per omnia priori similis. Sed hic spernitur quod dulciorem humorem emittit.

1 melitites] militinis *VF* melitine *L* militenes *BW* similis] similis est *LB* sed] *om. V* hic] propterea *W* **2** humorem] *om. L* emittit] emittit acceptum in omnia *W*

deest PC

cf. Diosc. V.133 (μελιτίτης). Pliny 36.140. Galen *Simp*. 9.2.2 (K.12, 196). Isidore 16.4.26.

172) Lapis morochthus

Lapis morochthus in Aegypto inuenitur et est pallidaster. Vires habet et ipse efficaciter stringentes.

1 morocthus] morochus *VW* morocos *F om. L* pallidaster *FLB*] paledaster *V* pallidus *W* pallidior *J*

deest PC

cf. Diosc. V.134 (λίθος μόροχθος). Pliny 27.103, 37.173

173) Lapis phloginus

Lapis phloginus et ipse in Aegypto inuenitur. Aureo uel flammineo colore, qui in fronte inlinitur somnium facere dicitur.

1 phloginus] fulligenus *V* foliginus *F* floginus *W* flammineo] flammeo *JC* **2** colore] oculorum *F* qui- somnium] illinitus senium *J*

170) Milkstone

Milkstone is almost ash-like in colour, sweet to the taste, and exudes a fluid that resembles milk. It is mildly astringent, and is therefore good for treating teary eyes.

171) Honey stone

Honey stone is similar to the previously mentioned (milkstone) in every respect, but it must be rejected if it also releases a fluid that is sweeter (than milkstone).

172) Pipe clay

Pipe clay is found in Egypt and is somewhat pale. It has effective astringent properties.

173) Flame stone

Flame stone is found in Egypt. It is golden or flame-coloured, and is said to induce sleep when smeared on the forehead.

deest P

cf. Pliny 37.179, 37.189. Isidore 16.14.9.

174) Lapis crystallus

Lapis crystallus est limpidus et perlucidus colore. Vires habet stringentes et lac prouocare dicitur.

1 crystallus] cristallus *VFW* limpidus] *om. F* perlucidus] prolucidus *J* colure] colore ad similis uiero *V* colore adsimilis *FL* 2 lac] lactae *VL* lac feminis *BW*

deest P

cf. Pliny 36.1–2, 37.23–9. Isidore 16.13.1

175) Lapis thytes

Lapis thytes, quem uidimus, in Aethiopia colligitur, colore subuiridis et ab humore remissus. Habet vires acriter mordentes propter quod caliginem et cicatrices oculorum extenuat.

1 thytes] ui *VL* sui *F* uidimus] *om. V* dicimus *L* 2 humore] ore *F*

deest PW

cf. Diosc. V.136 (θυΐτης). Galen *Simp.* 9.2.4 (K.12, 198–9). Isidore 16.4.30.

176) Lapis batrachides

Lapis batrachides ab Aegypto affertur, colore ranae rubra. Vires habet lenientes et relaxantes.

1 batrachides] batracidis *V* batracites *F* botrachites *W* affertur] deferetur *W* rubra] rube *V* rubea *L* rubetae *BW* 2 lenientes] leniter *W*

deest P

cf. Pliny 37.149. Isidore 16.4.20.

177) Lapis smiris

Lapis smiris potest uiscide stringere et pene nimie. Propter quod in caustica mittitur.

1 smiris] zemiris *V* cimeris *F* imiris *L* zimiris *B* potest] est parcus *VL* est parvus *FB* pene nimie] *om. W* bene nimie *F*

deest P

cf. Diosc. V.147 (σμύρις). Isidore 16.4.27.

174) Rock crystal

Rock crystal is clear and transparent in colour. It has astringent properties, and is said to stimulate lactation.

175) Thyite stone

The thyite stone, which as we saw[1] is mined in Ethiopia, is light green in colour and diminishes upon contact with moisture. It has properties that bite severely, and it therefore helps with dim eyesight and diminishes corneal abrasions.

1 Perhaps referring to #168 above. The description is very similar to Diosc. V.136, θυΐτης, translated as 'turquoise' by Goodyer (Beck 2005, 394 n.59), which seems unlikely.

176) Batrachid stone

Batrachid stone is exported from Egypt, and is the same colour as the red frog.[1] It has calming and loosening properties.

1 Pliny 37.149, the only author besides Isidore (whose brief comment seems independent) to mention this stone (from Coptos, in the Thebaid), says there are two varieties, one with veins, and a third that is 'red mixed with black.'

177) Emery

Emery can be used as a powerful astringent, though is mostly too strong. It is therefore used in caustics.

178) Lapis misy

Lapis misy prout chalcis et ipse in Cypro insula inuenitur et in eisdem metallis. Similiter et glebula est et aurosa fulgens ac fragilis. Cuius fragmenta sunt per minuta et in modum sabuli. Optimum autem misy dicimus quod in aurigineum colorem maxime splendet et dum laeuigatur
5   mollis sentitur et aequaliter fulget. Qui est idoneus non est pallidus et leue et aequalem tractamus et aqua tantum cito denigratur. Verum effectu uidetur exuperare chalcitem.

1 misy] mibum *L* miscum *B* missus *VF* (*cont. #177*) miseum *F* chalcis] calidis *F* ipse] ipse ad fauces *F* 3 per minuta] super minuta *J* pernuta *V* aurosa] aurora *JV* sabuli] sablune *V* sablonis *LB* autem] ipsum *B* misy] missus *F* 4 aurigineum] aginem *J* aurigenio *VB* eaurigineo *L* 6 et leue] *om. JC* aqua tantum] qua tacto *F* aequali actu *L* qui ad tactum *B* 7 effectu] *om. V*

deest PW

cf. Diosc. V.100 (μίσυ). Pliny 19.36, 34.121. Celsus 5.19.8. Scrib. 34, 240. Galen *Simp.* 9.3.21 (K.12, 226–9).

179) Mel

Mel succus est roris coelestis qui ab apibus colligitur. Cuius sunt genere duo, id est unum quod in cupiliones ab apibus quae sunt oblongae colligitur. Aliud est quod sub terra inuenitur et atticum appellatur, quod uariolae et lanuginosae et fortiores apes reponunt et habetur melius
5   ad oculorum claritatem faciens. Vtriusque uirtus est amborum calida. Solum enim mel quasi compositionem in se habet quia ex multarum flore uel succo herbarum colligitur. Propter quod omnibus ualetudinibus pro occursu citato prodesse potest. Ideo omnibus antidotis commiscetur. Vulnerum sorditiem adhibitus comedit adiuncto sale. Illinitur
10 extrinsecus faucibus exiccatis. Deducendo humores aperit. Mel solum fastidium tollit, orexin facit, nauseam prohibet. Vnde pulmoni et interioribus omnibus per totum suadendo medetur atque ea relaxat.

1 Mel] mellis *V* 2 id est] *om. J* unum soruile *B* sorbile *P* cupiliones] cupillionis *V* cupelinis *F* piliones *BP om. W* 3 aliud- inuenitur] *om. BP* appellatur] inuentur *C om. W* 4 uariolae] uariores *C* lanugosiae et fortiores] *om. BP* 5 faciens] faciei *J* amborum] *om. J* 7 ualetudinibus- *fin.*] antidotis miscetur *W* 8 pro- citato] prout curatio *BP* 9 adhibitus] aditus *V* sale] sale in uno exiccatis] siccant *VLBP* 11 orexin] horixin *V* osin *B* os *P om. C* 12 suadendo] leniendo *J om. C* ea] sanat uel *VF om. LC*

cf. Diosc. II.82 (μέλι). Pliny 11.30–45. Galen *Simp.* 7.12.9 (K.12, 70). Isidore 20.2.36.

178) Misy / Copper ore

Misy is just like rock alum[1] and likewise is found in the same mines on the island of Cyprus. And like rock alum it comes in the form of little lumps, sparkles with gold, and is delicate. Chunks are broken off and ground extremely fine until like sand. We declare the best misy to be that which shines with a resplendent, golden colour, and which feels soft when it is pulverized, yet still shines just as much. The type which is not yellow-green and light, and which quickly blackens upon contact with water, we consider just as useful. The effectiveness of misy seems to surpass that of rock alum.

1 #48. *VF* attach this entry to #177.

179) Honey

Honey is the juice of heavenly dew which is harvested by bees.[1] There are two types of honey:[2] one in which the containers used by the bees to harvest the honey are oblong; the other is that found under the ground and is called 'Attic,' which is stored by bees that are striped, more sprightly and energetic, and this type is considered better for clearing up vision. Both types have a warming property. Honey alone is something of a compound mixture in itself, because it is collected from many different flowers and the juice of different plants.[3] For this reason it can be useful as a fast-acting agent for all ailments, hence it is mixed with all types of antidotes. It is mixed with salt and used to consume the filth within wounds, and is smeared externally on dry throats. It reduces humours by thinning them. Honey on its own stops squeamishness, stimulates the appetite, and prevents nausea. It alleviates problems in the lungs and all internal organs by thoroughly soothing and thereby relaxing them.

1 The notion that honey is derived from bees collecting dew from the air, whereas pollen from flowers is collected for the comb, can be found in Aristotle, *Historia animalium* V.22, ed. Peck 1970, II:190–8. Hence Pliny 11.30 believed that early morning dew upon leaves was actually honey (*folia arborum melle roscida*), but was uncertain 'whether this is perspiration of the sky, or a sort of saliva of the stars, or the moisture of the air purging itself.' Nonetheless the term *ros coelestis* is rare: see above #12 n.1.
2 The description of two types here is unique: cf. Diosc. II. 82, and those cited by Balandier 1993. See also Byl 1999.

180) Myrrha

Myrrha lachryma est arboris eiusdem nominis quae maxime in Arabia inuenitur. Quae aliquo loco incisa hanc ipsam lachrymam emittit. Est ergo optima myrrha quae est mundissima et tacta quasi aspera et arida et fragilis et leuis et glebulis pusillis, extrinsecus uno colore, intrinsecus
5 deinde subrufa, et ueluti uenulas habens albicantes, odore suaui, gustu amara et excalfaciens. Haec igitur in montibus inuenitur. Colligitur autem altera myrrha in planitiis quae et ipsa est mollior et pinguior quam exprimere in liquorem solent. Quod est illius quasi liquidissimum in uascula reponitur et Graeci myrrham stacten appellant. Verum
10 est omnis myrrha styptica et acris et excalfactoria huius itaque effectus uehementer commouere potest.

1 Myrrha] murrah *V* mirra *FP* 2 quae- loco] cortex eius *C* est- igitur] *om. W* arida] a radice *F* 4 glebulis] gliuosis *VF* gleuibus *L* intrinsecus] extrinsecus *L* 5 uenulas] nebolas *V* nebulosa *F* 7 planitiis] planis *J* plano *P* planiis *W* mollior] melior *JCW* 8 liquorem] colleris *V* coris *F* hanc aliqui *BP om.C* 6 liquidissimum] recentissimum uel liquidissimum *V* recentissimum *F* liquiditate *W* 9 uascula] uascella *VC* uase *P* myrrham stacten appellant] smirna stactentis dicunt *V* 10 acris] *om. BP* huius- potest] *om. W*

cf. Diosc. I.64 (σμύρνα), I.60 (στακτή). Pliny 12.66. Galen *Simp.* 8.18.30 (K.12, 127). Isidore 17.8.4.

181) Mandragora

Mandragora herba est omnibus nota, sic et aliqui herbam ipsam apollinarem appellant. Cuius ergo succus colligitur foliis ipsius herbae penitus et improbe excisis, ut quasi concauus locus earum relinquatur, postea quidem humor illius effluit colligitur, aut ipsam radicem prout
5 panacis, aut mala eius uel totam radicem in pila tundunt, et sic omnis illius succus organo exprimitur quem graece mandragorochylon appellamus. Vires autem habet refrigerantes et hac ipsa ratione stringentes. Sed et grauem marcorem facere potest, unde et somnifera creditur. Compescit autem omnes dolores et uigilias.

1 omnibus] pluribus *J* apollinarem] apolinare *V* 2 cuius- appellamus] *om. W* 3 penitus] et intro *B* penitus- panacis] *om. P* et improbe] in pluribus *L om. C* concauus- panacis] *om. C* 4 postea- colligitur] *om. J* 5 in pila] *om. B* 6 organo] in uaso *B* in organeo *C* appellamus] appellant *V* 7 hac- ratione] sed et graue *B* ualde *P om. W* 8 somnifera] omnem ferrum *BP* 9 compescit- uigilias] *om. W* omnes] *om. J*

3 This statement is unique among the ancient literature on honey, but is quite accurate according to modern research: see ch. 2.H. In contrast, Pliny 11.31 laments that its heavenly nature is stained by contact with earthly foliage.

180) Myrrh (*Commiphora myrrha* Engl.; *Commiphora anglosomalie* Chiov.)

Myrrh is the sap from the tree of the same name which grows chiefly in Arabia, and any part of it may be incised to release this sap. The best myrrh is that which is purest, a little rough in texture, dry, delicate, light, and with tiny lumps in it, the one same colour on the outside, a little reddish on the inside, and with something like whitish, tiny veins running through it, has a sweet fragrance, and a bitter taste that warms. This is the myrrh that is found in the mountains: there is another type that is cultivated on the plains, which is more supple and more plump, and which is usually pressed to extract its liquid. That part which is very runny is stored separately in a small jar and is called by the Greeks *stacte*.[1] All myrrh is styptic, sharp, and heating, and its aggressive effect can be disturbing.

1 See Diosc. I.64.

181) Mandrake (*Mandragoras* sp. L.)

Mandrake is a plant known to everyone, and some call the same plant *apollinaris*.[1] Its juice is extracted from its leaves, which are cut out all the way around, leaving merely the outside arch of the leaf, and the fluid which pours out is then collected. Or its root, which is like that of opopanax,[2] or its fruit along with the whole root, is crushed with a mortar, and then its juice is squeezed out by a machine, and we call this in Greek 'mandrake-juice.'[3] Mandrake has cooling properties, and for the same reason also astringent properties. Mandrake can also induce a debilitating fatigue, and therefore is believed to be soporific. It does, however, reduce all pain and alertness.

1 *apollinaris*: also given as an alternative name for mandrake in *Ex herbis femininis* 15, and *DynVat* 2.119 (saec. VIII), but other texts apply it to different plants: Ps. Apul. 22, Scrib. 90 (= *Withania somnifera* Dun.: above #84 n.1), Pliny 26.140, 147 (= *hyoscyamos*, below #295). André, 1985, 21. On different names for mandrake, see Besnehard 1993. See also ch. 2.B and plate 4.

cf. Diosc. IV. 75 (μανδραγόρας). Pliny 25.147, 26.140. Galen *Simp.* 7.12.4 (K.12, 67). Ps. Apul. 131. *Ex herb. fem.* 15 (femina). Isidore 17.9.30.

182) Molybdaena

Molybdaena est quasi stercus auri et argenti quae reponitur quod conflatur colore spumae argenti. Est et alia quae foditur in Corintho. Verum optima est quae est colore quem diximus, non lapidosa, fragilis, splendida. Vires quoque habet assimiles spumae argenti.

1 Molybdaena] mollibdina *VF* molipdinis *P* quae- spumae argenti] *om. JF* 2 quae foditur] *om. V* quae datur *F* corintho] carento *F* 3 non] nunc *V*

deest W

cf. Diosc. V. 85 (μολύβδαινα). Pliny 34.173 [33.95, 34.159].

183) Malum granatum

Malum granatum quod graece rhoea dicitur, fructus est arboris eiusdem uocabuli, qui recens et maturus manducatur sine suo cortice, qui malicorium uel sidia dicitur. Et cum aliqua dulcedine succus suus uentrem stringit et stomachum erigit, magis tamen stringit quod est suba-
5   cidastrum. Cuius ipsius flos cytinus dicitur. Vehementissime stypticus est et efficacissime stringit. Vnde dysenterico sanat et uulnera recentia glutinat. Eadem et cortex illius praestat, quem malicorium diximus nominari.

1 quod- dicitur] *om. B* rhoea] quoa *V* ceoa *F* choa *LC om. W* 2 uocabuli] nominis *P* maturus] *om. P* manducatur] uocatur *L* sine suo cortice] sine sugus corticem *V om. B* 3 malicorium] malocorio *V* malum *B om. W* sidia] sicidia *L* sidicia *C* dicitur] dicitur eius cortex *B* succus] *om. J* 4 stomachum - tamen] *om. V* subacidastrum *VF*] acrum *L* subacidasterum *B* subaridum *P* subacidastera stomachi humorem corrigit *C* stomachum- stringit] *om. W* quod] quod magis *B* 5 cytinis] cytonori *V* cinidon *F* cihama *C om. BP* uehementissime stypticus] *om. V* 7 quem- nominari] *om. BPW* malicorium] malum corium *VFL* malum *C* 8 nominari] nomine *V* praestare *C*

cf. Diosc. I.110 (ῥόα). Pliny 13.113, 23.103–8, 23.110–12 (cytini). Scrib. 41. Galen *Simp.* 8.17.8 (K.12, 115). Garg. *Med.* 41. *DynSGall.* 92–6. Marc. *Med.* 10.7. 6.19.2.

184) Medulla

Medulla quidem omnis leniter mollire et relaxare durities et tumores potest. Verum tamen efficacissima uidetur ceruina, sequenter uitulina, deinde caprina numeratur.

2 #201, 220.
3 *mandragorochylon*: unattested elsewhere. The use of *quod* rather than *quem* in some manuscripts suggests this was the name of the machine (*organum*), that is, 'mandrake-juicer.'

182) Galena

Galena is the dross from molten gold and silver which is stored separately
and is the colour of litharge.[1] There is another type which is mined in Corinth.[2] The best is that which is the colour we mentioned, without any pebbles in it, delicate, and shiny. It has properties similar to those of litharge.

1 #256.
2 Cf. Diosc. V.85 'mined ... near Sebastes and Corycos.'

183) Pomegranate (*Punica granatum* L.)

Pomegranate is called in Greek *rhoa*, and is the fruit of the tree that goes by
the same name. It is eaten when ripe and fresh, and with its skin removed,
which is called 'pomegranate-rind,' or *sidia*.[1] The juice of the pomegranate
that is slightly sweet binds the bowels and stimulates the stomach, but the
pomegranate that is a little sour binds [the bowels] even more. The flower
of this sour type is called *cytinus*, which is extremely styptic and a very
effective astringent. It cures dysentery and agglutinates fresh wounds. Particularly good for these last is pomegranate skin, which as we said is called
'pomegranate-rind.'

1 *sidia* and *cytini*: likewise Diosc. I.110, Pliny 23.110–12. The specific term *malicorium* suggests an early date for the *AG*: the term is only used by classical writers such as Petronius
(*Satyr.* 47), Celsus (2.33.4, 4.23.2, 7.27.3, etc.), Pliny (23.107, who noted it was the term
used by doctors), and from him Marcellus (*Med.* 31.35, 32.26), though it is not found in
other works that borrowed from Pliny (e.g., Garg. *Med.* 41, *Med. Plin.*, 7 'coria mali in
aceto decocta'), or in any late antique or medieval text. See ch. 4.C.

184) Marrow

All marrow can gently soften and loosen indurations and swellings. The
most efficacious seems to be the marrow from bones of a deer, then that
from a calf, then that of goat is ranked next.

1 omnis] omnis medulla *VL* **2** potest] sedare potest *W* efficacissima] in effectu *W* **3** deinde] inde *JC* numeratur] nominatur *V om. BPC* postremo *W*

cf Diosc. II.77 (μυελός). Pliny 28.145. Galen *Simp.* 11.1.3 (K.12, 331–3). Isidore 11.1.87.

185) Menta

Menta herba est quam omnes nouerunt quae est apta stomacho ualde. Sanguinis quoque abundantiam stringit et capitis dolorem sedat. Item singultus et nimios uomitus reprimit, et conceptum uetare dicitur.

1 quam- quae] *om. P* apta] utilis *W* ualde] *om. FLBC* **2** quoque] *om. JW* **3** conceptum-dicitur] *om. W* uetare] uitia redigitur *F* uertere *C*

cf. Diosc. III.34 (ὑδύοσμον). Pliny 19.159–62, 20.147. Ps. Apul. 121. Garg. *Med.* 24. *Dyn. Vat.* 22/*SGall* 26. Isidore 17.11.6, 17.11.82 (agrestis).

186) Mastiche

Mastiche genus est resinae, quae in modum gummi ab abore manat quae schinus dicitur et in Chio insula nascitur locis asperis et petrosis. Est ergo mastiche aliqua subuiridis, aliqua albida. Verum optima est quae candidissima et perlucida et quae dum maturatur aliquatenus
5 mollescit. Vires autem habet euaporantes et cum suaui excalefactione relaxantes. Adulteratur resina a cypresso et a pino, sed ista odore sui deprehenduntur.

1 Mastiche] masticem *V* mastice *F* mastix *W* gummi ab] gemis *V* **2** schinus] scinus *VFLBC* lentiscus *W* chio] quo *F* cipro *LP* clio *C* chio- mollescit] *om. W* **3** aliqua subuiridis- albida] *om. J* **5** mollescit] mollis fit *LBC* euaporantes] *om. P* excalefactione] euaporatione *J* **6** cypresso- *fin.*] *om. W* pino] nudo pino *FL* nucleo pino *B* nucli *C* nucleis pinu *P* sui] soaui *V*

cf. Diosc. I.70 (σχῖνος). Pliny 12.72. Galen *Simp.* 7.11.6 (K.12, 68–9). Isidore 14.6.30, 17.8.7.

187) Marrubium

Marrubium herba est quod ramulos emittit oblongos uel minutos tres aut quatuor circa quos folia mollia et exalbida quasi rotunda. Flosculum rufum et assimilem pulegio. Nascitur in campis solidis et petrosis et est optimum creticum. Vires autem habet acres et efficaciter relaxan-
5 tes. Vnde succus illius cum melle ad tussim facit et in partibus neruosis optimum est.

1 Marrubium] marrubio *V* uel- creticum] *om. W* **2** quasi] fere *P* **3** solidis] audis *L* sudibus *BP* saudis *JC* **5** unde- est] *om. W*

185) Green mint (*Mentha* sp. L.)

Mint is a plant that everyone knows, and is very good for the stomach. It also staunches excessive blood and eases headaches. It subdues hiccups and excessive vomiting, and is said to prevent conception.

186) Mastic (*Pistacia lentiscus* L.)

Mastic is a type of resin which flows out of a tree like a gum,[1] and is called 'schinus.'[2] It grows in harsh, rocky places on the island of Chios. There is a type of mastic which is a little green in colour, and another type which is white. The best is that which is brightest in colour and transparent, and which softens a little as it matures. Mastic has dispersive properties, and also properties that loosen with a gentle warmth. It is adulterated with cypress and pine resin, but their presence is revealed by their distinctive fragrance.

1 #47.
2 Cf. #151, 228.

187) Horehound (*Marrubium vulgare* L.: *M. creticum* Miller)

Horehound is a plant which shoots out three or four small, lengthy, and slender branches that are covered in soft, whitish leaves that are almost rounded. It has a small yellow flower similar to pennyroyal.[1] It grows in dense[2] and rocky plains, and the best is that from Crete. It contains sharp and thoroughly loosening properties, and its juice mixed with honey is good for coughs, and is excellent for sinewy parts of the body.

1 #214.
2 Reading *solidis* (*VF*), though 'thorny/spinous' (*BP*) is possible: *saudis* (*JC*) is obscure.

cf. Diosc. III.105 (πράσιον). Pliny 20.242–4. Galen *Simp.* 8.16.35 (K.12, 107–8). Ps. Apul. 45. Isidore 17.9.58.

188) Myrta

Myrta genus est herbe lignosae duobus generibus est et nigra et alba, utraque est omnibus nota. Vires habet stypticas ualde et nigra tamen illa stringit potentius. Vnde et sanguinem excreantibus remedium est et dysentericis et omne profluuium stringit. Eius uero nigra, quae in
5　montuosis locis nascitur, grana sicca uel uiridia commanducanda offeruntur his qui sanguinem reiiciunt et uesicae taedia sustinent. Stomacho est conueniens, urinam mouet et scorpionum morsibus prodest si imponatur aut potui detur aut locus qui tactus fuerit ex eadem cataplasmetur. Et uiridis myrta similiter operatur. Coctura eius cum uino mix-
10　ta capillos tingit et ex uino decocta et in modum cataplasmatis imposita uulnera sanat. Cum polline tritici immista tumores oculorum compescit et aegylopibus utiliter facit. Grana eius contusa et expressa uino faciunt et ideo parum decoqui expedit ne acescat et uase piceo reseruatur. Facit autem contra ebrietatem et furfures capitis et uulnera curat si
15　frequenter inde lauentur et capillis fluentibus prodest. Fit etiam oleum de foliis eius contusis et in oleo diu decoctis quod in catismatis prodest et articulis uel fracturis utiliter imponitur. Sed et maculas perfricatum emendat. Aures pus reiicientes curat si tepefactum immittatur. Ipsa folia trita cum aqua imposita uulnerum rheumatismum sanant, uel
20　locis fluentibus prosunt. Fluxum uentris emendant et ignem sacrum, uel tumores testium curant. Sicca autem et cribrata panaritiis et locis uerendis prosunt. Commista autem folia eius tenue cum ceroto ignem sacrum curant. Folia eius trita pensus piperis medietate acetum quantum sufficiens maculis prosunt.

1 Myrta] myrtus *WP* lignosae] quasi lignum *B* lignosae- nota] *om. W* 3 potentius] potest *F* ualide *P* excreantibus] reicientibus *B* remedium- conueniens] ad bibendum datur *W* remedium- profluuium] profluuium sanguinis *F om. C* 4 eius- quae] *om. VBC* nigra] *om. F* 5 grana] grandia *VF* grandis *C* uiridia] uiscida *V* uel- commanducanda] *om. C* 6 his- taedia] etiam *V* uesicae] uiscide *JVC* uissicae *F* uesce *L* uesice *BP* 7 conueniens] commouens *F* 10 capillos tingit] constringit *V* stringit *FL* extringit *C* capillos- *fin.*] capillis ad errantibus prodest. commixta enim folia cum ceroto tenui ignem sacrum curat *W* imposita] imposita extrimitatum *L* imposita exaremitatum *B* imposita scarema *C* 11 uulnera- immista] *om. F* uulnera] extriniticota uulnera *V* immista] immista imposita *VLBC* 12 aegylopibus] igylopis *V* icylopis *F* egilopus *LB* egilopes *P om. C* eius] etiam *JC* contusa] *om. B* uino] *om. JB* 13 expedit] *om. F* ne acescat] *om. C* piceo] picito *V* picato *FLBP* picauo *C* 14 furfures] furoris *F* 15 fluentibus] fluentibus maxime *C* fit- prodest] *om. BP* 16 diu] diutissime *V* in catismatis] incuruationibus *J* 18 aures] *om.*

Cf. Diosc. III.105, 'grows around building lots and ruins [οἰκόπεδα καὶ ἐρείπια],' but his description differs considerably, and does not mention Crete.

188) Myrtle (*Myrtus communis* L.)

Myrtle is a woody type of plant, and there are two species, black and white, and both are well known. It contains strong styptic properties, and the black myrtle is a more powerful astringent. It is a cure for coughing up blood, dysentery, and staunches all haemorrhages. Either the fresh or dried seeds of the black myrtle, which grows in mountainous places, are chewed by those spitting blood, and revive a sluggish bladder.[1] They are also good for the stomach, stimulate urination, and are a remedy against scorpion stings when applied topically or when given as a drink, or when applied as a plaster directly onto the place where the scorpion struck. The fresh myrtle seeds achieve the same results. The liquid from the decoction of myrtle mixed with wine makes a hair-dye, and when stewed in wine and applied in the form of a plaster it cleanses wounds. Mixed with finely ground wheat flour, it eases swellings on the eyes and is useful against lachrymal fistulas.[2] Its seeds are also pounded, pressed, and stewed in wine, then separated from the wine before they turn sour and preserved in a pitch-sealed container. This is good against drunkenness, scaly infections on the head,[3] heals wounds when frequently washed with it, and promotes flowing hair. An oil is made from the crushed leaves of myrtle by stewing them in oil for a long time, and is beneficial used in sitz baths[4] and useful when applied topically on joints and fractures. It also removes blemishes when rubbed in. When warmed and placed inside the ear canal it stops ears oozing pus. The leaves of myrtle, when crushed and mixed with a little water, are applied topically to cleanse wounds of any morbid discharge, and are good against discharges anywhere else. They also cure diarrhoea, erysipelas, and swellings on the testicles. When dried and sifted, they are beneficial for paronychia,[5] and for sores on the genitals.[6] The leaves also, applied with a thin cerate, heal erysipelas. The crushed leaves, combined with about half their weight in pepper and then dipped in vinegar, help remove blemishes.

1 *uesicae*: following the similarities here with Diosc. I.122 (ἐπιδακνομένοις τὴν κύστιν) rather than *uiscide* (VJC).
2 *aegilops*: lit. 'goat-eye,' caused by an ulcer in the inner canthus of the eye: see Cels. 7. 7. 7, Pliny 35.34.
3 *furfures capitis*: cf. Pliny 20.101.
4 *catismatis*: cf. Diosc. I.122, 'τὰ ἐγκαθίσματα'.
5 *panaritium* (scil. παρωνυχία), whitlow or felon, a bacterial or fungal infection between nail and skin.
6 or 'or burning parts,' *J*.

P pus reiicientes] prurientes *J* plus reicientes *C* si] sed *J* tepefactum] *om. LB* denique si mittis *C* **19** imposita] *om. B* uel- prosunt] *om. F* **20** sacrum] acrum *VFL* **21** testium] tritium *BP* cribrata] cribellata trociscus factum uino ad disinteriam per bumunem *C* **22** uerendis] urentibus *J om. C* ceroto] ceroro *J* cerutum *V* cerom *L* **23** pensus- acetum] riperpensa *C* acetum] *om. VF* **24** maculis] masculis *J* maculis ueraciter prodesse sentitur *BP* in oculo prodest *C*

plerumque om. l. 4–23 W

cf. Diosc. I.112 (μυρσίνη ἡ ἥμερος). Pliny 15.118–26. Orib. *Syn.* 1.19A. Ps. Apul. 88.6. Isidore 17.7.50.

189) Meu

Meu radiculae sunt herbae assimilis anisi, quae radices inueniuntur in omnem partem exporrectae, tenues, oblongae, suauiter olentes et gustu linguam excalfacientes. Nascitur maxime in Macedonia. Vires habet ipsa radix relaxantes. Vnde potata omnem rigorem uesicae et renum remollit et inflationes dissipat. Item urinam et menstrua prouocat.

1 assimilis] familis *C* aspera *W* radices- exporrrectae] *om. W* **2** exporrectae] aspera *C* gustu- macedonia] *om. W* **4** potata] cataplasmata *BP* postumis *C* **5** item- prouocat] *om. W*

cf. Diosc. I.3 (μῆον). Pliny 20.253. Galen *Simp.* 7.12.20 (K.12, 78). Isidore 17.9.10.

190) Malabathrum

Malabathrum qui putant esse folium spicae falluntur. Licet autem sit similis odore tamen certe alterius est naturae. Inuenitur autem supernatans quibusdam paludibus in India in specie foliorum et sic habetur leuis, lino trahitur et siccatur. Est ergo optimum quod est recentissimum, odoratissimum, tenue, integrum, nigastrum, gustu non salsum, sed potius spicastrum. Potest autem eadem praestare quae spica et easdem uires habet quas et caetera aromata habere diximus.

1 Malabathrum] malabatrus *V* malabastrum *F* malabatrum *BPC* qui] alii *F* spicae] specie *F* **3** quibusdam] aliquibus *C* habetur- trahitur] colligitur *C* **4** leuis] foris *V* flores *F* foris litore *LBP* foras extractus et siccatus seruatur *W* **5** nigastrum *VFLC*] nigratum *BP* subnigrum *W* nigricans *J* non] *om. C* **6** spicastrum *VFBCP*] spicarum *L* picem referens *J* ut spica *W* **7** quas- diximus] *om. W*

cf. Diosc. 1.12 (μαλάβαθρον). Pliny 12.129. Galen *Simp.* 7.12.2 (K.12, 66)

191) Marrubiastrum

Marrubiastrum est subsimile marrubio et facile notum. Cuius uires faciunt ad serpentum morsus et ad suspiriosos. Item strumas dissoluit et menstrua prouocat.

189) Spignel / Baldmoney (*Meum athamanticum* Jacq.)

Spignel consists of the small roots of a plant that is similar to anise,[1] upon which the roots are found all over sticking up, and are thin, lengthy, sweetly fragrant and have a taste that heats the tongue. It is grown mostly in Macedonia, and its roots contain loosening properties. When drunk it dissolves stiffness in the bladder and kidneys, and eases swellings. It also stimulates urine and menstruation.

1 #15.

190) Malabar (*Cinnamomum* sp., *Pogostomon patchouli* Pell.)

Those who think that malabar is the leaf of folium spikenard[1] are mistaken: it may have an odour similar to such, but it is of an entirely different nature. It is found floating upon on the water in certain swamps in India like a type of leaf,[2] for it is so light that it is collected with a linen cloth upon which it is dried. It is the best when it is freshest and its fragrance powerful, and when thin, whole, a little black, with a taste not salty but rather like spikenard.[3] It can be used as a substitute for spikenard, and it has the same properties as the other aromatics[4] which we have mentioned.

1 #236, and see above #111. Cf. Diosc. I.12, 'Some think that malabar is the leaf of Indian spikenard, being misled by the similarity of their scent.'
2 *in specie foliorum* may here be referring to #111 (*folium*): see above.
3 #236.
4 *aromata*: see ch. 1.D.

191) Black horehound (~ *Ballota nigra* L.)

Black horehound is quite similar to horehound[1] and generally known about. Its properties heal snake bites and help those with respiratory problems. It also breaks down scrofulous tumours and stimulates menstruation.

1 cuius- faciunt] *om.* C 2 strumas] stricinas id est scrofulas uel hemorrhoidas caecas *J* extrumas *C*
deest W
cf. Diosc. III.103 (βαλλωτή). Pliny 20.244.

192) Millefolium

Millefolium thyrsum habet ex una radice mollem et subrufum et quasi manu lineatum. Folia quasi foeniculi, quae uulnera glutinant et sine tumore custodiunt.

1 Millefolium] milifolium *V* 2 lineatum] limatum *VFLBPC om.* W quae] uis eis *C* 3 tumore] timore *V*
cf. Diosc. IV.102 (στρατιώτης ὁ χιλιόφυλλος). Pliny 24.152, 25.42. Ps. Apul. 89.

193) Myrice

Myrice fructiculum affert mollem, fragilem et ueluti florem, in Aegypto qui idem similem gallae. Huius igitur succus et aqua in qua fuerit coctus pediculos occidit et culices et pulices fugat.

1 Myrice] mirice *FP* myrice id est tamarice *C* affert] habet *C* fragilem] *om.* C 2 gallae] gallem V huius- succus] *om.* W 3 culices] *om.* VFC pulices] pudices *P* fugat] fugat splenticis et colicis *C om.* W
cf. Diosc. I.87 (μυρίκη). Pliny 13.116, 24.67. Galen *Simp.* 7.12.28 (K.12, 80–1).

194) Morus

Morus arbor est spinosa quae affert mora similia rubo sed et fortiora et longiora quae et uentrem malaxant et indigestionem faciunt. Succus eorum cum melle et careno, uel uino austero pari mensura permixtus, et ad spissitudinem mellis decoctus, synanche laborantes curat. In sole
5 in uase eneo positus uehementior erit. Et ideo uulnera et attritiones curat, uel putredines. Sed, ut supradictum est, synanchen cum tumore uenientem emendat. Si cum alumine scisso, uel galla, croco et parum myrrhae commisceatur, non solum his rebus prodest sed et uitia oris abstergit. Cortex radicum eius in aqua decocta, in potione data, uen-
10 trem soluit et lumbricos eiicit. Folia eius quoque trita et imposita, in igne combustum sanant.

1 Morus] murus *V* mora] poma *W* 2 longiora] albidiora *J* longi molliora *F* malaxant] laxant *JC* indigestionem] digestionem *JCW* 3 careno] aceto *BPW* farina *F* pari] paris *J om.* W permixtus] promistus *JBC* 4 spissitudinem] similitudinem *C* synanche] sinances *VFL* sentuces *B* sinance *PC* sinanticos *W* curat] curatio *BP* dabitur *C* 5 eneo] enio *V* aereo

1 #187. The name *marrubiastrum* is unique to the *AG* (the *Alphita* [109, 24] later borrowed
   from it). It is another instance of the *-aster* suffix: see ch. 4.E and plate 6. Both André 1985,
   155, and Opsomer 1989, II:430, identify black horehound, but it may be some other variety: cf. Chiron 888, *marrubium montanum*; Pelagonius 341, *marrubium ponticum*, etc.

192) Milfoil / Yarrow (*Achillea millefolium* L.)

Milfoil has a soft, reddish stem that shoots from a singular root, and which
has lines on it like those on a hand. Its leaves are similar to fennel and they
agglutinate wounds and prevent them from swelling.

193) Tamarisk (*Tamarix* L.)

The tamarisk tree bears a small fruit like a flower which is soft, delicate,
and that from Egypt is very similar to nutgall.[1] Its juice, and the water in
which it is cooked, kills lice and drives away gnats and fleas.

1 #121. Likewise Diosc. I.87.

194) Mulberry (*Morus nigra* L.)

Mulberry is a thorny tree which bears berries that are similar to blackberries but are hardier and longer, and which loosen up the bowel, yet
cause indigestion. Their juice, when thoroughly blended with honey and
reduced sweet-wine, or with an equal measure of dry wine, and then
stewed until it achieves the consistency of honey, cures those suffering
from sore throat. It becomes even stronger when stored in a bronze container. It heals wounds and abrasions, and festering sores. It also, as said
above, cures sore throats that have become swollen. If mixed with split
alum, oak-gall, or saffron, and an equal amount of myrrh, it not only alleviates these problems but also cleanses sores in the mouth. The skin of its
root is stewed in water and, administered as a drink, loosens the bowel and
expels intestinal parasites. Its leaves, when crushed and applied topically,
heal burns received from fire.

LW hereo *B* ereo *C* erit] fit *C* attritiones] adstrictiones *BPC* constrictiones *W* 6 sed- est] *om. BPW* cum tumore] cunctos humores *BP* 7 uenientem] uehementer *FC* croco] *om. W* parum] pauco *J* in parum *C* 8 myrrhae] smyrnam *F* non- abstergit] malos humores emendat uicia oris constringit *W* prodest] *om. FLB* 9 abstergit] stringit *B* constringit *P* 10 lumbricos] uentrem *F* eiicit] dictitur *B*

cf. Diosc. I.126 (μορέα ἤ συκάμινον). Pliny 23.134–9, 24.120. Galen *Simp.* 7.11.23 (K.12, 78–9). *Dyn.SGall* 79. Isidore 17.7.19.

195) Mora syluatica

Mora quoque syluatica, id est quae de rubo colligitur, stringunt fortiter et synanchicos melius curant. Succus eorum uel decimas rubi succi cum melle aequali mensura coquatur ad tertias et sic gargarizetur frequentius. Si enim bibatur, uentrem stringit [apte].

1 Mora] mura *VL* quoque] etiam *J* id est] ratione *JB* idem *F om. P* rubo] rubro *J* rubi ad tercias *W* 2 synanchichos] *om. BW* succus] *om. B* decimas] alicuius *J* decienis *C* commisces *C* 4 uentrem] *om. F* apte] *om. VLBC* uehementer *F*

plerumque om. W, def. partim P

cf. Pliny 24.117, 120. Marc. *Med.* 34.24. Ps. Apul. 88. Isidore 17.7.19.

196) Nepita

Nepita est omnibus nota. Cuius sunt species tres. Est enim quae folia similia ocimo habet et est quae pulegio assimilatur. Est quae mentae habet speciem. Verum omnia folia gustu sunt acria et excalfactoria ualde sunt. Igitur coxiosis genere sinapismi prodest, succus eius ad uenerea mire et regium morbum emendat et lumbricos et tineas et partum expellit et uermes in auribus occidit.

1 Nepita] nepeta *V* nepeta herba *CW* 2 ocimo] ozimi *V* achimi *F* 3 acria] cara *F* 4 igitur- prodest] *om. JW* uenerea] ad uersus ueneriam *F* ad usus ueneria *C* uenerios excitat *W* 5 tineas] extenuata statim *B* tineas statim *L* 6 occidit] eiicit *WC om. F*

def. ex lacuna P

cf. Diosc. III.35 (καλαμίνθη). Garg. *Med.* 23. Ps. Apul. 94 [91?]. Isidore 17.9.82.

195) Blackberry (*Rubus fruticosus* L.)

Wild berries,[1] that is, those from the plant which produces blackberries, are powerfully astringent, and are excellent for curing those suffering from sore throat. Their juice, or the leftovers from making juice, is cooked with an equal measure of honey and reduced to a third of its mass and this is then regularly gargled. When drunk it tightens the bowels [considerably].

1 *mora syluatica* appears as a coda attached to *morus* (#194) in *VFLC*, yet clearly the blackberry (usually termed *rubus*) is intended, as Ps. Apul. 88.33 makes clear ('rubus ... Romani mora siluatica [appellant]),' and similarly Isidore 17.7.19 (*mora siluestris*). See further Halleux-Opsomer 1982a, 92, and André 1985, 164. The comparative entries in Diosc. I.126 for mulberry (as above #194), and bramble IV.37 (βάτος = *Rubus ulmifolius* Schott), differ considerably.

196) Catmint (*Calamintha* sp. Lmk.)

Catmint is known to everyone. There are three different types. There is that which has leaves similar to basil,[1] and another which looks like pennyroyal.[2] Then there is that which has the appearance of mint.[3] The leaves of all of these have a sharp taste, and are strong calefacients. Catmint is therefore excellent for people with hip ailments[4] when applied as a mustard-plaster. Its juice is a marvellous aphrodisiac, and it heals the royal disease.[5] It also drives out intestinal parasites and maggots, expels the foetus, and kills worms in the ears.

1 #205. See ch. 4.F and plate 5.
2 #214.
3 #185.
4 Cf. Diosc. III.35, 'applied on people with hip ailments to alter the state of their pores by burning the skin's surface.'
5 *regius morbus*: the term was used by ancient Latin writers for a number of ailments, most commonly jaundice (Langslow 2000, 123, 202, et passim), thought to be caused by the retention of bile. From the fourth century onwards, patristic authors used the term for leprosy, epilepsy, and other ailments that visibly affected the skin: see Barlow 1980, 3–27. The correspondence with Diosc. in this case, and for the term used again in #269 and #288, suggest that jaundice was intended, though #56 uses the Greek term *icteritium* for such.

197) Narcissus

Narcissus thyrsum habet oblongum, folia quasi porri sed breuiora et
plena, florem album inter croceastrum, aliquando purpureum. Fructum quasi quasdam membranas nigrum, oblongum. Radicem quasi
bulbi, quae maculas de facie tollit et suppurationem aperit et surculos
et si qua sunt aliena euocat et euaporat et intumescere non permitt.

2 plena] *om. C* inter] intus *V* intrinsecus *W* croceastrum *VFLBC*] croceum *J* croceo *W* 3
radicem] *om. BW* 4 aperit- euaporat] ponit *J* surculos] *om. W*

def. ex lacuna P

cf. Diosc. IV.158 (νάρκισσος). Pliny 21.25, 128. Galen *Simp.* 8.13.6 (K.12, 85–6). Ps. Apul.
55. Isidore 17.9.16.

198) Nascaphtum

Nascaphtum affertur ab India, et est quasi cortex elamulae non odoratus, et uiscide excalefaciens. Propter quod in antidotis et in acopis
excalefactoriis mittitur.

1 Nascaphtum] nagatmum *V* nastaticum *F* nascapthimum *LB* nascathimum *W* elamulae]
elamnolae *V* lamulae *F* lamnule *LB om. W* odoratus] duratus *V* 2 in acopis] *om. J*

def. ex lacuna P

cf. Diosc. I.23 (νάσκαφθον).

199) Opium

Opium succus est lactae quodammodo erractici papaueris quod sic
colligi solet. Postmodum cum hoc ipsum papauer flores suos amittit
et ampullulae succrescunt uirides et tenerae harum ipsarum capitella
ueluti stillae curae huius homines scalpello acuto leniter abscindunt
5 sic, ut intrinsecus non laedant, et deinceps in gyrum scarificant et sic
laticulum quod promanat digito conterunt et in conchulis reponunt et
siccant. Meconium autem fit contritis in unum foliis et ampullis uiridis
papaueris. Verum succus organo pressus et siccatus in pastillos redigitur et meconium appellatur. Erit ergo optimum quod est odore grau-
10 issimum et colore subrufum et gustu amarissimum et quod humorem
quo resoluiter quam citissime candidum facit et accensum splendidam

197) Narcissus (*Narcissus poeticus* L).

Narcissus has a lengthy stem, leaves like the leek, but shorter and plump, and a white flower with saffron-coloured hues, sometimes purplish. Its fruit is black, oblong, and has several skins. The root is like an onion, and removes blemishes on the face, facilitates suppuration, and draws out and disperses any abnormal growths and prevents them from swelling.

198) Nascaphthon

Nascaphthon[1] is imported from India, and is like the bark of elamulae,[2] non-fragrant and vigorously heats. It is therefore mixed in antidotes and in calefacient salves.

1 Diosc I.23 is the only other text to mention this spice, which remains unidentified: André 1985, 170.
2 *elamulae* (J, but note variants): unidentified (from *lamium*, 'dead nettle'? Pliny 21.83, 22.37). Diosc. ibid. describes nascapthon as 'like bast, resembling the bark of the mulberry tree [ἔστι δὲ φλοιῶδες, συκαμίνου λεπίσματι ἐοικός]' and that it is burned for its sweet scent.

199) Opium poppy (*Papaver somniferum* L.)

Opium is the milky juice, after some refinement, of a wild poppy which is harvested in the following manner.[1] After this particular poppy has flowered it sprouts little green, delicate capsules that look just like drops and are harvested by men who gently cut off the tips of these capsules with a small, sharp knife, carefully so as to not damage them on the inside, and then they place them on a rack to be scarified. The milky fluid which flows out is then wiped off with a finger and stored in small shells and dried. Meconium is made by crushing the leaves and green capsules of the poppy together at the same time. The juice of this is then squeezed through a press and dried out in the form of lozenges, and these are called meconium. The best opium has an extremely pungent fragrance, is slightly

et aliquatenus duracem flammam emittit et rursum extinctum pristinum odorem reseruat. Maxime hispanum tale inuenitur. Adulteratur autem et lentis et lupinorum farina, sed gummi nigro et lactucae erraticae succo, item glaucio. Sed et adulteratum quod est sic intelligimus eo quod non omnia coeant quae sinceritatem illius comprobare diximus. Potest autem syncerum opium lachrymam stringere collyriis admistum, ita et circumlinitum refrigerare et auricularum dolorem et totum corporis languorem mitigare, illa uidelicet ratione qua etiam somnum facere dicimus.

1 lactae] *om. JW* quodammodo- papueris] lacte quibusdam herbis et radicibus *B* erratici] radice *F* uidelicet *W* 2 colligi- siccant] ex tenerioribus capitellarum et in sole siccatur *W* postmodum- succrescunt] exprimitur succus papauerum *B* 3 ampullulae] ampullae *V* capsulles *C* 4 ueluti- huius] *om. J* inciduntur et ueluti stillas decurrunt *B* et ueluti stoillarum *L* sunt ueluti stillata quarum corium *C* leniter] leuiter *LB* 5 scarificant] scarifans *LBC* 6 lacticulum- meconium] permunt *B* conterunt] contergit *V* contergent *F* conchulis] oculis *F* concutis *C* 8 organo] origano *L om. B* pastillos] patella *C* rediguur] *om. C* 9 erit- diximus] *om. W* appellatur] ampullatur *F* reponitur *B* 11 citissime- facit] *om. C* accensum] adscissum *F* 12 duracem] durantem *JC* extinctum pristinum] extensus proprium *V* intinctum proprium *F* 14 erraticae] radices *LBC om. F* 15 quod- coeant] ominino non coherata *B* noeant *C def. L* 16 comprobare] improbare *C* 17 collyris- ita] *om. B* 18 ita] habet *J* item *F* circumlinitum] leniter *V* 19 languorem] *om. VFLCW* 20 facere] preparare *W* dicimus] dicitur *VFB* creditur *C* manifestum est *W*

def. ex lacuna P

cf. Diosc. IV.64 (μηκών). Pliny 20.198–209, 25.142. Garg. *Med.* 19. Ps. Apul. 53. *Ex herb. fem.* 45. *DynVat* 49/*SGall* 52. Isidore 17.9.27 (*cyreniacum*), 17.9.31 (*papauer*), 16.15.20 (*meconites*).

200) Omphax

Omphax succus est de radice morae syluaticae et immaturae uuae. Qui expressus siccatur in sole et in pastillis latis usque quo figi possit et sic reponitur. Optimum est illud quod est colore rubeastrum et fragile pastillo et gustu styptico ualde et linguam submordens. Vnde potenter stringit et omne profluuium confestim tenet.

1 Omphax] onfacio *V* omfacium *F* radice] racimis *V* morae] *om. V* uuae *C* uaae] ubique *F om. C* 2 pastillis] patellis *FC* figi possit] latis fieri potest formatur *J* 3 rubeastrum *VFLC*] rubido *J* 4 pastillo] pistillo *J* potenter] prudenter *C* 5 tenet] retinens *F*

deest BW, def. P

cf. Diosc.V.5 (ὀμφάκιον). Pliny 12.130, 14.98. Isidore 17.7.68.

reddish in colour, very quickly dissolves and turns white when in contact with moisture, and when ignited emits a flame that burns for a little while, and which when extinguished replenishes its fragrance to the same level of pungency as when fresh. It is found chiefly in Spain, and is adulterated with the flour made from lentils and lupins,[2] but is also mixed with black gum, or the juice of wild lettuce, or even glaucion.[3] But for the process of adulteration we understand that not all ingredients may work and some may compromise its effectiveness. Pure opium can be mixed into eye-salves for drying up teary eyes, or can be smeared on around areas that need cooling. It alleviates earaches, reduces all types of fatigue in the body, and for the same reason we find it also induces sleep.

1 The account of preparation is similar to Diosc. IV.64 (μηκώνιον), which otherwise differs considerably. On opium in the ancient world, see Scarborough 1995. The reference to Spain seems unique. See plate 2, ch. 1.C, ch. 5.C.
2 #152, #153.
3 #118.

200) **Omphacion** (Juice of blackberry root and immature grapes)

Omphacion is the juice from the root of wild blackberry plant and unripe grapes.[1] Once the juice is squeezed out it is dried in the sun until it can be fashioned into large lozenges for preservation. The best type is that which is reddish, and the lozenge itself is delicate, with a very strong styptic taste that slightly bites the tongue. It is a powerful astringent and can quickly staunch all haemorrhages.

1 The inclusion of (root of) wild blackberry (see #195) in omphacion is unique, though omitted in V, 'juice of wild and unripe grapes.' Cf. Diosc. V.5, juice of unripe Psithian or Aminnaian grapes; likewise Pliny 12.130, though he records another type made from unripe olives (cf. 14.98, from the wild vine, *labrusca*).

201) Opoponax

Opoponax succus est herbae quae panax dicitur, simillimae per omnia
ferulae, quae nascitur in Capadocia, Africa, Arcadia et Macedonia, locis
asperis et lapidosis. Colligitur ergo hic ipse succus hac ratione, mediis
aestatibus terra quae est circa radicem et herbam foditur et excauatur,
5   deinde foliis ipsius sternitur, et sic ipsa radix et ferula in oblongum et
in gyrum inciditur. Postque humor lacteus emanat et in folia suppo-
sita defluit et siccatur quasi exalbidae myrrhae colorem accipit. Ergo
optimus est opopanax qui ab Africa affertur. Hic est extrinsecus ueluti
crocatus, intrinsecus lacteus, mundus, odore graui et quidem uiscide.
10   Potest igitur opoponax et laxare et calefacere et mollire stricturas et
durities, fugare uiperas et omnem serpentum accensus.

1 Opoponax] opopanax *FB* quae- uiscide] *om. W* 2 arcadia] *om. J* arabia *C* 3 lapidosis]
petrosis *F* mediis aestatibus] est huis temporibus *C* 4 excauatur] cauatur *JC* 5 sternitur]
ternitur *L* tenes *C* 7 quasi- est] *om. C* exalbide] exadipe *V* myrrhae *V* myrte *J* morem *F*
murre *L* murra *B* 10 stricturas] *om. W* 11 accensus] accessum *J* facit *C om. W*
def. P

cf. Diosc. III.48 (πάνακες Ἡράκλειον). Pliny 20.264. Celsus 5.18.5. Scrib. 82.

202) Opobalsamum

Opobalsamum lachryma est arboris quam balsamum appellamus,
similis malo granato, ramosae, ramis tenuibus et exalbidis, qui xylobalsamum dicuntur. Quae folia habet similia erucae. Haec ipsa arbori
in India tantummodo crescit, aliqui tamen in Iudea dicunt et fructum
5   affert aliquatenus oblongum, contra medium uastiorem et quasi uentrosum, ad capitula utraque subito deductum, rubicundum ualde, grauem et plenum et si frangatur mollem et si masticetur similiter ut piper
mordentem et suauiter redolentem quem carpobalsamum dicimus.
Huius ergo arboris lachryma colligitur a uerno usque ad autumnum.
10   Inciditur autem ipse matrix in aliquot locis et ipsi ramuli et sic uascula
supponuntur quae lachrymam fluentem excipiant. Quod ipsum aliquot coloribus uidetur. Verum est optimum quod recens et sincerum et
exalbidum est et austerum et odore graui et acuto et non acidastrum et
quod aliqua lenitudine linguam submordet. Seqens est quod est colore
15   subrufo et quasi glutinum se distillat a digito, leue etiam aspectuque
gratum. Tertium autem quod est tenue ualde et perlucidum. Omne
autem opobalsamum quod uestutatem accipit fit subrubrum et resinastrum et siccum. Verum optimum est sincerum quod aquae mundae

201) Opopanax / Hercules' woundwort (*Opopanax hispidus* Grisb.)

Opopanax is the juice of the plant which is called panax,[1] which grows in Cappadocia, Africa, Arcadia, and Macedonia, and in rugged, rocky places, and which in every way is very similar to giant fennel.[2] Its juice is collected in the following way: in the middle of summer the earth around its root is dug away and the whole plant is dug out, then stripped of its leaves, upon which the root and stem are laid out and cut lengthwise and right around, so that the milky fluid which flows out then leaks onto the leaves below[3] and is then dried until it achieves a colour similar to the pale myrrh. The best opopanax is that imported from Africa, which is almost saffron-coloured on the outside, and milky on the inside, pure, very sticky and has a heavy fragrance. Opopanax can loosen, warm, relieve cramps, soften indurations, and when burned wards off vipers and all types of snakes.

1 #220.
2 See #4 n.1.
3 Similarly Diosc. III.48 (πάνακες Ἡράκλειον).

202) Mecca balsam (*Commiphora opobalsamum* Eng.)

Mecca balsam is the sap from the tree which we call balsam and which is similar to that of the pomegranate[1] in having many boughs with slender, pale branches that are termed 'xylobalsam.' Its leaves are similar to rocket.[2] The tree itself grows only in India, though some say in Judea,[3] and bears a fruit that is somewhat longish, much thicker in the middle as though pot-bellied, and then sharply tapered at both ends, and is heavy, plump, and a strong red colour. When broken the fruit goes soft, and when chewed it bites the tongue in a manner similar to pepper, but has a sweet fragrance, and we call this 'carpobalsam.'[4] The sap of this tree is collected from spring right up to autumn. The parent stem is cut in several places, and the tree's smaller branches are placed underneath to act as jars to collect the sap that pours out. Mecca balsam can come in a range of colours, but the best is that which is fresh, pure, whitish in colour, plain, and has a pungent, sharp fragrance that is not particularly sour, and which bites the tongue with a degree of gentleness. Ranked next is that with a reddish colour, and which trickles from your finger like a type of glue, yet is light and has an attractive appearance. Ranked third is that which is extremely thin and transparent. All Mecca balsam turns red, becomes a little resinous, and dries

instillatur, non protinus diffunditur, sed magis subsidet, postmodum
20 totam aquam exalbidam facit. Lacti autem admistum in caseum illud
adstringit ut coagulum. Item uestimento albo aut mundo instillatum
maculam non facit. Quod enim sincerum est in summa aqua diffunditur et supernatat et prout oleum quasi stillulas ostendit. Nam uires
habet aliquatenus acriores quae uiscide excalfacere et relaxare efficaciter possunt.

1 quam- appellamus] *om. C* 2 ramosae- arbori] *om. W* ramis- ipsa] *om. C* xylobal-
samum] chilobalsamum *F* 3 dicuntur] dicitur in egypto nascitur *C* 4 aliqui- dicunt]
om. J 5 contra] circa *J* conara *C* uastiorem] pastionem *V* 6 subito] *om. VC* 8 carpobal-
samum] ergo opobalsamum *B* 9 a uerno] uere *J* 10 matrix] truncus *J* marox *V* matrox *F*
matrix *L* arbor *B* cortex *C* in- locis] arboris *C* 11 excipiant] suscipiant *B* quod- uidetur]
et reponitur et habentur *C* 12 sincerum] sinceriter *VL* 13 acidastrum *VFLBC*] acri *J* 15
se distillat] glutinosum distillat *V* stillamim *C* digito] tacto digito *F* 16 aspectu gratum]
affectum ostendit *C* perlucidum] liquidum optime considerandum est *B* 18 resinastrum *VFLBC*] terrestre *J* 19 subsidet] desedet *VFL* 21 albo] mundam *V* 22 instillatum]
aut stellatum *F* sincerum] sincerum non *LB* 23 stillulas] quasi pilulas *J* quasi stellas *F*
stillam *C* ostendit] ostendit item in ouo cum canna scribe et in calida [..]uum durum
coque. molle t[er]esta[m] ouo quod scriperis interius inuenies si in minus adulteru[m]
e[st] *C*

3- 22] plerumque om. W, def. P

cf. Diosc. I.19 (βάλσαμον). Pliny 12.114–16. Scrib. 33, 38. Celsus 5.23.3. Isidore 17.8.14.

203) Oleum

Oleum uetus optimum est quod cum bona pinguedine perlucet, odore
et gustu simile uetusto pinguamini. Vires autem habet excalefactorias
quae penitus relaxant. Sed et ad caliginem oculorum facit si fuerit de
pluribus annis et ad diuturna et indurata uitia neruorum. In cibis uero
5 admistum omnibus conuenit et illinitum stomacho et uentri reddit et
si qua strictura aut uulneratio intrinsecus facta est leniendo relaxat et
per inferiores partes deducit et dum sit res thermantica. Accepta rosa
refrigerans inuenitur. Nardo decoctum metasyncreticam uirtutem
habere incipit. Chamaemelo uero itidem metasyncreticum esse uide-
10 tur. Cum bacca lauri decoctum plus quam oportet thermanticum esse
certum est. Semini lentisci mixtum sicut rosaceum per longum tempus
stypticam uirtutem affert. Mastiche admixtum tonoticum esse uidetur.
Anethum illic per tempus madefactum, thermanticum comprobatur.
Mala cydonia illic per tempora madefacta, paregoricum uel stypticum
15 aliquantum inuenitur. Cucumeris agrestis radice admixta et decocta
catharticam et thermanticam habet uirtutem et quidem proprie attri-
tis inunctio facit. Iridi admixtum anapleroticum inuenitur calidum

as it ages. The best is pure and when droplets are poured into clean water they do not dissolve straightaway but instead sink to the bottom and then after a while turn the water entirely white. When mixed with milk it binds cheese like rennet.[5] Similarly, it does not stain when spilled on a white or clean vestment. When pure it also spreads across the surface of water and floats like small droplets of oil. It possesses considerably sharp properties, which can vigorously heat and thoroughly loosen.

1 #183.
2 #100.
3 Cf. Diosc. I.19 (βάλσαμον) 'it grows only in Judea.' Berendes and Aufmesser both cited 'India' in their translations of Diosc. Beck 2005, 19 n.36.
4 Meaning 'fruit of balsam,' the use of a compound noun here seems to be a later Latin development: Diosc. I.19 simply speaks of 'the fruit' (καρπὸς) of Mecca balsam, but Chiron (827, 843), Cassius Felix (45), [Lat.] Alexander of Tralles (2.13), Isidore (17.8.14), and later glossaries all use the compound form. The Latin Diosc. pairs the term with xylobalsam to create a separate chapter heading (*De carpobalsamu et xylobalsamu*, I.19), but only *semen* is used in the text to refer to the fruit as well as the seed (likewise Pliny 12.118), which (Gr.) Diosc. saw as separate products. Both Diosc. and Pliny (12.114–16) record that it is another, different seed from Petra, with which the fruit of balsam is adulterated, that 'tastes like pepper.'
5 #87.

203) Olive oil (*Oliva europea* L.)

The best olive oil is old, gleams translucently from its rich oiliness, and has an odour and taste like aged fat. It has heating properties that loosen deeply within. But if taken daily for many years it causes cloudiness of vision and a hardening of any problems in the sinews. It is an agreeable addition to all types of food, and is smeared on the belly to relieve the bowels, or if there is any cramp or internal wound olive oil gently relieves them and draws them down to the lower parts where it becomes a heating agent.[1] Yet mixed with rose oil it becomes cooling in action. Stewed with nard-oil,[2] olive oil begins to acquire metasyncretic properties, and likewise the same effect is achieved if mixed with camomile.[3] When stewed with berries of the bay laurel[4] for an exceedingly long time it is sure to become thermantic. If mixed with the seeds of the mastic tree[5] or rose-oil and left for a good while it gains styptic properties. Mixed with mastic itself it becomes strengthening.[6] When dill[7] is soaked in it for a while the oil's thermantic properties are strengthened. When quinces[8] are soaked in it for a while the oil is found to be paregoric and a little styptic. Mixed and stewed with the root of wild cucumber,[9] it gains both cathartic and thermantic properties,

ideoque oleum multis speciebus admixtum pluribus ualetudinibus prodest.

1 quod- pinguamini] *om.* W **3** penitus] ponit *F* si fuerit- thermantica] *om.* W **4** diuturna] longa *V* olitana *F* ad olentana *LB* condilomata *C* **5** illinitum] et lenitiem *J* in linitia *V* liniciam *F* linit *B* reddit] sedit *V* reddit lenem *B* remollit *C* **7** per- thermantica] *om.* *J* deducit] obducat *C* **8** metasyncreticam] mittitur increticam *F* cretica *W* **9** chamaemelo] camimollau *V* camomillo *FW* metasyncreticum] admixtum *W* uero- uidetur] uiridis commixta meaurum in creticum est ualde si *C* **10** esse certum] *om.* *BC* necesse *F* **12** tonoticum] thoniticum *L* toniticon *B* tonaticon *C* toneticum *W* collecticum *J* **13** anethum- inuenitur] *om.* W **15** aliquantum] aliquantulum *BC* inuenitur] inueniri faciant *J* **16** thermanticum] farmaticam *F* quidem- speciebus] et arreacos sanat *W* quidem] quod *VFB* **17** iridi] hirius *V* yreos *FLBC* anapleroticum- ammixtum] *om.* *VJ* **18** speciebus] rebus *FL* pluribus- prodest] pluries inuenitur *V*

def. P

cf. Diosc. I.30 (ἔλαιον). Pliny 15, 13–17. Isidore 17.7.62, 17.7.68.

204) Ouum

Ouum partus gallinae est quod omnes nouerunt. Effectum habet duplicem. Si semicoctum bibatur uirtutem stomacho probet et non permittit corpus debilitari, eo quod naturae fit animantis. Lenit enim omnia quia leuem uicem habet. Crudum autem potatum raucos emendat. Lach-
5 ryma uero eius tepida in oculis dolentibus inuncta salutem restituet. Vitellus uero eius extrinsecus in aliqua parte corporis illinitus paregoricus uel stypticus inuenitur. Coctum uero in aceto ut sit perdurum dysentericis prodest. Virtute autem sua cibi et potus uicem praestat. Si crudum bibatur sitim prohibet.

1 omnes- **4** habet] *om.* W **2** semicoctum] semel coctum *VF* enim coctum *BC* probet] preuit *V* prebet *F* **3** debilitari] deuiari *L* deuiriari *LC* fit] *om.* *V* **4** leuem] lene *V* leneficem *F* potatum] si bibatur *F* raucos] raucorem *V* braucos *BC def.* *L* raucedinem *W* lachryma- *fin.*] et coli inflationes et tortiones remediat *W* **5** inuncta] iniectum *VFL* intinctum *B* **6** uero eius] quoque *J* **7** inuenitur] est *V*

def. P

cf. Diosc. II.50 (ᾠόν). Pliny 29.39–51. Sext. Plac. 29. Isidore 12.7.80–1.

and likewise especially if applied as an ointment and well rubbed in. Mixed with iris[10] [it becomes anaplerotic and warming, and when the oil is mixed with the different types of iris it] cures many illnesses.

1 Difficult passage, absent in *J.* On the Greek terms used only here in the *AG (thermanticum, metasyncreticum, catharticum, anapleroticum, paregoricum*; this last also used in the next entry #204), and their relation to the *Liber de Dynamidia,* see ch. 1.E, ch. 2.E, and below n.6.
2 Presumably spikenard #236.
3 #72.
4 #149.
5 #151.
6 *tonoticum.* Not used by Diosc., the term τονωτικῶς first appears in Soranus (c. 98–138) and Galen to mean 'bracing' or 'strengthening' in the context of drug actions, and only appears in three Latin writings directly dependent on Greek writings: Cassius Felix 42 qualified it as a 'confortatorium [medicamentum],' likewise Lat. Oribasius, Eup. 3.7, 'confortatiuus [medicamen]'; similarly Alex. Tral. 2.44.
7 #35.
8 *mala cydonia*: the quince tree is mentioned elsewhere in the *AG* (#148, 249), likewise Diosc. I.115 (μηλέα).
9 #78.
10 #137.

204) Egg

The egg is the offspring of a chicken, as everyone knows. It has a double application. If it is eaten when half-cooked, it fortifies the stomach and does not allow the body to become enfeebled because of the egg's nature as a life-giving organism. It moderates everything, because it has a gentle element. When raw egg is drunk it heals those who are hoarse. Its albumen when warmed and applied topically to inflamed eyes restores their health. Egg yolk; when applied externally by smearing on any part of the body, is found to be paregoric[1] and styptic. Eggs cooked in vinegar until they harden are helpful against dysentery. Their strength is greater when drunk rather than eaten. Raw egg when drunk prevents thirst.

1 See #203 n.1.

205) Ocimum

Ocimum herba est omnibus nota. Vires habet acriter excalfacientes et relaxantes et urinam prouocantes.

1 ocimum] ocum *F* et relaxantes] *om. BC*

deest W, def. P

cf. Diosc. II.141 (ὤκιμον). Pliny 19.119, 20.119. Galen *Simp* 8.24.1 (K.12, 158). Garg. *Med.* 22. Ps. Apul. 118. *DynVat* 37/*SGall* 41. Isidore 17.10.16.

206) Origanum

Origanum et ipsum omnibus notum est et habet uires acres et excalfactorias et relaxantes et euaporantes. Vnde et suspiriosis et tussientibus prodest et serpentes fugat. Item biles detrahit, uentrem temperat sed urinam et menstrua prouocat.

1 et ipsum] *om. V* 2 relaxantes] bene relaxantes *LBC* 3 biles detrahit] uires retrahit *BC* temperat] detrahit *J*

deest W, def. P

cf. Diosc. III.27 (ὀρίγανος). Pliny 20.175. Galen *Simp*. 8.15.13 (K.12, 91). Garg. *Med.* 37. Ps. Apul. 123. Isidore 17.19.76.

207) Orobum

Orobum omnes nouerunt. Huius igitur farina cum melle uulnera purgat et maculas tollit et alienationes tenet. Ipse autem orobus manducatus, caput replet et uentrem soluit et sanguinem per urinam prouocat. Et est optimus atticus est enim colore uiolaceo effectu uiscide stringens et in modum cyani.

1 omnes- huius] *om. P* 2 tenet ipse] *om. BPC* 3 replet] repit *V* soluit] purgat *V om. J* sanguinem- prouocat] *om. BPC* 3 prouocat] prouocat. Ochra (*etc. prout nov. cap.*) *VFL* 4 enim] non *J om. BC* effectu] effectu habet *BC* stringens] sistens *J* 5 cyani] cani *J* cyamini *V* ciminum *F* cyani *L* ciani *B* cyaini *C*

deest W

cf. Diosc. II.108 (ὄροβος). Pliny 18.57, 139, 22.151. *Dyn.Vat.* 15/*SGall* 19. Alex. Tral. 1.3, 144. Galen *Simp*. 8.15.14 (K12.91–2).

208) Psoricum

Psoricum hac ratione conficitur. Chalcitidis partes duae et cadmiae una, in mortario laeuigantur aestate maxime in sole calidissimo, infuso

205) Basil (*Ocimum basilicum* sp. L.)

Basil is a plant known to everyone. It has sharply heating and loosening properties, which also stimulate urination.

206) Oregano (*Origanum vulgare* L.; *Origanum heraclioticum* Richb.)

Oregano is very well known to everyone and contains sharp, heating, loosening, and dispersive properties. It is therefore good for those with trouble breathing or coughing, and it puts snakes to flight. It also draws away biles, calms the stomach, and stimulates urination and menstruation.

207) Bitter vetch (*Vicia ervillia* Willd. = *Ervum ervilia* L.)

Everyone knows bitter vetch.[1] The flour made from it mixed with honey cleanses wounds, removes blemishes, and curbs abnormal growths. When chewed, bitter vetch clouds the head, loosens the bowels, and stimulates blood to enter into urine. The best type is Attic,[2] which is the colour of violet and has a powerfully astringent effect like that of lapis lazuli.[3]

1 Already given above under the Latin name of [*h*]*erbum* #133. As with that entry, the first part here closely parallels Diosc. II.108.
2 Text seems corrupt: *VFL* all begin a new entry for 'ochre' (*ochra*) at this point, which better suits the following description, for Diosc. V.93 (ὤχρα) mentions that the best yellow ochre should 'come from Attic provenance,' but he makes no comparison with lapis lazuli and says that ochre is quince-yellow in colour.
3 #85.

208) Psoricum

Psoricum[1] is made in the following way. Take two parts of rock alum[2] to one of calamine,[3] and crush them in a mortar under a burning sun during

subinde feruente et acerrimo aceto et hoc fit donec res rubro colore apparent. Habet igitur psoricum uires quas chalcitis quidem uastiores.

1 Psoricum] psigon *V* psiricon *FL* ponitium *J* cadmiae] cani *J* catmia *F* casmie *L* cathmia *P* 2 aestate- sole] aestu *J* 3 subinde] inde *BC* in feruente *P* acerrimo] aqua agrissimo *V* acrissimi *LBC* fit] fieri habet *J* res] *om. BC* rubro] rubrice *VL* 4 apparent] *om. V* uires] *om. V* cures *L* quidem uastiores] maiores *P*

deest W

cf. Diosc.V.99 (χαλκῖτις). Pliny 34.119. Celsus 6.6.31. Scrib. 32. Galen *Simp.* 9.3.40 (K.12, 244) (ψωρικοῦ).

209) Piper

Piper fructus est arboris quae in India nascitur quod cum primum cre-scit oblongum et rotundum apparet, et longum piper appellatur. Quod ex minutis et quam pluribus granulis ueluti constipatis constat et quae tempore ipso accrescunt et dissipantur et adhuc alba inmatura. Album
5 piper nominatur maturata enim et nigra facta nigrum piper dicuntur. Eligimus ergo longum piper quod confrangitur strictissime granula sua coniuncta habet et circa suam ueluti medullam cohaerentia, quo-dque gustum uiscide mordet et calefacit. Album deinde quod sine ruga fit candidissimum, graue et non minutum et ad gustum simile longo.
10 Item nigrum quod praeter colorem in reliquis sit simile albo. Potest autem omne piper sinceriter stringere et excalfacere et totum corpus uehementer commouere. Nam adulteratur longum quidem, sinapi et gummi formatis in specie piperis longi. Album deinde laeuigatur cum spuma argenti et ex gummi in pilulas eiusdem magnitudinis redactum.
15 Item nigrum his ipsis pilulis de nigratis, sed et leue et cauum et farino-sum piper est et lapillulis [sibi similis] admistum.

1 primum crescit] primis *V* nigrescit *P* pernigrescit *C om. F* 2 longum] *om. BPC* appel-latur] assimilatur *J* 3 quam- constipatis] *om. F* constipatis] stipatis *VLB* 4 adhuc- enim] maturata non *J* 6 longum] *om. BPC* confrangitur] quoniam frangitur *J* 7 suam] se *P* 9 fit] sed *V* longo] oblongum *BPC* similiter locum *V* 11 sinceriter] syncere *J om. BC* 12 sinapi-in] informatum *BP* in farmatium *C* sinape et [..] fomato in *L* 14 ex gummi] excuditer *J* 15 cauum] in cauum *VLB* in cabo *F* 16 sibi similis] sui similibus *BPC om. J* admistum] admistum esse solet *BPC*

deest W

cf. Diosc. II.159 (πέπερι). Pliny 12.26–31. Celsus 2.21, 3.21.2. Scrib. 120. Isidore 17.8.8. Galen *Simp.* 8.16.11 (K.12, 97).

the hottest period of summer, and then immediately pour over it extremely sour boiling vinegar, and continue until it turns a ruddy colour. Psoricum contains the same properties as rock alum but they are more robust.

1 Similar prescriptions for making *psoricon* are found in Diosc (V.99 rock alum), Pliny, and Celsus, though these all add that the solution is to be buried in an earthenware jar for a number of days (20 in Celsus, 40 in Diosc. and Pliny, in dung and vinegar respectively).

2 #48.

3 #45.

209) Pepper, black and white (*Piper nigrum* L.; *Piper officinarum* DC.)

Pepper is the fruit from a tree that grows in India and which at first grows in a form that is long and round, and this is called long pepper. It contains many tiny granules that are packed together tightly, and then when these grow larger they begin to drop out though they are still white and unripe. When these mature they are called white pepper, and when they turn black are called black pepper. We select the long pepper which when broken still holds its granules tightly packed together because they stick to the flesh of the fruit, and their taste vigorously bites and heats the tongue. White pepper, however, should be very white, free of wrinkles, heavy and of considerable size, with a taste similar to that of the long pepper. Black pepper should be the same as white pepper in everything except its colour. All pepper can be used as a safe astringent, and can heat and agitate the entire body. Long pepper is adulterated with mustard seed and gum[1] and then fashioned again into the shape of a long pepper. White pepper is crushed with litharge,[2] then mixed with gum and shaped into little balls the same size of its original granules. Black pepper is also fashioned into similar little black balls, but this makes the pepper light, hollow inside, and powdery, and it is also mixed with small stones [the same size of its granules].

1 #261, #47.

2 #256.

210) Pompholyx

Pompholyx est quasi flos spodii similis autem utraque colliguntur circa metallas in calaminis id est in cryptis ubi cuprum et argentum conflatur. Verum pompholyx prout leuissima fuligo in ipsis cameris inuenitur spodium circa parietes. Adulteratur [stillatim] ficis tenuibus
5 combustis. Item cannis leuibus sed fiunt fuscidiora et non aequaliter leuia. Est enim optima pompholyx quae est leuissima et candidissima et ad tactum nullam habens asperitatem. Easdem enim et spodium uires habet stypticas et quae siccare possunt.

1 Pompholyx] ponfolex *V* pontofolix *F* ponfolix *PW* spodii] spondii *BC* similis] simul *J om. W* 2 calaminis- cryptis] calamis inscriptis *F om. W* in cryptis] ubi aes *J* in egyptum *BPCW* 3 uerum- parietes] in cameris et parietibus *W* fuligo] poligo *F* 4 stillatim ficis] stellis siccis *J* illis ficus *V* stilis ficus *F* stillicidus *BPC* tenuibus] teneribus *VF* cannis 5 leuibus] cannis quibuslibet *J* leuis *V* per quibus leue est *FBC* canne leuis est *P* de canita combusta et *W* fiunt fuscidiora] infuscatur nimium et graue fit *W*

partim def. L

cf. Diosc. V.75 (πομφόλυξ). Pliny 34.128–9. Galen *Simp.* 9.3.25 (K.12, 234–5).

211) Purpurissum

Purpurissum quod et fuscum aliqui dicunt, glebulae sunt quadratae et pusillae in modum tessellarum quae ab infectoribus tinguntur colore roseo. Tale est maxime canusinum, sequens puteolanum. Vires habet aliquatenus stypticas.

1 Purpurissum] porporiscum *V* puporissum *F* purpurismum *P* fuscum] sucum *JBC om. W* quadratae] rarae *W* 2 pusillae] post illum *F* infectoribus] inferioribus *F* colore] odore *VFLBPCW* 3 puteolanum] potiolanum *PW* 4 aliquatenus] aliquantulum *J om. W*

cf. Pliny 35.44. Isidore 19.17.15

212) Pyrethrum

Pyrethrum radix est oblonga, colore fuscastra, intus exalbida, crassitudine digiti pusilli. Gustu excalfaciens uehementer et saliuam copiosam prouocat. Cuius optima est quae est recens, sicca et plena. Potest autem efficaciter relaxare et calefacere acriter.

1 Pyrethrum] peretrum *F* piritrum *P* fuscastra *VLJ* fucastrum *F* fusco *BPC* fuscior *J* intus] trahitur *BC* crassitudine] grosituudine *VFB* 2 pusilli] *om. VF* uehementer] *om. P* 3 sicca et plena] *om. J* 4 acriter] celeriter *BPC*

210) Zinc oxide

Zinc oxide (*pompholyx*) is similar to the frothy slag of zinc (*flos spodii*)[1] in that both are collected from foundries on the Calamine islands[2] where copper and silver are smelted. Zinc oxide is like an extremely fine soot that is found on the walls of the foundry rooms, and it is adulterated [little by little] with lightly burned figs, or with slender reeds, which become quite blackened and have an unevenly smooth texture. The best zinc oxide is that which is lightest in weight, brightest in appearance, and has no coarseness on its surface. Zinc oxide has the same styptic and potentially drying properties as slag of zinc.

1 Two different forms of zinc oxide; *pompholyx* is the (pure) zinc oxide sublimated at the top of the furnace, *spodos* the (impure) oxide collected from the bottom: likewise treated together in Diosc. V.75 and Pliny 34.128. On the two types and their manufacture, see Riddle 1985, 149–52, Wilsdorf 1977; also Forbes 1955–64, VIII.203–4.
2 *in calaminis*: a conjectured reading for the floating islands in Lydia (*Calaminae isolae*, Pliny 2.209), though no other source mentions this location in relation to *pompholyx*. Another possibility *in calamis* (*-inis* from Gr. καλάμινος? 'reed-like,' 'toothbrush,' 'toothpick,' etc.), given the reference to reeds below, but no manuscript supports it. The first two-thirds of the text is corrupt, hence somewhat conjectural here.

211) Dark purple (a cosmetic)

Dark purple, [a cosmetic] which others call 'dusky,'[1] comes in the form of little lumps that are cubic like dice and are stained by dyers to achieve a rosy colour.[2] The best is that from Canosa, and after this that from Puteoli. Dark purple has somewhat styptic properties.

1 *fuscum*: unattested elsewhere. See ch. 3.C.
2 *colore J*: all MSS read *odore*, not impossible but less likely.

212) Pellitory (*Anacyclus pyrethrum* DC = *Anthemis pyrethrum* L.)

Pellitory is a root which is lengthy, the width of a little finger, darkish in colour, and whitish in the inside. It has a strongly heating taste that causes much salivation. The best is that which is fresh, dry, and plump. It can thoroughly loosen and sharply warm.

deest W

cf. Diosc. III.73 (πύρεθρος). Pliny 28.151. Celsus 5.4. Scrib. 9. Isidore 17.9.74. Galen *Simp*. 8.16.41 (K.12, 110).

213) Pumex

Pumex lapidis est genus qui in littoribus maris inuenitur quique assidue a fluctibus illisus fit leuissimus et exalbidus et undique concauus in similitudinem spongiae. Facit ergo pumex ad dentifricia et scriptores pennas uel calamos sibi in eisdem temperantes bene scribunt et mitti-
5   tur haec ipsa pumex in smegmata. Abradit autem et sua asperitudine laeuigat.

1 Pumex] pumix *P* illisus] conquassatus *W* 2 fit] sit *J* 3 scriptores pennas] scripturae pensas *V* scripturas pinnas *F* scriptores- scribunt] bene trita et cribrata *PW* 4 scribunt] cribrata *B* cubrata *C* 5 smegmata] migmata *V* smigmata *FP* megmata *L* sinigmata *BC* smiginate *W* malagmata *J* abradit- laeuigat] *om. P*

cf. Diosc. V.108 (κίσηρις). Pliny 36.154–6. Galen *Simp*. 9.2.15 (K.12, 205), 9.3.13 (K.12, 221–2). Isidore 16.3.7.

214) Pulegium

Pulegium herba est omnibus nota. Quod cum aliqua acrimonia excalefacit et est diureticum ualde. Sed et menstrua prouocat et partum expellit et olim tussientibus prodest et odor eius nauseam prohibet.

1 Pulegium] puleium *F* omnibus nota] *om. P* 3 olim tussientibus] olendo sitientibus *BPC*
deest W

cf. Diosc. III.31 (γλήχων). Pliny 20.152–6. Galen *Simp*. 6.3.7 (K.11, 857). Ps. Apul. 93. Isidore 17.9.29, 59.

215) Propolis

Propolis circa fauos et circa aluearia apicularum inuenitur et est quasi genus cerae. Cuius eligimus quae est colore subrubra, odore melastro autem quasi storacis. Malaxatur spissa et densa et per omnia munda. Vires habet similes cerae sed efficaciores.

1 Propolis] propulus *F* propulis *P* aluearia apicularum] aluearia apud *J* uiolas apicularum *V* uiolcis alpicolarum *F* uiolas specularum *BC* apicularum] *om. P* est- quae] *om. F* 2 melastro *VFBPC*] mellastro *L* melleo *J* 3 quasi- munda] quasimum *J*
deest W

cf. Diosc.II.84 (πρόπολις). Pliny 11.16. Celsus 5.3–4. Galen *Simp*. 8.16.36 (K.12, 108).

213) Pumice stone

Pumice is a type of stone which is found on ocean shores and which becomes extremely light and whitish in colour from being continually pounded by the waves, a process which also hollows it out in the same way as sea sponges. Pumice stone is used to make toothpaste, writers sharpen their quills[1] and reeds with it to write better, and pumice itself is placed in cleansing lotions.[2] It is also scraped and pulverized to smooth out its coarseness.

1 *pennas*: one of the earliest references to the use of a quill (feather) pen: see ch. 4.C.
2 *smegmata*: or possibly *migmata* (*VL*), 'mixtures,' 'compound medicines.' Cf. Preface (above), n.1. Note *malagmata* (*J*) matches Pliny 36.156.

214) Pennyroyal (*Mentha pulegium* L.)

Pennyroyal is a plant known to everyone. It heats with a degree of sharpness, is a powerful diuretic, and stimulates menstruation. It expels the foetus, is good against developing coughs, and its fragrance prevents nausea.

215) Propolis (Bee glue)

Propolis is found in the honeycomb and hives[1] of small bees and is like a type of wax. We prefer that which has a tinge of red, a honey-like smell, and resembles storax.[2] That which is thick, solid, and altogether pure mollifies, and has properties similar to wax, though more effective.

1 *aluearia* (*J*) is the better reading, corrupted to *uiolas* in most MSS.
2 #254. *melastrum*: see ch. 4.E.

216) Psyllium

Psyllium semen est herbae nigrum et minutum pulicis magnitudine, unde etiam cognominatum uidetur. Herba tota assimilis foeno, foliola longa et angusta, calamum minutum et oblongum habet, a quo medio incipit quasi spicula oblonga et rotunda in qua hoc ipsum semen inue-
5 nitur. Quod efficaciter refrigerare et stringere potest, unde oculis lachrymantibus cito prodest. Item dysentericos et podagricos remediat et capitis dolorem ex aqua laeuigatum et in fronte inlinitum confestim sedat. Et haec ipsa herba pulicaris dicitur.

1 Psyllium] psillon *V* psilion *F* 2 foliola] *om. J* 3 longa- minutum] *om. BC* angusta-incipit] *om. P* medio] modo *F* 4 et rotunda] *om. J* hoc] *om. J* 5 refrigerare et stringere] restringere *V* 6 podagricos remediat] uerumque emendat *V* remediat] emendat *F* 8 herba] *om. BC* pulicaris] policaria *F*

deest W

cf. Diosc. IV.69 (ψύλλιον). Pliny 25.140, 26.79. Galen *Simp.* 8.23.2 (K.12, 158). *Ex herb. fem.* 25. Isidore 17.9.54.

217) Plantago

Plantago similiter huius succus et folia efficaciter refrigerant et stringunt, quapropter succus illius sanguinem excreantibus utiliter datur. Item folia contrita cum pane et imposita ad eaque stringere et siccare uolumus utiliter faciunt.

1 similiter- et] *om. F* efficaciter] fricacitur *C* stringunt] restringunt *VL* 3 folia] foliola *L* contrita] arita *BC* pane] *om. C* eaque] quae *J*

deest W

cf. Diosc. II.126 (ἀρνόγλωσσον). Pliny 21.101, 25.80, 26.129. Ps. Apul. 1. *Ex herb.fem.* 50. Isidore 17.9.50. Galen *Simp.* 6.1.60 (K.11, 838–9).

218) Polium

Polium herba est oblonga exalbida folia habens et in summo quasi canitiem et comulam, semine pleno, odore graui et non suaui. Potest ergo omnem uiperam fugare et uenenosas extinguere, laxat quidem stomachum et caput purgat. Sed et urinam et uentrem mouet.

1 herba] herbula *JL* folia] foliola *C* 2 semine] simile *V* graui- fugare] *om. F* potest- fugare] *om. BPC* 3 uenenosas] potest res uenenosas *J* uenenosa esse *P* ueneriosas *F* 4 sed- mouet] *om. LC*

totum. om. W

216) Fleawort (*Plantago psyllium* L.; *Plantago cynops* L.)

Fleawort is the small black seed, about the size of a flea, which gives the plant its name. The entire plant is similar to hay, with long, narrow, and thin leaves, and a small, long reed for a stem, from the middle of which little long and rounded spikes begin to grow, and these contain the seed. Fleawort can thoroughly cool and bind, and is therefore good as a fast remedy for teary eyes. It also heals those suffering from dysentery and gout, and when crushed in a little water and smeared on the forehead it quickly eases headaches. This plant itself is also called *pulicaris*.[1]

1 This alternate name seems to have been known only to other late antique writers such as Caelius Aurelius *Acut.* 2.197, 3.96–7, Theodorus Priscianus 1.10 (*pulicaria*), *Ex herb. fem.* 25, *DynVat* II.114, the Latin Oribasius, Soranus, and Alex. Tral.: André 1985, 211. Diosc. IV.69 notes the likeness of the seeds to fleas but not the etymology (ψύλλιον, from ψύλλα, 'flea'), and gives alternative names of *cynocephalon*, *cynomyia*, and *crystallion* (a 'Sicilian' name).

217) Plantain (*Plantago major* sp. L.)

The juice and leaves of plantain thoroughly cool and contract in a similar way (to fleawort, mentioned above), and therefore its juice is good for administering to those coughing up blood. Its leaves pulped with bread are applied topically to whatever we wish to contract and dry.

218) Hulwort (*Teucrum pollium* L.)

Hulwort is a plant with oblong, pale leaves which at the ends have neat little gray hairs like those of an old man, and it has a plump seed with a heavy fragrance that is not pleasant. It can ward off all types of viper and act as an antidote against their poisons, and it also relaxes the stomach and clears the head. It stimulates urination and the bowel movement.

cf. Diosc. III.110 (πόλιον). Pliny 21.100, 21.145. Ps. Apul. 57. Isidore 17.9.63. Galen Simp. 8.16.30 (K.12, 106–7).

219) Pelecinus

Pelecinus herba est quae in segetibus nascitur, plures et tenues ramulos habet, foliola pusilla et granula in folliculis terna uel quaterna, subrufa et amara ualde. Quae possunt cum aliqua acrimonia uiscide stringere, propter quod et in antidota mittitur et menstrua prouocat.

1 Pelecinus] peleginus *F* pelecinus uel cepa canina *tit. J* herba] herbula *BPC* segetibus] setibus *def. V* egypto *BPC* itineribus *J* 2 foliola] folia *BJ* pusilla] postilla *V* quaterna] qua *V om. B* 4 quae- quod] *om. W*

cf. Diosc. III.130 (ἡδύσαρον). Pliny 18.155, 27.121. Galen *Simp.* 6.7.2 (K.11, 883–4). Marc. *Med.* 22.12. Orib. *Eup.* 4.116. Alex. Tral. 2.147.

220) Panax

Panax radix est herbae de qua opopanax conficitur, quae etiam et ipsa si infirmius eadem praestat quae opopanax.

1 herbae] *om. W* conficitur] constringitur *F* quae- opopanax] per se enim ispa radix opopanacis gistercon herba est folia habens modica uelut incisa radicem mollem et tursum stipticum tota hac herba uulnera glutinat alienationes cessat in potu data uenena extinguit tertianas et quartianas curat. huius sucus in colliriis mittitur *W*

cf. Diosc. III.48–50 (πάνακες Ἡράκλειον et al.). Pliny 12.127, 25.30–3, 26.113. Galen *Simp.* 8.16.4–6 (K.12, 94–5). Cael. *Acut.* 2.154, 3.42. *Ex herb. fem.* 63. Isidore 17.9.28.

221) Peucedanum

Peucedanum herba est thyrsum et ramos habens similiter ut foeniculum, radicem uastam et nigram, odore grauem, succi plenam et circa illam comulam densam, florem aurosum. Haec igitur radix aliquando propter succum suum exprimitur qui succus eius melior est si sit colore
5 subrufus et ad gustum excalfacit, aliquando siccatur et reponitur. Facit autem ad uulnera oletana et expurgat uiscide, propter quod in aliquibus emplastris et malagmatibus mittitur et radix eius incensa serpentes et uiperas fugat.

1 Peucedanum] peucidanum *F* similiter] similes *J* 3 illam] *om. BC* aurosum] aureo colore *J* 5 subrufus] subrubeos *V* excalfacit] renitum facit *V* aliquando- reponitur] *om. V* 6 oletana] antiquata *J*

deest W

cf. Diosc. III.78 (πευκέδανον). Pliny 25.117–18. Celsus 5.18.29. Scrib. 56. Galen *Simp.* 8.16.17 (K.12, 99–100). Ps. Apul. 95.

219) Axe weed / Crown vetch (*Securigera securidaca* L.)

Axe weed is a plant which grows in grain fields and has many small and slender branches, tiny little leaves, and little pods in which red and extremely bitter seeds grow in bunches of three and four together. The seeds can vigorously contract with a degree of causticity, and are therefore used in antidotes, and they stimulate menstruation.

220) Panax / Sweet myrrh (*Opopanax* L.)

Panax is the root of the plant from which opopanax[1] is made, and has the same properties as opopanax but is less effective.

1 #201.

221) Hog's fennel / Sulphurwort (*Peucedanum officinale* L.)

Hog's fennel is a plant that has a stalk and branches like fennel,[1] and a large, black root that is full of juice and has a heavy fragrance. Around the root it bears dense foliage and a golden flower. The root is sometimes squeezed for its juice, which is best when it shows a slightly red colour and heats the mouth when tasted; and sometimes the root is dried and preserved. It is good for foul-smelling wounds and is a powerful cleansing agent. It is therefore used in particular plasters and emollients, and its root when burned wards off snakes and vipers.

1 #110.

222) Pastinaca

Pastinaca omnibus nota est. Nam sumitur in cibo et in reliqua uiridia. Vires habet excalfactorias et diureticas.

1 omnibus] herba breuior est omnibus *W* omnibus-uiridia] *om. P* in cibo- uiridia] *om. W* reliqua] aliqua *V* uiridia] uirtutibus *J* **2** excalfactorias] acres et excalfactorias *JL*

cf. Diosc. II.113 (σίσαρον). Pliny 20.29–32. Galen *Simp.* 8.18.19 (K.12, 124). Garg. *Med.* 33. Ps. Apul. 81. *Dyn Vat* 35/*SGall* 39. Isidore 17.10.6.

223) Petroselinum

Petroselinum nascitur circa petras et est optimum Macedonicum, enim est odoratu subdulce, diureticum ualde et excalfactorium. Vnde et stomachi et colli inflationes et tortiones remediat.

1 Petroselinum] petrosilino *F* macedonicum] *om. C* **2** odoratu] odoriferum *P* **3** colli] gulae *F* tortiones] rosiones *L* tensiones *J* remediat] emendat *P*

deest W

cf. Diosc. III. 66 (πετροσέλινον). Pliny 20.118. Celsus 4.21.2. Scrib. 106. Ps. Apul. 128. *Dyn Vat* 44. Isidore 17.11.2. Galen *Simp.* 8.16.16 (K.12, 99).

224) Peristereon

Peristereon herba est [quae] folia habet modica ueluti incisa, radicem mollem, thyrsulum stypticum. Tota herba uulnera glutinat et alienationes stringit et ad omnia uenena potui data optime facit. Simul ad tertianas et quartanas disposita est. Huius succus in collyriis mittitur. Haec autem et columbina et uerbena dicitur.

1 Peristereon] peristereon uel uerbena *J* est] est columbaria *J* ueluti] quasi *J* **3** stringit] extinguit *BP* uenena] *om. V* simul] similiter *JL* **4** disposita est] curat *BPC* collyris] gliria *F*

deest W

cf. Diosc. IV 59 (περιστέριον), IV.60 (ἱερὰ βοτάνη). Pliny 25.105, 25.125–6. Galen *Simp.* 8.16.14 (K.12, 98). Ps. Apul. 66 (et columbina), Cael. *Acut.* 3.160. Marc. *Med.* 8.28 (1.46 uerbena). *Med. Plin.* 1.18 (uerbena) *Ex herb. fem.* 54. Alex. Tral. 1.6 (1.1 uerbena). Isidore 17.9.55 (uerbena).

225) Polypodium

Polypodium herba est similis filici, quae in petris et circa radices arborum uetustarum et in querceis arboribus crescit, haec melior habetur.

222) Parsnip (*Pastinaca sativa* L.)

Parsnip is known to everyone.[1] It is consumed by mixing with food and with left-over greens. It has heating and diuretic properties.

1 On possible alternative identifications, see Andrews 1958c; André 1985, 190; and below #273.

223) Parsley (*Petroselinon hortense* Hoffm.)

Parsley[1] grows among rocks, and the best is Macedonian, which has a mildly sweet fragrance and is a powerful diuretic and a calefacient. It therefore alleviates swellings and cramps in the stomach and throat.

1 On the different identifications among ancient writers, see Andrews 1949a.

224) Vervain (*Lycopus europaeus* L.)

Vervain[1] is a plant that has moderate-sized leaves that are sort of jagged, a soft root, and an astringent stalk. The entire plant agglutinates wounds and contracts abnormal growths, and is an excellent antidote to all poisons that have been drunk. Likewise it is used to settle tertian and quartan fever. Its juice is placed in eye-salves. It also goes by the names of *columbina* and *uerbena*.[2]

1 Or possibly holy vervain (*Verbena officinalis* L.): the description contains aspects of the separate entries for both in Diosc. IV.59–60.
2 *columbina*, a direct translation of Gr. περιστέριον, from περιστέρα/*columba*, pigeon or dove. Also recorded as an alternative name by Ps. Apul. 66, and for holy vervain, 3 (*herba uerbenaca*). The term *uerbena*, meaning simply 'foliage' or 'herbs,' often those associated with ancient religious rituals (laurel, olive, myrtle, etc.), is used by Celsus 2.22, 8.10.7, as a category for 'cooling' plants; in *Med. Plin.* 1.18, Alex. Tral. 1.1 and Isidore 17.9.55 it seems to refer to holy vervain: André 1985, 269.

225) Polypody (*Polypodium vulgare* L.)

Polypody is a plant that is similar to the fern, and grows among rocks, around the roots of old trees, and upon oak trees; this last is considered to

Cuius radix quasi pilosa et hirsuta uidetur, gustu styptica cum aliqua dulcedine. In foliis uero habet binos ordines quasi punctorum aureorum. Vires habet uentrem soluentes et bilem detrahentes.

1 Polypodium] pulipodium *V* polipodium *FLBP* petris] petris exit *BPC* 2 querceis arboribus] quercibus *PW* 3 quasi] *om. J* hirsuta] sitatula *VFLBPC om. W* 4 aureorum] arborum cubrum *V* arborum aureorum *F om. BPC* 5 bilem] *om. BPCW* detrahentes] distrahentes *J* extrahentes *BC om. W*

cf. Diosc. IV.186 (πολυπόδιον). Pliny 26.58, 26.105. Galen *Simp.* 8.16.34 (K.12, 107). Marc. *Med.* 30.44, 31.11. Ps. Apul. 84. Isidore 17.9.62.

226) Quolocynthis (Colocynthis)

Colocynthis similis est cucurbitae rotundae uel malogranato. Huius uero medulla, id est quod intus, habet carnosum et semen uiscide uentrem purgare potest et in plura cathartica mittitur.

1 Colocynthis] quoluquintidas *V* quoloquintida(s) *FP* quoioquintions *L* *qtum *B* quod aut coloquintida *C* similis] *om. W* rotundae- mittitur] nascitur maxime in aegypto gustu amarissima haec igitur amaritudine sua et acrimonio totum corpus mouet et uentrem purgat unde catarticis adhibetur *W* 2 carnosum] et quod quasi carnosum *F* 3 plura] gyra *V* hiera *C* gera *FB* ieras *L* uera *P*

cf. Diosc. IV.176 (κολόκυνθα ἀγρία). Pliny 19.70, 20.14. Marc. *Med.* 20.13, 30.9. *Ex herb. fem.* 46. Isidore 17.9.32. Galen *Simp.* 7.10.38 (K.12, 34).

227) Ruta

Ruta herba est omnibus nota, utilis ualde. Haec enim stomachum erigit et tussientibus prodest et colli dolorem sedat. Item urinam et menstrua prouocat, partum quoque si plus sumatur expellit.

1 nota] *om. P* herba- enim] *om. W* 2 item- prouocat] *om. W* 3 plus] multum *P* plurimum *W*

cf. Diosc III.45 (πήγανον). Pliny 19.156, 20.131–5. Galen *Simp.* 7.16.18 (K.12, 100–1). Garg. *Med.* 3. Marc. *Med.* 20.64. Ps. Apul. 90 (116 siluatica, 126 erifion). *DynVat* 42/ *SGall* 46. Alex Tral. 1.107. Isidore 17.11.8.

228) Resina

Resina est quasi lachryma arboris et locis quam pluribus colligitur. Verum sicca quae est inuenitur circa nuclipinum et pityn. Alia liquida circa terebinthum et peucem et cypressum et mastichem quam [lenti- scum aliqui esse dicunt et] in Chio insula colligitur. Aliae prout diximus
5 pluribus locis inueniuntur. Communiter enim omnis resina laudatur si sit munda, perlucida et si odorem suum pristinum reseruat. Specialiter

be the best type. Its root is hairy, so that it looks a little shaggy,[1] and has a styptic taste with a degree of sweetness to it. On its leaves it has two rows of what look like gold puncture-marks. Polypody contains properties that loosen the bowels and reduce bile.

1 *hirsuta*: following the conjecture of *J*. All MSS read *sitatula* (?). Cf. Diosc. IV.186 'thick, having tentacles like an octopus.'

226) Colocynth / Bitter Apple (*Citrullus colocynthis*, Schrader)

Colocynth[1] looks like a round gourd[2] or a pomegranate.[3] Its marrow, or the inner part, is fleshy. Its seed can vigorously purge the bowels and is used in many purgatives.

1 Rendered with initial 'q[u]' in all manuscripts except *W*, hence its position in the text. See also plate 8.
2 *cucurbita*: see #54, where *cucurbita syluatica* is an alternative name for colocynth, doubtless modelled on the Greek κολόκυνθα ἀγρία (e.g., Diosc. IV.176).
3 #183.

227) Rue (*Ruta graveolens* L.; *R. halepensis* L.)

Rue is a well-known plant that is very useful. It fortifies the stomach, helps cure coughs, and eases pain in the neck. It also stimulates urination and menstruation, and if eaten in large quantities it expels the foetus.

228) Resin

Resin is mostly the sap of a tree and is harvested in many different places. Dry resin is derived from stone pine[1] and from *pitys* pine.[2] Other resins are those of liquid form which are derived from terebinth, the *peuce* pine,[2] the cypress, and mastic tree, [which some call *lentiscus*],[3] and which is harvested on the island of Chios. The other types we mentioned are found in many places. In general all resins are recommended when they are pure,

tamen liquidarum optima est terebinthina. Haec enim et candorem et munditiem et eo quod perlucet et uitro assimilis esse debet. Postquam schinine recipitur, colore non dissimili ponticae cerae. Item siccarum
10 receptissima est pityina, postquam numeratur peucena et nuclipinea, quam graece strobilinam dicimus. Verum omnis resina efficaciter relaxare et remollire potest et in usu aliquando crudae accipiuntur, aliquando friguntur.

1 arboris] herbae *J* colligitur] nascitur *V* 2 nuclipinum] nuglupinio *V* nucleos pini *BC* cluneos pini *P* pini nucleum *W* pityn] pece *V* pici *F* pytina *L* pituin *BC* pituina *P* alia] *om. VL* aliquid *F* 3 peucem] peucine *V* mastichem] masticem *VLBC* mastichinam *J* stacten *W* lentiscum- dicunt] *om. VFLBCW* cirenen dicunt *P* 4 chio] istintine et cio *V* scyninem *L* in cirinen et chio *BC* aliae] aliae ille *VLC* aliae- inuenitur] *om. PW* prout] *om. V* ille ut *F* 6 suum pristinum] sucus prestinare *F* reseruat] seruet *J* seruat *P* 7 liquidarum] aliqua eorum aliqua harum *BPC* aliqua earum *FL om.W* terebinthina] terouentina *V* teruentina *BC* haec] habet *J* candorem] cum odore *F om.* W 8 eo quod] *om. J* quod in ipsa lucet *BC* postquam] post *J* 9 schinine] scinen *V* scinum *F* schyninem *L* scini *W* in se *BPC* mastichina *J* recipitur- friguntur] laudabilem *ce* manifestum est *W* non dissimili] similis *V* ponticae] ponticalem *F* 10 pityina] pitia *V* pithya *F* pituina *P* numeratur] nominatur *V* 11 strobilinam] struuolinas *V* tropilanan *F* id est robolina *L* 12 in usu] in uaso *F* aliquando· friguntur] *om. F*

multum om. W

cf. Diosc. I.71 (τέρμινθος). Pliny. 13.54, 14.122–4, 14.127, 16.38, 16.54, 16.58, 24.27–8, 24.34. Galen *Simp.* 8.19.1 (K.12, 137–8). Garg. *Med.* 58 (pinea). *DynSGall.* 88 (nucleus pinorum). Isidore 17.7.1.

229) Rosmarinus

Rosmarinus herba est fructiculus et semen unum cachrys adsimilis spondylo, quae herba radicem et folia habet quasi foeniculum, sed asperiora et exalbidiora et crassiora et ueluti thyrsum oblongum et ramosum. Huius igitur semen quod est breuius et uastius, cachrys dici-
5 tur quod inde tenerius et oblongum. Rosmarinus uerum tamen efficaciter stringere et relaxare potest. Vnde et urinam mouet et antidotis et malagmatis admiscetur et quidem cachry acrius et uiscidius.

1 herba- fructiculus] quaedam arbuscula *W* unum] dicta *J* cachrys] acris *F* cacreos dicitur *W* adsimils- tamen] *om.* W 2 spondylo] herbae *P* 3 crassiora] grossiora *VFP* oblongum] longum *BC* 4 cachrys] cacreus *F* cacreos *P* 5 tenerius et oblongum] interius et oblongius *V* uerum tamen] *om. BC* 6 relaxare] excalfacere *LBC* urinam] omnia *F* 7 malagmatis] contra maligna *J* et malignis *F* et quidem- uiscidius] *om.* W cachry acrius] corius *F*

cf. Diosc. III.75 (λιβανωτίς). Pliny 19.187, 24.99. Galen *Simp.* 7.11.14 (K.12, 60–1). Isidore 17.9.81.

translucent, and retain their initial fragrance. In particular the best liquid resin is that from the terebinth tree, and this should be bright in colour and pure to the point of transparency, so that it resembles glass. Next after this ranks *schinine*,[4] when it has a colour not unlike Pontic wax. The most popular of the dry resins is that from the *pitys* pine, and after this ranks those of the *peuce* pine and the stone pine, which we call in Greek *strobilina*.[1] All resins can thoroughly loosen and mollify, and are sometimes used when raw, other times after they have been roasted.

1 *nuclipinum* (*VF*): a contraction of *nucleus pineus*, pine nut or pine cone. Its use here, otherwise unattested in Latin, to denote stone pine (*Pinus pinea* L., or Swiss or arolla pine, *Pinus cembra* L.) seems to follow the similar development in Greek which we find in Diosc. (I.68.5, I.69.2, 4, I.71), who likewise used the word στρόβιλος, 'pine cone' (e.g., Theophrastus, *HP* 3.9.1) to mean stone pine. The term *strobilina* (resin) is otherwise unattested in Greek or Latin.

2 *Pitys* can refer to several different types, most commonly Corsican pine (*Pinus laricio* Poir.), but also Aleppo pine (*Pinus halepensis* Miller) and even stone pine (*Pinus pinea* L.). *Peuce* is another name for Corsican pine. Beck 2005, 50 n.91, notes the 'hopelessly confusing' use of names for pines in Diosc. and Greek literature in general, and leaves the Greek terms without translating: likewise here for *pitys* and *peuce*.

3 *lentiscus*: unattested in the MSS, it is probably a later interpolation culled from its use in #151.

4 Diosc. I.70 (σχῖνος, mastic) mentions that the resin is called σχινίνη, or alternatively μαστίχη.

229) Frankincense rosemary (*Lecokia cretica* Lam.; ~ *Cachyris libanotis* sp. Koch; ~ *Rosemarinus officinalis* L.)

Frankincense rosemary[1] is a plant the size of a little shrub and bears seed inside a capsule that looks like a vertebra,[2] while the root and leaves of the plant look a little like fennel,[3] but the frankincense rosemary itself is more coarse, more pale, more dense, and has a lengthy stem that bears many branches. Its seed is short and fat, and the part that is called its capsule is quite frail and oblong in shape. Frankincense rosemary can both thoroughly contract and loosen. It therefore induces urination, and is mixed into many antidotes and emollients. The capsule itself is quite severe and powerful in its effect.

1 The mention of *cachrys* confirms the identification of frankincense rosemary (*Lecokia cretica* Lam., or *Cachyris libanotis* sp. Koch) rather than rosemary (*Rosmarinus officinalis* L.). Both Theophrastus, *HP*.9.11.10, and Diosc. III.74, distinguish two types of frankincense rosemary (λιβανωτίς), one without fruit (ἄκαρπος), and another bearing fruit (κάρπιμος) which is called κάχρυ. Likewise Pliny 24.99 and elsewhere (19.187) he notes that *libanotis* was 'called rosmarinum by some.' Diosc. III.75 gives a separate entry for the λιβανωτίς

230) Rubus

Rubum omnes nouerunt. Huius igitur et folia et flos et mora leuiter stringunt. Vnde ad uuluam et fauces proprie euidenter faciunt, quamuis omne profluuium idonee stringant.

1 omnes- igitur] *om. P* leuiter] leniter *FBCW* 2 stringunt] exstringunt *VBC* uuluam] uuam *VFLBPCW* proprie] *om. J* euidenter] proprietatem uidentes *F* uidetur *BPCW om. L* faciunt] *om. L* 3 indonee stringant] ualiter dantur *W*

cf. Diosc. IV.37 (βάτος). Pliny 24.117. Galen *Simp.* 6.2.4 (K.11, 848–9).

231) Rhu

Rhu, quod ab aliquibus radix dicitur, aliquatenus uasta et nigrastra, fragilis, leuissime, odore et gustu amara et leniter stringens et si fricetur et masticetur fit ueluti glutinosa et croceato colore. Vnde optimum rhu est ponticum. Vires autem habet efficaciter stringentes, propter quod cum uino laeuigatum et quasi fomentum impositum, inflationem siccat et uulnera glutinat et sanguinis abundantiam reprimit et mittitur in antidotis.

1 Rhu] reu *VFP* rheu *L* reum *B* reu ponticum *W (idem Diosc. Lat. III.2) def. C* quod] quod aliqui radicola dicunt *VL* dicitur] est *LBC om. WP* aliquatenus] aliqualiter *L* aliquatenus omnibus nota *BC* aspera *P* nigrastra *VFL*] nigra aspera *P* nigra astra *BC* nigra *W* nigricans *J* 2 fragilis] fragilis necnon *BC* odore] sine odore *FLW* gustu- ponticum] *om. W* fricetur] frangitur *F* 4 propter- impositum] *om. W* 5 fomentum] fromento *V* fruentum *F* fumotium *L* frumentum *P* 6 reprimit- mittitur] praemittitur *F* sistit *W* in antidotis] *om. W*

cf. Diosc. III.2 (ῥᾶ). Pliny 27.128–30. Galen *Simp.* 8.17.9 (K.12, 115–16). Orib. *Eup.* 2 1R.2. Marc. *Med.* 22.31. Cael. *Chron.* 4.72. Isidore 17.9.40 (reubarbara).

'which the Romans call rosemary [ρουσμαρῖνον]' (though he says the same for betony, IV.1, below #290), probably *Rosmarinus officinalis* L.

2 Likewise Diosc. III.74, 'ἐοικὼς σφονδυλίῳ, but σφονδύλιον can also mean hogweed (*Heracleum sphondylium* L.: below #264), which is the meaning Diosc. employs only a few lines down (σπέρμα ... μέλαν ὡς σφονδύλιον) when discussing λιβανωτὶς ἄκαρπος (note above), and which he describes in full (III.76, σφονδύλιον) immediately after the entries on λιβανωτίς. It can also mean the flesh or muscle of an oyster: Pliny 32.154, *spondylus*: cf. 12.128, hogweed.

3 #110.

230) Bramble (*Rubus ulmifolius* Schott)

Everyone knows bramble.[1] Its leaves, flower, and berries are gently astringent. It is therefore rightly used for sores in the vulva and in the pharynx, and it is equally suitable for staunching all types of haemorrhages.

1 *Rubus* can also mean the blackberry (*Rubus fructicosus* L.), as suggested by Halleux-Opsomer 1982a, 95, but blackberry is already described in #195 (*mora sylvatica*). The term 'bramble' can be used for several different fruit-bearing species of the genus *Rubus*, used here to avoid certainty, as it might for most ancient authors on these berries: see André 1985, 83. Beck 2005, 264, identifies Diosc. IV.37 (βάτος) as *Rubus ulmifolius* L., elmleaf blackberry, and translates as 'bramble.' Cf. Pliny 24.117.

231) Rhubarb (*Rheum ribes* L.)

Rhubarb, which some call 'root,'[1] is rather large and somewhat black, delicate and very smooth, with a bitter fragrance and taste that is mildly astringent. When scoured and chewed it becomes sticky and turns saffron-coloured. The best rhubarb is Pontic. Rhubarb contains astringent properties, hence is crushed with wine and applied in the form of a compress to drain swellings, agglutinate wounds, and prevent discharges of blood. It is also used in antidotes.

1 Or 'little root' (*radicola VL*). Pliny refers several times to a plant called *radicula* which 'struthion vocant Graeci' (19.48; also 24.96, 25.52, 29.39), that is, soapwort, #246, but it is more likely weld (*Reseda luteola* L.). Cf. Diosc. III.2, 'but some call it rheon.' André 1985, 217. Cf. below #234, radish, 'also known as "root,"' and #269.

232) Rosa

Rosam omnes nouerunt, quae et ipsa suauiter stringit et refrigerat.

1 omnes nouerunt] *om. P* suauiter] *om. F* refrigerat] uires refrigerantes et so oleo admis-
cceatur et in puteum per quinqeginta dies ut maneat dimittitur nimium febris aufert
ardorem et si mastix adhibeatur intrinsecus reddit frigdorem et febrem eicit *W*

cf. Diosc. I.99 (ῥόδα). Pliny 21.16–20. Isidore 17.9.17. Galen *Simp*. 8.17.5 (K.12, 114).

233) Rapa et napus

Rapa omnibus nota est. Huius radix comesta inflationes facit et col-
lectum carnis, sed inutilem nutrit humorem. Aqua uero in qua fuerit
cocta, podagricis et ad pernionem optimum fomentum est. Sed et ipsa
contrita et imposita idem operatur. Si excauetur et accipiat oleum rosa-
5 ceum et cerae modicum et sic in cinere calido coquatur ita ut cerotum
faciat, uulnerosos perniones et curat. Est etiam rapae similis napus qui
similiter inflat, sed semen huius napi contra uenena potui cum uino
optimo datur, uel ex mulsa contritum.

1 omnibus- est] *om. P* comesta] comixta *V* collectum] multum *J* ocitum *L* ulcum *BC*
ulcera *P* cultum *W* 2 sed] sedat *LBPCW* inutilem] utilem *FBPC* humorem] *om. W* 3
cocta] *om. F* pernionem] praegnationem *J* praegniones *F* pregnionem *L* perinionibus
BP perioionibus *C* 4 excauetur] excoquatur *BC* coquitur *P* 5 sic] si *FLBC* sic in cera
V cerotum] crustum *F* 6 faciat- contritum] fiat *W* perniones] promouet *J* (c)eniotis
V pregnione(s) *FL* periniones *PC* curat] sedat *BPC* 7 napi] huius *V* potui] potatum
BPC

cf. Diosc. II. 110 (γογγύλη), II.111 (βουνιάς). Pliny 18.131, 19.77, 20.21. Garg. *Med*. 34,
35. *DynVat*. 36/*SGall* 40, 40b. Isidore 17.10.7–8.

234) Raphanus

Raphanus herba omnibus nota est quae radix appellatur. Radicem
habet similem rapae sed longiorem. Quae comesta a ieiunio prout antidotum, magnum est adiutorium sanitatis et tutamentum corporis, sed
inflationem operatur, ructus foetidos facit et urinam mouet, uentrem
5 malaxat. Decocta autem in cibo sumpta phthisicis prodest et thora-
cem a phlegmate laborantem emendat. Cortex uero eius cum aceto et
melle acceptus, uomitus mouet, hydropicis et spleniticis prodest. Si
in modum cataplasmatis imponatur cum melle commista putredines
curat et liuores tollit et contra uiperarum morsus prodest. Alopecias
10 capitis iuuat, fungos malos aut in potu aut in cibo sumpta compescit.
Semen uero eius cum aceto datum splenem minuit. Si in aceto et melle

232) Rose (*Rosa* sp. L.)

Everyone knows the rose, which is a gentle astringent and cools.

233) Turnip (*Brassica rapa* L.) and Rapeseed (*Brassica napus* L.)

The turnip is well known. Its root when eaten causes flatulence, makes the flesh flabby, and nourishes harmful humour. The water in which it is cooked makes an excellent compress for gout and chilblain, or the root when crushed and applied topically achieves the same results. If it is hollowed out and filled with rose-oil and a little wax and placed on hot ashes to cook, it makes a wax-salve that heals ulcerated chilblains. The rapeseed[1] is similar to the turnip, and likewise causes flatulence, but the seed itself of rapeseed is excellent against poisons when drunk with wine, or if crushed and mixed with honeyed wine.

1 Cf. Diosc. II.110–11, and ch. 3.B.

234) Radish (*Raphanus sativus* L.)

Radish[1] is a well-known plant, which is also called 'root.' Its root is in fact similar to the turnip, but longer. When eaten after a period of fasting, like an antidote, it is an excellent aid for restoring health and protecting the body, though it also causes flatulence, odiferous belching, urination, and loosens the bowels. When stewed and consumed with food it alleviates consumption, and helps to clear the chest of excess phlegm. Its peel taken with vinegar and honey is an emetic, and helps with dropsy and spleen disease. If mixed with honey and applied in a poultice it heals putrefaction and clears bruises, and is good against viper bites. It helps with baldness of the head, and when drunk or eaten with food it counteracts the harmful effects caused by consuming bad mushrooms whether eaten or drunk. Its

radix ipsa coquatur et sic gargarizetur synanchen curat. Folia eius tunsa et imposita omnibus articulis dolentibus prodest.

1 Raphanus] rafanus *FPW* herba- quae] radix ieiunio *W* herba] *om. JFW* omnibus nota] *om. P* 2 longiorem] *om. VF* comesta] comixta *V* 3 tutamentum] et aumentum est *V* tota metentum est *L* augmentum *B* totum argumentum est *C* 4 ructus foetidos] ruptus fetitus *V* ructuationes fastidios *F* fructus eius *W* 5 malaxat] relaxat *JBC* laxat *P* sumpta] sumatur *V* phthisicis] siccis *V* ptisici *F* tyssicis *LC* ptisicis *B* 6 cortex- eius] *om. F* 7 acceptus-spleniticis] *om. V* prodest si] *om. BPC* 8 commista] mixtus *V* commista etiam prodest *J* 10 capitis iuuat] berteis *V* bisteis *F* uestis *L* habentes *BC* 11 minuit] mouit *V* 12 folia-prodest] *om. W* 13 dolentibus] *om. F*

multum breviatum W

cf. Diosc. II.112 (ῥαφανίς). Pliny 20.13. Garg. *Med.* 1. *DynVat* II.4/*SGall* 59. Isidore 17.10.10. Galen *Simp*. 8.17.2 (K.12, 111–12).

235) Sarcocolla

Sarcocolla est, prout aliqui aiunt, lachryma arboris quae in Perside nascitur. Aliqui fieri dicunt ex suffimentis quae similiter resinae per parietes laeuigando. Alii enim sarcocollam agrimoniam herbam uocant. Quam optimam iudicamus quae est subalbida, fragilis, sine odore, flo-
5   re subrufo, foliis quasi rotundis, radice longa quarum sunt infirmiores, quae subrubicant et ad gustum commista dulcedine aliquam habent amaritudinem et dentibus pressae agglutinantur. Potest igitur stringere lachrymam et quodammodo hilarem uisum facere. Adulteratur autem gummi et resina sed tactu facile deprehenditur.

1 Sarcocolla] sarcocola *F* 2 fieri] igitur ea ex suffimenti *V om. BC* alii quidem *J* suffimentis] in suffusio mentis *F* subfigmentis *L* figmentis *C* compositionem *W* quae] aquae *J* 3 parietes] partes *FLBPCW* agrimoniam] argeni *V* agrimonia *FPW* argimonia *LBC* argemonem *J* 4 sine- agglutinatur] *om. W* 5 foliis] foliola *LP* rotundis] subrotunda *V* rotundula *LPB* 6 gustum- aliquam] cedinam cum aliqua *C* commista] commita *V* mista *F* commentasa *B* pressa *V* prestat *BC* pressa *L* 8 hilarem uisum] iularem uim *L om. W* 9 facile] aliquid *V*

cf. Diosc. III.85 (σαρκοκόλλα) Pliny 13.67, 24.72, 24.128. Galen *Simp*. 8.18.4 (K.12, 118). Ps. Apul. 31.27. Marc. *Med.* 8.163.

236) Spica nardi

Spica nardi ad similitudinem cognominata est. Similis est enim spicae frumenti et haec plurimas habet uelut aristas et quasi setarum uel

seed when given with vinegar reduces the spleen. The root itself is cooked in vinegar and honey and used as a gargle to cure sore throats. The leaves when pulped and applied topically relieve aching joints.

1 Halleux-Opsomer 1982a, 95, identifies this as horseradish (*Armoraica rusticana* Gaertn., et al.), but the term usually means radish (André 1985, 218), an identification confirmed by the similarities to the entry for such in Diosc. II.112 (ῥαφανίς).

235) Sarcocolla (*Astragalus fasciculifolius* sp. Boiss.)

Sarcocolla, according to some, is the sap from a tree that grows in Persia. Others say it is a resin-like substance that is scraped from walls where incense has been burnt. Yet others call the plant agrimony[1] 'sarcocolla.' We consider the best type[2] to be whitish, delicate, without fragrance, producing a reddish flower, leaves that are almost round, and long roots, the least effective of which are those that are turning slightly red and have a bitter taste but with a fair degree of sweetness mixed in, and which make your teeth stick together. It can staunch the flow of tears, and can help put people in a happy mood. It is adulterated with gum[3] and resin,[4] but this is easy to detect by touch.

1 The confusion among manuscript readings (agrimonia/argimonia, agrimony/windrose) reflects the same confusion among authors recorded above for #13 (and n.1), though here it seems resolved by Ps. Apul. 31, who notes that sarcocolla is an alternative name for agrimonia (common agrimony, *Agrimonia eupatoria* L., not in *AG*), and subsequently Marc. *Med.* 8.163, though using *argimonia*: see André 1985, 227.
2 The following description of the plant – unique among ancient and medieval sources, which focus exclusively upon its sap – may be corrupt, given the confusion of *agrimonia/argemone* (see note above). The description holds up well enough, but oddly does not mention its primary renown as an agent for agglutinating wounds, whence its name (σάρξ, flesh, κόλλα, glue): see Innes Miller 1969, 101. Cf. Diosc. III.85, and Pliny 24.72, and 13.20 ('most useful to doctors and painters').
3 #47.
4 #228.

236) Spikenard (*Nardostachys jatamansi* DC.; *Patrinia scabiosifolia* Fisch.)

Spikenard is so-named because of its appearance, for it is similar to an ear

capillorum fasciculus est ad modum patens. Nascitur enim in India ex
una minima radicula sine foliis. Dicitur tamen et Syriaca non quia in
5 Syria legitur sed quia montis illis in India in qua plurima inuenitur pars
ad Syriam expectat. Verum optimam dicimus quae est odoratissima et
colore subrufa, gustu subamara, recens et ad tactum non grauis, nec
uasta sed magis densa. Haec ergo uires habet suauiter stypticas et lente
calfactorias.

1 Spica] spiga *V* 2 plurimas] similis *V* 3 est] *om. J* patens] patet *J* patentis *BC* paruis-
simus *W om. P* 4 minima] *om. J* minuta *F* parua *W* 5 plurima] abundantia *W* 6 expectat]
aspectat *J* spectat *PW* odoratissima] odorifera *W* 7 gustu subamara] *om. J* ad- densa] *om.*
W 8 lente] leniter *JP*

cf. Diosc. I.7–9 (νάρδος, κελτικὴ –, ὀρεινὴ –). Pliny 12.42–3. Celsus 3.21.8. Scrib. 120.
Isidore 17.9.3 (nardus). Galen *Simp.* 8.13.1–3 (K.12, 84–5).

237) Syricum

Syricum fit de cerussa plurimum siccata et cribrata. Coquitur autem
in patella fictili super carbones usquequo sandaricae habeat colorem,
quod et ipsum cerussae uires repraesentat. Est optimum quod ignis
habet colorem.

1 Syricum] siricum *V* sirico *W* cribrata] criuellantur *V* 2 sandaricae] sandycis *J* sandarace
V sandarice *FL* candescat *BPC* ui candescat *W* habeat] *om. BPC*

cf. Pliny 33.120, 35.40. Isidore 19.17.5.

238) Saliunca

Saliunca herbulae genus est quae radiculas et folia subrufa, floribus
similibus uiolae quam plurima habet quae tota prodest, haec a terra
refossa in modum scapularum stringitur. Nascitur autem in alpibus
summis et in Norica regione, quam ipsam optimam iudicamus et odo-
5 ratissimam cui tamen et haec esse debet recens, et sine sabulo sit, et
quam plurimas cum ramulis suis radices habeat. Illae autem sunt uiscidissimae. Potestas igitur saluiculae est quae et spicae et si infirmior.

1 Saliunca] saluicula *J* saliuncola *V* salincincula *L* saliunca *FP* saliuncula *W* herbulae]
herba *V* subrufa] sub floribus *J* 2 quam- modum] *om. W* a terra] haec *J* aurea *L* 3 scapu-

(*spica*) of grain in having plenty of downy beard which forms a bundle of bristly hairs that are quite visible from the outside. It grows in India from a singular, small root and has no foliage. That which is called Syrian is not named so because it is from Syria, but rather because it is mostly found upon the side of mountains in India which face Syria.[1] We say the best is that which is the most fragrant, reddish in colour, is fresh, has a slightly bitter taste, is light in weight, and which is thick rather than large. Spikenard has gently styptic and slowly warming properties.

1 Likewise Diosc. I.7, though otherwise his description differs considerably. Cf. Pliny, 12.42.

237) Syrian (minium)

Syrian (minium)[1] is made from cerussite[2] that has been well dried and sifted. It is then cooked on an earthenware plate over coals until it turns the colour of sulphide of arsenic,[3] and it has the same properties as cerussite. The best is that which is the colour of fire.

1 Unique description. Pliny (33.120, 35.40) records *syricum* as a type of colouring made from mixing Sinopic red ochre (*sinopis*, #276) and sulphide of arsenic (*sandyx*, cognate with *sandaracha*, #242) and used to adulterate minium (cinnabar or red-lead: see #58). Unknown to Dioscorides, the only other writer to mention it is Isidore (19.17.5), who records it as a red pigment 'with which the chapter-headings in books are written,' and who adds that it was also known as 'Phoenecian' because it was collected on the shores of the Red Sea. Halleux-Opsomer 1982a, 95, identifies it as lead oxide, but if made from cerussite as claimed it was more likely a lead carbonate.
2 #46.
3 #242. *sandyx*, J.

238) Celtic spikenard (*Valeriana celtica* L.)

Celtic spikenard is a species of shrub which has small roots, reddish leaves, and many flowers similar to the violet when it is in full bloom, which is when it is dug out from the ground and bound by its shoulders, as it were.[1] It grows on the summits of the Alps and in the region of Noricum,[2] and we judge this the best, for it is the most fragrant and should be fresh, without any gravel attached, and it should have many roots with little branches sprouting from them: these roots are extremely potent. The strength of Celtic spikenard is the same as that of spikenard,[3] though it is less reliable.

1 Difficult passage, the translation is tentative. Diosc. I.8 says it has a quince-yellow flower

larum] stabularum *J* spicula rarum *F* **4** norica] orica *W* odoratissima] odorifera *W* **5** esse
debet] adesse oportet *J* **7** potestas] potest *BPC* si infirmior] sic esse infirmior *BC* si est
infirmior *VL* si sit infirmior *P* tantum inde ualet quantum et spica ita enim apparet infir-
mat *W*

cf. Diosc. I.8 (κελτικὴ νάρδος). Scrib. 153, 166. Pliny 21.43, 21.144. Galen *Simp.* 8.13.2
(K.12, 85). Marc. *Med.* 8.15.

239) Sal

Salem omnes nouerunt. Colligitur autem in quibusdam littoribus
et partibus maris, aut foditur aut paratur, prout aliquibus in locis in
Gallia, maxime tamen in Italia uel in diuersis prouinciis. Verum est
optimus sal qui ex mare fit mundus et candidus, cuius uirtus sapore
5 ipso probatur, potens comedere, siccare et purgare. Multis itaque rebus
misceri praecipitur. Cuius salis uirtus omnibus cibis admistus conuenit
et orexim facit, fastidium auertit, nauseam prohibet. Qui autem foditur
graece alasorexaton dicitur est uelut uena alba et perlucida. Venas habet
prout schiston quem et nos salem amoniacum diximus.

1 omnes nouerunt] *om. P* **2** partibus] *om. P* horas *W* aut foditur aut paratur] *om. B*
aut foditur- prouinciis] *om. W* aliquibus] aliquod *V* **4** cuius- facit] *om. W* **5** probatur]
prouocatur *V* potens] posse *LBC* possit *V* siccare] *om. J* itaque] quidem *J* **7** orexim]
orixin *VF* oresin *L* ad orexin *BP* ad orecin *C* auertit] atque *BC* **8** graece- dicitur] *om. JP*
lasor et tunditur *F* sal frictus dicitur quod etenim foditur *W* alasorexaton] αλλασωριχεων
L αλασφριχθων *BC* uenas] henas *V* uini *F* uenas- schistos] uerum est aliud *BPC* **9** amo-
niacum] armeniacum *J*

cf. Diosc. V.109 (ἄλς). Pliny 31.73–92. Galen *Simp.* 6.2.5 (K.12, 372–4). Isidore 16.2.3–6.

240) Succus

Succus tritici duplex est uel triplex uirtute. Habetur easdem etsi inertius praestare quae et ouum diximus facere, uicem enim cibi et potus
praestat. De hordeo ptisanas faciunt quae frigidam et humectam habent
uirtutem et relaxare possunt. Si oleum rosaceum admisceas febricitan-
5 tes refrigerat et reparat. De hordeo immaturo succus, qui graece alphiton, uentrem stringit et infrigidat et uires acquirit et alicae uel orizae
succus similiter operatur. Siligo enim tritici uentrem stringit et ardorem et corruptionem facit.

1 Succus tritici] succi uirtus *J* sucum succi *F* succi aritici *BC* succus triticis *W* sugum suci

and that it is pulled up in 'bundles that are as big as a handful.' Pliny 21.43 describes the plant as 'more a grass than a flower, matted as though squeezed by hand' and as a 'unique kind of turf.'

2 A region that covers part of today's Austria and Slovenia. Cf. Pliny 21.43, 'grows in Pannonia and in the sunny places among the Norican Alps, and in Ivrea'; Diosc. I.8, 'grows in the Alps around Liguria. The locals call it *saliunca*, but it grows also in Istria.' See ch. 1.F. For different forms *saliunc[ul]a*, André 1985, 224.

3 #236.

239) Salt

Everyone knows about salt. It is collected from particular shores and regions of the sea, or it is mined or processed, as is the case in certain places in Gaul, but especially in Italy and various other provinces. The best salt is that derived from the sea and which is pure, white, and whose potency can be tested by its very taste, and by the degree to which it is edible, can dry, and can cleanse. It is recommended for seasoning many things. The qualities of salt enhance all foods when it is mixed with them, it stimulates the appetite, alleviates squeamishness, and prevents nausea. The salt which is mined is called in Greek *alasorexaton*[1] and has the appearance of a white, transparent vein. Likewise the salt which we have called ammoniac salt has veins like *schiston*.[2]

1 *alasorexaton* (*V*): unattested elsewhere (nor in the variants in later manuscripts *LBC*), the name seems to derive from the words for 'salt' (ἅλς, -ός) and 'mined' (ορυκτος), or alternatively 'desire/appentency/appetite' (ὄρξις, as used in line above, + superlative *-aton*), hence a corruption of the notion that mined salt is 'the most desired': cf. Diosc. V.109, 'The most effective [ἐνεργέστατον] salt is that which is mined [τὸ ὀρυκτόν].'

2 A type of alum, see #7 n.1. Ammoniac salt was named because it was found under sand (ἄμμος), and was mined near the temple of Jupiter Ammon in Libya. Pliny, 31.79, likens its colour to *schiston*.

240) Juices (barley water, rose-oil, unripe barley, spelt, rice, wheaten-flour)

The juice from wheat has a double and even triple application, and is considered to possess the same properties, though a little weaker, as those which we said for the egg,[1] hence it is better to drink wheat juice than eat wheat. From barley they make barley-water (*ptisanas*),[2] which has a cool and moist property that can loosen. If you mix it with rose-oil, it cools down those with fevers and revives them. Juice from immature barley, which in Greek is called *alphiton*,[2] contracts the bowels, cools, and restores strength, and likewise the juice from spelt or rice does the same.

uentus *V* habetur] hic *J* habetur- praestare] potest facere *P* inertius] incertius *J* inhertibus *VL* certius *BC om. W* **2** praestare] praestare potest *LBCJ* cibi] cui *F* **3** ptisanas] hisanas *V* tysanes *L* humectam] humidam *JBP* **4** admisceas] admista sit *V* admisceatur *WP* **5** qui- alphiton] *om. J* alphiton] dfita *V* alfita *FW* δαynan *L* δαφιταν *B* δαφνταν *C om. P* **6** infrigidat- stringit] *om. VF* infrigidat- acquirit] *om. C* acquirit] tribuit *P* orizae] ut ordei *L* orodis *BPC* **7** siligo- facit] *om. W* siligo] sucus *BPC* ardorem] hudorem *V* **8** corruptionem] correptionem *V* facit] *om. J*

cf. Diosc. II.85 (πυροί), 86 (κριθή), 93 (τράγος), 95 (ὄρυζα). Pliny 18.63–71, 81–96 (triticum, 91–5 siligo), 18.72–5 (hordeum; 75 tisanae; 78–80 farina), 109–16 (alica), 18.71, 75 (oryza). *Dyn. Vat./S. Gall.* 4. Isidore 20.3.21.

241) Stibium

Stibium lapis est fulgens et splendens [nullo lapidi similis] qui in metallis inuenitur colore assimilis plumbo. Cuius est optimum quod late refulget, fragilissimum et quod in se nihil lapidosum uel terrenum continet. Vires autem habet stypicas, unde refrigerare et stringere et lachrymas uelut oppilare potest.

1 Stibium] stibeus *V* stibeos *FW* stiben *BPC om. L* est] enim *VJ* nullo lapidi similis] alia similis *def. V* et mollis similis *F* habet similis *BC def. L* nulli habet similiter *P* nulli lapidi silis *J om. W* **3** late] ueluti aes *BPC* latere *L om. W* quod] quod est *J* nihil lapidosum] mundum *J* **4** continet] non continens *J* refrigerare] *om. W* stringere] *om. VF* **5** oppilare] retinere *W*

cf. Diosc V.84 (στίβι). Pliny 33.101–2. Celsus 6.6.5–6, 12–13. Scrib. 23–4. Marc. *Med.* 8.4–6. Cass. Fel. 29.8–12. Alex. Tral. 1.96–7, 101.

242) Sandaracha

Sandaracha affertur maxime a Ponto et Cappadocia et Sicilia. Est a primo lapis igneo colore qui in minuta comminuitur. Est ergo optima sandaracha, quam florem appellamus, quae est roseo colore, munda, idem nihil exterum admistum habens. Effectu et ipsa fortiter stringens.

1 Sandaracha] sandaracae *V* sandaraca *FP* sicilia] cilicia *J* **2** in minuta] ita *BCP* **4** idem] id est *FLC* uel *J* ipsa] *om. P*

deest W

cf. Diosc. V.105 (σανδαράκη). Pliny 34.177, 35.39. Galen *Simp.* 9.3.26 (K.12, 235). Isidore 19.17.11.

Wheaten flour also contracts the bowels, though also causes heartburn and disease.

1 The opening sentence is very corrupt, the reconstruction here is tentative. Both aloe (#5) and egg (#204) were said to have 'double application' (*effectus duplex*), which seems to refer to being eaten or drunk for different purposes: see ch. 4.H. The readings of *inertibus* (*VL*) and *praestare potest* (*LBCJ*) suggest a possible meaning of 'properties that are good to administer to the weak.'

2 πτισάνη can also mean peeled barley or barley gruel, but here refers to barley water, as with Pliny (*tisanae*): ἄλφιτον is properly 'barley groats' (Diosc. II.86), though Hippocrates, *Epid.* 5.10, mentioned that a type of barley water (ἄλφιτα λεπτὰ ἐφ᾽ ὕδατι) was made from them. The initial 'd' among most manuscript readings (*dfita, daphitan*, etc.) probably derives from scribal corruption of 'ἀλ-', though some derivation from διαλφιτόω ('fill up with barley meal') is also possible.

241) Antimony

Antimony is a shining, glittering stone, [similar to no other], which is mined and has a colour similar to lead. The best is that which glitters brightly, is very delicate, and is free of any rubble or dirt. It contains styptic properties, and hence can cool and contract, as well as block tears.

242) Sulphide of arsenic

Sulphide of arsenic is imported mainly from Pontus, Cappadocia, and Sicily. The stone from which it is derived is a fiery colour, and this is then crumbled into small pieces. The best sulphide of arsenic is that which we call 'flower,'[1] which is rose-coloured and pure, having absolutely no foreign matter in it. It is used as a powerful astringent.

1 Otherwise unattested, and the entire description above is unique: Pliny 35.39 mentions that an adulterated version 'ought to be flame-coloured'; elsewhere, 35.43, he describes that which sticks to bronze pans and is used to make a dye as 'black florescence' (*flos niger*).

243) Sagapenum

Sagapenum est lachrymae genus quod in aliquibus locis colligitur. Cuius optimum dicimus quod est glebulis minutis durum, colore subsimillimum myrrhae, gustu acri et quasi glutinosum, odore cepastrum, recens, mundum et intus album et extrinsecus rubeastrum. Vires habet
5 quae cum acrimonia uiscide laxare possunt. Hoc ipsum galbano aliqui adulterant, formatis ex eo guttulis eiusdem magnitudinis.

1 Sagapenum] sagapino *VF* sagapinum *P* lacrhymae genus] *om. BC* lachrymae- quod est] *om. P* locis] *om. V* 2 minutis] ex inmutis *BC* subsimillimum] subsimile *V* 3 myrrhae] murra *V* cepastrum *VJLBPC*] cipastrum *F* 4 album] albidum *J* rubeastrum *VFLBC*] rubeastum *P* rubidum *J*

deest W

cf. Diosc. III.81 (σαγάπηνον). Pliny 20.197. Celsus 5.23.3A. Scrib. 69. Galen *Simp*. 8.18.1 (K.12, 117).

244) Sepum

Sepum omne mollire potest duricies. De quadrupedibus optimum est leoninum, melius est leopardinum. Item melius ursinum quam taurinum uel ceruinum. Caprinum uero spissius est et siccius et magis hircinum uel asinium. De uolatilibus uero optimum est anserinum, melius pullinum. Item melius phasianinum et omnibus melius uulturinum.

1 Sepum] seuum *V* sibum *FP* duricies] duritias *VF* 2 item] nunc *J def. V* taurinum] *om. F* 3 caprinum- hircinum] *om. F* siccius] *om. V* 4 uolatilibus] uolatilibus omnibus *V* melius] *om. BPC* 5 phasianinum] fasianino *V* fasininum *L* fasaninum *FBPC* melius] *om. P*

cf. Diosc. II.76 (στέαρ). Pliny 11.212–13. Celsus 5.18–19. Galen *Simp*. 11.1.2 (K.12, 323–31).

245) Sanguis

Sanguis omnis humidus est. Foris in corpore illinitus stypticus inuenitur, sed palumbinorum omnibus siccior inuenitur et oculis prodesse comprobatur.

1 omnis] omnibus *JBCW* foris] a forissis *F* sed *BC om. P* inuenitur] stringit ualde *W* 2 sed- inuenitur] *om. J* siccior] sicci dicitur *F* oculis- comprobatur] indefricio est optimum *W* 3 comprobatur] et in prouatur *V*

cf. Diosc. II. 79 (αἷμα). Galen *Simp*. 10.2.1 (K.12, 253). Isidore 4.5.4, 4.5.6, 11.1.122–3.

243) Sagapenon (*Ferula persica* Willd.)

Sagapenon is a type of sap[1] which is harvested in several places, and we reckon the best to be that which comes in hard, small globules, with a colour that is somewhat similar to myrrh,[2] a sharp taste, a sticky consistency, smells a little like an onion, and is fresh, pure, white on the inside, and reddish on the outside. It contains properties which can vigorously loosen with some stinging. Some adulterate it with galbanum[3] and then shape it into little droplets the same size as the sagapenon globules.

1 Diosc. III.81 describes it as juice (from a plant in Media), but it is more like a gum: Innes Miller 1969, 100; Dalby 2003, 29.
2 #180.
3 #119.

244) Fat

Fat can soften all types of induration. From among the quadrupeds the best is that derived from a lion, then that from a leopard. After these rank fat from a bear, then a bull, then a stag. She-goat fat is thicker, drier, and better than fat from a he-goat or a donkey. From among birds the best fat is that from a goose, then chicken fat. After these rank pheasant fat, and fat from all types of vultures.

245) Blood

All types of blood are moist. It is found to be styptic when applied topically to the outside of the body. Blood of wood-pigeons dries more than any other type,[1] and has been proven to relieve eye problems.

1 Cf. Diosc. II.79, 'pigeon blood in particular stays haemorrhages from membranes.'

246) Struthium

Struthium herbulae radix est nota quam pluribus, exalbida et spissula. Quae acrimonia sua et asperitate urinam et cauculum prouocat, sternutare copiose facit tunsa et cribrata et in naribus imposita.

1 Struthium] sirucitio *V* syratium *F* struthium *L* serietium *P* syricium *J* nota- pluribus] *om. P* spissula] spissa *P* 2 cauculum] calculum *J* canculum *P* sternutare copiose] sternuta compescit *BPC* 3 tunsa] tusa *BC* cribrata] criuellata *V* imposita] missa *BPC*

deest W

cf. Diosc. II.163 (στρούθιον). Pliny 19.48–9, 24.96. Galen *Simp.* 8.18.41 (K.12, 131). *Ex herb. fem.* 55. Isidore 17.9.56.

247) Staphisagria

Staphisagria semen est herbae quae florescit similiter ut uites erraticae et fructum affert in foliculis uiridibus quasi ciceris, triangularem, album, subrufum, gustu acerrimum et amarum. Verum optima est quae est recens, nigra, plena. Haec ipsa cum maximis uexationibus uiscidissime relaxat, pediculos quoque ex oleo laeuigata et corpori illinita occidit.

1 Staphisagria] stafisiagria *V* staphideagria *J* simisagrie *L* stafisagria *FBPC* florescit similiter] follis floribus et in radice similis *BPC* erraticae] *om. BPC* 2 ciceris] ceteris *F* quasi- triangularem] gleba *BPC* 3 acerrimum] acrissimum *VL* amarissimum *BPC* amarum] *om. BPC*

deest W

cf. Diosc. IV.152 (σταφὶς ἀγρία). Pliny 23.17. Scrib. 8, 166. Marc. *Med.* 1.8 *Ex herb. fem.* 37. Isidore 17.9.87.

248) Sepia

Sepia est marinum animaliculum cuius testa et item os siccatum et repositum, leue et candidum et oblongum uidetur et ad usum medicinae facit. Admiscetur autem collyriis et solum in mortario laeuigatum dentifricium est optimum.

1 Sepia] sippia *V* sipia *W* animaliculum] animal *W* cuius- ad] ossum *W* item] id est *VLBPC* os] ossa *F* 3 admisceatur- optimum] *om. W* in mortario] tantum modo *FLBC* tantum *P*

cf. Diosc. II.21 (σηπία). Pliny 32.141. Galen *Simp.* 6.1.27 (K.12, 347–8). Isidore 12.6.46.

249) Sulphur

Sulphur optimum est quod uiuum appellatur, enim est splendidissimum et perlucidum et ueluti perunctum. Vires habet uiscide relaxantes et plurima uitia remedicantes.

246) Soapwort (*Saponaria officinalis* L.)

Soapwort[1] is the root of a shrub that is known to many, and which is pale in colour and a little dense. Its tang and severity stimulate urination and the emission of bladder and kidney stones, and when pulped, sifted, and poured into the nostrils it causes a great deal of sneezing.

1 *Struthium*, following *L*: the name given in *J, syricium*, is otherwise unattested according to André 1985, 253, who rightly identified it as a corruption of Gr. στρούθιον: on which Diosc. II.163, who also notes the sneezing. Cf. Pliny, 19.48 (*struthion*), and below #269 n. 1.

247) Stavesacre (*Delphium staphisagria* L.)

Stavesacre is the seed of a plant which flowers just like the wild grapevine and bears in green pods a seed like a chickpea but triangular, white, a little reddish, and with an extremely sharp and bitter taste. Stavesacre is best when fresh, dark, and plump, and when like this it is powerful enough to relieve the most violent type of shaking, and when crushed with oil and applied topically on the body it kills lice.

248) Cuttlefish

Cuttlefish is a sea creature whose and head and mouth are dried and stored, so that it is light in weight, white in colour, and is an oblong shape, and this is then used for medical purposes. It is mixed with eye-salves and when ground in a mortar alone it makes an excellent toothpaste.

249) Sulphur / Brimstone

The best sulphur is that which is called 'live,'[1] for it glitters brilliantly and is transparent, as though it has been oiled. It has loosening properties and is a remedy for most ailments.

1 uiuum] uinum *B* 2 perlucidum] lucet *P* ueluti perunctum] uel unctum *F*
plerumque def. V, deest W
cf. Pliny 35.175. Galen *Simp*. 6.3.9 (K.12, 217–18). Isidore 16.1.9.

250) Silphium

Silphium radix herbae est quae graece appellatur mastiton cuius lachryma laser esse diximus, haec est radix pluribus nota. Virose et acriter
excalfacit et uesicam laedit. Verum admiscetur antidotis et purgatoriis
et sale condita suauiter facit.

1 Silphium] silper *V* silfor *F* silber *L* silfer *P* silfera *W* herbae] *om. JC* herba multis nota
W quae- mastiton] *om. JP* mastiton] maseton *L* ag[er]ico itata *W om. JBPC* 2 laser]
lasaris *B* laxaris *B* lasar *PW* esse] sepe *BPC* sic iam sepe *W* pluribus nota] *om. P* uirose-
laedit] uires habit sed et acriter uiscide dit *def. V* uirose- excalfacit] uires habet acriter
calfacientes *W* radix acriter calefacit *P*

cf. Diosc. III.80 (σίλφιον). Pliny 19.38–45, 22.100–1. Galen *Simp*. 8.18.16 (K.12, 123).

251) Seseli

Seseli semen est herbae quae similatur foeniculo et anetho crescit et
grauat et grana affert uastiora et oblongiora quam utrumque. Verum
et hoc est diureticum et in plurima antidota mittitur. Proprie tamen ad
colicum dolorem facit.

1 Seseli] siseli *V* siceli *F* siselos *W* similatur] *om. P* anetho] nitro *F* crescit- utrumque] fert
maiora et longiora *P* 2 grauat] granat *L om. FB* grana] granula LB 3 proprie- facit] *om.*
VW 4 colicum] coli *L* colicum dolorem] colorem *P*

deest C

cf. Diosc. III.53 (σέσελι). Pliny 8.112, 20.238. Galen *Simp*. 8.18.9 (K.12, 120). Marc. *Med*.
20.88, 101.

252) Scammonium

Scammonium, quod colophonium uel diagridium dicitur, succus est
herbae quae a sua radice plures ramos uelut uiticulas mittit, quae nascitur et in Mysia et in Colophone et similiter mandragorae premitur,
aut quasi panacis radix illius inciditur et foliis suis suppositis lachryma
5 quae emenat excipitur. Verum optimum diximus scammonium quod
est aetate medium, leue, perlucidum, intus non spissum, sed magis
prout spongia fistulosum et splendidum et quod spongia tangitur cito
humorem lactentem et amarum ostendit. Hoc igitur ipsum stomachum
quidem euertit, sed uentrem uiscide purgat. Adulteratur autem quod

1 *uiuum*: also known to Pliny 35.175, 'vivum [sulphurem], quod Graeci apyron (that is, 'untouched by fire') [appellant],' a type which doctors mostly use.

250) Laserwort (~ *Ferula tingitana* L.)

Laserwort is the root of the plant which in Greek is called *mastiton*, the sap of which we called above *laser*,[1] which comes from the plant's well-known root. It heats in a sharp and fetid way, but injures the bladder. It is mixed into antidotes and purgatives, and becomes tasty when seasoned with salt.

1 #147. *Mastiton* (*VF*, *maseton L*) otherwise unattested, but Diosc. III.80 mentions that 'some call the stalk *silphion*, the root *magydaris*, and the leaves *maspeta*,' though he also mentions in this entry that *maspeton* is the name of the stem, and that another similar plant goes by the name of *magydaris*. Likewise Pliny 19.38–45, though qualifying that *magydaris* is the juice from the root of *silphium/laserpicium* (cf. 22.101).

251) Hartwort (*Tordylium officinale* L.) ~ Massilian hartwort (*Seseli tortuosum* L.)

Hartwort is the seed of the plant that resembles fennel,[1] yet grows like dill[2] in that it becomes heavy and bears seeds that are quite wide and equally lengthy. It is a diuretic and is put in many antidotes. It is particularly good for pain associated with colic.

1 #110.
2 #35.

252) Scammony (*Convulvulus scammonia* L.)

Scammony, which is also called *colophonium* and *diagrydium*,[1] is the juice of the plant which sprouts many branches like vine tendrils, commencing low down near its roots, and which grows in Mysia and Colophon. It is pressed in the same manner as mandrake, and just as is done with panax[2] the root is cut and the leaves are placed under to catch the sap that flows out. The best scammony is that which is middle-aged, light, transparent, not dense in the middle, but rather porous like a sponge, and shiny, and which releases a milky, bitter fluid upon touching it, like a full sponge. Scammony really churns up the stomach, but forcefully purges the bow-

10   est tithymalli succo, sed ita deprehenditur, quod tithymalli succus ad
gustum minime excalefacit quodammodo, sed urit, quod abest ab sin-
cero colophonio.

1 Scammonium] scamonia *VF* scamonea *P* quod] quod est *V* quod est quasi *BC* colopho-
nium] colufona *V* colifinia *F* colofonia *P om.* W dicitur] quia *F* 2 uelut uiticulas] intra se *J*
ueluti uiticlis *V* uelut uiticulas *F* ueluticlas *L* quasi uiticulas *BPC* 3 mysia- mandragorae]
om. F colophone] colonia *BPC* premitur] ut creditur *J* 5 emenat] manat *LBC* excipitur]
accipitur *V.* suscipitur *P* 6 aetate] *om. JBPC* medium] *om. BPC* non] *om. BPC* magis
prout] ut pro *F* 7 spongia] spungia *V* simul ac saliua *J* contexta *C* contacta *BP* cito] *om.*
JB 8 lactentem] lacteum *J* lactantem *BC* 9 quod est] *om. JBC* 10 tithymalli] tytimalis *V* ex
titimalli *BPC* sed- succus] *om. VFP* sed ita- colophonio] sed urici quod est sincere colure
colufoniae *V* habet sincere ut colofonia *BP* habet *C* 11 sincero] sentire *F*

2–5, 7–8, 9–11 plerumque om. W

cf. Diosc. IV.170 (σκαμμωνία). Pliny 26.59–60. Celsus 3.20.6. Scrib. 137, 140, 254. Cael.
Acut. 1.180–1, 2.155. Isidore 17.9.64.

253) Singiber

Singiber herbulae radix est pusilla, exalbida, leuis et ueluti trunca.
Inuenitur maxime in Troglotydis et Arabiae partibus et aliqui illam
piperis radicem putant. Huius igitur optimum est, quod est recentissimum et sine fistula ulla, quod atreton graece dicimus, odore suaui,
5   gustu et sapore similis piperi. Cuius uires plurimae sunt, sed proprie
digestionem facit et stomachum confirmat et corroborat et uentrem
relaxat.

1 Singiber] sengiber *PW* herbulae] herba *V* pusilla] sincere ut colofonia pusilla *C* leuis]
leuissima *W* trunca] flusca *BPC om.* W 2 troglotydis] troncloditiditis *V* trocliditis *F* tra-
contidis *BC* droconitidis *P def. L* intra mediis regionibus *W* 3 huius igitur] hoc ergo *J* 4
ulla] alba *V om. F* quod- dicimus] *om. BPCW* atreton] ἄτρητον *J* aypum *V* athepum *F*
def. L odore suaui] *om. VF* 6 stomachum confirmat] *om. F* 7 relaxat] laxat *J*

cf. Diosc. II.160 (ζιγγίβερι). Pliny 12.28. Celsus 5.23.3B. Galen *Simp.* 6.4.2 (K.11, 880–2).
Isidore 17.9.8.

254) Styrax

Styrax est quasi lachryma arboris quae et ipsa styrax appellatur, similis
tota cydonia. Huius igitur arboris ramuli et uirgulae ut uermiculi quae
circa ipsas medullas nascuntur, a quibus incipiuntur qui foras terebrant
et siccantur caccabello et foramina ipsorum scobem quasi furfur inue-
5   nitur, quae cum lachryma confiscatur, quod in modum resinae pityinae
manat maxime aestibus mediis et sole calidissimo. Verum optimum
dicimus styracem qui est subrufus et qui multum resinaster permanet,

els. It is adulterated with the juice of spurge,[3] but this is detected by the way that the spurge juice upon tasting does not just mildly heat the tongue but rather burns, which does not happen with pure *colophonium*.

1 Pliny, 26.60, mentions that the scammony from Colophon 'is recommended [*laudatur*],' and elsewhere (14.123) refers to a resin called *colophonia*; likewise Scrib. 137. Caelius Aurelius, *Acut.* 1.179, 2.155 (also *Chron.* 2.189, 3.11) mentions that *diagrydium* is the juice of scammony. Diosc. IV.170 does not mention these names, or Colophon.
2 #220, but see #201 (*opopanax*).
3 #280. The spurge taste test is also in Diosc. IV.170, Pliny 26.61.

253) Ginger (*Zinziber officinale* L.)

Ginger is the little root of a shrub, and is pale, light, and seems as though it is deprived of limbs. It is found mostly in the land of the Troglodytes,[1] and in parts of Arabia, although some people think it is the root of pepper. It is best when freshest, is free of holes, which we call in Greek *atreton*,[2] and has a sweet fragrance and a flavour similar to pepper. Its properties are many, but it is especially good for digestion, for it strengthens and fortifies the stomach yet loosens the bowels.

1 A cave-dwelling people in Ethiopia. Cf. Diosc. II.160, grows 'in Troglodytic Arabia.'
2 ἄτρητος, 'without aperture/not perforated.' Cf. Diosc. ibid., 'choose that which is not worm-eaten [τὰ ἀτερηδόνιστα].' The only other attested use of the term in Latin is Theodorus Priscianus, *Gyn.* tit. 4, *De atretis*, a woman's complaint, the same meaning found in Latin translations of Soranus, *Gyn.* tit. 33 (*de atretoe*).

254) Storax (*Styrax officinalis* L.)

Storax is the sap of the tree which itself is called storax and resembles the quince tree in every way.[1] The tree has branches and twigs that breed larvae within their core, from which the larvae begin to bore their way out and are then dried in a cooking pot, and within the little holes they leave is found a bran-like sawdust which is then mixed in with the sap and left under the burning sun in the hottest period of summer until it achieves the consistency of pine resin.[2] We declare the best storax to be

odore suaui et quasi melleo et subausterus, sine humore et leuis dum
fricatur. Viribus autem potest leniter relaxare, calefacit enim et leniter
10 durities emollit. Sed adulteratur copiosa scobe ipsius arboris, uel etiam
alterius et inde tunsa [in] et cribrata et cum melle decocto subacta et
rursus cera et adipe reuoluta.

1 Styrax] storax *VFBP* sthorax *L* arboris] arboria *J* similis- cydonia] *om.
F* 2 cydonia]
citonia *V* cidonia *BPC* scidonia *L* arboris] arboria *J* 3 incipiuntur- caccabello] inciduntur
J terebrant] terebant *L* 4 siccantur] siccat *V* siccis *LBC* cacabello] capillus *V* scobem]
ex cubiis *P* inuenitur] emittunt *J* 5 confiscatur] confegatur *V* confringatur *F* confricatur
L confrigatur *P* 6 mediis] meridis *BPC* calidissimo] candidissimo *J* 7 multum] *om. V*
resinaster *FLBC*] r(i)sinaster *V* resinastrum *P* resinae specie *J* 8 subausterus] austerus
VF subalterus *L* sine] siue *P* 9 leniter] lenit *J* lente *FL* leuiter *P* 10 emollit] soluit *P* adul-
teratur] ulceratur *F* copiosa scobe] gumme *BPC def. V* copiosa cupa *F* 11 tunsa] inlirita
V inligat[ur] *F* in lyrica *L* ylliri *B* illirica tunsa *P* yllyri *C* 12 cera] siccat *V* in siccata *F* in
sicca *LBPC* adipe] pridie *VLPC* die *F* reuoluta] resoluta *P*

deest W

cf. Diosc. I.66 (στύραξ). Pliny 12.124. Scrib. 88–9. Galen *Simp.* 8.18.42 (K.12, 131–2).
Marc. *Med.* 35.24–5. *Phys. Plinii* 57.44–6. Isidore 17.8.5.

255) Symphytum

Symphytum, quod aliqui anagallicum dicunt, herba est pluribus nota.
Cuius radix uidetur oblonga, cortice nigra ad gustum subamara et glu-
tinosa, propter quod ad restringendum sanguinem excreantibus datur.
Viscide autem stringit et glutinat.

1 Symphytum] sinfitum *VFP* anagallicum] gallicum *V* alum gallice *L* alumen *B* alumen
gallie *P* lumen gallie *C* pluribus nota] *om. P*

deest W

cf. Diosc. IV.10 (σύμφυτον ἄλλο). Pliny 14.108, 26.45, 81, 137, 27.41. Scrib. 83. Marc.
Med. 10.68, 17.21. Ps. Apul. 59, 127 (album). *Dyn.* 2.74. Isidore 17.9.61.

256) Spuma argenti

Spuma argenti sit in metallis et purgamentis argenti cum conflatur. Item
ex scoria quam Graeci molybdenam dicunt. Aliqua ex plumbo et ipsa
coctura in speciem pusillorum tabularum redigitur. Verum optimam
dicimus quae est colore assimilis auro et usque grauis. Potest autem
5 aliquatenus restringere et refrigerare et eadem praestat quae et cerusam
facere dicimus. Ab aliquibus uero lithargyrus appellatur.

1 Spuma argenti] spuma argenti uel lythargyrum *tit. J* sit] fit *FLBP* purgamentis] expur-
gantis *F* 2 ex scoria] exalbida *F* est aliquam *B* est alia *C* molybdenam] molibitidem *V*

that which is reddish, continues to remain very resinous, has a sweet, honey-like fragrance, is somewhat coarse, contains no moisture, and becomes smooth when you rub it. In terms of its properties it can gently loosen, for it warms and gently softens indurations. It is heavily adulterated with sawdust from the tree itself, and there is another type which is pulped[3] and then sifted, and then cut with stewed honey, and then again with wax and then rolled in fat.[4]

1 Likewise Diosc. I.66 verbatim, but the rest of the description differs considerably. The *AG* contains no entry for quince: cf. Diosc. I.115, Pliny 15.37.
2 See #228. The text concerning the preparation process is corrupt and difficult to construe: see n.4
3 *tunsa*: following *J* (and *P*), though the manuscripts all record some other now irrecoverable word (*inlirita, inligatur, in lyrica*, etc.), possibly *inlisa* (beaten, crushed) or *inlita* (spread, smeared, etc.).
4 *wax- fat*: a restoration, following the sense of Diosc. I.66.

255) Comfrey (*Symphytum officinale* L.)

Comfrey, which some call 'Gallic garlic,'[1] is a well-known plant. Its root is oblong, has black bark, and a taste that is slightly bitter and a little sticky. Because of this it is administered to stop the coughing up of blood. It vigorously binds and agglutinates.

1 *anagallicum*: an alternative name also noted by Vegetius, *Mulomed.* 2.6.10, 2.9.6, and Ps. Apul. 59, 127 (*alum gallicum*); the latter is probably the source for its attestation in Anglo-Saxon England: Amalia d'Aronco 1988, 29. Cf. Diosc. IV.10, 'some call it pecte [πηκτή]': André 1985, 253, notes that the common variety in Greece (and hence Diosc.) was *Symphytum bulbosum* Schimp. Pliny 27.41, 'Alum nos vocamus, Graeci symphyton petraeum,' is referring to low pine (*Symphytum tuberosum* L.), for which see Diosc. IV.9.

256) Litharge

Litharge ('silver foam') is mined and also formed from the refuse of smelting silver. It is also obtained from dross, which the Greeks call *molybdaena*.[1] There is another type made from lead, which is baked after it has been fashioned into small tablets. We declare the best litharge to be that which has a golden colour and is evenly weighted. It can act as a mild astringent, can cool, and it can be used for same purposes as cerussite.[2] It is all called 'litharge' by some.

1 #182.

oliuiae *F* MWΛYB *L* MWΛHBYTIN *BC* 3 tabularum] tribolorum *JL* tubulorum *F* redigitur] dicuntur *V* esse dicitur *BC* 4 usque] quasi *BC om. P* 6 lithargyrus] litargero *V* litargire *F* litargirum *P* litarguina *C*

deest W

cf. Diosc. V.87 (λιθάργυρος). Pliny 33.106–9, 34.173. Galen *Simp.* 9.3.17 (K.12, 224–5). Isidore 16.19.4.

257) Sori

Sori glebula est uiridis quae foramina [in se] habet quam plurima. Dicitur et aliud [eius genus] inueniri colore argenteo. Verum optimum iudicatur Aegyptium et est quod ab Aegypto affertur, quod plurima habet foramina et est fragile et dum comminuitur infuscatur, gustu
5 stypticum et odore nauseosum sic, ut et uomitum quibusdam moueat. Coquitur autem usquequo sit colore rufum et habet uires ualde stypicas et acres. Vnde et alienationes continet et uulnera quae luxuriantur stringit et siccat.

1 Sori] suri *P* foramina] foramina in se *FLBPC* foraminibus multis *W* 2 aliud] alia *VWBC* eius genus] *om. FLBPCW def. V* 3 iudicatur] dicimus *BPC* indicatur *F* et est] id est *FL om. BC* quod- affertur] *om. P* 5 nauseosum] rolusiosum *P* sic ut] sicut *L def. V om. BC* 6 habet- siccat] *om. W* 7 alienationes] aliginationes *V* 8 stringit et siccat] exiccat *J* stringit et desiccat *F*

cf. Diosc.V.102 (σῶρι). Pliny 34.118, 120–1. Celsus 6.9.23.

258) Stoechas

Stoechas herbula est quae ramulos emittit tenues et oblongos, circa quos habet folia oblongiora quam thymum sed non dissimilia et semen minutum. Tota herba est odore suaui, gustu acris et subamara. Vires habet acriter relaxantes. Vnde menstrua prouocat et uitia thoracis sedat
5 et in antidotis miscetur. Nascitur plurimum in insulis quae stoechades dicuntur.

1 Stoechas] sticadu(e) *V* sticados *F* sthicas *W* sticas *P* herbula] herba *VL* 2 habet] *om. LBC* thymum] tymi *LBPC* cymas *F* thimum *W def. V* thymbra *J* 3 odore- habet] marubiores herba acriter *F* 4 acriter] *om. CJ* 5 nascitur- dicuntur] *om. P* plurimum] primum *L* stoechades] stigadus *V* sticadus *F* stichades *W*

cf. Diosc. III.26 (στοιχάς). Pliny 20.247, 27.131. Celsus 8.9.1E. Scrib. 106. Galen *Simp.* 8.18.39 (K.12, 130–1). Marc. *Med.* 1.106. *Ex herb. fem.* 13. Isidore 17.9.88.

2 #46.

257) Melanterite

Melanterite is a green blob that has many small holes in it. Another type is said to exist which is silver in colour. The Egyptian type is judged to be the best, along with that which is said to be Egyptian, which has many holes, is delicate, when crumbled turns dark, and which has a styptic taste and an odour so nauseating it causes some people to vomit. It is heated until it turns red, and has very styptic and sharp properties. It restricts abnormal growths, and contracts and dries festering wounds.

258) French lavender (*Lavendula stoechas* L.)

French lavender is a shrub which shoots small, slender, and lengthy branches bearing leaves that are more oblong in shape than those of (Cretan) thyme[1] but otherwise are similar, and has a small seed. The whole plant has a sweet fragrance, and a sharp, slightly bitter taste. It contains properties that severely loosen, and therefore it stimulates menstruation, eases chest complaints, and is mixed with antidotes. It grows mostly on islands which are called *stoechades*.[2]

1 #282. Note *thymbra J*: likewise Pliny 20.247.
2 Today Iles d'Hyeres, off the French coast east of Marseilles. Likewise, Diosc. III.26, whose entry is very similar.

259) Sapo

Sapo coquitur ex sebo bubulo uel caprino aut ueruecino et lixiuio cum calce. Quod optimum iudicamus germanicum est enim mundissimum et ueluti pinguissimum, deinde gallicum. Verum omnis sapo acriter relaxare potest et omnem sordem de corpore abstergere uel de pannis et exsiccare similiter ut nitrum, uel aphronitrum, mittitur et in caustica.

1 Sapo] sapone *VFW* coquitur] conficitur *J* bubulo] bubalino *F* bubillino *P* ueruecino] ouino *P om. W* lixiuio] lixiua *V* cum calce] calcis sugo *W* 2 optimum- sapo] alii capitellum alii cursiam gr(ec)i u(er)o putren appellant *W* iudicamus] dicimus *FBPC* 3 deinde] sed inde *L* uel ut *BC* ut *P* acriter] *om. F* 4 de pannis] quin immo de pannis *W* 5 exsiccare] *om. W* nitrum] merum *P*

cf. Diosc. I.128 (σῦκα: 6. κονία). Pliny 28.101.

260) Sampsuchus

Sampsuchus herba est folia et ramos habet plures et pusillos circa terram diffusos non dissimiles nepitae, tota herba olens suauiter. Viribus acris et excalfactoria et ad omnem duritiem uel rigorem est ualde idonea.

1 Sampsuchus] samsugum *V* sumsucus *P* ramos] ramulos *P* pusillos] plure postillum *V* 3 omnem duritiem] *om. BPC*

deest F, W

cf. Diosc. III.39 (σαμψούχινον). Pliny 21.63, 163. Galen *Simp.* 8.18.2 (K.12, 118). *Ex herb. fem.*10. Isidore 17.9.14.

261) Sinapi

Sinapi omnes nouerunt, cuius efficacissimum est alexandrinum. Habet enim uires acriter et uiscide stringentes, propter quod uitia longinqua eliminat, eo quod loca languentia exsiccat et exstringit et corroborat simul stomachum, quod ipsum sinapismo sano uidimus contingere.

1 Sinapi] sinape *VW* senape *F* omnes- cuius] *om. P* 2 stringentes] distinguentes *J* destringentes *F* 3 eliminat] elimat *L* limat *FBPC* languentia] sanguine manantia *J* exstringit] extringit *VJ* constringit *W* asstringit *LBC* 4 quod- contingere] *om. W* sinapismo sano] sinapi mosano *J* a sinapismo sanum *FL* sinapismum *BPC* contingere] solum contingere *BPC*

desinit V

cf. Diosc. II. 154 (σίνηπι ἢ νᾶπυ). Pliny 12.28, 19.171, 20.236–8. Garg. *Med.* 29. Marc. *Med.* 15.87. *DynVat* II.8/*SGall* 62. Isidore 17.10.9.

259) Soap

Soap is made by cooking beef, she-goat, or wether fat, mixed in with lye[1] and quicklime.[2] We judge the best soap to be Germanic, for it is the purest and the creamiest; after this ranks Gallic soap. All types of soap can severely loosen and remove all filth from the body and from clothing. It can also dry things out in the same manner as soda or foam of soda,[3] and is put in caustics.

1 #150.
2 #67.
3 #9.

260) Marjoram (*Majorana hortensis* Moench)

Marjoram is a plant that has many small branches and leaves which spread along the ground, not unlike catmint.[1] The whole plant emits a pleasant fragrance, has sharp and heating properties, and is excellent against all types of induration and stiffness.

1 #196. Similarly Diosc. III.39. On the different types of marjoram see Andrews 1961a.

261) Mustard (seed) (*Sinapis alba* L.)

Mustard seed is known to everyone, and the most effective is that from Alexandria.[1] For it has severe and vigorous astringent properties, and therefore eradicates chronic ailments, dries out and contracts enfeebled parts, and also fortifies the stomach, which we have also seen happen by application of a clean mustard-plaster.

1 Mentioned in #94. Diosc. (II.154) does not mention the Alexandrian type, and Pliny (12.28) only in passing.

262) Sertula

Sertula herbula est humilis, odore suaui, exalbida, flosculum habens subrufum. Cuius melior habetur quae est recens et uastior et non nimis exalbida. Vires autem habet leniter relaxantes, propter quod et oculorum tumorem sedat et partibus neruosis et delicatis ex aqua mulsa cocta in modum fomenti trita et imposita facit.

1 herbula] herba *JFPL* 4 tumorem] dolorem *BPC* neruosis] *om. WBPC* delicatis] ligatis *F* mulsa] multa *JL* 5 modum- facit] et inposita contrita pro fomento est *W* fomenti trita] frumenti *F*

cf. Diosc. III.40 (μελίλωτος). Pliny 21.53. Celsus 5.11, 6.5.3. Scrib. 220, 258. Galen *Simp.* 7.12.8 (K.12, 70).

263) Scinum

Scinum nascitur in Africa et in India, optimum tamen habetur Arabicum et huius ipsius quod est recentissimum et plures et breues calamos habet et si findatur rubicundum intus apparet, redolens suauiter si fricetur inter manus et gustu linguam submordens. Quidem uires habet
5 acriter relaxantes, proprie quidem uesicam petit et urinam prouocat. Mittitur et in antidota quae aduersus uenena faciunt.

1 Scinum] syrum *J* india] iudea *BPC* habetur] est *BPC* 2 est- breues] *om. F* calamos] calamulos *L* ramulos *BPC* 3 findatur] frangitur *J* fundatur *F* funditur *P* 5 uesicam petit] per uesicam *BPC*

deest W

cf. Diosc. I.17 (σχοῖνος). Cato Agr. 113. Scrib. 70. Galen *Simp.* 8.18.50 (K.12, 136–7).

264) Scilla

Scilla radix est herbae pluribus nota. Est enim colore et corticibus uasta et cepae uel bulbo assimilis. Vires habet acerrimas et excalfacientes. Vnde et tussientibus et hydropicis utiliter datur et si fricetur in corpore magnam et intolerabilem pruriginem facit.

1 Scilla] squilla *P* pluribus nota] *om. P* colore] dolore *P* colore- assimilis] *om. W* 2 bulbo] uulno *L* uulbis *BPC* acerrimas] acrissimas *FLBCW* 3 tussientibus- datur] ydropicis prodest *W* utiliter datur] prodest *F* 4 pruriginem] prodiginem *F* perdignem *L*

cf. Diosc. II.171 (σκίλλα). Pliny 19.93–5, 20.97. Galen *Simp.* 8.18.23 (K.12, 125). Marc. *Med.* 19.10, 28.66. Ps. Apul. 42. *Ex herb. fem.* 52. Isidore 17.9.85.

262) King's clover (*Melilotus* sp. Adans.)

King's clover[1] is a small, low-growing, pale plant that has a pleasant fragrance and bears a small yellow flower. That which is fresh, larger, and not too pale is considered the best. It contains gently loosening properties, therefore reduces swellings on the eyes, and is cooked with honeyed wine and then ground down to be applied topically as a compress to treat sinewy and delicate areas.

1 *Sertula* (*serta*): Pliny 21.53 notes that Romans call clover (*melilotus*) 'sertula campana'; Diosc. III.40 mentions that there is a type with a weaker scent that grows in Campania 'around Nola.'

263) Camel hay (*Cymbogopon schoenanthus*, Spreng.)

Camel hay[1] grows in Africa and India, but that from Arabia is considered to be the best, and this type is best when freshest and when it has many reeds which turn red inside when you split them open, smell sweetly if you rub them in your hands, and have a taste that gently bites the tongue. It has severely loosening properties, and in particular it acts upon the bladder to stimulate urination. It is placed in antidotes that combat poison.

1 *scinum*: the name derives from Gr. σχοῖνος (e.g., Diosc. I.17). Despite the odd reading of *J* (*syrum*) it was identified correctly by Meyer 1855, IV:495.

264) Squill (*Scilla maritima* L. = *Urginea maritima* Baker)

Squill is the root of a plant known to many. It is the same colour as an onion or bulb, and likewise has numerous layers of skin. It contains very sharp and heating properties, and is therefore usefully administered to those with coughs or dropsy,[1] and when rubbed on the body is causes tremendous and unbearable itching.

1 Squill was commonly thought to cure dropsy in the ancient and medieval periods: see Stannard 1974, and Aliotta et al. 2004.

265) Scordium

Scordium herbula est uirgulas mittens quadratas, rectas et oblongas et
folia stringentia ad gustum subamara, odore oleastro, flore rubicundo.
Vires habet acres et diureticas et aduersus uiperas facientes.

1 Scordium] scordeon *F* scordium uel allium agrestis *tit. J* herbula] herba *BPC* uirgulas]
om. L mittens] emittit *BC* quadratas] quadratulas *BC* quadnutulas *L* 2 oleastro *LBC*]
oliuastrum *F* aliastro *P* oleaginosa *J* 3 aduersus] ad *J*

deest W

cf. Diosc. III.111 (σκόρδιον). Pliny 25.63, 26.77. Marc. *Med.* 20.19. Scrib. 177. Ps. Apul.
71. Galen *Simp.* 8.18.25 (K.12, 125–6).

266) Scincus

Scincus est animaliculum quod in India inuenitur, quadrupes et simile
lacerto, sed multo maius et longius et uentosum hoc igitur salitur. Et
ad uenerea facit drachma una carunculae illius ex uini cyatho potata. Si
enim plus sumatur affert periculum.

1 Scincus] sticas *B* animaliculum] animal paruulum *J* animaculum *P* india inuenitur] *om.*
FP 2 longius] oblongum *BPC* uentosum] uentrosum *LBPCJ* 3 ad] *om. J* drachma] dra-
gis *BC om. F* drachma– periculum] ac tussim caro illius exiunciato *W* una] in *J om. BC*
carunculae] caracule *JP* caracullae *F*

cf. Diosc. II.66 (σκίγκος). Pliny 28.119–20. Theod. *Log.* 34.3.

267) Sedum

Sedum herba est quam aliqui senetion uocant quae et in petra nascitur et super tecta. Folia enim habet uiridia, densa et quercui similia
sed grossiora et pinguia. Thyrsulum emittit oblongum et subrufum,
capitella habet oblonga, in summo habens in modum uerticilli. Vires
5 habet stringentes et refrigerantes propter quod similis est plantagini et
ad eadem proficit.

1 aliqui] *om. BC* senition] senetionem *J* senition *W* senecion *BP* senicion *FC* 2 super
tecta] super textis *B* in tectis *PW* 3 grossiora] crassiuscula *J* grossula *FLPC* grossa *L*
grossiora *W* thyrsulum] solum *F* oblongum] oblongum floscellum *F* 4 capitella] capitula
FLBC habet- habens] in sumitate tursi habens *W* in summo habentia *P* habet oblonga]
om. FLBPC uerticelli] uiticelle *BPC* uicellae *W* 5 refrigerantes] *om. F* similis- proficit]
easdem praestat quod plantago *W*

cf. Diosc. IV.88 (ἀείζωον μέγα). Pliny 25.160. Galen *Simp.* 6.1.8 (K.11, 815). Ps. Apul.
124. *Ex herb. fem.* 11 (cestros), 33 (aizos).

265) Garlic germander (*Teucrium scordium* L.)

Garlic germander is a shrub that shoots small, lengthy, and straight quadrangular stems, with contracted leaves that are slightly bitter in taste, smell like the wild olive, and it bears a red flower. It contains sharp and diuretic properties, and properties that are good against vipers.

266) Skink

The skink is a little creature which is found in India. It has four legs and is similar to a lizard, but is larger, longer, and jumps about very quickly.[1] A drachma of its inner flesh when drunk with a twelfth of a pint of wine is an aphrodisiac,[2] but consuming more than this is dangerous.

1 Or is 'pot-bellied yet jumps around,' *LBPCJ*.
2 Similarly Diosc. II.66, 'one drachma of the part that surrounds its kidneys'; cf. Pliny 28.119, 'its muzzle and feet' or 'two *oboli* of the flesh of the flanks.'

267) Houseleek (*Sempervivum arboreum* L.), and Common houseleek (*Sempervivum tectorum* L.)

Houseleek is a plant which some call *senition*,[1] and which grows among rocks and upon roofs. It has thick foliage of green leaves similar to those of the oak, but thicker and quite fat. It shoots up a stalk that is oblong and yellow and bears little oblong pods, having one at the top also like a small crown. Houseleek has astringent and cooling properties, and is therefore similar to plantain[2] and can be used in its place.

1 *senition/senetio*. Unattested elsewhere for houseleek: André 1985, 238. Both Pliny 25.160 and Diosc. IV.88 give several other names for houseleek, but not *senecio/senetio*, a term which other authors use to refer to different plants, mostly *erigeron* or groundsel (*Senecio vulgaris* L.): Pliny 25.167, Ps. Apul. 76.
2 #217.

268) Satureia

Satureia fere easdem uires habet quas et thymus et eadem facit et si est aliquatenus infirmior.

1 Satureia] saturea *F* fere] herba *C* caldes uires habet *W* et] etiam *BC* si- infirmior] quam ius (?) fit infirmior parum *W*

cf. Diosc. III.37 (θύμβρα). Pliny 19.107, 165. Scrib. 124. Garg. *Med.* 20. Marc. *Med.* 22.42. *DynVat* 53/*SGall* 65.

269) Spondylium

Spondylium locis humidis crescit et habet radicem albam et assimilem radici quam in cibos sumimus et thyrsus oblongus quasi foeniculi, folia quasi plataginis, semen quasi seseleos, duplex sed candidius et latius et ueluti paleastrum, florem album. Huius igitur semen potatum bilem
5 detrahit et uentrem soluit acrimonia sua, regium morbum sedat et ad tussem et suspiriosis uidetur prodesse.

1 Spondylium] spondilium *WBC* spondilion *FP* assimilem] similis est *BC* 2 sumimus] mittimus *JL* 3 plataginis] platani *J* seseleos duplex] elippidum *BPC om. W* silu duplex *F* 4 paleastrum *BPC*] palestrum *F* palidastrum *L* latum *W* pilastrum *J* album] album uires habet trahentes *W* bilem detrahit] utile extrahit *BPC om. W* 5 acrimonia] acradine *W* sedat] *om. JLBC* ad tussem] discitit *F* 6 suspiriosis] suspiris *BP*

cf. Diosc. III.76 (σφονδύλιον). Pliny 12.128, 24.25. Scrib. 2, 5. Orib. *Eup.* 2. 1S.11. *Med. Plin.* 1.5.4. Cael. *Acut.* 2.37, *Chron.* I.49.

270) Sisymbrium

Sisymbrium est assimile mentae, latoria quidem folia habet et odore suaui et excalfactorium est et diureticum et ad apicularum et similium animalculorum puncturam confert praesidium. Semen illius potatum ad eadem facit.

1 Sisymbrium] sisimbrium *FP* sisinbrium *W* assimile mentae] assimilitudine *C* 2 apicularum- facit] semen eius ad dolores et punctas optime facit *W* apicularum] dolorem *BPC* similium- animalculorum] *om. BPC* 3 confert praesidium] *om. BC* possit *F* potatum ad eadem] datum optime *BC* optime *P*

cf. Diosc. III.41 (σισύμβριον). Pliny 19.172, 20.247. Galen *Simp.* 8.18.21 (K.12, 124). Ps. Apul. 106. *DynVat.* 23/*SGall* 27.

268) Savory (*Satureia hortensis* L.)

Savory has nearly the same properties as (Cretan) thyme,[1] and is used for the same purposes, although it is a little less reliable.

1 #282.

269) Hogweed / Cow parsnip (*Heracleum sphondylium* L.)

Hogweed grows in moist places, and has a white root that is similar to the root we consume as food.[1] Its stalk is long and like that of fennel,[2] its leaves almost the same as those of plantain,[3] and its seed is like that of hartwort,[4] but is doubled, wider, whiter, a little chaffy,[5] and it blooms a white flower. Its seed when drunk draws down bile, its bitterness loosens the bowels, it alleviates the royal disease,[6] and seems to help coughs and those with respiratory problems.

1 Probably radish, given that this is the alternative name mentioned above #234, though it was also given for rhubarb #231 (and n.1). Cf. Diosc. III.76, whose description mirrors that above, and says that it has a white root which 'resembles that of the cabbage' ('ὁμοίαν ῥαφάνῳ'): some confusion between ῥάφανος (cabbage) and ῥαφανίς (radish) may lie behind the use of *radix* here.
2 #110.
3 #217.
4 #251.
5 *paleastrum*, see ch. 4.E.
6 See #196 n.5.

270) Bergamot mint (*Mentha sylvestris* L.; *Mentha viridis* L.)

Bergamot mint is like green mint,[1] but its leaves are wider and it has a sweet fragrance. It is both a diuretic and calefacient, and it protects against the stings of small bees and similar insects.[2] Its seed when drunk does the same.

1 #185. On the different types and uses of mint in the ancient world, see Andrews 1958a.
2 Similarly Diosc. III.41, *DynS.Gall.* 22/ *Vat.* 26.

271) Smyrnium

Smyrnium est assimile apio, sed folia habet latiora et subaureo colore et quasi pinguia et fortiora, redolentia suauiter cum aliqua acrimonia. Ramulum et capitellum prout anethi, semen rotundum, nigrum,
fere gustu myrrastrum, radicem sucosam, mollem, leuem, odore
5　suaui, cortice nigro, intus uiridem aut exalbidam et ad gustum fauces
mordentem. Vires autem habet acres et sudorem suscitantes. Vnde et
tussientibus et suspiriosis et hydropicis et coxiosis prodest et ad serpentum morsum facit.

1 Smyrnium] smyrnionem *F* smyrnion *W* simirnium *P* apio sed] *om. BPC* subaureo]
aurosa *LB* subaurosa *WP* colore] *om. BC* 2 fortiora] *om. W* suauiter] *om. J* acrimonia]
acrimonia coctum *BPC* acredine *W* 3 rotundum] in capite rotundum *W* 4 myrrastrum *F*]
mirastrum *J* murrastrum *L* morastrum *BC* marastrum *P om. W* mollem leuem] amulum
F 6 suscitantes] prouocantes *W* 7 coxiosis] *om. W* 8 facit] similiter facit *BPC*

cf. Diosc. III.68 (σμύρνιον). Pliny 27.133 (cf. 19.162, 187). Scrib. 126. Galen *Simp.* 8.18.32
(K.12, 128). Ps. Apul. 107 [?].

272) Satyrium

Satyrium herba est ramulum habens oblongum et purum, florem purpurastrum et exalbidum. Folia quasi lilii, subbreuiora et subrufastra,
tria uel quatuor ad terram iacentia. Radicem duplicem, quasi testiculos,
quae radix potata uenerea prouocat. Succus ipsius inunctus lippitudi-
5　nes optime curat. Est et aliud satyrium erythraicum, quod assimilatur
semini lini. Vastius quidem est et leue et splendidum. Cuius radicis
cortex tenuis est leuis ipsa penitus alba et dulcis. Tertius est quidem
satyrion aliquid herbam basiliscam appellant, cuius folia sunt priori satyrio similia, sed grauiora et leuiora, punctis asperis nigrioribus,
10　radix eius pedis ursi habens similitudinem, flore aureo uel croceo et si
cum aqua uel uino fuerit tritus croceatum efficit colorem. Vires habet
excalfactorias et aduersus uenena facientes, quam herbam siquis secum
habuerit ab omni genere serpentum erit securus.

1 Satyrium] satirion *FP* purum] purpureo *BC* purpurastrum *FLP*] (flos) purpuraster *BC*
purpureum *J* 2 subbreuiora] et breuiora *BC* sed breuiores *F* sed breuiora *L* subrufastra
BPC] subrufa *FLJ* 3 tria] *om. BC* duplicem] simplicem *F om. P* 4 ipsius] illius *J* ipsius
radicis *BC* enim radicis ipsius expremitur et *F* 5 assimilatur] similat *LBC* simulans *P* 6
radicis- aliquid] *om. J* 7 leuis] rubeis *FL* lenis *P* tertius- *fin.*] *om. F* 8 basiliscam] basilicam
JPC 9 grauiora] graciliora *LCP* 10 ursi] moris *BPC* aureo] *om. BC* croceo] cryso *L* coc-
cum *B* coctum *P* 11 colorem] *om. BPC* 13 genere] generationes *L* genere male fricorum
et ueneno *BC* serpentum] maleficorum et ueneno *P*

deest W

271) Cretan alexanders (*Smyrnium perfoliatum* Mill.)

Cretan alexanders looks like celery,[1] but the leaves are wider, thicker, and tougher, are slightly golden in colour, and emit a sweet scent with a degree of pungency. It has a small stalk and little umbel like dill,[2] and a round, black seed that has an almost myrrh-like[3] taste. Its root is juicy, soft, smooth, has a sweet smell, and a black peel which is green or white on the inside and has a taste that stings in the throat. It has properties that are sharp and which cause perspiration, and therefore helps those suffering from coughs, respiratory problems, dropsy, and hip diseases, and is good for snake bites.

1 #14.
2 #35.
3 *myrrastrum*: privileging *FL*, though *P* 'fennel-like' (*marastrum/marathrum*) and *J* 'remarkable' (*mirastrum/miratum*) are also possible readings. See ch. 4.E.

272) Man orchis (*Fritillaria graeca* L.) and Heart-flowered orchid (*Serapias cordigera* L.)

Man orchis is a plant with a long, branchless stalk bearing a purplish and off-white flower. Its leaves are like those of the lily,[1] but are slightly shorter and reddish, and three or four of them hang down to the ground. Its root is a doubled bulb, like a pair of testicles, and when drunk it is an aphrodisiac. The juice from the root is an excellent cure for ophthalmia when smeared on [the eyes]. There is another type of man orchis called *erythraicum*, which resembles linseed,[2] and is considerably larger, smooth and shiny. The bark of its root is thin and smooth, and on the inside is white and has a sweet taste. There is a third type of man orchis that they call *basilisca*,[3] whose leaves are similar to the man orchis just mentioned, but are heavier and have jagged, slightly dark holes through them. Its root resembles a bear's paw, and it bears a golden or yellow flower which when crushed in water or wine turns it the same saffron colour. It contains properties that heat and which are useful against poisons, so that anyone who has this plant with them is safe from all types of snakes.

1 #155.
2 That is, flax (*Linum usitatissimum* L.), not mentioned elsewhere in *AG*, which here follows Greek nomenclature of flax-seed (λινόσπερμον) to describe the plant. Cf. Pliny 19.2–25. This type, the heart-flowered orchid (*Serapias cordigera* L: André 1985, 228), is also mentioned by Diosc. III.128 (ἐρυθραϊκὸν σατύριον) as an aphrodisiac 'when held in the hand but more so when drunk with wine'; likewise Pliny 26.98 (also used on animals).

cf. Diosc. III.128 (σατύριον). Pliny 19.2–25, 26.96–8. Galen *Simp*. 8.18.5 (K.12, 118). Ps. Apul. 130 (basilisca). Isidore 17.9.43.

273) Sisarium

Sisarium notum est omnibus et huius radix uires habet diureticas.

1 notum- omnibus] *om. L* et] ergo *BC* uires] *om. LB* habet] est *WC*

deest P

cf. Diosc. II.113 *(σίσαρον)*. Pliny 19.90–2, 20.35. Celsus 2.21. Galen *Simp*. 8.18.19 (K.12, 124).

274) Scolopendria

Scolopendria similis est scolopendriae animali, in petris humidis nascitur. Nec florem habet [nec semen], folia ipsius in superficie uiridia sunt, subter rufa et lanosa. Cum aceto potata splenem curat et extrinsecus cum uino posita iuuat et calculos expellit. Vires habet diureticas.

2 nec] nec semem nec florem *F* 3 subter] sed *J* lanosa] limosa *J* 4 uino posita] uino cocta mirabiliter operatur. Nam et ictericas cum uino potata *F* calculos] cauculosos *F* diueticas] *om. F et add*. tertius autem satirio etc. *ex #272 supra*.

deest LBPCW

cf. Diosc. III.134 (ἄσπληνος). Pliny 27.34. Galen *Simp*. 6.1.67 (K.11, 841). Ps. Apul. 56. *Ex herb. fem*. 40. Isidore 17.9.87.

275) Scopa regia

Scopa regia facile est nota omnibus et habet radicem tenuem et oblongam et uirgulas oblongas et folia ex interuallo assimilia lactucae, sed fixiora, lanuginosa et fortiora. Vires autem habet ipsa acres et diureticas.

1 Scopa regia] scupareia *F* facile] *om. W* facile- omnibus] *om. LP* est- omnibus] potest *F* 3 lanuginosa] *om. JL* fortiora] *om. L* habet] habet in ipsa *JW*

cf. Diosc. III.154–7 (vide #299), IV.144 (μυρσίνη ἀγρία). Pliny 21.28, 23.166. Scrib. 153. Galen *Simp*. 8.20.5 (K.12, 148). Marc. *Med*. 26.1.

3 Not mentioned by any other source except Ps. Apul. 130, *herba basilisca*, 'which Italians call *regia*' (or *regula*, l.12), but it remains unidentified: Maggiulli and Buffa Giolito 1996, 173. It shares the description of the leaves, the similarity of the root to a bear's paw, and the protection from snakes: see ch. 3.F.

273) Skirret (*Sium sisarum* L.)

Skirret[1] is well known. Its root has diuretic properties.

1 Though *sisarium* is merely a Latinized version of the Greek word for parsnip (σίσαρον, Diosc. III.113, Diosc. Lat. sisarum), parsnip has already been given above, #222 (*pastinaca*); hence I follow here the suggestion of Halleux-Opsomer 1982a, 97, for skirret (*Sium sisarum* L.), though see Andrews, 1958c, 146–9. André 1985, 241, opts for *Pastinaca sativa* L., citing Diosc. and *AG*.

274) Miltwaste / Rustyback fern (*Ceterach officinarum* Willd.)

Miltwaste[1] looks like the insect known as the centipede, and grows among damp rocks. It bears no flower [or seed], its leaves are green on the top surface, but red and woolly underneath. When drunk with vinegar it heals the spleen, is beneficial to the same when applied topically with wine, and it expels bladder and kidney stones. It contains diuretic properties.

1 Eng. Miltwaste (lit. 'spleen-reducer'), alternatively 'spleenwort,' preserves the etymology of the Gr. ἄσπληνος: Diosc. III.134 also mentions its resemblance to the centipede (*scolopendra*), and that it is also called *scolopendrion*: likewise Ps. Apul. 56. Cf. Pliny 27.34 'Asplenon sunt qui hemionion vocant.' See André 1985, 230.

275) Royal broom (~ Butcher's broom, *Ruscus aculeatus* L.)

Royal broom[1] is generally known to everyone. It has a slender, oblong root and oblong branches, its leaves are spaced out like those of a lettuce, but are more sturdy, downy, and tougher. It has sharp and diuretic properties.

1 Halleux-Opsomer 1982a, 97, identified this as St John's wort (*Hypericum perfoliatum* L.), another species of which is given below #299 (*hypericum* = *Hypericum crispum* L.); André 1985, 231, records the uniqueness of *AG*'s description and declares its identification 'indéterminée.' Other sources use the name *scopa regia* for butcher's broom (*Ruscus aculeatus* L.), *Bunium pumilum* Sm., and camomile (*Anthemis nobilis* L.). The equally mysterious *scopa regia* of Pliny 21.28 possibly refers to *Chenopodium scoparium* L. or Wilkin's bellflower (*Camphorosma monspeliaca* L.), and may be what was intended here.

276) Sinopis

Sinopis optima est quae in Lemno insula inuenitur et huius ipsius quae sphragis appellatur. In terris autem sinopis quod in quibusdam spelaeis colligitur et caprino sanguine in modum fermenti subigitur et in pastillos redigitur. Adhuc optima sinopis dicitur cuiuscumque loci erit, si
5   bene fuerit cocta, quae est rubicundissima et spissa, grauis, per omnia eodem colore et sine lapillis, quae ex minima mica plurimum humoris in similitudine sanguinis inficit. Vires habet efficaciter stringentes.

1 Sinopis] sinopidum id est minium *F* sinodope *W* sinopide *P* 2 sphragis] fragilis *F* fragis *BP* fragile *W om. L* in terris] in quores *F* in choris *L* integrum *BPC om. W* spelaeis] *om. L* 3 caprino sanguine] exprimitur *BP* primitur *C* expressus *W* in modum- subigitur] *om. W* fermenti] frumenti *F* pastillos] patellis *BW* 4 redigitur] positus et decoquuunt *W* adhuc] ob hoc *FLBC* ad hoc *P* illud *W* sinopis- est] *om. W* dicitur] dicimus *LB* 5 grauis- colore] *om. W* 6 quae- inficit] quod in modum sanginis inficit *W* minima] una *J* minica *L* 7 efficaciter] acriter *B om. W*

cf. Diosc.V.96 (μίλτος Σινωτική), V.97 (Λημνία γῆ). Pliny 35.31.

277) Sinon

Sinon semen est quod maxime de Assyria defertur, fusculum, oblongum, acre et uiscide excalfactorium, propter quod ad inflationes et rosiones stomachi et coli proprie facit.

1 Sinon] sium *J* sinonus *F* sinon *L* sinonos *B* sinonos *P* defertur] affertur *F* fertur *P* 3 rosiones] *om. FL* torsiones *BC* tortiones *P* coli] cholerae *J*

deest W

cf. Diosc III.55 (σίνων). Pliny 27.136. Theod. *Log.* 29, 89. Alex.Tral. 2.35.

278) Tragacantha

Tragacantha est lachryma herbae lignosae et spinosae, cuius radix incisa hanc ipsam lachrymam copiose emittit. Quae aliquatenus siccata, prius tamen quam indurescat, cultello in tenues laminas secatur. Est ergo optima tragacantha illa mundissima et candidissima, tenuis et lucens,
5   recens et subdulcis. Vires autem habet quales sarcocolla et aliquatenus efficaciorẹs magis enim stringit plus et lachrymam reprimit.

1 Tragacantha] tracacantum *F* trachanta *L* tragagantum *W* tracacanta *B* tragacanta *PC* 2 hanc] unde *F* copiose] *òm. J* quae- secatur] *om. W* siccata] *om. L* 3 laminas] lactucidis *F* latigulis *BP* secatur] siccactus est *F* sic per siccit *L* siccatur *B* siccamur *C* secatur et siccatur *P* 4 candidissima] *om. F* tenuis- subdulcis] *om. W* subdulcis] succis *F* 6 reprimit] *om. L* lacrimas stringere potest *W*

cf. Diosc. III.20 (τραγάκανθα). Pliny 13.115. Celsus 6.6.7. Scrib. 22. Galen *Simp.* 8.18.19 (K.12, 143). Alex. Tral. 1.88–9.

276) Sinopic red ochre

The best Sinopic red ochre is found on the island of Lemnos, and is called *sphragis*.[1] Sinopic red ochre is collected from the ground and in caves. It then mixed with goat's blood and left to leaven, and then is moulded into lozenges. It is said that excellent Sinopic red ochre can be from any place so long as it has been well leavened, is richly red, and is thick, heavy, with no other colours in it, free of rubble, and merely a few grains of it mixed with a lot of liquid can dye things the colour of blood. It has effective astringent properties.

1 Diosc. V.97 (Λημνία γῆ) states that Lemnian earth is mixed with goat's blood and stamped with the figure of a goat and called *sphragis* (lit. 'seal,' 'signet'), while Sinopic red earth V.96 (μίλτος Σινωπική) is collected in caves in Cappadocia and is sold in Sinope (today Sinop, Turkey, on the Black Sea coast). Pliny 35.31 says the ochre *sinope* was first discovered in Sinope, hence its name, and also grows in Egypt, Africa, and the Balearic Islands, but the best is mined from caves in Lemnos and Cappadocia; the red ochre (*rubrica*) of Lemnos is particularly prized and is sold stamped and called *sphragis*.

277) Stone parsley (*Sison amomum* L.)

Stone parsley[1] is a seed which is mostly imported from Assyria and is quite dark, oblong, pointed, and is a powerful calefacient, and is therefore good for gas and gnawing sensation in the stomach and in the colon.

1 The reading of *J* (*sium*: see #158) hindered previous attempts at identification.

278) Tragacanth (*Astragallus gummifer* Labill. : *Astragallus microcephalis* Willd.)

Tragacanth is the sap of a woody, thorny plant, the root of which is cut to release copious amounts of this sap, which is then dried for a little while until it has nearly hardened, and then cut with a knife into thin strips. The best tragacanth is that which is really pure and brightest in colour, thin and clear, fresh and slightly sweet. It has the same properties as sarco-colla[1] though is a little more effective, for it is more astringent and better at blocking tears.

1 #235.

279) Thus

Thus lachryma est arboris quae in Arabia et in India nascitur quae graece libanus dicitur. Quod ergo de Arabiae arbore manat candidius est,
quod de India subrufum. Verum utrunque masculum appellatur. Est et
tertium minutum ualde et rufum de quo maxime manna conficitur. Est
5   ergo optimum thus quod masculum diximus, subrufum, rotundum et
naturaliter lene, fragile et si massetur in miculas minutas et ut sabulum
dissipatur et quod ad ignem cito accenditur odorem suauem et copiosum emittit. Vires autem habet leniter excalfacientes. Sed adulteratur
resina pityina et gummi cum more commistis et sic in speciem sinceri
10   thuris formatur. Idem et manna adulteratur aliqui, laeuigata resina pityina ut farina et cortice thuris et cortice nuclei pinei. Sed deprehenduntur ita. Thus non quod resinam ut gummi habet si masticetur, fit quasi
mastiche et ad ignem non cito accenditur, nec talem odorem habet.
Similiter et manna et cortex, si sincera non sint tardius et fumosius
incenduntur.

1 Thus] tus *W* india] indica *BC* 2 libanus] *ΛΙΒΑΝΩΝ BC om. P* arabiae] arabica *BC*
arbore manat] *om. W* 3 utrunque- rufum] *om. L* 4 tertium] tertiarum *B* tertiarium *C*
etiam *P* de- conficitur] quod mannis dicitur *W* manna] annis *F* est] erit *JL* 6 naturaliter-
emittet] *om. W* lene] leue *L* massetur] pincatur *J* frangitur *F* 7 cito] *om. BC* 8 leniter]
uelociter *W om. F* sed- *fin.*] *om. W* 9 cum more] et thure *J* commistis- formatur] idem *P*
sic] fit *J* 10 formatur] firmatur *F* formans *L* fortis *J* adulteratur- pityina] *om. F* 11 cortice]
om. F ut] uel *JC* 12 ita- non] status ut *F* statim enim *BC* statim si massetur *P* 13 mastiche]
masasta *L* assata *BC* mastiche assata *P* non] *om. L* nec] nec non *BC* 14 fumosius] *om. PC*
15 incenduntur] concipit *L def. F* accenditur *BPC*

cf. Diosc. I.68 (λίβανος). Pliny 12.52–64, 29.119. Scrib. 81. Isidore 17.8.2–3.

280) Tithymallum

Tithymallum fere tota habet uocabula quot et species, sed non refert
eadem autem omnium uis est. Colligitur ergo et quasi lachryma et
quasi succus, quasi lachryma sic. Singulae uirgulae prope ipsa capitella
franguntur et sic lachryma ueluti colum quod manat in conchulas quasi
5   opium colligitur et siccatur. Similiter et hoc est uirosissimus. Succus
deinde ab ipsis foliis contusis et expressis excipitur et eadem ratione
siccatur et est infirmior. Verum eadem facere tithymallum potest, quae
et elaterium facere diximus, et extra quasi psilothrum pilos tollit et
si incidat cutem qua causticum urit, maxime cum est recens et dum
colligitur.

1 Tithymallum] titimalum theretho *F* titimallum *W* titmalus *P* refert] fert *L om. BPC* 2 uis]

279) Frankincense (*Boswellia carterii* Birdw.).

Frankincense is the sap of a tree which grows in Arabia and India, and which in Greek is called *libanus*. The sap which flows from the Arabian tree is whitish in colour, while that from India is a little reddish: both of these trees are called masculine. There is a third type which is very small and red, from which frankincense powder[1] is chiefly made. The best frankincense, which as we said is the male of the species, is slightly reddish, rounded, naturally smooth, delicate, and when chewed it breaks apart into small particles and disperses like sand, and when placed in a fire it ignites quickly, and releases a sweet and abundant fragrance. It has gently heating properties. It is usually adulterated by mixing it with pine resin and gum,[2] which gives it the appearance of pure frankincense. Others adulterate the powder in a similar way, using pine resin that is ground down to a fine flour, along with the bark of the frankincense tree and the shell of the pine nut.[2] The adulteration is revealed by the following traits: if frankincense does not have a resinous consistency like gum when it is chewed, and is like mastic; or it does not ignite quickly when placed in fire, or emit the fragrance mentioned. Likewise, if the powder and the bark used to adulterate it burn very slowly and very smokily, they are not pure.

1 *manna*: also referred to by Diosc. I.68.6, but he does not mention a third type of frankincense as here. Cf. Pliny 29.119 (*manna turis*), Scribonius 81.
2 See #228 n.1; #47.

280) Spurge (*Euphorbia* L.)

Spurge has almost as many names as there are types, but is it said that not all have the same property.[1] It is prepared both as a sap and as juice which is almost like a sap. The tips appearing on individual branches are crushed and sap very much like glue flows out and, just like opium,[2] is collected in shells and then dried, and likewise this method produces the most powerful sap. The juice is made by pounding and pressing the leaves and is then dried by the same method, but it is less effective. Spurge can do everything that we said the squirting cucumber[3] can do, and moreover can remove hair like *psilotron*,[4] and when it is collected, is very fresh, or recently cut, it burns the skin like a caustic.[5]

1 Diosc. IV.164 mentions that there are seven types (likewise W, uniquely among the manuscripts), but none of his descriptions matches the above (except see nn.4, 5).

uisa *F* et quasi lachryma- succus] *om. BPC* **3** uirgulae] *om. BPC* **4** sic] *om.
BPC* colum] solum *J* culum *L* colatum *BP* colami *C* manat] emanat *P* **5** uirosissimus] uiridissimum *J*
uirosissimus- siccatur] *om. P* **6** ab] alii *BC* **8** extra- psilothrum] qua ispi potum *L om. PC*
9 incidat] sedat *F* insidat *BPC* qua] quasi *LB* cum] quod *F* quando *J*.

totum **W**: Titimallum omnibus nota est et species cuius sunt septem quorum nota in
herbarii libro repperiuntur. sunt omnium uirtus una esse dignoscitur prouocatur enim
uentrem superius riferit.

cf. Diosc. IV.164 (τιθύμαλλος). Pliny 26.62. Galen *Simp.* 8.19.7 (K.12, 141–3). Ps. Apul.
109. *Ex herb. fem.* 33, 41. Isidore 17.9.77.

281) Thapsia

Thapsia succus est herbae eiusdem uocabuli quae folia habet similia
foeniculo, sed rectiora et uastiora. Thyrsulum quasi ferula, florem
uiolaceum, semen latulum et ferulae seminis simile. Radicem albam et
uastam, quam qui uolet effodere succum illius exprimere, prius cor-
5   poris sui quicquid erit nudum ceroto liquido perungere debebit. Sin
minus, inflationes et pustulas circa faciem et manus patietur. Exprimitur ergo succus thapsiae lota ipsa radice et in pila contusa et sic organo
pressa. Qui succus in mortario laeuigatur et in sole siccatur. Aliqui et
folia cum radice iungunt et in unum tundunt et exprimunt, quod est
10 inertius. Alii autem, quod est optimum et uiscidissimum, ipsam radicem refossa terra nudant, ita tamen quae una in humo maneat et sic
eam cum ferula sua acuto scalpello incidunt et ueluti scarificant et lachrymam quae manat colligunt. Vires autem habet thapsia acerrimas et
in modum ignis causticas. Prodest tamen longinquis uitiis, illa ratione
15 qua et sinapismus.

1 Thapsia] tapsia *FP* **2** rectiora] recentiora *L* florem] *om. BC* **3** latulum- simile] latum
J **4** uolet] uult *BC* **5** liquido- sin] *om. F* debebit] *om. P* sin] *om. L* ut *BPC* **6** pustulas]
pustulas faciat *BC* faciem et manus] facinus *F* patietur] *om. BPC* **7** organo] *om. BPC* **9**
in unum] nimium *J* **10** inertius] interius *L* quod] ut *L* ipsius *BC* **11** in] semen *BC* humo]
terra *J* huma *B* hima *V* una- incidunt] *om. F* **13** manat] habeat *L* acerrimas] acrissimas *LB*
15 sinapismus] sepissimus *F* sinapis *BPC*

totum **W**: tapsia sucus est herbae similis feniculo. sunt folia uastiora habet et rectiora
tursum quasi ferula florem uiolatum et latum. uires habet acerrimas et in mundum ignis
caustias prodest tamen longinquis uiciis

cf. Diosc IV.153 (θαψία). Pliny 13.124, 19.173. Galen *Simp.* 6.8.2 (K.11, 885).

282) Thymum

Thymum herbula est humilis et ramosa et ramulis duris et lignosis, qui
folia habet pusilla et angusta et oblonga et subalbida. In summo capi-

2 #199.

3 #97.

4 See #8 n.4. Diosc. ibid. 'smeared on with olive oil in the sun it removes hair and makes the regrowth fine and blond, but in the end it destroys the hair entirely.'

5 Diosc., ibid., warns (without really saying why) that for its collection 'one must neither stand against the wind nor touch his eyes with his hands, but even before extraction one must anoint his body with suet or with oil combined with wine, especially the face, scrotum, and neck.'

281) Deadly carrot (*Thapsia garganica* L.)

Deadly carrot is the juice of the plant of the same name which has leaves similar to fennel,[1] but straighter and larger. It has a thin stalk like giant fennel,[2] a violet-coloured flower, and a broad seed also like that of giant fennel. Its root is white and large, and anyone who wishes to dig it out in order to extract its juice should first smear cerotic liquid on the parts of the body that are uncovered, for if they do not they will suffer swellings and blisters on the face and hands.[3] The juice of deadly carrot is extracted from its root after it has been washed, crushed with a pestle, and then pressed through a machine. This juice is then smoothed out over a mortar board and dried in the sun. Some add leaves to the root then crush them together to extract the juice, but the juice is less effective. Others – and this creates the best and strongest form – dig out the earth from around the root so that it remains in the ground exposed, and then cut it and its stem with a very sharp knife, scarring them in a way that makes the sap flow out, which they then collect. Deadly carrot has extremely sharp properties, and caustic properties that burn like fire. It is good for chronic illnesses, and for the same reason can also be used in a mustard-plaster.

1 #110. On the depiction of *thapsia* in Munich 377 (fol. 123v), see Riddle 1985, 199–200.

2 See #4 n.1.

3 Similarly Diosc. IV.153, adding that there must be no wind about.

282) Cretan thyme (*Satureia thymbra* sp. L).

Cretan thyme[1] is a low-growing shrub that has many boughs with small, hardy, and woody branches that bear small, narrow, oblong, and pale

tula in speciem formicarum in se implicitarum et concurrentium. Vires habet acres et excalfactorias et stringentes, propter quod omnibus uitiis stomachi et thoracis aptissimum est.

1 Thymum] tymum *F* thimus *W* herbula] herba *P* 2 subalbida] exalbida *B* subexaluida *C* in summo- concurrentium] *om. W* 3 se implicitarum] semplicitarum *C* concurrentium] coherentium *FL* currentium *BP* 4 stringentes- est] uicia stomachi et toracis curat *W* 5 thoracis] ptoracis *BC* est] habetur *L* habetur ualde *BPC*

cf. Diosc. III.36 (θύμος). Pliny 21.57, 21.154–7. Galen *Simp.* 6.8.7 (K.11, 887–8). Garg. *Med*. 36. Isidore 17.9.12–13.

283) Thlaspi

Thlaspi herbula est quae folia habet magnitudine digiti, thyrsulum subalbidum, tenuem et oblongum, semen assimile cardamo et quasi battutum et dilatatum. Vires quoque habet acres et relaxantes et ueluti mordentes. Vnde asperitudines oculorum et cicatrices extenuat et
5 potatum uentrem sursum et deorsum purgat. Cum oleo laeuigatum pro unctione frigora et horrores in accessione discutit.

1 Thlaspi] lapis *F* thalpi *BW* thalp *C* thalpa *P* folia] foliola *LPC* 2 tenuem] *om. F* semen] *om. BC* cardamo] cardamomum *F* et quasi- dilatatum] *om. W* 3 battutum] patulum *J* baptudum *F* battumum *C* 4 extenuat] extemiat *C* 5 sursum] *om. F* 6 pro unctione] punctiones *J* in unctioni *W* accessione] accidentes *W*

cf. Diosc. II.156 (θλάσπι). Pliny 19.118, 27.139. Scrib.129. Galen *Simp*. 6.8.5 (K.11, 886–7). Ps. Apul. 20 (cardamon). *Ex herb. fem*.16.

284) Trifolium

Trifolium herba est omnibus nota. Huius igitur folia et potata et pro fomento imposita uiperae morsus remediant. Item semen illius in emplastrum mittitur et eius radix in antidota. Vtraque ad uenenosorum serpentum morsus faciunt.

1 omnibus nota] *om. P* potata] pota *J* puto *F* poto *L* 2 fomento] frumento *F* 3 eius] *om. JL* antidota] antidota intraque *BC* utraque] intrat *P* quae *W* 4 faciunt] facientia *J* facit *F*

cf. Diosc.III. 109 (τρίφυλλον). Pliny 21.54, 26.89. Scrib. 163. Galen *Simp*. 8.19.13 (K.12, 144–5). Veg. *Mulomed*. 4.21.2. Marc. *Med*. 26.64 Isidore 17.9.72.

285) Veratrum

Veratrum radiculae herbae sunt eiusdem uocabuli, exalbidae et oblongae et ab uno capitulo copiosae et directae et quae per aetatem maiores quidem quam porri capitula sed non dissimiles. Quarum optimae sunt

leaves. The top of the plant bears little heads that are like ants which have entangled themselves and are locked together. It has sharp, heating, and astringent properties, and is therefore most suitable for all ailments of the stomach and chest.

1 On the different types and names in the ancient world, see Andrews 1958b, 1961a: on medicinal uses, Bonet 1993. See also #104. Similar description in Diosc. III.36, though 'on top it has very many little heads teeming with purple flowers.'

283) Shepherd's-purse (*Capella bursa pastoris* L).

Shepherd's-purse is a shrub which has leaves the size of a finger, and a thin, pale stem which is slender and long, and a seed that is similar to garden cress,[1] in that it seems bruised and widened out. It has both sharp and loosening properties, and properties that sort of bite. It therefore diminishes trachoma and corneal abrasions and when drunk it purges the bowels at both ends. Crushed with olive oil and used as an ointment it relieves frost-bite and the onset of shivering.

1 *cardamum* (not to be confused with *cardamomum*, #71, as does *F*), garden cress (*Lepidum sativum* L.), does not appear in *AG*. Diosc. II.156 also likens its small seed to garden cress ('[σπερμάτιον] δισκοειδές, οἱονεὶ ἐντεθλασμένον'), from which the plant gets its name (θλάω, bruise, crush). *Cardamum* (or *nasturtium*) was, however, fairly well known: André 1985, 49, and below #289.

284) Treacle clover / Trefoil (*Psorolea bituminosa* L.)

Treacle clover is a plant known to everyone. A drink is made from its leaves or it is applied in a compress as a remedy against viper bites. Likewise its seed is used in plasters, and its root is used in antidotes. Both the leaves and seed are good for poisonous snake bites.

285) White hellebore (*Veratrum album* L.) and Black hellebore (*Helleborus cyclophyllus* L.)

White hellebore[1] consists of the roots of the plant of the same name, which are pale and oblong, numerous and straight at one end, while the older

quae aspectu unctulae uidentur et non rugosae et quae medullam tenu-
5 issimam habent et gustu magis excalfaciunt et uomitum prouocant. Tale
est creticum, nam gallicum multum est inertius. Verum omne ueratrum
est uiscidissimum medicamentum et fortissimum purgatorium, in tantum ut uenenosum esse dicatur. Est et aliud ueratrum nigrum, quod
magis purgat, aliqui ambo et elleborum dicunt.

1 Veratrum] uaratrum *F* uelatrum *BPCW* uereatrum uel elleborum *tit. J* 2 ab- capitulo]
rugosae *W* directae] directae ab uno capite proedentes *W* quae] quasi *FL* quae- dissimi-
les] *om. BPC* quae- quae] *om. W* maiores quidem] minores sunt *F* 3 capitula] *om. L* non]
om. BPC 4 tenuissimam] tenuem *J* tenuissimam instrinsecus *W* 5 magis] *om. P* uomitum]
om. F iste uomitum *L* uomicam *B* prouocant] faciunt *L* tale- inertius] *om. W* tale] tale
namque *J* 6 gallicum] creticum *F* 7 in tantum] quoniam *W* 9 magis] magis de uisum *F def.*
L aliqui] aliqui secundum ambo purgant ueratra *J* aliqui utrumque uelatrum elleborum
dicunt album scilicet atque nigrum *W* aliqui utrum elleborum uocant *P*

cf. Diosc. IV.162 (ἐλλέβορος ὁ μέλας). Pliny 25.47–61. Celsus 3.18.17. Scrib. 10. Galen
Simp. 6.5.9 (K.11, 874). Isidore 17.9.24.

286) Viscum

Viscum succus herbae est quae de ramis arborum excrescit [et] quae
herba ramulos diffusos et lignosos habet, folia assimilis buxo sed fortiora modice, fructum rotundulum in specie uuae, pallidastrum. Qui
fructus tunditur et commodatur ad aquam frigidam copiose et diligen-
5 ter, usquequo nihil asperum de cortice fructiculus habeat. Sic enim euadit glutinosissimum, alii quoque in aqua coquunt et eadem diligentia
expurgatur. Verum optimum est quod in quercu arbore crescit et quod
est mundum, subuiride, purum et quod si extendatur non rumpitur
sed in specie membranae extenuatur. Vires autem habet acres et quae
durities remollire possunt.

1 Viscum] uiscus *W* herbae] eius quasi herbae *F* 2 ramulos- lignosos] *om. J* diffosos] effusos
CW folia- modice] *om. J* 3 modice] *om. P* fructum] fructulum *F* rotundulum] rotundum
JPW uuae] acinae uuae *F* acini uuae *L* pallidastrum *FLBCP*] palludiusculum *J* pallidum
W qui- uerum] *om. F* 4 fructus] fractus *F* commodatur] conmundatur *L* commedatur *B*
frigidam] *om. P* 5 cortice] cortice eius *J* 6 in aqua] ex dua *F* ex aqua *P* 8 est] est uiscum *FL*
mundum] colore *W* purum] putidum *FLPW* puridum *BC* 9 sed- extenuatur] *om. W*

cf. Diosc. III.89 (ἰξός). Pliny 16.245. Celsus 5.18.6–7. Galen *Simp*. 6.9.2 (K.11, 888–9).

287) Viola

Viola species habet tres. Est enim alba et purpurea et aurosa, flore-
sque de radice eius excrescunt, quarum melior habetur quae est purpurea. Verum omnes uiscide relaxant, sic et urinam prouocat aliquando

roots look very similar to the top parts of a leek. The best roots are those without wrinkles and which seem as though they have been lightly oiled, their marrow is extremely tender, their taste heats the mouth, and they cause vomiting. Those from Crete are like this, but those from Gaul are much weaker.[2] Every type of hellebore makes very powerful medicines, and purgatives so strong that they are claimed to be poisonous. There is another hellebore which is black and is an even stronger purgative, and some people call both [black and white] types 'hellebore.'[1]

1 Notably using the Latin name (*ueratrum*), which according to Pliny was that used in Italy, 25.52. On Pliny's otherwise somewhat fantastical account, see Mudry 2000.
2 No source mentions a Cretan variety; on Gallic cf. Diosc. IV.148 ('more likely to cause choking'), and Pliny 25.61 (used by Gauls when hunting).

286) Mistletoe (*Viscum album* L.)

Mistletoe is the juice of the plant which grows upon the branches of trees and has small but extensive, woody branches, leaves that are like boxthorn but are a little more hardy, and a somewhat pale fruit that is slightly rounded like a type of grape. The fruit is crushed and then thoroughly rinsed in very cold water until the remainder of the fruit has a totally smooth skin. Others, in order to rid it of stickiness, also cook it in water, which cleans it to the same degree. The best mistletoe grows in oak trees, and is pure, green, clear, and does not break when it is stretched, but thins out like a type of skin. It has sharp properties which can also soften indurations.

287) Violet (*Viola odorata* L.)

There are three types of violet[1] – the white, the purple, and the golden violet. Its flowers grow directly from its root, and among these the purple is considered the best. They all are powerful relaxants, and therefore can

et semen illarum partum expellit et lumbricos occidit. Odorem enim habet flos ipse suauem.

1 aurosa] aureo colore *J* 2 quarum- omnes] *om. W* habetur] *om. B* 3 sic] sicut *JC om.W* aliquando] aliquantum *JP* 4 semen illarum] *om. W* expellit] excludit *W* occidit] *om. P* 5 habet] habens est *J om. W*

cf. Diosc. IV.121 (ἴον). Pliny 21.27, 64, 130. Galen *Simp.* 6.9.3 (K.11, 889). *Ex herb. fem.* 58, 65 (purpurea). Isidore 17.9.19.

288) Vrtica

Vtrica est omnibus nota, et est calefactoria ualde, quapropter contrita ex oleo uel axungia inuncta contra frigus facit. Item potata ex uino morbum regium emendat et antidotis quae colicis sunt propria admiscetur. Item trita cum melle singula cochlearia diurno potui data tus-
5   sem ueterem ex frigore in pulmonibus conceptam sedat et calore suo omnes humores dissoluit.

1 est- et] *om. W* omnibus nota] *om.P* quapropter- 2 item] *om. W* contrita] attrita *J* 2 axungia] auxungis *B* inuncta] pro unctius *F* per unctrone *L* in unctione *BP* 3 antidotis] semen eius antidotis *FL* sunt propria] prosunt *BC* prodest *W* 4 singula] *om. F* singula-sedat] elactiquarum diximus tussem ueterem ex frigore conceptam emendat *W* tussem ueterem 5 ex frigore] tumorem uentris uel frigus *JBPC*

cf. Diosc. IV.93 (ἀκαλήφη). Pliny 21.92–4. Marc. *Med.* 27, 134. Isidore 17.9.44.

289) Vella

Vella herba est foliis quidem minor quam eruca sed non dissimilis, flore uiolaceo. Semen habet uelut in folliculis oblongis, ordinis stipati prout nasturtii et gustu calefactoria et submordens et uiribus relaxans. Vnde ad tussem ueterem et ad regium morbum cum melle optime facit.

1 Vella] ueia *F* uaeoia *L* uela *PW* non] *om. W* 2 uelut] constipatum *W* ordinis- nasturtii] uelut nasturtium *W* 4 optime facit] facit *F* mixta *W*

cf. Diosc. II.158 (ἐρύσιμον). Pliny 22.158. Galen *Simp.* 6.5.21 (K.11, 877–8). *Ex herb. fem.* 17. *Dyn.Vat.* 19/*SGall* 23.

sometimes stimulate urination, and their seed expels the foetus and kills intestinal parasites. The flower has a pleasant fragrance.

1 Diosc. IV.121 only mentions one type or colour: Riddle 1981, 80, identifies *Ex herb. fem.* 58 (*viola*), which likewise gives three types (*purpureum, album, melinum*) as wallflower (*Cheiranthus cheiri* L.), and points to the similar description of Diosc. III.123 (λευκόϊον), but Beck 2005, 236, identifies this as gilliflower (*Matthiola incana* L.). Pliny 21.130 says there are several types of violet but restricts comments to the purple, yellow, and white violets.

288) Stinging nettle (*Urtica* sp. L.)

Stinging nettle is known to everyone. It is a strong calefacient, and therefore ground up and smeared on with olive oil or axle grease it helps against the cold. When ground up and drunk with wine it cures the royal disease,[1] and it is mixed with antidotes which help with colic. Ground up and taken daily with a spoonful of honey it eases chronic coughs[2] derived from cold which has settled in the lungs, and its heat breaks down all humours.

1 See #196 n.5.
2 Similarly Diosc. IV.93.

289) Hedge mustard (*Sisymbrium officinale* Scop.)

Hedge mustard[1] is a plant with leaves that are smaller than those of rocket[2] but are otherwise quite similar, and it has a violet-coloured flower. It produces its seed in long pods together in rows just like the nasturtium.[3] The seed has a warming and somewhat biting taste and has loosening properties. It is therefore excellent for treating chronic coughs, and likewise the royal disease[4] when taken with honey.

1 Following the identification of Halleux-Opsomer 1982a, 96, and André 1985, 268. Meyer 1856, IV:494, proposed that this was the first known description of wards weed (*Carrichtera vella* DC.). Pliny 22.158 points out that *uela* is the Gallic name.
2 #100.
3 *Nasturtium* was used for different types of watercress of the Tropaeolaceae family (André 1985, 49) and is not mentioned in *AG*. Diosc. II.185 likens the seed to watercress (κάρδαμον), hence *nasturtium* may refer to the *cardamum* mentioned above #283. Cf. Pliny 19.155, 20.127.
4 #196 n.5.

290) Vetonica

Vetonica herba est quae radices habet aliquatenus grossitudine digiti et ramulos longos, quadratos, folia quasi quercus, suauiter olentia, florem purpureum, semen circa ramulos in circuitu in modum spicae nigrum. Nascitur in pratis et montibus. Vires habet leniter acres et diureticas ualde.

1 Vetonica] uetionica *L* uittonica *BP* aliquatenus] ad *W* grossitudine] crassitudine *J* 2 ramulos] *om. F* quadratos] *om. F* quadra folia *W* quadratos et oblongos *L* florem- montibus] *om. W* 3 modum] medium *P* 4 nascitur- montibus] *om. P* leniter] *om. W* 5 ualde] *om. JBPC*

cf. Diosc. IV.1 (κέστρον). Pliny 25.84, 122. Celsus 5.27.10. Galen *Simp.* 7.10.21 (K.12, 23–4).

291) Vitis

Vitis est, quam omnes nouerunt, ex cuius fructu uinum exprimitur. Huius igitur pampini id est folia uiridia in frontem frequenter imposita et mutata oculis ex rheumate dolentibus prosunt, eo quod stypticum et austerum habent. Botrus uero eius maturus calidus est, infundit et pro-
5 curat uentrem et maxime si albus sit. Passa autem uua ardorem facit et uentrem procurat. Immaturus uero botrus uuae calidus est et uentrem stringit, unde et dysentericis facit.

1 Vitis] uites *W* quam- nouerunt] *om. P* ex cuius fructu] ex qua *W* 2 pampini - est] pamputa *W om. J* imposita] exposita *JC* 3 mutata] immutata *JB* prosunt] propter *P* eo- habent] stipticae enim uiratas sunt et austere *W* 4 maturus] matrum *C* 5 uua] uua id est febricis- simam facit sicccat *F* uua id est fabriam siccam *L* 6 calidus] frigidus *J*

cf. Diosc. V.1 (ἄμπελος οἰνόφορος), V.3 (σταφυλή). Pliny 14.8–43 et passim. Ps. Apul. 67.10. *DynVat* II.4/*SGall* 77 (uva). Isidore 17.5.1–33.

292) Vinum

Vinum omne natura calidum est. Vires atque motum corpori praestat, eo quod calefacit et resumit et reficit omnia membra quae ex frigore et dolore constricta tenentur. Vinum et oleum si calidum ex unguento ponatur, oportet diu perfricare quo usque totum corpus illud perbibat.
5 Similiter et his qui in humeris dolore tenentur. Vinum recens calidum minus est uetusto et uentrem procurat. Vinum album diureticum est et toto corpori aptum. Vinum nigrum et dulce sanguinem nutrit et oculis contrarium est et capitis uertiginem, id est scotosin, et corpori grauitu- dinem et stomacho turbationem et somnos graues infert. Vinum maxi-

290) Betony (*Stachys officinalis* L.)

Betony is a plant that has roots about the thickness of a finger, long but slender square-shaped branches, leaves like the oak that are pleasingly fragrant, a purple flower, and black seed that grows in a circle around the branches like some sort of spike.[1] It grows in fields and upon mountains. It has properties that are gently sharp and vigorously diuretic.

1 Cf. Diosc. IV.1, at the top of a quadrangular stalk 'lies the the seed growing in a spike just like that of savory.'

291) Grapevine (*Vitis vinifera* L.)

The grapevine, which everyone knows, is the plant whose fruit is crushed to produce wine. Its tendrils, that is, its young leaves, when applied topically and frequently changed, treat painful discharges of fluid from the eyes because the tendrils have styptic and drying qualities. Its grape clusters, when mature, are warming: they are good for the stomach when plastered over it, and even more so when this is done with white grapes. When dried the grapes produce heat which is also good for the stomach. A cluster of immature grapes is warming, contracts the bowels, and is good for those suffering from dysentery.

292) Wine

All wine is naturally warming. It supplies the body with both strength and movement, because it heats, repairs, and restores any part of the body which is constrained by cold or pain. When applying wine with oil to the body as an ointment it should be well rubbed in until the body totally absorbs it. The same should be done for those suffering pain in the shoulders. Fresh wine is less warming than old wine, and it is good for digestion. White wine is a diuretic and is good for the entire body. Dark[1] and sweet wines nourish the blood, though harm the eyes, cause dizziness in the head, that is *scotosis*,[2] and heaviness in the body, as well as churning the

10 me si cum modica aqua bibatur saluti corporis prodest, eo quod nutrit
et custodire potest sanitatem. Primo quod album est magnum remedium corpori facit. Vnde et de ipsa aqua ratio est reddenda. Omnis
uirtus aquae frigida est et dulcis et humida et febribus acutis aliquoties
facit. Vulneribus cancerosis si sola adhibeatur perniciem infert, quia
15 omne uulnus in corpore dulce esse uidetur. Enim tunc gaudet cancer
et crescit, dum sibi consentiens et amplius nutriens incrementum acce-
perit. Nam cum melle et aqua iuncta aut lixiuia cancer semper lauari
praecipitur, aut cum uino et melle simul. Vinum autem acetum factum
incipit frigidum esse et stypticum et non nutrit sed magis compescit.
20 Fiunt enim de uino diuersis modis propomae corporibus aptae. Vnde
mel et uinum solum uentrem procurat et calefacit. Piper uero adiun-
ctum idem conditum plus calefacit et siccat. Conditum ex passo similiter et urinam deducit et corroborat stomachum et infrigidat maxime
si crocus modice admiscetur. Defretum calefacit et infundit et uentrem
25 procurat. Sambucatum similter ut mastichatum sed et uentrem procurat et choleram rubram temperat. Rosatum infrigidat et urinam et uentrem procurat, absinthium uentrem et choleram emundat, uiolacium
urinam procurat, plus stomachum corroborat et lumbricos excludit.

1 natura] *om. J* naturale *F* motum] multum *FL* 2 reficit] refrigit *B* 3 si] *om. F* ex unguento
4 ponatur] pro unguere *FL* per unguatur *BPC* illud] *om. LB* 5 humeris] humorem *F*
calidum- uestusto] uetustis calidiora sunt *BPCL* 6 minus est] *om. F* 8 scotosin] *om. FP*
scosin *L* grauitudinem] grauidinem *BC* 10 modica] *om. B* sicut maxime sicut modica *F* eo
quod- potest] *om. C* 11 est] sit *L* fit *B* 12 de ipsa- reddenda] danda est aqua rationem *F* 16
amplius] alius *F* consentiens] consentientem *LB* conspetientem *C* acceperit] accipiet *P* 17
nam] aut *LC* 18 simul] similiter *F* 19 non] *om. F* magis] magis humore *F* magis tumores
L 20 propomae] proponie *F* propome *LBC* propone *P* proportiones *J* corporibus aptae]
om. F 22 idem] id est *FC* in *J* conditum ex passo] *om. J* passo] expresso *F* 23 infrigidat]
om. J infrigide *F* frigidat *L* 24 modice] *om. F* defretum calefacit] *om. J* de frigido calefacit
F de fructu calefacit *P* 25 sambucatum- procurat] *om. F* sed] *om. J* 26 choleram] colorem
F rubram] rubeam *P* rosatum] anesatum *F* uentrem] *om. J* 27 procurat] prouocat *JP*
absinthium] absinthicatum *P* absinthium- procurat] *om. J* choleram] cholorem *F*

deest W

cf. Diosc. V.6 (οἶνοι), V.7 (μελιτίτης), V.8 (οἰνόμελι). V.10 (ὕδωρ). Pliny 14.59–70, 23.31–
53. Galen *Simp.* 8.15.2 (K.12, 88). Isidore 20.3.2–12.

293) Xylobalsamum

Xylobalsamum ramuli sunt arboris quae balsamus dicitur, tenues
[et] exalbidi, suauiter olentes, gustu et uiribus acres et calefactorii et
relaxantes.

stomach and inducing heavy sleepiness. Wine is excellent for maintaining a healthy body when it is drunk with moderate amounts of water, for then it both nourishes and can protect one's health against illness. White wine is the best, and constitutes an important medicine for the body. And here something needs to be said about water itself. All water has cold, soft, and moist properties, and hence is administered many times when treating acute fevers. But if it alone is applied to malignant wounds the result is disastrous, for all wounds to the body themselves are soft in nature. Then the malignancy thrives and grows, finding an agreeable environment in which it can nourish itself and develop further. But when water is mixed with honey or lye,[3] the malignancy should be repeatedly washed with it, or with wine and honey at the same time. Wine that has turned to vinegar instead becomes cooling and styptic, and does not nourish but rather staunches. Aperitifs[4] which are excellent for the body are made from wine in various ways. Honey and wine alone maintains regularity of the bowels, and warms them. When pepper is added to wine as a method of seasoning it warms even more, as well as dries. When seasoned with raisins wine draws down urine, fortifies the stomach, and cools it if a little saffron is added. Boiled down must warms, coats, and maintains healthy bowels. Wine that has been seasoned with marjoram[5] or mastic[6] maintains the bowels, and calms red bile. Rose-wine cools, regulates micturition, and maintains healthy bowels, while wine tinctured with wormwood[7] purges the bowels and clears cholera, and violet-wine maintains regular micturition, strongly fortifies the stomach, and prevents intestinal parasites.

1 *nigrum*: used for red wines, likewise Pliny 20.82, 23.39, as in Greek, Diosc. V.6 (μέλας οἶνος), Homer *Od.* 5.265.

2 σκότωσις: vertigo. See Galen, K.19, 417. Not used by Diosc., though cf. V.34 (wine flavoured with resin) 'σκοτωματικός' (again in II.70).

3 #146.

4 *propomae*, a rare term in Latin writings: see ch. 4.C. On medicinal uses of spiced wines, see Magdelaine and Fournet 2001.

5 #260.

6 #186.

7 #22.

293) (Twigs of) Mecca balsam (*Commiphora opobalsamum* Endl.)

The twigs of Mecca balsam[1] are the small, pale, and slender branches of the tree called balsam. They emit a sweet fragrance, are sharp in taste and in properties, which are both heating and loosening.

1 #202.

1 Xylobalsamum] xilobalsamum *PW* ramuli] ramusculi *W* dicitur] *om. W* tenues] tenues et *PC* tenuisse *F* 2 olentes] redolentes *W* gustu] *om. W* acres] *om. W*

cf. Diosc. I.19.3 (βάλσαμον). Pliny 12.118. Scrib. 110. Celsus 5.18.7B. Galen *Simp*. 6.2.2 (K.11, 846–7). Marc. *Med*. 15.99. Isidore 17.8.14.

294) Xylocinnamum

Xylocinnamum est assimile aspectu cinnami, sed grossius et longius et quodammodo lignosum, sed uiribus et odore infirmius et tamen ad eadem sumitur et eadem praestat.

1 Xylocinnamum] xylocynnamum *L* xilocinnamomum *FW* xilocinamum *P* aspectu] *om. W* cinnami] cinnamomi *F* grossius] crassius *J* grossior *F* gloriosus *W* 2 quodammodo lignosum] *om. F* infirmius] infirmius fere *F* infirmius fert *L* inferius fert *W* tamen] *om. J*

cf. Diosc. I.14 (κινάμωμον). Pliny 12.91. Scrib. 271. Galen *Simp*. 7.10.25 (K.12, 26). Marc. *Med*. 35.9. Alex. Tral. 2.44.

295) (H)Yosciamus

Hyosciamus est quem aliqui laterculum dicunt. Herba omnibus nota quae species habet tres. Est enim quae semen et florem album habet quaeque inertius est, quae utrumque nigrum, quaeque uelut purpureum maxime. Eligimus omnes quae quidem foliis et cauliculis non multum differunt. Vires habet refrigerantes et stringentes et marcorem facientes, quapropter succus illorum ex foliis et cauliculis dum adhuc sunt uirides expressus et siccatus et similiter ut opium in antidotis et collyriis mittitur et eodem genere omnem dolorem compescit. Aliqui uero eum interficere dicunt si bibatur et esse uenenosum.

1 Hyosciamus] yiusquiami *F* yosciamum *LB* yiusquiamus *P def. C* laterculum] altercum *J* herba- nota] *om. P* omnibus- multum] *om. W* habet] est *LBC om. F* 3 quaeque- est] *om. J* inertissima est *FL* hertius est *BC* quaeque] quaeque uehementius *J* purpureum] purpureum omnes *BC* 4 eligimus] quod eligimus *J* omnes] *om. BCW* quod *J* cauliculis] calcellissius *F* calicellis *BPC* calicellis suis *L* 5 differunt] *om. J* marcorem facientes] *om. W* 6 ex foliis- cauliculis] foliorum aut folliculorum *W* illorum] *om. J* 7 expressus et siccatus] exprimitur et siccatur *J* 8 collyriis] *om. F* genere] modo *JC* aliqui- uenenosum] *om. W* 9 interficere] insania facere *F* inficere *P*

cf. Diosc. IV. 68 (ὑοσκύαμος). Pliny 25.35–8, 26.89. Scrib. 181. Galen *Simp*. 8.20.4 (K.12, 147–8). Marc. *Med*. 26.33. Ps. Apul. 4. Alex.Trall. 2.183, 188. Isidore 17.9.41.

296) [Oes]Ypus

Oesypus est quasi sordes et succus lanarum sucidarum. Qui hac ratione colligitur. Lanas sucidas sordidissimas et mox detonsas mittimus in

294) (Twig of) Cinnamon (*Cinnamomum aromaticum* Nees)

Twig of cinnamon[1] is similar in appearance to cinnamon,[2] but is larger, longer, and a little woody. Its properties and fragrance are weaker than cinnamon, but it is used for the same purposes, and can be substituted for cinnamon.

1 There seems no precise modern identification: André 1985, 278. Diosc. I.14 notes that 'some say *xylocinamomon* is even generically different from cinnamon, being of a different nature': similarly Pliny 12.91.
2 #77, 75.

295) Henbane (*Hyoscyamus* sp. L.)

Henbane is that which others call *laterculum*.[1] It is a plant known to everyone, and there are three types. There is the weaker kind that has a white flower and seed; there is another type that has a black flower and seed, and another that has a purple flower and seed, which is the more powerful. We recommend all of those whose leaves and stems are nearly the same colour. Henbane has properties that are cooling, astringent, and induce languor. The juice of its leaves and stems is therefore extracted while they are still green and is then dried. This is used in antidotes and eye-salves in the same way as opium, and likewise eases all types of pain. Some say that it is fatal if drunk, and that it is poisonous.[2]

1 *laterculum*, as all MSS read: but André 1985, 139, specifically citing the *AG* (though for *latriculus*), believes it to be a corruption of *alterculum* (recognized by Pinzi, hence *altercum* in *J*), as found in Scribonius 181 and Ps. Apul. 4. Pliny 25.35 states that *altercangenum* is the Arab word for henbane. Cf. Diosc. IV.68 'but some call it *adamanta*,' and also refers to three types: white flower (superior), yellow flower, and purple flower (weakest).
2 See ch. 2.C.

296) Greasy wool fat

Greasy wool fat is obtained from the debris and juice of greasy wool. It is collected in the following way: take the dirtiest greasy wool as soon as

uase, quod habeat calidam aquam et aquam succendimus ut aliquantulum ferueat. Deinde refrigeramus et quod supernatat in modum pin-
5 guaminis abradimus manu et in uase stagneo abstergimus et sic ipsum
uas aqua pluuiali replemus et opertum tenui linteolo in sole ponimus
et rursus delimpidamus et tunc oesypum reponimus. Vires enim habet
cum aliqua acrimonia mollientes et relaxantes. Adulteratur ex pin-
guamine et ceroto molli, sed statim deprehenditur, eo quod sincerus
10 oesypus odorem reseruat succidae lanae et si manibus fricetur in similitudinem cerusae efficitur.

1 Oesypus] ysopum *F* ysopo cerotis *L* ysopocerotus *W* ysopum cerotis *BPC* quasi]
om. W qui- ratione] *om. W* 2 sucidas] *om. F* mox] iam mox *J* 3 et aquam] et aquam sic
aliquando *LBPC* simili quando *F* aliquantulum] aliquid *L* diu *W* 5 abradimus manu]
aufertur *W* in uase- reponimus] *om. W* stagneo] stanneo *J* abstergimus] absternimus *J*
tergimus *P* 6 replemus] implemus *J* opertum- linteolo] oportet tenuere uirente oleo *F*
linteolo] lineo *J* ponimus] ponimus biduo sic tertio die de aqua plena limpidamus et
aliam recentem mittimus et similiter relinquimus *F* ponimus biduo et sic ayacdie aquam
primam delimpadamus et aliam recoltem supermittimus et similiter reliquimus et rurus
L 8 cum- acrimonia] *om. W* pinguamine] pinguedine *J* 9 sed- efficitur] *om. W* statim] *om.*
J ista *F* sincerus] in sincere *L* non sincere *BPC* 10 odorem] succidum *J* odorem succidum
F fricetur] fricetur candidum uidetur *FL* 11 efficitur] *om. FL*

cf. Diosc. II.74 (οἴσυπος). Pliny 29.35, 30.28.

297) (H)Yssopum

Hyssopum herba est uero erraticum et hortulanum. Verum omne hyssopum facile notum est et habet uires acres et calefactorias. Vnde tussientibus et suspiriosis prodest et lumbricos expellit.

1 Hyssopum] ysopum *FLBPW* est- notum est] quam omnes nouerat *W* 2 facile- est] *om.*
P 3 et lumbricos expellit] *om. W*

def. C

cf. Diosc. III.25 (ὕσσωπος). Pliny 25.136. Galen *Simp.* 8.20.8 (K.12, 149). Marc. *Med.*
17.52. Isidore 17.9.39.

298) (H)Ypocisthis

Hypocisthis herba est quae nascitur exalbida et subrufa, huius succus exprimitur et reponitur non dissimilis acaciae sed mollior tactu et
effectu uiscidior. Stringit enim uehementissime. Vnde uentris profluuium et sanguinis abundantiam reprimit et sine mora praebet effectum.

1 Hypocisthis] ypoquistidis *F* ypocistis *L* ypoquistidas *W* ypochistis *BC* ypoquistis *P*
herba- subrufa] uocant quandam herbam *W* exalbida] *om. BPCW* subrufa] subrufa quam

it is shorn off and place it in a jar containing hot water and then heat the water until it boils for a little. Then let it cool and scrape off with your hand the greasy substance which floats upon the surface, place it in a tin container, fill the container with rain water, and cover it with a strip of linen and place in the sun. Rinse it once again, and upon storing it becomes greasy wool.[1] Its has properties that soften with degree of pungency, and loosening properties. It is adulterated with fat and soft wax-salve, but this is immediately revealed because pure greasy wool retains its smell of sappy wool and when rubbed between the hands it begins to resemble cerussite.[2]

1 The addition found in *FL* (waiting two or three days and re-rinsing) has a parallel in Diosc. II.74.
2 #46. Similarly Pliny 29.35. Var. 'it appears white like cerussite,' *FL*.

297) Hyssop (*Hyssopus officinalis* L.)

Hyssop[1] is a plant that has both wild and garden varieties. But all types of hyssop are generally well known, and have sharp and warming properties. Hyssop is therefore good for those with coughs and respiratory problems, and it expels intestinal parasites.

1 On the different types see Andrews 1961b. André 1985, 129, points out that Greek sources (e.g., Diosc. III.25) usually refer to the species *Satureia graeca* L. = *Micromeria graeca* Benth, and cites the *AG*'s two types as *Hyssopus officinalis* L., the variety 'pour Italie,' as with Pliny 25.136.

298) Hypocist (*Cytinus hypocistis* L.)

Hypocist[1] is a plant that is either a white or a tawny-coloured variety, the juice of which is extracted and stored just like that from the shittah tree,[2] but is softer in texture and more powerful in effect. It is an extremely powerful astringent, and therefore staunches diarrhoea and discharges of blood, and its effect is immediate.

1 The hypocist (lit. 'under the rockrose') is a parasitic plant (André 1985, 128), noticeable only when it blooms, that grows on the roots of the rockrose (*Cistis* sp. L.), noted by

Graece γποχιcταιω appellant *BC* ypocistidos *F* ypoquisados *L* 2 mollior] melius *F* mollior- uehementissime] melioris uirtutis *W* 4 effectum] *om. FLW*

cf. Diosc. I.97.2 (κίσθος). Pliny 24.81, 26.49, 81. Galen *Simp.* 7.10.28 (K.12, 28–9).

299) (H)Ypericum

Hypericum herba est, quae ramulos habet plures, tenues et oblongos colore subrufo, folia similia rutae, florem tenuem et aurosum, ut est similis uiolae, semen nigrum, forma hordei. Vires autem habet acriter stringentes, potatumque coxiosos emendat et urinam prouocat, uentrem stringit et pro fomento posita combusta sanat.

1 Hypericum] yppiciricum *F* ypyricum *L* ypericum *W* ypiricum *BPC* plures- subrufo] *om. W* 2 aurosum] aureo colore *J* aurosom- hordei] *om. W* 3 forma] farinam *F* 4 coxiosos] ischiadicos *J* coctum si manducetur *F* noxios emendat humores *L* 5 combusta] *om. B* composita *F* combustura *W*

cf. Diosc. III.154 (ὑπερικόν). Pliny 26.85. Galen *Simp.* 8.20.5 (K.12, 148). *Ex herb. fem.* 47.

300) (H)Yacinthus

Hyacinthus thyrsulum habet leuem, florem pupureum, folia et radices quasi bulbi. Tota herba uiribus et effectu styptica est ualde.

1 Hyacinthus] yhacantum *F* yancintus *W* yacinthum *BC* yacinctus *P* habet] herba *F* 2 uiribus] uiribus imirnum est assimile appu sed folia habet latiora et subaurosa quasi pinguia et fortiora sed leniora *B* effectu] affectu *W* ualde] *om. F* in ziber est radix aromatici acer gustu et calida ualde *W*

cf. Diosc. IV.62 (ὑάκινθος). Pliny 21.170. Galen *Simp.* 8.20.1 (K.12, 146–7). Isidore 17.9.15.

301) Zmyrnium

Zimyrnium est herba similis apio, sed folia habet latiora et subaurosa et quasi pinguia et fortiora redolentia suauiter. Vires habet acres et sudorem suscitantes, unde tussientibus et suspiriosis et hydropicis prodest et serpentium morsus facit.

1 Zmyrnium] zimium *P* zimirnium *BC* zimirinum *Pinzi* zmirnion *W* herba] *om. BW* 3 suscitantes] prouocantes *W* suspiriosis] suspirium patientibus *W*

deest JFL

cf. Diosc. III.68 (σμύρνιον). Pliny 27.133 (**cf.** 19.162, 187). Scrib. 126. Ps. Apul. 107.6 (cf. 128.9).

Diosc. I.97.2, which contains similar information, though says that there is one type that is part white and part tawny: cf. Pliny 26.49, who also notes them as separate types.
2 #2. Similarly Diosc. I.97.2.

299) St John's wort (*Hypericum crispum* L.)

St John's wort[1] is a plant which has many small, slender, and lengthy branches that are a tawny colour, leaves that are similar to rue,[2] a delicate, golden flower that resembles the violet,[3] and a black seed that is the shape of barley.[4] It has severely astringent properties, and therefore when drunk it heals hip disease, stimulates urination, contracts the bowels, and heals burns when applied topically by means of a compress.

1 On the identification, see above #275, and ch. 3.D.
2 #227.
3 #287.
4 Cf. #240.

300) Wild hyacinth (*Scilla bifolia* L.)

Wild hyacinth has a smooth stalk, a purple flower, and leaves and roots like purse tassels.[1] The whole plant is vigorously styptic in terms of its properties and use.

1 Similarly Diosc. IV.62. Purse tassels (*Muscari comosum* Miller: Diosc. II.170 βολβὸς ἐδώδιμος, Pliny 21.170) does not appear in the *AG*.

301) Cretan alexanders

Cretan alexanders[1] is a plant similar to celery, but its leaves are wider, of a slightly golden colour, and are thick, a bit more tough, and emit a pleasant fragrance. It has properties that are sharp and which cause perspiration, and therefore is good for treating coughs, respiratory problems, and dropsy, and for snake bites.

1 This is an abbreviated version of the entry given above #271 (*smyrnium*), which appears in four manuscripts (*BPCW*; unfortunately *V* is defective from #261 onwards) and the Pinzi edition of 1490, but not the Junta edition of 1522 (*J*). The name *smyrnium/zmyrnium* was also used for alexanders (*Smyrnium oluasatrum* L., cf. Diosc. III.67, ἱπποσέλινον, Pliny 19.162, 187, 20.186), which perhaps explains the repetition.

Epilogus

Haec sunt pater carissime Paterniane, quae memoriae nostrae subuenire potuerunt et quae per nos ipsos probauimus et quae experti sumus et apud antiquos et receptissimos auctores medicinae inuenimus. Nunc pro tua examinatissima diligentia curauimus omnia intimare et etiam
5   singulas res cum sua scriptura comparare, ideo ut qui post auctorum lectiones ad hanc fuerit reuersus scripturam ne erret et in nullo possit reprobus inueniri.

1 Epilogus] prologus *BP om. C* pater] frater *JC* carissime] karissime *PC* charissime *J*
2 per nos] partim *J* per mei (?) *P* ipsos] ipsi *J* ipsum *P* probauimus] probaui *P* quae]
om. P experti sumus] expertus sum *P* 3 apud] quae apud *B* inuenimus] inueni *P* nunc]
non quod *P* 4 examinatissima] examinantissima *C om. P* curauimus] curabo *P* curabis
J curabimus *C* intimare] examinare *J* et etiam] etiam que *C om. P* 5 scriptura] captura
BP comparare] comparere *C* ideo] ne erret *B om. PC* ut qui] ut si quis *J* auctorum lectiones] lectionis auctorem *BP* anticorum lectiones *C* 6 reuersus] conuersus *J* scripturam]
scriptura *B* ne erret et] *om. JC supra B* 7 reprobus] reprobi *P* inueniri] inuenire *C Addit*
C ex Isidoro 17.9.1, 'exant quaranda[m] herbarum notaque ex aliqua sui causa resonant
habentes nominu[m] explanationem non aut[e]m o[mn]ium herbarum ethimologiam
inuenies. Nam p[ro] locis mutantur eti[am] nomina. (*Idem M2, sed sine hoc epilogo uel*
prologo initiali).

Epilogue

These, dear father Paternianus, are the things we could recall from memory, along with that which we ourselves have tried and tested, and which we have found among the ancient and most popular authorities. Now, because of your most exacting attentiveness, we have endeavoured to describe everything with each entry accorded its own article, so that anyone who turns to this work after reading the authorities will not commit an error or be faulted for anything.

Bibliography

✤

A. Ancient and Medieval Sources

Note: Anonymous sources are listed under their common title; if edited in a collection or in a journal article, the full reference is provided under the name of the editor(s) listed among *Modern Studies*.

Agnellus (*iatrostophista*) of Ravenna.
– ed. and trans. L.G. Westerink, D.O. Davies, K.M. Dickson, A. Kershaw, J.P. Peters, B.K. Robbins, T.A. Virginia, and J.J. Walsh. 1981. *Agnellus of Ravenna. Lectures on Galen's De sectis.* Buffalo: State University of New York.
– ed. and trans. N. Palmieri. 2005. *Agnellus de Ravenne. Lectures Galénique: Le De pulsibus ad tirones* (Centre Jean Palerne Mémoires 28) Saint-Étienne: Université de Saint-Étienne.
Aetius (Aetios) of Amida. *Aetii Amideni libri medicinales.* Ed. A. Olivieri. 1938, 1950. 2 vols. CMG VIII. 1–2. Leipzig.
Alexander of Tralles. *Therapeutica.* Ed. and German trans. Th. Puschmann. 1879. 2 vols. Vienna: W. Braumüller. Repr. Amsterdam 1963.
– (Latin) *Practica Alexandrri Yatros greci cum expositione glose interlinearis Jacobi de Partibus et Januensis in margine positi.* 1504. Lyon: F. Fradin.
– (Latin). *See* Langslow 2006.
Anthimus (physician). *De obseruatione ciborum: On the Observance of Foods.* Trans. and ed. M. Grant. 1996. Blackawton: Prospect Books.
Pseudo Apuleius. *Herbarius.* Ed. E. Howald and H.E. Sigerist. 1927. CML 4. *Antonii Musae de herba vettonica liber. Pseudo Apuleius herbarius. Anonymi de taxone liber, Sexti Placiti liber medicinae ex animalibus.* Leipzig and Berlin: Teubner.
– Facsimile edition. Ed. F.W. Hunger. 1935. *The herbal of Pseudo-Apuleius from the ninth century manuscript in the Abbey of Monte Cassino [Codex Casinensis*

97], together with the first printed edition of Joh. Phil de Lignamine [Editio princeps Romae 1481] both in facsimile described and annotated by F.W. Hunger. Leiden: E.J. Brill.

Aristotle. *Historia animalium.* Ed. and trans. A.L. Peck. 1965–91. 3 vols. Loeb. Cambridge, MA: Harvard University Press.

Caelius Aurelianus. *Celerum Passionum Libri III, Tardarum Passionum Libri V.* Ed. G. Bendz, with Germans trans. I. Pape. 1990. 2 vols. CML 6.1. Berlin: Akademie der Wissenschaft der DDR, Akademie Verlag.

– *De morbis acutis, De morbis chronicis.* Ed. and trans. I.E. Drabkin. 1950. *Caelius Aurelianus, On Acute Diseases and On Chronic Diseases.* Chicago: University of Chicago Press.

– *Gynaecia.* Ed. and trans. M.F. Drabkin and I.E. Drabkin. 1951. *Caelius Aurelianus. Gynaecia, Fragments of Latin version of Soranus' Gynaecia from a thirteenth century manuscript.* Baltimore: Johns Hopkins University Press.

Cassiodorus. *Institutes.* Ed. R.A.B. Mynors. 1937. *Cassiodorus: 'Institutiones divinarum et saecularium litterarum.* Oxford: Clarendon Press.

– Eng. trans. J. Halporn. 2004. *Institutions of Divine and Secular Learning. On the Soul.* Liverpool: Liverpool University Press.

Cassius Felix. *Cassius Felix De la Médicine.* Ed. A. Fraisse. Collection des universités de France. Série latine 366. Paris: Les Belles Lettres.

– *Cassii Felicis De medicina: ex graecis logicae sectae auctoribus liber translatus, sub Artabure et Calepio consulibus (anno 447).* Ed. V. Rose. 1979. Berlin: Teubner.

Cato (Marcus Porcius) the Elder. *On agriculture [De agri cultura].* Trans. W. Davis Hooper. 1979. Loeb. Cambridge, MA: Harvard University Press; London: Heinemann.

Celsus (Aulus Cornelius). *De Medicina.* Trans. W.G. Spencer (Latin text ed. Marx). 1935–8. 3 vols. Loeb. Cambridge, MA: Harvard University Press.

– Bks 1 and 2. Ed. and French trans. G. Serbat. 1995. *See* Modern Studies.

De observantia ciborum. Ed. I. Mazzini. 1984. *De observantia ciborum. Tradizione tardo-antica del Περὶ διαίτης pseudo-hippocratico l.II.* Università di Macerata. Pubblicazioni della facoltà di lettere e filosofia istituto fi filologia classica 18. Rome: G. Bretschneider.

Diaeta Theodori. Ed. Sudhoff. 1915.

Dioscorides. *Pedanii Dioscuridis Anazarbei De materia medica libri quinque.* Ed. M. Wellmann. 1907–14. 3 vols. Berlin: Weidmann (repr. 1958).

– Latin trans. ('Dioscorides Lombardus'). The editions below were based primarily on the recension in Munich 337 (saec. IX), in which book I is defective: Stadler's editions of books II–V benefited from the discovery of the recension in Paris, BN, lat 9332, also used by Mihaescu to reedit book I.

Book I, ed. K. Hoffman and T.M. Auracher. 1882. *Romanische Forschungen* 1: 49–105.

– ed. H. Mihaescu. 1938. *Dioscoride Latino. Materia medica libro primo.* Iasi.

Book II, ed. H. Stadler. 1899. *Romanische Forschungen* 10: 181–247.

Book III, ed. H. Stadler. 1899. *Romanische Forschungen* 10: 369–446.

Book IV, ed. H. Stadler. 1901. *Romanische Forschungen* 11: 1–121.

Book V, ed. H. Stadler. 1902. *Romanische Forschungen* 13: 161–243.

– Eng. trans. L.Y. Beck *Dioscorides. De Materia Medica.* 2005. Altertumswissenschaftliche Texte und Studien 38. Hildesheim: Olms-Weidmann.

– facsimile edition ed. F. Unterkircher. *Der Wiener Dioskurides. Coc. Med. gr. I. der Österreich Nationalbibliothek.* 1965. Graz.

Dynamidia (Ps. Hippocratic). *Vat.* = ed. Mai 1835 (ex Vat. Pal. lat. 1088, saec. IX ex., and Vat. Reg. lat 1004, saec. XII), *SGall* = ed. Rose 1864–70, II, 131–150 (ex St Gall 762, saec. IX).

Ex herbis femininis. (Ps. Dioscorides). Ed. Kästner. 1896.

Galen. *Claudii Galeni Opera Omnia.* Ed. C.G. Kühn. 20 vols in 22. 1821–33 (Medicorum Graecorum opera quae exstant). Leipzig: C. Cnobloch (repr. Hildesheim, 1964–5).

– *De simplicium medicamentorum temperamentis et facultatibus,* K.11 (1826), 379–892; 12 (1826), 1–377.

Gargilius Martialis. *Gargilii Martialis quae extant.* Ed. S. Condorelli. 1978. Roma: L'Erma di Bretschneider.

– Eng. trans. in R.M. Tapper. 'The Materia Medica of Gargilius Martialis.' 1980. Unpublished PhD dissertation. Classics, University of Wisconsin, Madison.

– *Les Remèdes tirés des légumes et des fruits.* Ed. and French trans. B. Maire. 2002. Collections des Universités de France. Paris: Les Belles Letters.

Hippocrates (Latin versions or translations are listed under the editor's name in Modern Studies below).

– *Oeuvres complètes d'Hippocrate.* Ed. E. Littré. 1839–61. 2 vols. Paris: Ballière (repr. 1961 Amsterdam: M. Hakkert).

Isidore of Seville. *Isidori Hipalensis Episcopi Etymologiarum siue Originum libri XX.* Ed. W.M. Lindsay. 1911. Oxford: Oxford University Press.

– *The Etymologies of Isidore of Seville.* Trans. S.J. Barney, W.J. Lewis, J.A. Beach, and O. Berghof. 2006. Cambridge: Cambridge University Press.

Marcellus (Empiricus) of Bordeaux. *Marcellus de medicamentis.* Ed. E. Liechtenhan and M. Niedermann. 1968. 2 vols. CML 5. Berlin: Akademie Verlag.

Oribasius. *Collectionum medicarum reliqiae.* Ed. J. Raeder. 1928–33. 4 vols. CMG 6.1–2. Leipzig: Teubner.

– (Latin). *Synopsis* (I–II). *Oribasius Latinus.* Ed. J. Mørland. 1940. Symbolae Osloenses. Suppl. X. Oslo: A. Brøgger.

– (Latin) *Synopsis* (III–IX). *Oeuvres d'Oribase*. Ed. U.C. Bussemaker, C. Daremberg, and A. Molinier. 1873–6. Paris: Imprimerie Nationale.

– (Latin) *Euporista*. *Oeuvres d'Oribase* VI. Ed. A. Molinier. 1876. Paris: Imprimerie Nationale.

Medicina Plinii, Plinii Secundii Iunioris qui feruntur de medicina libri tres. Ed. A. Önnerfors. 1964. CML 3. Berlin: Akademie der Wissenschaften.

Paul of Aegineta. *Paulus Aegineta*. Ed. J.L. Heiberg. 1921–4. 2 vols. CMG 9. Leipzig and Berlin: Teubner.

– Eng. trans. F. Adams. *The Seven books of Paulus Aeginata*. 1844–7. 3 vols. London: Sydenham Society.

Pelagonius. *Pelagonii ars veterinaria*. Ed. K.-D. Fischer. 1980. Leipzig: Teubner.

Pliny. *Natural History*. Ed. W.H. Jones, H. Rackham, and D.E. Eichholz. 1938–63. 10 vols. Loeb. Cambridge, MA: Harvard University Press.

Physica Plinii Bambergensis (Cod. Bamb. Med. 2, fol. 93v–232r). Ed. A. Önnerfors. 1975. Hildesheim: G. Olms.

Plinii secundi iunioris qui feruntur de medicina libri tres. Ed. A. Önnerfors. 1964. CML 3. Berlin: Academische Verlag.

Scribonius Largus. *Compositiones*. Ed. S. Sconocchia. 1983. Bibliotheca scriptorum graecorum et latinorum Teubneriana. Leipzig: Teubner.

Stephanus of Alexandria. *Stephanus the Philosopher and Physician*: *Commentary on Galen's Therapeutics to Glaucon*. Ed. and trans. K.M. Dickson. 1998. Leiden: Brill.

– *Stephanus of Athens, Commentary on Hippocrates' Aphorisms*, sects. 1–2. Ed. L. Westerink. 1984. Berlin: Akademie-Verlag.

– *Stephanus the Philosopher: A Commentary on the Prognosticon of Hippocrates*. Ed. J. M. Duffy. 1983. Berlin: Akademie-Verlag.

Theodorus Priscianus. *Euporiston libri III cum physicorum fragmento et additamentis pseudo-Theodoreis*. Ed. V. Rose. 1894. Leipzig: Teubner.

Varro (Marcus Terentius). *De re rustica*. Trans. W. Davis Hooper. 1979. Loeb. Cambridge, MA: Harvard University Press; London: Heinemann.

Vitruvius Pollio. *On Architecture*. Ed. and trans. F. Granger. 1931–4. Loeb. London: Heinemann.

B. Modern Studies

Bibliotheca casinensis (1873–94) = *Bibliotheca casinensis seu, Codicum manuscriptorum qui in tabulario casinensi asservantur series per paginas singillatim enucleata; notis, characterum speciminibus ad unguem exemplatis aucta, cura et studio monachorum ordinis S. Benedicti, abbatiae Montis Casini*. Tomes 1–5. Monte Cassino.

Adams, J.N. 1976. *The Text and Language of a Vulgar Latin Chronicle (Anonymous Valesianus II)* Bulletin supplement – Institute of Classical Studies no. 36. London: University of London.

– 1995. *Pelagonius and Latin Veterinary Terminology in the Roman Empire.* Leiden: Brill.

– 2003. *Bilingualism and the Latin Language.* Cambridge: Cambridge University Press.

– and M. Deegan. 1992. 'Bald's *Leechbook* and the *Physica Plinii.*' *Anglo-Saxon England* 21: 87–114.

– and M. Janse and S. Swain, eds. 2002. *Bilingualism in Ancient Society: Language Contact and the Written Text.* Oxford: Oxford University Press.

– 2007. *The Regional Diversification of Latin 200 BC–AD 600.* Cambridge: Cambridge University Press.

Akhondzadeh, S., M. Noroozian, M. Mohammadi, S. Ohadinia, A.H. Jamshidi, and M. Khani. 2003. 'Salvia Officinalis Extract in the Treatment of Patients with Mild to Moderate Alzheimer's Disease: A Double Blind, Randomized and Placebo-Controlled Trial.' *Journal of Clinical Pharmacy and Therapeutics* 28.1: 53–9.

Alexanderson, B. 1963. *Die hippocratische Schrift Prognostikon. Überlieferung und Text.* Studia Graeca et Latina Gothoburgensia 17. Stockholm: Almqvist and Wicksell.

Aliotta, G., et al., eds. 2003. *Le piante medicinali del Corpus Hippocraticum.* Naples: Istituto italiano per gli studi filosofici. Milan: Guerini e associati.

Aliotta, G., N.G. De Santo, A. Pollio, J. Sepe, and A. Touwaide. 2004. 'The Diuretic Use of Scilla from Dioscorides to the End of the 18th Century.' *Journal of Nephrology* 17: 342–7.

Amalia d'Aronco, M. 1988. 'The Botanical Lexicon of the Old English Herbarium.' *Anglo-Saxon England* 17: 15–34.

Amouretti, M.-C., and G. Comet, eds. *Des hommes et des plantes. Plantes Méditerranéennes, vocabulaire et usages anciens.* Cahier d'histoire des techniques 2. Université de Provence: Service des Publications.

André, J. 1949. *Étude sur les termes de couleur dans le langue latine.* Paris: Gap, impr. L. Jean.

– 1963. 'Remarques sur la traduction des mots grecs dans les textes medicaux du V siècle.' *Revue Philologique* 89: 47–67.

– 1985. *Les noms des plantes dans la Rome antique.* Paris: Belles Lettres.

– 1987. 'Sur quelques noms latins de maladies.' *Revue de philologie* 61: 7–12.

– 1991. *Le vocabuliare latin de l'anatomie.* Études anciennes 59. Paris: Belles Lettres.

– 1998. 'Chronologie des noms latins de trois maladies.' In Sabbah, ed., 1998, 9–18.

Andrews, A.C. 1941. 'Alimentary Use of Lovage in the Classical Period.' *Isis* 33: 514–18.

– 1942–3. 'Alimentary Use of Hoary Mustard in the Classical Period.' *Isis* 34: 161–2.

– 1948. 'Orach as the Spinach of the Classical Period.' *Isis* 39: 169–72.

– 1949a. 'Celery and Parsley as Foods in the Graeco-Roman Period.' *Classical Philology* 44: 91–9.

– 1949b. 'The Carrot as a Food in the Classical Era.' *Classical Philology* 44: 182–96.

– 1951. 'Alkannet and Borage in the Classical Period.' *Classical Weekly* 4: 165–6.

– 1956a. 'Sage as a Condiment in the Graeco-Roman Era.' *Economic Botany* 10: 263–6.

– 1956b. 'Melons and Watermelons in the Classical Era.' *Osiris* 12: 370–5.

– 1956c. 'Acclimatization of Citrus Fruits in the Mediterranean Region.' *Agricultural History* 35: 35–46.

– 1958a. 'The Mints of the Greeks and Romans and Their Condimentary Uses.' *Osiris* 13: 126–49.

– 1958b. 'Thyme as a Condiment in the Graeco-Roman Era.' *Osiris* 13: 150–6.

– 1958c. 'The Parsnip as a Food in the Classical Era.' *Classical Philology* 53: 145–52.

– 1961a. 'Marjoram as a Spice in the Classical Era.' *Classical Philology* 56: 73–82.

– 1961b. 'Hyssop in the Classical Era.' *Classical Philology* 56: 230–48.

Artelt, W. 1968. *Studien zur Geschichte der Begriffe 'Heilmittel' und 'Gift.'* Darmstadt: Darmstadt Wissenschaftliche Buchgesellschaft.

Auberger, J. 1999. 'Le buerre dans la Grèc antique. Une énigme dans l'histoire de la consommation.' *Histoire et Sociétés Rurales* 11: 15–30.

Baader, G. 1970. 'Lo sviluppo del linguaggio medico nell'antichità e nel primo medioevo.' *Atene e Roma* 15: 1–20.

– 1971. 'Lo sviluppo del linguaggio medico nell'alto e basso medioevo.' *Atti dell'accademia La Columbaria* 36: 59–109.

– 1972. 'Anfänge der medizinischen Ausbildung im Abendland bis 1100.' In *La scuola nell'occidente latine dell'alto medioevo*. Spoleto: CISAM, 669–772.

Balandier, C. 1993. 'Production et usages du miel dans l'Antiquité gréco-romaine.' In Amouretti and Comet, eds., 93–125.

Barlow, F. 1980. 'The King's Evil.' *English Historical Review* 95: 3–27.

Barnes, J. 1997. 'Logique et pharmacologie. À propos de quelques remarques d'ordre linguistique dans le *De simplicium*.' In Debru, ed. 1997, 3–33.

Bartoshuk, L.M. 1978. 'History of Taste Research.' In Canterette and Friedman, eds. 1978: 3–18.

Bates-Smith, E.C. 1973. 'Haemanalysis of Tannins: The Concept of Relative Astringency.' *Phytochemistry* 12: 907–12.

Beagon, M. 1992. *Roman Nature: The Thought of Pliny the Elder*. Oxford.

Beavis, I.S. 1988. *Insects and Other Inverterbrates in Classical Antiquity*. Exeter: University of Exeter.

Beccaria, A. 1956. I *codici di medicina del periodo presalernitano (secoli IX, X, XI)*. Rome: Edizioni di Storia e letteratura

– 1959. 'Sulle tracce d'un antico canone latino di Ippocrate e di Galeno I.' *Italia medioevale e umanistica* 2: 1–56

– 1961. 'Sulle tracce d'un antico canone latino di Ippocrate e di Galeno II.' *Italia medioevale e umanistica* 4: 1–75

– 1971. 'Quatrro opere di Galeno nei commenti della scuola di Ravenna all'inizio del medioevo.' *Italia medieovale e umanistica* 14: 1–23.

Beck, L.Y. 2005. *See above under* Dioscorides in Ancient and Medieval Sources.

Becker, G. 1885–7 (repr. 1973). *Catalogi Bibliothecarum Antiqui*. Bonn and Leipzig: Max Cohen et filium. Repr. Hildesheim, New York: Georg Olms Verlag.

Benassai, L. 1998. Per una lettura del testo delle cure ex animalibus. Dottorato di ricerca in filologia mediolatina. Unpublished. Università di Firenze. Roma, Biblioteca Nazionale Centrale, Diss. 90/235.

Bennett, B.C. 2007. 'Doctrine of Signatures: An Explanation of Medicinal Plant Discovery or Dissemination of Knowledge?' *Economic Botany* 61.3: 246–55.

Bensky, D., S. Clavey, and E. Stöger. 2004. *Chinese Herbal Medicine. Materia Medica*. 3rd ed. Seattle, WA: Eastland Press.

Berendes, J. 1902. (repr. 1970) *Des Pedanios Dioskurides aus Anazarbos Arznei-mittellehre in fünf Büchern. Übersetzung und mit Erklärungen versehen*. Stuttgart.

Bertelli, C., S. Lilla, and G. Orofino, eds. 1992. With introduction by G. Cavallo. *Dioscurides Neapolitanus, Biblioteca nazionale di Napoli, Codex ex Vindobonensis Graecus 1: commentarium*. Rome: Salerno Editrice.

Besnehard, P. 1993. 'Nommer la mandragore.' In Amouretti and Comet, eds. 1993, 127–34.

Bischoff, B. 1974. *Lorsch im Spiegel seiner Handschriften*. Münchener Beiträge zur Mediävistik und Renaissance-Forschung. Munich: Arben-Gesellschaft.

– 1990. *Latin Palaeography. Antiquity and the Middle Ages*. Trans. D. Ó Cróinín and D. Ganz. Cambridge: Cambridge University Press.

Bischoff, B., M. Budny, G. Harlow, M.B. Parkes, and J.D. Pheifer, eds. 1988. *The Epinal, Erfurt, Werden and Corpus Glossaries. Early English Manuscripts in Facsimile* 22. Copenhagen: Rosenkilde and Bagger. Baltimore: Johns Hopkins University Press.

Bodson, L. 1986. 'Aspects of Pliny's Zoology.' In French and Greenaway, eds. 1986, 98–110.

– 1991. 'Le vocabulaire latin des maladies pestilentielles et épizootiques.' In Sabbah, ed. 1991, 215–41.

Boehm, I. 2003. 'Toucher du doigt: le vocabulaire du toucher dans les textes médicaux grecs et latins.' In F. Gaide and F. Biville, eds., *Manus Medica*. Aix-en-Provence: Université de Provence, 229–40.

Bolling, G. Melville. 1897. 'Latin -astro-.' *American Journal of Philology* 18: 70–3.

Bonacelli, B. 1925. 'L'ammoniaco dell'antica Cirenaica.' *Bollettino di Informazioni economiche* 4: 1–12.

Bonet, V. 1993. 'Le thym médicinal antique, un cadeau divin.' In Amouretti and Comet, eds. 1993, 11–22.

– 2003. 'Les applications dans la pharmacopée végétale de Pline l'Ancien.' In Gaide and Biville, eds. 2003, 131–47.

Boon, H., and M. Smith. 2009. *55 Most Common Medicinal Herbs: The Complete Natural Medicine Guide*. Toronto: Robert Rose.

Boscherini, S. 1970. *Lingua e scienza greca nel de agri cultura di Catone*. Roma: Edizioni dell' Ateneo.

– 1993. 'Termini medici negli scritti di M. Porcio Catone.' In idem, ed. 1993, 31–44.

– ed. 1993. *Studi di lessicologia medica antica*. Bologna: Pàtron Editore.

– 2000. 'La dottrina medica communicata per epistulam. Struttura e storia di un genere.' In Pigeaud and Pigeaud, eds. 2000, 1–12.

Bosworth, C.E. 1981. 'A Medieval Islamic Prototype of the Fountain Pen?' *Journal of Semitic Studies* 25: 229–34.

Bozin, B., and N. Mimica-Dukić. 2007. 'Antibacterial and Antioxidant Properties of Rosemary and Sage (Rosmarinus officinalis L. and Salvia officinalis L.) Essential Oils.' *Planta Medica* 73: 164–5.

Bozzi, A. 1981. *Il trattato ippocratico Peri aeron, hydaton, topon, e la sua traduzione latina tardo-antica: concordanze contrastive con il calcolatore elettronico e commento linguistico-filologico al lessico tecnico latino*. Orientamenti linguistici 16. Pisa: Giardini.

Brown, M. 1998. *The British Library Guide to Writing and Scripts: History and Techniques*. Toronto: University of Toronto Press.

Brown, T.S. 1984. *Gentlemen and Officers, Imperial Administration and Aristocratic Power in Byzantine Italy, A.D. 554–800*. Rome: British School at Rome.

Brush. *See under* Dana.

Buffa Giolito, M.F. 2000. 'Topoi della tradizione letteraria in tre prefazioni di testi medici latine.' In Pigeaud and Pigeaud, eds. 2000, 13–32.

– and G. Maggiulli. 1996 *L'altro Apuleio: problemi aperti per una nuova edizione dell'Herbarius* I. Naples: Loffredo.

Byl, S. 1994. 'Les mentions d'Hippocrate dans Histoire naturelle de Pline.' In Vázquez Buján, ed. 1994, 163–70.

– 1999. 'La thérapeutique per le miel dans le *Corpus Hippocraticum.*' In I. Garofalo et al., eds., *Aspetti della terapia nel Corpus Hippocraticum.* Florence: Leo S. Olschki, 119–24.

Cain, W.S. 1978. 'History of Research on Smell.' In Canterette and Friedman, eds. 1978, 197–229.

Cameron, M.L. 1982. 'The Sources of Medical Knowledge in Anglo-Saxon England.' *Anglo-Saxon England* 11: 135–55.

– 1983. 'Bald's *Leechbook*: Its Sources and Their Use in Its Compilation.' *Anglo-Saxon England* 12: 153–82.

– 1993. *Anglo-Saxon Medicine.* Cambridge: Cambridge University Press.

Canterette, E.C., and M.P. Friedman, eds. 1978. *Handbook of Perception: Tasting and Smelling.* Vol. VIA. New York: Academic Press.

Capitani, U. 1991. 'I Sesti e la medicina.' In P. Mudry and J. Pigeaud, eds., *Les écoles médicales à Rome: actes du 2ème colloque international sur les textes médicaux latins antiques (Lausanne, septembre 1986).* Geneva: Université de Nantes, 95–123.

Cappelli, A. 1961. *Lexicon abbreviaturarum: Dizionario di abbreviature latine ed italiane usate nelle carte e codici specialmente del Medio-Evo.* Milan: Hoepli

Carnoy, A. 1959. *Dictionnaire étymologique des noms grecs de plantes.* Louvain: Publications universitaires

Centani, M. 1988–9. 'Nomi del male. "Phrenitis" e "epilepsia" nel *corpus Galenicum.*' *Museum Patavium* 5: 47–79.

Chandelier, J., L. Moulinier-Brogi, and N. Nicoud. 2006. 'Manuscrits médicaux latins de la Bibliothèque Nationale de France. Un index des oeuvres et des auteurs.' *Archives d'histoire doctrinale et littéraire du Moyen Age* 73: 63–163.

Chen, J.K., and T.T. Chen. 2004. *Chinese Medical Herbology and Pharmacology.* City of Industry, CA: Art of Medicine Press.

Cilliers, L. 2005. 'Vindicianus's *Gynaecia*: Text and Translation of the Codex Monacensis.' *Journal of Medieval Latin* 15: 153–63.

Cimolai, N. 2007. 'Sweet Success? Honey as a Topical Wound Dressing.' *British Columbia Medical Journal* 49.2: 64–7.

Clackson, J., and G. Horrocks. 2007. *The Blackwell History of the Latin Language.* Oxford: Blackwell Publishing.

Collins, M. 2000. *Medieval Herbals: The Illustrative Traditions.* London, Toronto: University of Toronto Press.

Conde Salazar, M., and M.J. López de Ayala. 2000. 'Recursos literarios en la obra de Teodoro Prisciano.' In Pigeaud and Pigeaud, eds. 2000, 33–46.

Condorelli, S. 1977. *See above under* Gargilius Martialis in Ancient and Medieval Sources.

Contreni, J. J. 1990. 'Masters and Medicine in Northern France during the Reign of Charles the Bald.' In M.T. Gibson and J.L. Nelson, eds., *Charles the Bald. Court and Kingdom*. 2nd rev. ed. Aldershot: Variorum, 267–82.

Corleto, L.M. 1992. 'The Concept of Mania in Greek Medical and Philosophical Literature.' *Medicina nei secoli* 4.3: 33–42.

Cosentino, S. 1997. 'La figura del medicus in Italia tra tardoantico e altomedievo. Tipologie sociali e forme di rappresentazione culturale.' *Medicina nei secoli, arte e scienze* 9.3: 361–89.

Court, W.E. 1999. 'The Ancient Doctrine of the Signatures or Similitudes.' *Pharmaceutical Historian* 29.3: 41–8.

Cox, P.A. 2007. 'Biodiversity and the Search for New Medicines.' *Planta Medica* 73.9:2.

Dalby, A. 2003. *Food in the Ancient World from A to Z*. London, New York: Routledge.

Daly, Lloyd William. 1967. *Contributions to a History of Alphabetization in Antiquity and the Middle Ages*. Brussels: Latomus.

Dana, J.D., and G.J. Brush. 1868. *A System of Mineralogy*. Oxford: J. Wiley and Son.

Daryaee, T. 2003. 'The Persian Gulf Trade in Late Antiquity.' *Journal of World History* 14.1:1–16.

Debru, A., ed. 1997. *Galen on Pharmacology: Philosophy, History, and Medicine: Proceedings of the Vth International Galen Colloquium, Lille, 16–18 March 1995*. Leiden, New York: Brill.

– 1997. 'Philosophie et pharmacologie: la dynamique des substances *leptomères* chex Galien.' In Debru, ed. 1997, 85–102.

– and Sabbah, G., eds. 1998. *Nommer la maladie: recherches sur le lexique gréco-latin de la pathologie*. Saint-Étienne: Publications de l'Université de Saint-Étienne.

– and Palmieri, N., eds. 2001. *Docente Natura. Mélanges de médecine ancienne et médiévale offerts à Guy Sabbah*. Saint-Étienne: Publications de l'université.

Deichgräber, K. 1965. *Die griechische Empirikerschule*. 2nd ed. Berlin, Zurich: Weidmann.

Dendle, P. and Touwaide, A., eds. 2008. *Health and Healing from the the Medieval Garden*. Woodbridge, UK: Boydell and Brewer.

Denniger, H.S. 1930. 'A History of Substances Known as Aphrodisiacs.' *Annals of Medical History* 2: 383–93.

De Renzi, S. 1852. *Collectio Salernitana* I. Naples: Filiatre-Sebezio. *See also* Jacquart and Paravicini Bagliani, eds. 2008.

Derolez, A. 1979. 'Observations sur la catalogographie en Flandre et en Hainaut aux 11e et 12e siecles.' In P. Cockshaw, M-C. Garand, and P. Jodogne, eds., *Miscellanea codicologica F. Masai dicata 1979*. Vol. 1 Ghent: E. Story-Scientia S.P.R.L., 229–35.

Deroux, C., ed. 1998. *Maladie et maladies dans les textes latins antiques et médiévaux. Actes du Ve Colloque International 'Textes médicaux latin.'* Brussels: Brepols.

Deroux, C., and R. Joly. 1978. 'La version latine du livre I du traité pseudo-hippocratique Du régime (*editio princeps*).' In G. Cambier, C. Deroux, and J. Préaux, eds., *Lettres latines du moyen âge et de la Renaissance.* Collection Latomus 158. Brussels: Latomus.

Desilve, J. 1890. *De schola Elnonense Santi Amandi a saeculo IX ad XII usque.* Dissertatio, Université Catholique de Louvain.

Diederich, N.J., and C.G Goetz. 2008. 'The Placebo Treatments in Neurosciences: New Insights from Clinical and Neuroimaging Studies.' *Neurology* 71: 677–84.

Diels, H.A. 1905. *Die Handschriften der Antiken Ärzte.* Leipzig: Königliche Akademie der Wissenschaften.

Duffy, J.M. 1983. Ed. and trans. *Stephanus the Philosopher: A Commentary on the Prognosticon of Hippocrates.* Berlin: Akademie-Verlag: 1983.

Durling, R.J. 1961. 'A Chronological Census of Renaissance Editions and Translations of Galen.' *Journal of the Warburg and Courtauld Institutes* 24: 230–305.

– and F. Kudlien, eds. 1991. *Galen's Method of Healing: Proceedings of the 1982 Galen Symposium.* Leiden: Brill.

– 1993. *A Dictionary of Medical Terms in Galen.* Studies in Ancient Medicine, 5. Leiden: Brill.

Elliott, C. 2004. 'Purple Pasts: Colour Codification in the Ancient World.' *Law and Social Inquiry* 33.1: 173–94.

Elsakkers, M. 2008. 'The Early Medieval Latin and Vernacular Vocabulary of Abortion and Embryology.' In Goyens, Leemans, and Smets, eds. 2008, 377–414.

Englert, K., J.G. Mayer, and C. Staiger 2005. 'Symphytum officinale L. – der Beinwell in der europäischen Pharmazie- und Medizingeschichte' ('Symphytum officinale L.: Comfrey in European Pharmacy and Medical History'). *Zeitschrift für Phytotherapie* 26. 4 (August): 158–68.

Evans, W.C., and G.E. Trease. 2002. *Pharmacognosy.* 15th ed. rev. with the assistance of D. Evans. Edinburgh, New York: W.B. Saunders.

Everett, N. 2003. *Literacy in Lombard Italy.* Cambridge: Cambridge University Press.

– 2009. 'Literacy from Late Antiquity to the Early Middle Ages.' In D. Olson and N. Torrance, eds., *The Cambridge Handbook of Literacy.* Cambridge: Cambridge University Press, 362–85.

– 2010. 'New Evidence for the Use of the Latin Suffix -aster from the *Alphabetum Galieni.* Contributions to Tracing a Linguistic Phenomenon from Latin to Romance and Lost Ancient Medical Texts.' *Galenos* 4: 169–82.

Fabricius, C. 1972. *Galens Exzerpte aus älteren Pharmakologen.* Berlin: de Gruyter.

Faraone, C.A. 1999. *Ancient Greek Love Magic.* Cambridge, MA: Harvard University Press.

Feemster Jashemski, W.M., and F.G. Meyer. 2002. *The Natural History of Pompeii: A Systematic Survey.* Cambridge: Cambridge University Press.

Ferraces Rodrígez, A. 1994. 'El Pseudo-Dioscórides *De herbis feminis,* los *Dynamidia* e Isidoro de Sevilla, *Etym.* XVII, 7–11.' In Vázquez Buján, ed. 1994, 183–203.

– 1999. *Estudios sobre textos latinos de fitoterapia entre la antigüedad tardía y la Alta Edad Media.* La Coruña: Universidade da Coruña, Servicio de Publicacións.

– 2000. 'Le *Ex herbis feminis*: traduction, réélaboration, problèmes stylistiques.' In Pigeaud and Pigeaud, eds. 2000, 77–90.

– 2004. 'Las *Curae herbarum* y las interpolaciones dioscorideas en el *Herbario* del Pseudo-Apuleyo.' *Euphrosyne. Revista de Filologia Clássica* 32: 223–40.

– ed. 2005. *Isidorus Medicus. Isidoro de Sevilla y los textos de medicina.* La Coruña: Universidade da Coruña, Servicio de Publicacións.

– 2006. 'Antropoterapia de la Antigüedad Tardía: *Curae quae ex hominibus fiunt.*' *Les Études Classiques* 74: 219–52

– 2007. 'El manuscrito de Lucca' (pp. 17–25), 'Un corpus altomedieval de materia médica' (pp. 43–53), 'La formación del texto del manuscrito de Lucca' (pp. 55–61) In A. Touwaide, ed., *Herbolarium et materia medica. Libro de Estudios (Biblioteca Statale de Lucca, ms. 296).* Facsimile y estudio del manuscrito 296 de la Biblioteca Civica de Luca. Madrid: Arte y Naturaleza Ediciones, & Lucca: Biblioteca Statale di Lucca.

– ed. 2009. *Fito-zooterapia antigua y altomedieval: textos y doctrinas.* Coruña: Universidade da Coruña.

– 2009a. 'Problemas de edición y límites en la enmienda de recetarios de tradición difusa: el "De herbis femininis" y las Curae herbarum.' In Ferraces Rodríguez, ed. 2009, 61–78.

– 2009b. Unité, réélaboration des sources et composition d'un réceptaire du haut moyen âge: *Curae que ex hominibus atque animalibus fiunt.*' In F. Le Blay, ed., *Transmettre les savoirs dans les mondes hellénistique et romain.* Rennes: Presses Universitaires de Rennes, 207–22.

Fichtner, Gerhard. 1985. *Corpus galenicum: Verzeichnis der galenischen und pseudogalenischen Schriften.* Tübingen: Institut für Geschichte der Medizin.

Fischer, K.-D. 1987. '*Universorum ferramentorum nomina.* Frühmittelalterliche Listen chirurgischer Instrumente und ihr griechisches Vorbild (mit einem Beitrag von Joseph A.M. Sonderkamp).' *Mittellateinisches Jahrbuch* 22: 28–44.

– 1988a. 'Ancient Veterinary Medicine. A Survey of Greek and Latin Sources and Some Recent Scholarship.' *Medizinhistorisches Journal* 23: 191–209.

– 1988b. 'Anweisungen zur Selbstmedikation von Laien in der Spätantike.'
Akten des 30. Internationalen Kongresses für Geschichte der Medizin, Düsseldorf 1986. Düsseldorf, 867–74.

– and F. Kudlien. 1989a. 'Die sogenannte Medicina Plinii.' *Handbuch der lateinischen Literatur der Antike*, Bd. 5, R. Herzog and P.L. Schmidt, eds. Munich:
C.H. Beck, 512.

– 1989b. 'Die sogenannte Mulomedicina Chironis.' *Handbuch der lateinischen
Literatur der Antike*, Bd. 5, R. Herzog and P.L. Schmidt, eds. Munich: C.H.
Beck, 513.

– 1997a 'Q. Gargilius Martialis.' *Handbuch der lateinischen Literatur der
Antike*, Bd. 4, R. Herzog and P.L. Schmidt, eds. Munich: C.H. Beck, 452.1
(269–73).

– 1997b. 'Pseudo-Gargilius Martialis, *Curae boum*.' *Handbuch der lateinischen
Literatur der Antike*, Bd. 4, R. Herzog and P.L. Schmidt, eds. Munich: C.H.
Beck, 452.2 (269–73).

– 1998. 'Der *Liber Byzantii*, ein unveröffentlichtes griechisches therapeutisches
Handbuch in lateinischer Übersetzung.' In Deroux, ed. 1998, 276–94. – ed.
2000. *Bibliographie des textes médicaux latins. Antiquité et haut moyen
âge. Premier supplément 1986–1999*. Mémoires du Centre Jean Palerne. 19.
Saint-Étienne

– 2000a. 'Theodorus Priscianus.' *Handbuch der lateinischen Literatur der
Antike*, Bd. 6, R. Herzog, P.L. Schmidt, and J. Fontaine, eds. Munich: C.H.
Beck, 607.2.

– 2000b. 'Vindicianus.' *Handbuch der lateinischen Literatur der Antike*, Bd. 6, R.
Herzog, P.L. Schmidt, and J. Fontaine, eds. Munich: C.H. Beck, 607.1.

– 2000c. 'Zusätze zu Theodorus Priscianus.' *Handbuch der lateinischen Literatur
der Antike*, Bd. 6, R. Herzog, P.L. Schmidt and J. Fontaine, eds. Munich: C.H.
Beck, 607.3.

– 2000d. 'De sanguinem reicientibus.' *Handbuch der lateinischen Literatur der
Antike*, Bd. 6, R. Herzog, P.L. Schmidt, and J. Fontaine, eds. Munich: C.H.
Beck, 607.5.

– 2000e 'Marcellus.' *Handbuch der lateinischen Literatur der Antike*, Bd. 6, R.
Herzog, P.L. Schmidt, and J. Fontaine, eds. Munich: C.H. Beck, 608.2.

– 2000f. 'Ps.Hippocrates epist. ad Maecen. und ad Antioch.' *Handbuch der lateinischen Literatur der Antike*, Bd. 6, R. Herzog, P.L. Schmidt, and J. Fontaine,
eds. Munich: C.H. Beck, 608.3.

– 2000g. 'Die frühesten Belege für lat. *cataracta* als Bezeichnung einer Augen-
rankheit.' In S. Sconocchia and L. Toneatto, eds., *Lingue tecniche del greco e
del latino. Atti del III Seminario internazionale sulla letteratura scientifica e
tecnica greca e latina*. Bologna: Pàtron, 69–79.

– 2002. 'Die pseudohippokratische Epistula de uirginibus. Bemerkungen zu ihrer

Textüberlieferung und zu ihrem Vokabular.' *Les Études Classiques* 70: 101–22.
- 2003. 'Galeni qui fertur ad Glauconem Liber tertius.' In I. Garofalo and A. Roselli, eds., *Fonte greche, latine e arabe*. Naples, 101–32, text 283–346.
- 2005. 'Neue oder vernachlässigte Quellen der Etymologien Isidors von Sevilla (Buch 4 und 11).' In Ferraces Rodrígez, ed. 2005, 95–174.
- and Sabbah and Corsetti, eds. 1987: *see under* Sabbah.
Flammini, G. 1998a. 'L"Epistula" Pseudogalenica *de febribus*.' In Santini, Scivoletto, and Zurli, eds. 1998, 239–57.
- 1998b. 'Celio Aureliano e le prefazioni ai *Gynaecia* e ai frammenti di *Medicinales responsiones*.' In Santini, Scivoletto, and Zurli, eds. 1998, 145–76.
- 1998c. 'La *praefatio* ai *Libri quinque de simplicium virtutibus* di Pseudo-Oribasio.' In Santini, Scivoletto, and Zurli, eds. 1998, 285–312.
Flobert, P. 1975. *Les verbes déponents latins des Origines à Charlemagne*. Paris: Publications de la Sorbonne, Série N.S. Recherches, 17.
Forbes, R.J. 1955–64. *Studies in Ancient Technology*. 9 vols. Leiden: Brill.
Fortenbaugh, W.W., and R.W. Sharples. 1991. *Theophrastus of Eresus: Sources for His Life, Writings, Thought, and Influence*. Leiden: Brill.
Fortuna, S., and A. Raia. 2006. 'Corrigenda and Addenda to Diel's Galenica by Richard J. Durling: III. Manuscripts and Editions.' *Traditio* 61: 1–30.
Fraisse, A. 2000. 'Observationes littéraires sur la Préface du livre I des Euporista de Théodore Priscien.' In Pigeaud and Pigeaud, eds. 2000, 91–100.
- 2003a. 'Médicine rationelle et irrationelle dans le livre I des Euporista de Théodore Priscien.' In Palmieri, ed. 2003, 183–92.
- 2003b. 'Place et statut des pratiques magiques dans les texts médicaux tardifs. Le cas de Cassius Felix et de Théodore Priscien.' In Gaide and Bivelle, eds. 2003, 161–72.
French, R.K. 1994. *Ancient Natural History: Histories of Nature*. London and New York: Routledge.
French, R.K., and F. Greenaway, eds. 1986. *Science in the Early Roman Empire: Pliny the Elder, His Sources and Influence*. London: Croom Helm.
Fronimopoulos, J., and J. Lascaratos. 1991. 'The Terms Glaucoma and Cataract in the Ancient Greek and Byzantine Writers.' *Documenta Ophptalmologica. Advances in Ophthalmology* 77.4: 369–75.
Fry, D.K. 1992. 'Exeter Riddle 31: Feather-Pen.' In J.M. Foley, J.C. Womack, and W.A. Womack, eds., *De gustibus: Essays for Alain Renoir*. New York: Garland.
Gaide, F., and F. Biville, eds. 2003. *Manus medica. Actions et gestes de l'officiant dans les texts médicaux latins. Questions de thérapeutique et de lexique. Actes du Colloque tenu à l'Université Lumière-Lyons II, les 18 et 19 septembre 2001*. Aix-en-Provence: Publications de l'Université de Provence.
Gaillard-Seux, P. 2003. 'La crémation des remèdes dans les texts médicaux latins.' In Gaide and Bivelle, eds. 2003, 69–86.

Gamble, H. 1995. *Books and Readers in the Early Church: A History of Early Christian Texts.* New Haven and London: Yale University Press.

Garofalo, I., ed. 1997. *Anonymi Medici De morbis acutis et chroniis.* With Eng. trans. B. Fuchs. Leiden: Brill.

Garofalo, I., and A. Rosselli, eds. 2003. *Galenismo e medicina tardoantica. Fonti greche, latine e arabe. Atti del Seminario Internazionale di Siena, Certosa di Pontignano, 9 e 10 settembre 2002.* Napoli: Istituto Universitario Orientale.

Gemmill, C.L. 1966. 'Silphium.' *Bulletin of the History of Medicine* 40.4: 295–313.

Gennadios, P.G. 1914. Λεξικόν Φυτολογικόν. Athens: Academy of Sciences (repr. 1997).

Gertsch, J. 2009. 'How Scientific Is the Science in Ethnopharmacology? Historical Perspectives and Epistemological Problems.' *Journal of Ethnopharmacology* 122: 177–83.

Giacosa, P. 1886. *Un ricettario del secolo XI (Archivio capitolare d'Ivrea no.87).* Turin: Fratelli Bocca.

– 1901. *Magistri Salernitani nondum editio* Turin: Fratelli Bocca.

Gilani, A.H., and A.U. Rahman. 2005. 'Trends in Ethnopharmacology.' *Journal of Ethnopharmacology* 100: 43–9.

Glaze, F.E. 2000. 'The Perforated Wall: The Ownership and Circulation of Medical Books in Europe, c. 800–1200.' PhD dissertation, Duke University.

– 2005. 'Galen Refashioned: Gariopontus in the Later Middle Ages and Renaissance.' In E. Lane Furdell, ed., *Textual Healing. Essays on Medieval and Early Modern Medicine.* Leiden: Brill, 53–76.

– 2008. 'Gariopontus and the Salernitans: Textual Traditions in the Eleventh and Twelfth Centuries.' In Jacquart and Paravicini Bagliani, eds. 2008, 149–90.

Gobeau, R. 1993. 'Le crocus, usages médicaux antiques.' In Amouretti and Comet, eds. 1993, 23–6.

Gourevitch, D. 2003. 'Fabriquer un médicament composé, solide et compact, dur et sec: formulaire et réalités.' In Gaide and Bivelle, eds. 2003, 49–68.

Goyens, M., P. De Leemans, and A. Smets, eds. 2008. *Science Translated: Latin and Vernacular Translations of Scientific Treatises in Medieval Europe.* Leuven: Leuven University Press.

Grant, M. *See* Anthimus under Ancient and Medieval Sources.

Green, M. 2005. 'Flowers, Poisons, and Men: Menstruation in Medieval Western Europe.' In *Menstruation: A Cultural History*, ed. A. Shail and G. Howie. New York: Palgrave, 51–64.

– 2008a. 'Gendering the History of Women's Healthcare.' *Gender and History* 20.3: 487–518.

– 2008b. *Making Women's Medicine Masculine: The Rise of Male Authority in Pre-Modern Gynaecology.* Oxford: Oxford University Press.

– and Hanson 1994. *See* Hanson.

Grensemann, H. 1996. *De aeribus aquis locis: interlineare Ausgabe der spätlatei-nischen Übersetzung und des Fragments einer hochmittelalterlichen Überset-zung.* Bonn: Habelt.

Grigson, G. 1974. *A Dictionary of English Plant Names (and Some Products of Plants).* London: A. Lane.

Grmek, M. 1984. 'Les vicissitudes des notions d'infection, de contagion et de germe dans la médecine antique.' In Sabbah, ed. 1984, 53–70.

– 1998. '"Albule oculorum": cataracte ou taies de la cornée.' In Deraux, ed. 1998, 422–33.

Guillaumont, A. 1950. *Les sens des noms du coeur dans l'antiquité.* Le Coeur, Études Carmélitaines 29. Paris: Desclée de Brouwer, 41–81.

Halleux-Opsomer, C. *See under* Opsomer (Halleux-).

Halleux, R., and J. Schamp, eds. 1985. *Les lapidaires grecs.* Paris: Les Belles lettres.

Hankinson, R.J. 1995. 'The Growth of Medical Empiricism.' In D. Bates, ed., *Knowledge and the Scholarly Medical Traditions.* Cambridge: Cambridge University Press.

Hanson, A.E., and M. Green. 1994. 'Soranus of Ephesus: *Methodicorum prin-ceps.*' In W. Haase and H. Temporini, eds., *Aufstieg und Niedergang der römischen Welt*, Teilband II, Band 37.2 Berlin and New York: Walter de Gru-yter, 968–1075.

Harborne, J.B., et al., eds. 1993. *Phytochemical Dictionary: A Handbook of Bio-active Compounds in Plants.* London and Washington: Taylor and Francis.

Harig, G. 1974. *Bestimmung der Intensität im medizinischen System Galens: ein Beitrag zur theoretischen Pharmakologie, Nosologie und Therapie in der Galenischen Medizin.* Tübingen: Akademie-Verlag.

Hartlich, O., ed. 1923. *Galeni De sanitate tuenda, De alimentorum facultatibus. De bonis malisque sucis, De victu attenuante. De ptisana.* Ed. K. Koch, G. Helmreich, C. Kalbfleisch, and O. Hartlich. *CMG* V.4.2. Leipzig and Berlin: Teubneri, 1923.

Haverling, G. 2008. 'On Variation and Syntax in Late Latin Texts.' In Wright, ed. 2008, 351–60.

Healy, J.F. 1986. 'Pliny on Mineralogy and Metals.' In French and Greenaway, eds. 1986, 111–46.

– 1999. *Pliny the Elder on Science and Technology.* Oxford: Oxford University Press.

Hedfors, H. 1932. *Compositiones ad tingenda musiva, herausgegegen übersetzt und philologish erklärt.* Uppsala: Almquist and Wiksells boktryckeri-a.b.

Herman, J. 2000. *Vulgar Latin.* Trans. R. Wright. University Park: Pennsylvania State University Press.

Hirth, W. 1980. 'Popularisierungstendenzen in der mittelalterlichen Fachliteratur.' *Medieval History Journal* 15: 70–89.

Hohmann, J., I. Zupko, D. Redei, M. Csanyi, G. Falkay, I. Mathe, and G. Janicsak. 1999 (August). 'Protective Effects of the Aerial Parts of *Salvia officinalis, Melissa Officinalis* and *Lavandula angustifolia* and Their Constituents against Enzyme-Dependent and Enzyme-Independent Lipid Peroxidation.' *Planta Medica* 65.6: 576–8.

Holland, B.K., ed. 1996. *Prospecting for Drugs in Ancient and Medieval European Texts: A Scientific Approach.* Amsterdam: Harwood Academic Publishers.

Hrobjartsson, A., and P.C. Gotzsche. 2001. 'Is the Placebo Powerless? – An Analysis of Clinical Trials Comparing Placebo with No Treatment.' *New England Journal of Medicine* 344: 1594–1602.

Hurtado, L. 2006. *The Earliest Christian Artefacts: Manuscripts and Christian Origins.* Grand Rapids, MI: Eerdmans.

Huguet-Termes, T. 2008. 'Islamic Pharmacology and Pharmacy in the Latin West: An Approach to Early Pharmacopoeias.' *European Review* 16.2: 229–39.

Hunger, F.W. 1935. *See above under* Pseudo-Apuleius in Ancient and Medieval Sources.

Ieraci Bio, A.M. 1991. 'Un témoinage grec à propos des Dynamidia.' In *Le latin médical. La constitution d' un language scientifique. Actes du IIIe Colloque International 'Textes médicaux latins antiques,' Saint-Étienne 11–12–13 septembre 1989*: 63–73. Saint-Étienne: Université de Saint-Étienne, 63–73.

Innes Miller, J. 1969. *The Spice Trade of the Roman Empire 29 B.C.to 641 A.D.* Oxford: Oxford University Press.

Irvine, J.T., and O. Temkin. 2003. 'Who Was Aki\la\o\s? A Problem in Medical History.' *Bulletin of the History of Medicine* 77: 12–24.

Jacquart, D. 2008. 'Islamic Pharmacology in the Middle Ages: Theories and Substances.' *European Review* 16.2: 219–27.

– and A. Paravicini Bagliani, eds. 2008. *La Collectio Salernitana di Salvatore de Renzi*. La Scuola Medica Salernitana 3. Florence: Sismel.

Jensen, L.B. 1963. 'Royal Purple of Tyre.' *Journal of Near Eastern Studies* 22: 104–18.

Joly, R. 1975. 'Les versions latines du Régime Pseudo-Hippocratique.' *Scriptorium* 29: 3–22.

Johnson, R.P. 1937. 'Some Continental Mss of the Mappae Clavicula.' *Speculum* 12: 90–1.

Jones, C. W. 1939. *Bedae pseudepigrapha: Scientific writings falsely attributed to Bede*. Ithaca, NY: Cornell University Press. Oxford: Oxford University Press.

Jorgensen, E. 1926. *Catalogus Codicum Latinorum Medii Aevi Bibliotecae Regiae Hafniensis*. Hafnia: in aedibus Gyldendalianis.

Jörimann, J. 1925. *Frühmittelalterliche Rezeptarien*. Beiträge zur Geschichte der Medizin, Heft 1. Zurich: Orell Füssli (repr. 1977, Vaduz: Topos Verlag).

Jouanna, J. 2003. 'Sur la dénomination et le nombre des sens d'Hippocrate à la médecine impériale: réflexions à partir de l'énumération des sens dans le traité hippocratique du Régime, c. 23.' In I. Boehm and P. Luccioni, eds., *Les cinq sens dans la médecine de l'époque impériale: sources et développements. Actes de la table ronde du 14 juin 2001*. Lyon and Paris: Diffusion de Boccard, 9–20.

– and V. Boudon. 1997. 'Remarques sur la place d'Hippocrate dans la pharmacologie de Galien.' In Debru, ed. 1997, 213–34.

Kästner, H. 1896. 'Pseudo-Dioscorides *De herbis feminis*.' *Hermes* 31: 578–636.

Kibre, P. 1945. 'Hippocratic Writings in the Latin Middle Ages.' *Bulletin of the History of Medicine* 18: 371–412.

– 1968. 'Further Addenda and Corridgenda to the Revised Edition of Lynn Thorndike and Pearly Kibre A Catalogue of Incipits of Medieval Scientific Writings in Latin (1963).' *Speculum* 43: 78–114.

– 1985. *Hippocrates Latinus: Repertorium of Hippocratic Writings in the Latin Middle Ages*. New York: Fordham University Press.

Klebs, A.C. 1938. *Incunabula scientifica et medica*. Bruges: G. Olms (repr. from *Osiris* 4).

Kollesch, J. 1973. *Untersuchungen zu den pseudogalenischen Definitiones medicae*. Schriften zur Geschichte und Kultur der Antike 7. Berlin: Akademie-Verlag.

Kühlewein, H. 1882. 'Mittheilungen aus einer alten lateinischen Übersetzung der Aphorismen des Hippokrates.' *Hermes* 18: 484–8.

– 1884. 'Beiträge zur geschichte und beurheilung der hippokratischen schriften. I. Zu Hippocrates' Prognosticon.' *Philologus* 42: 120–1.

– 1890. 'Die Handschriftliche Grundlage des Hippokratischen Prognostikon und eine lateinische Übersetzung desselben.' *Hermes* 25: 113–40.

Kudlien, F. *See under* Durling.

Langslow, D.R. 1989. 'Latin Technical Language: Synonyms and Greek Words in Latin Medical Terminology.' *Transactions of the American Philological Society* 87: 33–53.

– 1991. 'The Formation of Latin Technical Vocabulary with Special Reference to Medicine.' In R.G.G. Coleman, ed., *New Studies in Latin Linguistics: Selected Papers from the 4th International Colloquium on Latin Linguistics, Cambridge, April 1987*. Amsterdam: J. Benjamins Pub. Co., 187–200.

– 1991–92. 'The Development of Latin Medical Terminology: Some Working Hypotheses.' *Proceedings of the Cambridge Philological Society* 37: 106–30.

– 1994. 'Some Historical Developments in the Terminology and Style of Latin Medical Writings.' In M.E. Vázquez Buján, ed., *Traditión e innovación de la medicina latina de la antigüedad y de la alta edad media: Actas del IV coloquio internacional sobre los 'textos médicos latinos antiguo.'* Santiago de Compostela: Servicio de Publicacións e Intercambio Científico da Universidade de Santiago de Compostela, 225–40.

– 1999. 'Late Latin Discourse Particles in "Medical Latin" and "Classical Latin."' In H. Petersmann and R. Kettemann, eds., *Latin vulgaire-latin tardif. Actes du 5e colloque international sur le latin vulgaire et tardif (Heidelberg, 5–8, September 1997).* Heidelberg: Universitätsverlag C. Winter, 169–82 (published also in *Mnemosyne* 53 (2000): 537–82).

– 2000. *Medical Latin in the Roman Empire.* Oxford: Oxford University Press.

– 2003. 'The Doctor, His Action and the Terminology.' In F. Gaide and F. Biville, eds., 2003, 25–36.

– 2006. *The Latin Alexander Trallianus: The Text and Transmission of a Late Latin Medical Book.* Journal of Roman Studies Monographs 10. London: Society for the Promotion of Roman Studies.

– 2007. 'The *Epistula* in Ancient Scientific and Technical Literature, with Special Reference to Medicine.' In R. Morello and A.D. Morrison, eds., *Ancient Letter: Classical and Late Antique Epistolography.* Oxford: Oxford University Press.

Laux, R. 1930. '*Ars medicinae*: ein frühmittelalterliches Kompendium der Medizin.' *Kyklos* 3: 417–34.

Lee, M.R. 1999. 'Colchicum autumnale and the Gout: Naked Ladies and Portly Gentlemen.' *Proceedings of the Royal College of Physicians of Edinburgh* 29.1: 65–70.

Leisinger, H. 1925. *Die lateinischen Harnschriften Pseudo-Galens.* Zurich: Füssli.

Leslie, D.D., and K.H.J. Gardiner. 1996. *The Roman Empire in Chinese Sources.* Rome: Bardi.

Leung, A.Y., and S. Foster. 1996. *Encyclopedia of Common Natural Ingredients Used in Food, Drugs and Cosmetics.* 2nd ed. New York: Wiley and Sons.

Leven, K.-H. 2005. *Antike Medizin: Ein Lexikon.* Munich: Beck.

Levey, M. 1973. *Early Arabic Pharmacology. An Introduction Based on Ancient and Medieval Sources.* Leiden: Brill.

Lloyd, G.E.R. 1964. 'The Hot and the Cold, the Dry and the Wet in Greek Philosophy.' *Journal of Hellenic Studies* 84: 92–106.

– 1979. *Magic, Reason and Experience: Studies in the Origin and the Development of Greek Science.* Cambridge: Cambridge University Press.

– 1983. *Science, Folklore and Ideology.* Cambridge: Cambridge University Press.

– 1987. *The Revolutions of Wisdom: Studies in the Claims and Practices of Ancient Greek Science.* Sather Classical Lectures 52. Berkeley: University of California Press.

Louhilala, P., and R. Puustinen. 2008. 'Rethinking the Placebo Effect.' *Medical Humanities* 34: 107–9.

Lowe, E.A. 1914. *The Beneventan Script: A History of the South Italian Minuscule.* Oxford: Oxford University Press.

– ed. 1935–71. *Codices Latini Antiquiores.* 11 vols and suppl. Oxford and Princeton: Clarendon.

Maciocia, G. 1989. *The Foundations of Chinese Medicine: A Comprehensive Text for Acupuncturists and Herbalists.* Edinburgh, London, Melbourne, and New York: Churchill Livingstone.

Magdelaine, C., and J.L. Fournet. 2001. 'Liste de vins aromatisés à usage médical (réédition du PAlex 36).' In I. Andorlini, ed., *Greek Medical Papyri I.* Florence: Istituto papirologico G. Vitelli, 163–70.

Maggiulli, G. 2000. 'Dinamidia come genere letterario.' In Pigeaud and Pigeaud, eds. 2000, 141–52.

– and Buffa Giolito 1996. *See under* Buffa Giolito.

Mai, A. 1835. *Classicorum auctorum e vaticanis codicibus editorum,* VII. Rome: Typis vaticanis.

Makato, M. 2005. 'The Three *juan* Edition of *Bencao jizhu* and Excavated Sources.' In V. Lo and C. Cullen, eds., *Medieval Chinese Medicine: The Dunhuang Medical Manuscripts.* London and New York: Routledge Curzon, 306–21.

MacKinney, L.C. 1934. 'Tenth-Century Medicine As Seen in the *Historia* of Richer of Rheims.' *Bulletin of the Institute of the History of Medicine* 2: 347–58.

– 1935–6. '*Dynamidia* in Medieval Medical Literature.' *Isis* 24: 400–14.

– 1937. *Early Medieval Medicine with Special Reference to France and Chartres.* Publications of the Institute of the History of Medicine, Johns Hopkins University, ser. 3 vol. III. Baltimore: Johns Hopkins University Press.

– 1946. 'Animal Substances in Material Medica: A Study in the Persistence of the Primitive.' *Journal of the History of Medicine and Allied Sciences* 1:149–70.

– 1952a. 'Medical Ethics and Etiquette in the Early Middle Ages: The Persistence of Hippocratic Ideals.' *Bulletin of the History of Medicine* 26: 1–31.

– 1952b. 'Multiple Explicits of a Medieval *Dynamidia*.' *Osiris* 10: 195–205.

Maire, B. 1997. 'La *variatio* dans le lexique des *Medicinae ex oleribus et pomis* de Gargilius Martialis.' *RFIC* 125.3: 306–18.

– 2000. 'Les Medicinae de Gargilus: un manuel pratique aux ambitions littéraires?' In Pigeaud and Pigeaud, eds. 2000, 153–64.

– 2003. 'Actions thérapeutiques ou gestes littéraires: le lexique des *Medicinae* de Gargilius Martialis.' In Gaide and Biville, eds. 2003, 147–60.

Majno, G. 1975. *The Healing Hand: Man and Wound in the Ancient World.* Cambridge, MA: Harvard University Press.

Maltby, R. 2008. 'The Language, Style and Origins of Ps. Dioscorides De herbis femininis.' In Wright, ed. 2008, 392–9.

Mann, J. 2000. *Murder, Magic and Medicine*. Oxford: Oxford University Press.

Mantello, F.A.C., and A.G. Rigg, eds. 1996. *Medieval Latin: An Introduction and Bibliographical Guide*. Washington, DC: Catholic University of America Press.

Manzanero Cano, F. 1996. Liber Esculapii. Edición crítica y estudio. Thèse. Universidad Complutense de Madrid. Unpublished.

Marganne, M.H. 1980. 'Glaucome ou cataracte? Sur l'emploi des dérivés de glaukos en opthalmologie antique [emplois de *glaukoma* et de *hypochyma*].' *History and Philosophy of the Life Sciences* 1: 199–214.

– 1982. 'Nouvelles perspectives dans l'étude des sources de Dioscoride.' In Sabbah, ed. 1982, 81–4.

Martindale, J.R., A.H.M. Jones, and J. Morris, eds. 1971–92. *Prosopography of the Later Roman Empire*. Cambridge: Cambridge University Press.

Mattei, S. 1995. Curae Herbarum: Tesi di Dottorato. Università degli Studi di Macerata. Facoltà di lettere e filosofia. Unpublished. Roma, Biblioteca Nazionale Centrale, Diss. 96/2455.

– 1998. *Curae herbarum*: elementi di cristianizzazione in un erbario di età tardoantica.' In P. Gatti and L. de Finis, eds., *Dalla tarda latinità agli albori dell'Umanesimo: alla radice della storia Europea*. Labarinti 33. Trent: Università degli Studi di Trento, 383–98

Maurel, H. 2000. 'La catharsis dans la culture et la médecine grecques.' *Cahiers du Centre d'étude d'histoire de la médecine* 8: 6–23.

Mazzini, I. 1977. 'Il *De observantia ciborum*. Tradizione tardo-antica del Περὶ διαίτης pseudo-hippocratico l.II.' *Romanobarbarica* 2: 287–357.

– 1978. 'Il greco nella lingua tercnica medica latina (Spunti per un "indagine sociolingustica").' *Annali della Facoltà di lettere e filologia Università Macerata* 11: 543–56.

– 1981. 'Il latino medico in Italia nei secoli V e VI.' *La cultura in Italia fra tardo antico e alto medio evo. Atti del convegno tenuto a Roma 12–16 nov. 1979*. Rome: Herder, 433–41.

– 1982–4. 'Le accuse contro i medici nella lettatura latina ed il loro fondamento.' *Quaderni linquistici et filologici* (Università di Macerata) 2: 75–90.

– 1984. *De observantia. See under* Ancient and Medieval Sources.

– 1985. 'Caratteri comuni a tutto l'Ippocrate latino tardo-antico e conseguenti considerazioni su alcuni emendamenti al testo.' In Mazzini and Fusco, eds. 1985, 63–74.

– 1991. 'Il lessico medico latino antico: caratteri e strumenti della sua differenziazione.' In Sabbah, ed. 1991, 175–85.

– 1992. 'Dynamidia Yppocratis: esempio di problematiche ecdotiche tardoantiche.' In A. Garzya, ed., *Tradizione e ecdotica dei testi medici tardoantichi e bizantini. Atti del Convegno Internazionale Anacapri 29–31 Ott. 1990.* Naples: M. D'Auria, 257–69.

– 1993. 'Il linguaggio della ginecologia latina antica: lessico e fraseologia.' In Boscherini, ed. 1993, 45–92.

– 1998. 'La malattia conseguenza e metafora del peccato nel mondo antico, pagano et cristiano.' In E. Dal Covolo and I. Gianetto, eds., *Cultura e promozione umana. La cura del corpo e delle spirito nell'antichità classica e nei primi secoli cristiani. Un magistero ancora attuale ? Atti del II Convegno Internazionale (Troina, 29 ottobre–1 novembre 1997).* Troina: Oasi, 159–72.

– and F. Fusco, eds. 1985. *I testi di medicina latini antichi. Problemi filologici e storici. Atti del Convengo Internazionale Macerata-San Severino M., 26–28 aprile 1984.* Rome: G. Bretschneider.

– and N. Palmieri. 1991. 'L'école medicale de Ravenne.' In Deroux, ed. 1991, 286–93.

McCormick, M. 2001. *Origins of the European Economy: Communications and Commerce 300–900.* Cambridge: Cambridge University Press.

McManus, J.P., K.G. Davis, T.H. Lilley, and E. Halsam. 1981. 'The Association of Proteins with Phenols.' *Journal of the Chemical Society. Chemical Communications* 7: 309–11.

McVaugh, M. 1975. *Arnaldi de Villanova Aphorismi de Gradibus, Arnaldi de Villanova Opera medica omni,* II. Barcelona, Granada: Seminarium Historiae Medicae Granatensis.

Meyer, E.H.F. 1854–7. *Geschichte der Botanik.* 4 vols. Königsberg: Gebrüder Bornträger.

Michler, M. 1993. 'Principis medicus: Anontius Musa.' *ANRW* 2.37.1: 757–85.

Migliorini, P. 1993. 'Alcune denominazioni della malattia nella letteratura latina.' In Boscherini, ed. 1993, 93–132.

– 1997. *Scienza e terminologia medica nella letteratura latina di età neroniana: Seneca, Lucano, Persio, Petronio.* Frankfurt-am-Main and New York: P. Lang.

Mihaescu, H. 1938. 'La versione latina di Dioscoride, tradizione manoscritta, critica del testo, cenno linguistico.' *Ephemeris Dacoromana* 8: 298–348.

Miller, G.L. 1990. 'Literacy and the Hippocratic Art: Reading, Writing, and Epistemology in Ancient Greek Medicine.' *Journal of the History of Medicine and Allied Sciences* 45.1: 11–40.

Miller, H.W. 1952. '*Dynamis* and *Physis* in On Ancient Medicine.' *Transactions and Proceedings of the American Philological Association* 83: 184–97.

– 1959. 'The Concept of *Dynamis* in *De victu.*' *Transactions and Proceedings of the American Philological Association* 90: 147–64.

Miles, B. 2004. 'The *Carmina Rhythmica* of Aethilwald: Edition, Translation, Commentary.' *Journal of Medieval Latin* 14: 73–117.

Moncrieff, R.W. 1967. *The Chemical Senses.* London: Leonard Hill.

Mørland, J. 1932. *Die Lateinische Oribasiusübersetzungen. Symbolae Osloenses* Suppl. Fasc. V. Oslo: Brogger.

– 1952. 'Theodorus Priscianus im Lateinischen Oribasius.' *Symbolae Oloenses* 29: 79–91.

Morton, A.G. 1986. 'Pliny on Plants: His Place in the History of Botany.' In French and Greenaway, eds. 1986, 86–97.

Mouget, C. 1993. 'Les utilisations antiques du ciste et du ladanum et leur héritage.' In Amouretti and Comet, eds. 1993, 27–44.

Mras, K. 'Sprachliche und textkritische Bemerkungen zur spätlateinsichen Übersetzung der hippokratischen Schrift von der Siebenzahl.' *Wiener Studien* 41: 61–74.

Mudry, P. 2000. 'L'ellébore ur la victoire de la littérature (Pline Nat. 25, 47–61).' In Pigeaud and Pigeaud, eds. 2000, 105–202.

Mudry, P., and G. Sabbah. 1994. *La médecine de Celse: aspects historiques, scientifiques et littéraires.* Université de Saint-Étienne.

Muzzioli, G., ed. 1954. *Mostra storica nazionale della miniatura.* Rome and Florence: Sansoni.

Nairn, J.A. 1899. 'The Meaning of Hellespontus in Latin.' *Classical Review* 13.9: 436–8.

Needham, J. 1954–. *Science and Civilization in China.* 6 vols. Cambridge: Cambridge University Press.

Newman, D.J., and G.M. Cragg. 2007. 'Natural Products as Sources of New Drugs over the Last 25 Years.' *Journal of Natural Products* 70: 461–77.

Newton, F. 1999. *The Scriptorium and Library at Monte Cassino, 1058–1105.* Cambridge: Cambridge University Press.

Niermeyer, J.F. 1997. *Mediae Latinitatis lexicon minus: lexique latin médiéval-français/anglais.* Leiden, New York: Brill.

Nockels Fabbri, C. 2007. 'Treating Medieval Plague: The Wonderful Virtues of Theriac.' *Early Science and Medicine* 12: 247–83.

Norton, S. 2006. 'The Pharmacology of Mithridatum: A 2000-Year-Old-Recipe.' *Reflections* 6.2: 60–6.

Nutton, V. 1969. 'Medicine and the Roman Army.' *Medical History* 12: 260–70.

– 1981. 'Continuity or Rediscovery: The City Physician in Classical Antiquity and Medieval Italy.' In A.W. Russell, ed., *The Town and the State Physician in Europe from the Middle Ages to the Enlightenment.* Wolfenbütteler Forschungen Bd. 17 Wolfenbüttel: Herzog August Bibliothek.

– 1984. 'From Galen to Alexander: Aspects of Medicine and Medical Practice in Late Antiquity.' *Dumbarton Oaks Papers* 39: 1–14.

- 1985. 'The Drug Trade in Antiquity.' *Journal of the Royal Society of Medicine* 78: 18–145.
- 1986. 'The Perils of Patriotism: Pliny and Roman Medicine.' In R. French and F. Greenaway, eds. 1986, 30–58.
- 1988. '*Archiatri* and the Medical Profession in Antiquity.' *Papers of the British School at Rome* 45: 191–226.
- 1995. 'Scribonius Largus, the Unknown Pharmacologist.' *Pharmaceutical Historian* 25.1: 5–8.
- 1997 'Did the Greeks Have a Word for It? Contagion and Contagion Theory in Classical Antiquity.' In L.I. Conrad, ed., *Concepts of Contagion and Infection in Premodern Society*. Aldershot: Ashgate, 137–62.
- 2004. *Ancient Medicine*. London and New York: Routledge.
- 2008. 'Ancient Mediterranean Pharmacology and Cultural Transfer.' *European Review* 16.2: 211–17.
- and Scarborough: *see* Scarborough.
Oken, B.S. 2008. 'Placebo Effects: Clinical Aspects and Neurobiology.' *Brain* 131: 2812–23
Önnerfors, A. 1993. 'Das medizinische Latein von Celsus bis Cassius Felix.' *Aufstieg und Niedergang der römischen Welt* 2.37: I.227–392.
Opsomer (Halleux-), C. 1982a. 'Un herbier médicinal du haut moyen âge: L'Alphabetum Galieni.' *History and Philosophy of the Life Sciences* 4.1: 65–96.
- (Halleux-) 1982b. 'Prolégomènes à une étude (informatique) des recettes médicales latines.' In G. Sabbah, ed., *Médicins et Médecine dans l'Antiquité*. Saint-Étienne: Université de Saint-Étienne: 85–104.
- 1986. 'The Medieval Garden and Its Role in Medicine.' In E.B. MacDougall, ed., *Medieval Gardens*. Washington, DC: Dumbarton Oaks Research Library and Collection, 93–113
- 1989. *Index de la pharmacopée di Ier au Xe siècle*. 2 vols. Hildesheim: Olms-Weidmann.
- and R. Halleux. 1985. 'La lettre d'Hippocrate à Antiochus et la lettere d'Hippocrate à Mécène.' In Mazzini and Fusco, eds. 1985, 339–64.
Orlandini, A. 1998. 'Parmi les noms latins de l'épilepsie: *morbus maior*.' In Debru and Sabbah, eds. 1998, 83–91.
Orofino, G. 1994. *I codici decorati di Montecassino. I*. Rome: Istituto Poligrafico e Zecca dello stato.
Osman O.F, I.S. Mansour, and S. El-Hakim. 2003. 'Honey Compound for Wound Care: A Preliminary Report.' *Annals of Burns and Fire Disasters* 16.3 (September).
Osol, A., R. Pratt, and A.R. Gennaro. 1973. *The United States Dispensatory*. 27th ed. Philadelphia and Toronto: J.B. Lippincott.

O'Sullivan, T.D. 1978. *The* De excidio *of Gildas. Its Authenticity and Date.* Leiden: Brill.

Palmieri, N. 1981. 'Un antico commento a Galeno della scuola medica di Ravenna.' *Physis* 22: 197–296.

– 1989. *L'antica versione latina del De sectis di Galeno (Pal. lat. 1090).* Pisa: ETS.

– 1993. 'Survivance d'une lecture Alexandrine de l'Ars medica en latin et en arabe.' *Archives d'histoire doctrinale et littéraire du Moyen Age* 60: 57–102.

– 1994. Il commento latino-Ravennate all' Ars medica di Galeno e la tradizione alessandrina.' In Vásquez Buján, ed. 1994, 55–75.

– 1997. 'La théorie de la médecine des Alexandrins aux Arabes.' In D. Jacquart, ed., *Les voies de la science grecque. Études sur la transmission des textes de l'Antiquité au dix-neuvième siècle.* Geneva: Droz; Paris: Champion, 33–133.

– 2000. 'Rhétorique et pédagogie dans les commentaires à Galien d'Agnellus de Ravenne.' In Pigeaud and Pigeaud, eds. 2000, 203–20.

– 2001. 'Nouvelles remarques sur les commentaires a Galien de l'école médicale de Ravenne.' In A. Debru and N. Palmieri, eds. 2001, 209–46.

– 2002. 'La Médecine alexandrine et son rayonnement occidental (VIe–VIIe s. ap.J.-Ch.).' *Lettre d'Informations du Centre Jean Palerne. Médecine antique et médiévale* 1: 5–23.

– 2003 ed. *Rationnel et irrationnel dans la médicine ancienne et médiévale. Aspects historiques, scientifiques et culturels.* Actes du colloque international organisé par le centre Jean Palerne à l'Université Jean Monnet de Saint-Étienne les 14 et 15 novembre 2002. Saint-Étienne: Publications de l'université.

– 2005. *Agnellus of Ravenne. Lectures Galéniques: le 'De pulsibus ad tirones'*: See *above under* Agnellus in Ancient and Medieval Sources.

Panzer, G.W. 1793–1803. *Annales typographici ab artis inventae origine ad annum MD (ab anno M DI, ad annum MCXXXVI continuati).* 11 vols. Nuremberg: J.E. Zeh (repr. 1963, Hildesheim: G. Olms).

Pardon, M. 2003. '*In medicinis uenena.* Celse et la défense de la médecine pharmaceutique.' In Gaide and Biville, eds. 2003, 103–16.

Parfitt, K., ed. 1999. *Martindale: The Complete Drug Reference.* 32nd ed. London: Pharmaceutical Press.

Pazzini, A. 1967. 'La letteratura medica salernitana e la storia della scuola di Salerno.' *Salerno-Civitas Hippocratica* 1.1–2: 5–18.

Paxton, F.S. 1993. '*Signa mortifera*: Death and Prognostication in Early Medieval Monastic Medicine.' *Bulletin of the History of Medicine* 67: 631–50.

Pearce, J.M.S. 2008. 'The Doctrine of the Signatures.' *European Neurology* 60: 51–2.

Pelicier, Y. 1982. 'Stability in the Meaning of the Latin Vocabulary for "Folie"

(Insanity).' *Zeitschrift für Klinische Psychologie und Psychotherapie* 30.1: 68–76.

Perdok, E.A., ed. 1968; 2nd ed. 1982. *A Multilingual Glossary of Common Plant-Names*. 2 vols. Wageningen: International Seed Testing Association.

Peterson, D.W. 1974. 'Galen's *Therapeutics to Glaucon* and Its Early Commentaries.' Unpublished PhD thesis, Johns Hopkins University.

Petit, C. 2009. *Galien. Le Médecin. Introduction.* Paris: Les Belles Lettres.

Petrucci, A. 1986. 'Alfabetismo ed educazione grafica degli scribi altomedievali.' In P. Ganz, ed., *The Role of the Book in Medieval Culture.* Turnhout: Brepols, 109–32.

Pharies, D. 2002. *Diccionario etimológico de los sufijos españoles. Madrid: Gredos.*

Pigeaud, A., and J. Pigeaud, eds. 2000. *Les texts médicaux latins comme literature. Acts du VIe colloque international sur les texts médicaux latins du 1er au 3 septembre 1998 à Nantes.* Université de Nantes: Institut universitaire de France.

Pilsworth, C. 2009. 'Could You Just Sign This for Me John? Doctors, Charters and Occupational Identity in Early Medieval Northern and Central Italy.' *Early Medieval Europe* 17.4: 363–88.

Plamböck, G. 1964. *Dynamis in Corpus Hippocraticum.* Akademie der Wissenschaften und der Literatur, Abhandlung geistes- und sozialwissenshafltiche Klasse. no. 2. Wiesbaden.

Pollan, M. 2008. *In Defense of Food. An Eater's Manifesto.* New York: Penguin Press.

Puhlmann, W. 1930. 'Die lateinische medizinische Literatur fes fruhen Mittelalters: Ein bibliographischer Versuch.' *Kyklos* 3:12: 395–416.

Raschke, M. 1978. 'New Studies in *Roman Trade* with the East.' *Aufstieg und Niedergang der römischen Welt* 2.9.2: 604–1378

Reiche, R. 1973. 'Einige lateinische Monatsdiätetike aus Wiener und St. Galler Handschriften.' *Sudhoffs Archiv für Geschichte der Medizin und der Naturwissenschaften. Zeitschrift für Wissenschaftsgeschichte* Bd 57 Heft 2: 113–41.

Reuter, J., A. Jocher, S. Hornstein, J.S. Mönting, and C.M. Schempp. 2007. 'Sage Extract Rich in Phenolic Diterpenes Inhibits Ultraviolet-Induced Erythema in Vivo.' *Planta Medica* 11: 1190–1.

Reveal, J.L. 1996. 'What's in a Name: Identifying Plants in Pre-Linnaean Botanical Literature.' In Holland, ed. 1996, 57–90.

Riddle, J. 1974. 'The Latin Alphabetical Dioscorides Manuscript Group.' *Actes du XIIIe congrès international d'histoire des sciences (Moscou 18–24 août 1971)*, vol. 5. sect. IV (Histoire des sciences et des techniques au Moyen Âge, org. B.A. Rosenfeld, secrétaire V. Gaidenko). Moscow: Nauka, 204–9.

– 1980. *Dioscorides.* In F.E.Cranz and P.O. Kristeller, *Catalogus translationum et commentarium IV.* Washington, DC: Catholic University of America Press, 1–143

– 1981. 'Pseudo-Dioscorides ex herbis femininis.' *Journal of the History of Biology* 14: 43–81.

– 1984. 'Gargilius Martialis as a Medical Writer.' *Journal of the History of Medicine and Allied Sciences* 39: 408–29.

– 1985. *Dioscorides on Pharmacy and Medicine.* Austin: University of Texas Press

– 1992a. *Contraception and Abortion from the Ancient World to the Renaissance.* Cambridge, MA: Harvard University Press.

– 1992b. *Quid pro quo: Studies in the History of Drugs.* Aldershot: Ashgate.

– 1996. 'The Medicines of Greco-Roman Antiquity as a Source of Medicines for Today.' In Holland, ed. 1996, 7–18.

– 1997. *Eve's Herbs: A History of Contraception and Abortion in the West.* Cambridge, MA: Harvard University Press.

– 2004 'Kidney and Urinary Therapeutics in Early Medieval Monastic Medicine.' *Journal of Nephrology* 17: 324–8.

Roberts, C.H., and T.C. Skeat. 1983. *The Birth of the Codex.* London and Oxford: Oxford University Press.

Roberts, M. 1989. *The Jewelled Style: Poetry and Poetics in Late Antiquity.* Ithaca and London: Cornell University Press.

Roberts, M.F., and M. Wink, eds. 1998. *Alkaloids: Biochemistry, Ecology, and Medicinal Applications.* New York: Plenum Press.

Rohlfs, G.R. 1966–9. *Grammatica storica della lingua Italiana e dei suoi dialetti.* Turin: Einaudi.

Roscher, W.H. 1911. *Über Alter, Ursprung und Bedeutung der Hippokratischen Schrift Von der Siebenzahl: ein Beitrag zur Geschichte der ältesten griechischen Philosophie und Prosaliteratur.* Sächsische Akademie der Wissenschaften zu Leipzig. Philologisch-Historische Klasse. Abhandlungh. Bd. 28, no. 5 Leipzig: Teubner.

Rose, V. 1874. 'Über die medicina Plinii.' *Hermes* 8: 18–66.

– 1864–70. *Anecdota graeca et graeco-latina: Mitteilungen aus Handschriften zur Geschichte der griechischen Wissenschaft.* 2 vols. Berlin: Duemmler (repr. Amsterdam, 1963).

Ross, I.A. 1998–. *Medicinal Plants of the World: Chemical Constituents, Traditional and Modern Medicinal Uses.* 3 vols (and in progress: vol. 1: 1998, 2: 2001: 3, 2005). Totowa, NJ: Humana Press.

Sabbah, G., ed. 1984. *Textes médicaux latins antiques.* Saint-Étienne: Publications de l'Université.

– ed. 1991. *La constitution d'un langage scientifique: le latin medical. Realités et*

language de la médicine dans le monde romaine. Actes du 3 Colloque Inter-
nationale Les texts medicaux latins antiques. Saint-Étienne: Université de
Saint-Étienne.

– ed. 1998. *Études de médecine romaine.* Saint-Étienne: Publications de
l'Université.

– and Debru, eds. 1998. *See under* Debru.

Sabbah, G., P.-P. Corsetti, and K.-D. Fischer. 1987. *Bibliographie des textes
médicaux latins. Antiquité et haut moyen âge.* Saint-Étienne: Publications de
l'Université.

Samama, E. 2002. 'Empoisonné ou guérie? Remarques lexicologiques sur les
pharmaka et venena.' In F. Collard and E. Sammama, eds., *Le corps à l'épreuve.*
*Poissons, remèdes et chirurgie: aspects des pratiques médicales dans l'Antiquité
et au Moyen-Âge.* Reims: D. Guéniot, 13–27.

Santamaría Hernández, M.T. 1994. 'Le denominación verbal de las cualidades
humorales.' In Vázquez Buján, ed. 1994, 297–317.

Santini, C., and N.I. Scivoletto, eds. 1990–2. *Prefazioni, prologi, proemi di opere
tecnico-scientifiche latine.* Vols. I–II. Rome: Herder.

– and N.I. Scivoletto and L. Zurli, eds. 1998. *Prefazioni, prologi, proemi di opere
tecnico-scientifiche latine.* Vol. III. Rome: Herder.

Scarborough, J. 1969. *Roman Medicine.* London and Southampton: Thames and
Hudson.

– 1975. 'The Drug Lore of Asclepiades of Bithynia.' *Pharmacy in History* 17:
43–57.

– 1977. 'Nicander's Toxicology: I, Snakes,' *Pharmacy in History* 19.1: 3–23.

– 1978. 'Theophrastus on Herbals and Herbal Remedies.' *Journal of the History
of Biology* 11: 353–85.

– 1979. 'Nicander's Toxicology II. Spiders, Scorpions, Insects and Myriapods.'
Pharmacy in History 21.2: 3–33, 72–92.

– 1982. 'Roman Pharmacy and the Eastern Drug Trade: Some Problems As Illustrated by Aloe.' *Pharmacy in History* 24: 135–43.

– and V. Nutton. 1982. 'The Preface of Dioscorides' Materia medica: Introduction, Translation, Commentary.' *Transactions and Studies of the College of
Physicians of Philadelphia* 4: 187–227.

– 1983. 'Theoretical Assumptions in Hippocratic Pharmacology.' In F. Lassere
and P. Mudry, eds., *Formes de pensée dans la collection hippocratique: actes du
IVe Colloque international hippocratique (Lausanne, 21–26 septembre 1981).*
Publications de la Faculté des lettres, Université de Lausanne 26. Geneva:
Librairie Droz.

– 1984. 'Early Byzantine Pharmacology.' *Dumbarton Oaks Papers* 38: 213–
32.

– 1985. 'Criton, Physician to Trajan: Historian and Pharmacist.' In J.W. Eadie and J. Ober, eds., *The Craft of the Ancient Historian: Essays in Honour of Chester G. Starr.* Lanham, MD: University Press of America, 387–405.

- 1986. 'Pharmacy in Pliny's Natural History: Some Observations of Substances and Sources.' In French and Greenaway, eds. 1986, 59–95.
- ed. 1987. *Folklore and Folk Medicine.* Madison, WI: American Institute of the History of Pharmacy.
- 1989. 'Contraception in Antiquity: The Case of Pennyroyal.' *Wisconsin Academy Review* 35.2 (March): 19–25.
- 1991a. 'The Pharmacology of Sacred Plants, Herbs and Roots.' In C.A. Faraone and D. Obbink, eds., *Magika Hiera: Ancient Greek Magic and Religion.* Oxford: Oxford University Press, 138–74.
- 1991b. 'The Pharmacy of Methodist Medicine: The Evidence of Soranus "Gynacology."' In P. Mudry and J. Pigeaud, *Les écoles médicales à Rome: actes du 2ème colloque international sur les textes médicaux latins antiques : Lausanne, septembre 1986.* Geneva: Université de Nantes, 204–16.
- 1995. 'The Opium Poppy in Hellenistic and Roman Medicine.' In R. Porter and M. Teich, eds., *Drugs and Narcotics in History.* Cambridge: Cambridge University Press, 5–23.
- 2005. Introduction, in L. Beck, trans. *Dioscorides' De materia medica*, xiii–xxi.
- 2006. 'Drugs and Drug Lore in the Time of Theophrastus: Folklore, Magic, Botany, Philosophy and Rootcutters.' *Acta Classica* 49: 1–29.
- forthcoming. 'Thornapple in Greco-Roman Pharmacology.' *Classical Philology.*
Schiefsky, M.J. 2005. *Hippocrates* On Ancient Medicine. *Translated with introduction and commentary.* Leiden and Boston: Brill.
Schmitz, R. 1998 (with Franz-Josef Kuhlen). *Geschichte der Pharmazie Band 1: Von den Anfängen bis zum Ausgang des Mittelalters.* Eschborn: Govi-Verlag Pharmazeutischer Verlag.
Schöner, E. 1964. *Das Viererschema in der antiken Humoralpathologie.* Wiesbaden: Steiner.
Schottus, J. 1544. *Experimentarius medicinae.* Strasbourg.
Sconocchia, S. 1985. 'Le fonti e la fortuna di Scribonio Largo.' In Mazzini and Fusco, eds. 1985, 151–213.
- 1993. 'Alcuni remedi nella letteratura medica latina del I sec. d. C.' In Boscherini, ed. 1993, 133–59.
- 1998. 'La lettera di Diocle ad Antigono e le sue traduzioni latine.' In Mazzini, Scivoletto, and Zurli, eds. 1998, III:113–32.
Scivoletto, N.I. *See under* Santini.
Seck, F., and H. Schnorr von Carolsfeld. 1884. 'Das lateinischen Suffix -aster, astra, astrum.' *Archiv für lateinische Lexicographie und Grammatik* 1: 390–407.
Segoloni, M.P. 1990a. 'Il prologus della *Medicina Plinii.*' In Santini and Schivoletto, eds. 1990–2, I: 363–6.

– 1990b. 'L'epistola dedicatoria e l'appendice in versi del De medicamentis liber di Marcello.' In Santini and Schivoletto, eds. 1990–2, I: 369–79.

Serbat, G. 1995. *Celse, De la médecine, i: Livres i–ii.* Paris: Les Belles lettres.

Shapiro, A.K. 1959. 'The Placebo Effect in the History of Medical Treatment: Implications for Psychiatry.' *American Journal of Psychiatry* 116: 298–301.

– and E. Shapiro. 1997. *The Powerful Placebo: From Ancient Priest to Modern Physician.* Baltimore: Johns Hopkins University Press.

Schiffman, S. 2000. 'Taste Quality and Neural Coding: Implications from Psychophysics and Neurophysiology.' *Physiology and Behavior* 69: 147–59.

Sigerist, H.E. 1923. *Studien und Texte zur fruhmittelalterlichen Rezeptliteratur.* Studien zur Geschichte der Medizin fasc 13. Leipzig: J.A. Barth.

– 1930. 'Fragment einer unbekannten lateinischen Übersetzung des hippokratischen Prognostikon.' *Sudoffs Archiv für Geschichte der Medizin* 23: 87–90.

– 1934. 'The Medical Literature of the Early Middle Ages.' *Bulletin of the Institute of the History of Medicine* 2: 26–50.

– 1934b. 'A Summer of Research in European Libraries.' *Bulletin of the Institute of the History of Medicine* 2: 559–613.

– 1943. 'Early Medieval Medical Texts in Manuscripts of Vendôme.' *Bulletin of the History of Medicine* 14: 78–89.

Singer, C. 1927. 'The Herbal in Antiquity and Its Transmission to Later Ages.' *Journal of Hellenic Studies* 47: 1–52, and tav. I–X

– 1957. 'Medical Science in the Dark Ages.' In Z. Cope, ed., *Sidelights on the History of Medicine.* London: Butterworth, 37–46.

Singhal, G.D., and T.J.S. Patterson. 1993. *Synopsis of Ayurveda.* Oxford: Oxford University Press.

Skard, S. 1946. 'The Use of Colour in Literature, a Survey of Research.' *Proceedings of the American Philological Society* 90: 163–249.

Skinner, P. 1997. *Health and Medicine in Early Medieval Southern Italy.* Leiden: Brill.

Skoda, F. 1994. 'L'eau et le vocabulaire de la maladie.' In R. Ginouvès, A.M. Guimier-Sorbets, J. Jouanna, and L. Villard, eds., *L'eau, la santé et la maladie dans le monde grec. Actes du colloque (25–27 novembre 1992, Paris).* Athens: École française d'Athènes, 249–64.

– 'Désignations de l'antidote en grec ancien.' In Debru and Palmieri, eds. 2001, 273–91.

Spiegel, D., and A. Harrington. 2008. 'What Is the Placebo Worth?' *British Medical Journal* 336: 967–8.

Sprecher, E., and W. Caesar, eds. 2003. *Society for Medicinal Plant Research – Gesellschaft für Arzneipflanzenforschung: 50 Years 1953 – 2003 A Jubilee Edition.* Stuttgart: Wissenschaftliche Verlagsgesellschaft mbH.

Simonini, R. 1936. 'Herbolarium et materia medica: cod. ms: 296 della Bib. governativa di Lucca.' In *Atti e memorie della Reale. Accademia di scienze, lettere ed arti di Modena*, s. 5a I, 188–90, tav. I–IX.

Stadler, H. 1898. 'Lateinische Pflanzennamen im Dioskurides.' *Archiv für lateinische Lexikographie und Grammatik* 8: 83–115.

– 1902. 'Epistola Pseudohippocratis.' *Archiv für lateinische Lexikographie und Grammatik* 12: 21–5.

Staiger, C. 2005. 'Beinwell – eine moderne Arzneipflanze (Comfrey, a Modern Medicinal Plant).' *Zeitschrift für Phytotherapie* 26.4 (August): 169–74.

Stannard, J. 1961. 'Hippocratic Pharmacology.' *Bulletin of the History of Medicine* 35.6: 497–518 (repr. in idem 1991a, I)

– 1965. 'Pliny and Roman Botany.' *Isis* 56.186: 420–5.

– 1966. 'Benedictus Crispus, an Eighth-Century Medical Poet.' *Journal of the History of Medicine and Allied Sciences* 21.1: 24–46 (repr. in idem 1999a, X).

– 1968. 'Lost Botanical Writings of Antiquity.' *Actes du XIe Congrès international d'histoire des sciences, Varsovie-Cracoviue 24–31 Août 1965 Warsaw*, 319–22 (repr. in idem 1999a, VII).

– 1969. 'The Herbal as Medical Document.' *Bulletin of the History of Medicine* 4: 212–20 (repr. idem 1999b, II).

– 1973. 'Marcellus of Bordeaux and the Beginnings of Medieval *Materia Medica*.' *Pharmacy in History* 15.2: 47–53 (repr. in idem 1999a, VI).

– 1974. 'Squill in Ancient and Medieval Materia Medica with Special Reference to Its Employment for Dropsy.' *Bulletin of the New York Academy of Medicine* 50: 684–713 (repr. in idem 1999b, XVI).

– 1977. 'Magiferous Plants and Magic in Medieval Medical Botany.' *Maryland Historian* 8: 33–46 (repr. in idem 1999b, V).

– 1979. 'Medieval Herbals and Their Development.' *Medica* 9: 23–33 (repr. in idem 1999b, III).

– 1982a. 'Medical Plants and Folk Remedies in Pliny, *Historia Naturalis*.' *History and Philosophy of the Life Sciences* 4.1: 3–23 (repr. in idem 1999a, III).

– 1982b. 'The Multiple Uses of Dill (*Anethum graveolens* L.) in Medieval Medicine.' In *'Gelêrter der arzenîe, ouch apotêker': Beiträge zur Wissenschaftsgeschichte; Festschrift zum 70. Geburtstag von Willem F. Daems*, ed. G. Keil (Würzburger medizinhistorische Forschungen 24). Hannover: 411–24 (repr. in idem 1999b, XVII).

– 1984. 'Aspects of Byzantine *Materia medica*.' *DOP* 38: 205–17 (repr. in idem 1999a, IX).

– 1985. 'The Theoretical Basis of Medieval Herbalism.' *Medical Heritage* 1: 186–98 (repr. in idem 1999b, IV).

– 1999a. *Pristina Medicamenta*. Ed. K. Stannard and R. Kay. Aldershot: Ashgate.

– 1999b. *Herbs and Herbalism in the Middle Ages and the Renaissance*. Ed. K. Stannard and R. Kay. Aldershot: Ashgate.

Stearn, W.T. 1966. *Botanical Latin. History, Grammar, Syntax, Terminology and Vocabulary*. New York: Hafner Publishing Co.

Stiennon, J. 1973. *Paléographie du Moyen-Age*. Paris: Armand Colin.

Stok, F. 2000. 'Il lessico del contagio.' In P. Radici-Colace and A. Zumbo, eds., *Letteratura scientifica e tecnica greca e latina: atti del Seminario internazionale di studi. Messina, 29–31 ottobre 1997*. Messina: EDAS, 55–89.

Stoll, U. 1992. *Das 'Lorscher Arzneibuch': ein medizinisches Kompendium des 8. Jahrhunderts (Codex Bambergensis medicinalis 1): Text, Übersetzung und Fachglossar*. Stuttgart: F. Steiner.

Stotz, P. 1996–2004. *Handbuch zur lateinischen Sprache des Mittelalters*. 5 vols. Handbuch der Altertumswissenschaft 2. Abt., 5. T., 2–5. Bd. Munich: Beck.

Sudhoff, K. 1915. 'Diaeta Theodori.' *Sudoffs Archiv für Geschichte der Medizin* 8.6: 377–403.

Svennung, J. 1934. 'Lat. *uiscum* und *uirus*.' *Zeitschrift für Vergleichende Sprachforschung* 62: 17–22.

– 1935. *Untersuchungen zu Palladius und zur lateinischen Fach- und Volkssprache*. Uppsala: Almquist & Wiksells

Sweet, H. ed. 1885. *The Oldest English Texts* 83. Oxford: Oxford University Press.

Tameanko, M. 1992. 'The Silphium Plant: Wonder Drug of the Ancient World Depicted on Coins.' *Celator* 6.4: 26–8.

Tecusan, M. 2004. *The Fragments of the Methodists: Methodism outside Soranus. Volume 1: Text and Translation*. Leiden: Brill.

Temkin, O. 1935. 'Studies on Late Alexandrian Medicine I. Alexandrian Commentaries on Galen's *De sectis ad Introducendos*.' *Bulletin of the Institute of the History of Medicine* 3: 405–30.

– 1973. *Galenism: Rise and Decline of a Medical Philosophy*. Ithaca, NY: Cornell University Press.

– 2006. 'History of Hippocratism in Late Antiquity: The Third Century and Latin West.' In idem, *The Double Face of Janus and Other Essays in the History of Medicine*. Baltimore: Johns Hopkins University Press, 167–77 (1st publ. 1977).

– and Irvine 2003. *See under* Irvine.

Thielmann, Ph. 1885. 'Habere mit infinitive und die Enstehung des romanischen Futurums.' *Archiv für lateinische Lexicographie und Grammatik* 2: 49–98, 157–202.

Thivel, A., and A. Zucker, eds. 2002. *Le normal et le pathologique dans la Collection hippocratique: actes du Xème colloque international hippocratique (Nice, 6–8 octobre, 1999)*. 2 vols. Nice: Publication de la faculté des lettres, arts et sciences humaines de Nice-Sophia Antipolis.

Thomas, F. 1940. 'Le suffixe Latin -aster/ -astrum.' *Revue des Études Anciennes* 48: 520–8.

Thorndike, L. 1946. *The Herbal of Rufinus*. Chicago: University of Chicago Press.

Tjäder, J-O. 1955–82. *Die nichtliterarischen lateinischen Papyri Italiens aus der Zeit 445–700*. 3 vols. Lund: C.W.K. Gleerup, 1955

Totelin, L.T. 2009. *Hippocratic Recipes: Oral and Written Transmission of Pharmacological Knowledge in Fifth- and Fourth-Century Greece*. Leiden: Brill.

Touwaide, A. 1992. 'Le traité de matière médical de Dioscoride en Italie depuis la fin de l'Empire romain jusqu' aux débuts de l'école de Salerne: essai de synthèse.' *PACT (Journal of the European Study Group on Physical, Chemical, Biological and Mathematical techniques Applied to Archaeology)*, 18: 291–335.

– 1997. 'La Thérapeutique médicamenteuse de Dioscoride à Galien.' In Debru, ed. 1997, 255–82.

– 2008. 'The Legacy of Classical Antiquity in Byzantium and the West.' In Dendle and Touwaide, eds. 2008, 15–28.

Trager, G.L. 1959. *The Use of Latin Demonstratives (Especially* ille *and* ipse*) up to 600 A.D. as the Source of the Romance Article*. New York: Publications of the Institute of French Studies.

Trease, G.E., and W.C. Evans. 2002. *Pharmacognosy*. 15th ed. Edinburgh, London, and New York: W.B. Saunders.

Tribalet, J. 1938. *Histoire médicale de Chartres jusqu'au XII siècle*. Paris: Vigot ères.

Ullmann, M. 1978. *Islamic Medicine*. Edinburgh: Edinburgh University Press.

Unschuld, P. 1985. *Medicine in China: A History of Ideas*. Berkeley: University of California Press.

– 1986. *Medicine in China: A History of Pharmaceutics*. Berkeley: University of California Press.

Urso, A.M. 1998. 'Destinazione e finalità nella praefatio delle passiones celeres di Caelio Aureliano.' In Santini, Scivoletto, and Zurli, eds. 1998, 177–98.

Váczy, C. 1980. *Lexicon Botanicum Polyglottum*. Bucharest: Stiintifica si Enciclopedica.

van der Eijk, P.J. 1997. 'Galen's Concept of "Qualified Experience" in His Dietetic and Pharmacological Works.' In Debru, ed. 1997, 35–58.

– 1999. 'Antiquarianism and Criticism: Forms and Functions of Medical Doxography in Methodism (Soranus, Caelius Aurelianus).' In idem, ed., *Ancient Histories of Medicine*. Leiden: Brill, 397–452.

Van Wyk, B.-E., and M. Wink. 2004. *Medicinal Plants of the World*. Pretoria: Briza Publications.

Vázquez Buján, M.E. 1984a. 'Remarques sur la technique de traduction des anciennes versions latines d'Hippocrate.' In Sabbah, ed. 1984, 153–63.

– 1984b. 'Problemas generales de las antiguas traducciones médicas latinas.' *Studi Medievali* 3rd ser. 25: 641–80.

– 1998. 'Quelques remarques lexicales sur l'ancienne traduction latine des Apho-
rismes hippocratiques.' In Deroux ed. 1998, 354–65.

– ed. 1994. *Tradición e innovación de la medicina latina de la antigüedad y de la
Alta Edad Media: actas del IV Coloquio Internacional sobre los "Textos Médicos Latinos Antiguos."* Santiago de Compostella: La Coruña.

Ventura, I. 2008. 'Translating, Commenting, Re-Translating: Some Considerations on the Latin Translations of the Pseudo-Aristotelian Prolemata and Their
Readers.' In Goyens, Leemans, and Smets, eds. 2008, 123–54.

Villard, L., ed. 2002. *Couleurs et visions dans l'Antiquité classique.* Rouen: Publications de l'Université de Rouen.

Voigts, L.E. 1978. 'The Significance of the Name Apuleius to the *Herbarium
Apulei.' Dumbarton Oaks Papers* 52.2: 214–27.

von Staden, H. 1975. 'Experiment and Experience in Hellenistic Medicine.' *University of London, Institute for Classical Studies Bulletin* 22: 178–99.

– 1989. *Herophilus: The Art of Medicine in Early Alexandria. Edition, Translation and Essays.* Cambridge: Cambridge University Press.

– 1996. 'Liminal Perils: Early Roman Receptions of Greek Medicine.' In F. Jamil
Ragep, S.P. Ragep, and S.J. Livesey, eds., *Tradition, Transmission, Transforma-
tions: Proceedings of Two Conferences on Pre-Modern Science Held at the
University of Oklahoma.* Leiden: Brill, 369–418.

– 1997. 'Inefficacy, Error and Failure: Galen on δόκιμα φάρμακα.' In Debru, ed.
1997, 59–84.

Voultsiadou, E. 2007. 'Sponges: An Historical Survey of Their Knowledge in
Greek Antiquity.' *Journal of the Marine Biological Association of the UK* 87:
1757–63.

Wallis, F. 1995. 'The Experience of the Book: Manuscripts, Texts, and the Role
of Epistemology in Early Medieval Medicine.' In D.G. Bates, ed., *Knowledge
and the Scholarly Medical Traditions.* Cambridge: Cambridge University Press,
101–26.

– 2000. 'Signs and Senses: Diagnosis and Prognosis in Early Medieval Pulse and
Urine Texts.' *The Year 1000: Medical Practice at the End of the First Millennium,*
ed. P. Horden and E. Savage-Smith = *Social History of Medicine* 13: 265–78.

Watson, P. 1989. 'Filiaster: Privignus or "Illegitimate Child"?' *Classical Quarterly*
39.2: 536–48.

Wattenbach, W. 1896. *Das Schriftwesen im Mittelalter.* Leipzig: Hirzel.

Wellmann, M. 1888. 'Zur Geschichte der Medizin im Alterthum.' *Hermes* 23:
556–66.

– 1889. 'Sextius Niger: Eine Quellenuntersuchung zu Dioscorides.' *Hermes* 24:
530–69.

- 1897. *Krateuas*. Abhandlüngen der königlichen Gesellschaft der wissenshaften zu Gottingen, Phil.- Hist. Klasse, Neue Folge 2. no. 1 Berlin.
- 1898. 'Die Planzennamen des Dioskurides.' *Hermes* 33: 360–422.
- 1907. 'Xenocrates aus Aphrodisias.' *Hermes* 42: 614–29.
- 1913. *A. Cornelius Celsus. Eine Quellenuntersuchung*. Berlin: Weidmann.
- 1916. 'Pamphilos.' *Hermes* 51: 1–64.
- 1924. 'Beiträge zur Quellenanalyse des Älteren Plinius.' *Hermes* 59: 129–56.
Westerink, L.G., with D.O. Davies, K.M. Dickson, A. Kershaw, J.P. Peters, B.K. Robbins, T.A. Virginia, and J.J. Walsh, eds. and trans. 1981. *Agnellus of Ravenna: Lectures on Galen's De sectis*. Buffalo: State University of New York.
- 1984. *Stephanus of Athens, Commentary on Hippocrates' Aphorisms*, sects. 1–2. Berlin: Akademie-Verlag.
Wichtl, M., ed. 2004. *Herbal Drugs and Phytopharmceuticals*. Trans. J.A. Brinckmann and M.P. Lindenmaier. Stuttgart and Boca Raton, FL: Medpharm GmbH and CRC Press.
Wickersheimer, E. 1953. 'Textes médicaux chartrains des IX, X et XI siècles.' In *Science, Medicine and History: Essays Written in Honour of Ch. Singer*. Oxford: Oxford University Press. I: 166–9.
- 1966. *Les manuscrits latins de médicine du haut moyen âge dens les bibliothèques de France*. Paris: Éditions du Centre national de la recherche scientifique.
Wiedemann, W. 1976. 'Untersuchungen zu dem frühmittelalterlichen medizinischen Briefbuch des Codex Bruxellensis 3701–15.' PhD dissertation, Freie Universität, Berlin.
Wilsdorf, H. 1977. 'Die architektonische Rekonstruktion antiker Produktionsanlagen für Bergbau und Hüttenwesen.' *Klio. Beiträge zur alten Geschichte* 59: 11–24.
Wink, M. *See under* Roberts, M.F, and van Wyk, B-E.
Wise, P.M., M.J. Olsson, and W.S. Cain. 2000. 'Quantification of Odor Quality.' *Chemical Senses* 25: 429–43.
Wright, R. 1982. *Late Latin and Early Romance in Spain and Carolingian France*. Liverpool: Francis Cairns.
- ed. 1991. *Latin and the Romance Languages in the Early Middle Age*. London and New York: Routledge.
- ed. 2008. *Latin vulgaire – latin tardif VIII: actes du VIIIe Colloque international sur le latin vulgaire et tardif, Oxford, 6–9 septembre 2006*. Hildesheim: Olms-Weidmann.
Wright, W.S. 2001. 'Silphium Rediscovered.' *Celator* 15.2: 23–4.
Zimmermann, F. 1987. *The Jungle and the Aroma of Meats. An Ecological Theme in Hindu Medicine*. Trans. J. Lloyd. Berkeley: University of California Press.

– 1989. *Les discourse des remèdes au pays épices. Enquête sur la médecine hindoue*. Paris: Éditions Payot.

Zumla, A., and A. Lulat. 1989. 'Honey – a Remedy Rediscovered.' *Journal of the Royal Society of Medicine* 82: 384–5.

Zurli, L. 1992a. 'L'epistola prefatoria dell'Herbarius dello ps. Apuleius.' In Santini and Schivoletto, eds. 1990–2, II: 445–51.

– 1992b. 'L'epistola a Pentadio (e altre reliquie) di Vindiciano.' In Santini and Schivoletto, eds. 1990–2, II: 455–62.

– 1992c. 'L'epistola dello Ps. Antonius Musa.' In Santini and Schivoletto, eds. 1990–2, II: 433–42.

– 1992d. 'Il pensiero medico di Teodoro Prisciano nelle prefazioni ai suoi libri.' In Santini and Schivoletto, eds. 1990–2, II: 465–97.

– 1998. 'Trilogia medica: I. Antimo: una dietetica per le nazioni agli albori del VI secolo; II. *L'Epistula ad Marcellinum*; III. Sulle tracce del gemello cortonese di Harley 4986.' In Santini, Scivoletto, and Zurli, 1998, III: 313–420 (I = 315–364, II = 365–390, III = 391–420).

Zysk, K.G. 1996. *Medicine in the Veda*. 3rd ed. Delhi: Motilala Barnardsidass.

Index

꙳

All numbers refer to *AG* entry numbers (#).

A. Plants and Plant Products

B. Minerals and Mineral Products

C. Animals and Animal Products

D. Places and Place-Names

Africa, 69, 137, 201, 263
Alexandria, 163, 261
Alps, 238
Amista, 8
Ammon, lands of, 4
Arabia, 39, 64, 71, 75, 156, 180, 253,
    263, 279
Arcadia, 201
Armenia, 8, 28, 66
Asia, 5, 8
Assyria, 277
Athens, 207

Babylonia, 11
Black Sea: *see* Pontus

Calamine islands, 210
Canosa, 211
Cappadocia, 8, 201, 242
cave, 276
Chios (island), 151, 186, 228
Cilicia, 156
Colophon, 252
Corinth, 182
Crete, 15, 49, 64, 91, 124, 127, 187,
    258, 285
Cyprus, 48, 66, 156, 178
Cyrene, 4, 97, 147

dry places, 152

Egypt, 2, 7, 10, 12, 15, 54, 64, 76 n.6,
    161, 168, 172, 173, 176, 257
Ethiopia, 26, 69, 95, 95, 168, 175

field, 36, 104, 120 (grain), 122, 219
    (grain), 290

Gage (river), 166
garden, 14, 84, 99, 104
Gaul, 15, 17, 27, 29, 94, 239, 259, 285
Germany, 259
gravelly place, 122

Hellespont, 8

Iacinthus (Zakynthos), 11
Illyricum, 137
India, 5, 64, 73, 95, 111, 135, 161, 190,
    198, 202, 209, 236, 263, 266, 279
islands (*stoechades*/Iles d'Hyeres), 258
Italy, 63 n.1, 94, 98 n.1, 103 n.1, 135,
    239, 285 n.1

Judea, 5, 11, 202

lagoon, 11
Lemnos (island), 276
Lycia, 166

Macedonia, 7, 66, 137, 189, 201, 223
Mauritania, 94
Media, 28
Melos, 7
mine, 48, 51, 52, 84, 89
moist places, 269
mountain, 8, 62, 98, 180, 188, 236, 290
mud (found in), 89
Mysia, 252

Nicomedia, 119
Nile, 10
Noricum, 238

ocean, 165 (shore), 213 (shore)

E. Medical

digestion, 5 (disturbed), 139 (aids),
148 (useful for), 153, 253, 292
discharges, 48, 126
disease, 240
diuretic, 13, 14, 104 n.1, 105, 143, 157,
158, 214, 223, 251, 270, 292
diuretic (properties), 15 (qualities), 23,
35, 100, 114, 133, 139, 142, 222, 265,
273, 274, 275
dizziness, 292
dropsy, 107, 165, 234, 264, 271, 301
drunkenness, 188
dysentery, 82, 101, 102, 106, 125, 126,
151, 183, 188, 204, 216, 291

ear, 188 (oozing pus), 196 (worms in),
199 (ache)
emetic, 234
emollients, 10, 43, 59, 156, 221, 229
enfeebled parts, 261
epilepsy, 12
eruptions, 109 (on skin)
erysipelas, 68, 132, 152, 188
eye (salve), 6, 10, 44, 45, 56, 58, 66, 95,
136, 175 (scar tissue), 199, 224, 248,
295
eyelid, 37
eyes, corneal abrasions (*cicatrices oculorum*), 52, 116, 175, 283; cataracts
(*suffusiones*), 63; dim-sightedness
(*caligo*), 56, 96, 116, 152, 175, 203;
clears vision (*claritas*), 25, 30, 93, 164,
179; rough eye (*asperitudo oculorum*) 283; teary eyes, 6, 37, 66, 106,
170, 199, 216; ophthalmia (*lippitudo*),
118, 272; 'goat eye' (*aegylopis*), 188;
other eye complaints, 110, 151, 152,
204, 245, 262, 291, 292. *See* ch. 4.H.

face, 281

fatigue, 181, 199
fetus (dead), 88
fetus (expels), 22, 55, 71, 74, 91, 99,
123, 153, 155, 196 (pushes out), 214,
227, 287
fever, 5, 109, 153, 224 (tertian, quartan), 240, 292
filth, 9, 109 (in chest), 116, 179 (in
wounds), 259
flatulence, 69, 233, 234. *See also* gas
flesh, 233 (makes flabby)
fomentation, 153
forehead, 173
fractures, 151, 188
freckles, 70, 109, 133
frost-bite, 283

gargle, 109, 195, 234
gas, 35, 71, 103 (in diaphragm), 148,
152 (in bowels), 277. *See also*
flatulence
genitals, 151 (sores), 152 (sores)
gout, 78, 115, 152, 216, 233
gums, 151 (swollen)

haemorrhage, 86, 87, 125, 188, 200, 230
hair, 8, 102, 121, 151 (loss; ingrown),
156 (frizzy), 188, 280
hand, 281
head, 12, 63 (clears), 104, 188 (scaly
infections), 207 (clouds), 218
('clears'), 292 (dizziness in)
headaches, 12, 133, 185, 216
heartburn, 240
hiccups, 151, 185
hip, 99 (ailments), 196 (ailments), 271
(diseases), 299 (disease)
humour, 48, 94, 160, 165, 179, 288

indigestion, 194

F. General

G. Interesting or Rare Words (see also ch. 4.I)

H. *Materia medica* (General)